T0296907

Energy, Sustainability and the Environment:

Technology, Incentives, Behavior

Energy, Sustainability and the Environment:

Technology, Incentives, Behavior

Edited by
Fereidoon P. Sioshansi
Menlo Energy Economics

AMSTERDAM • BOSTON • HEIDELBERG • LONDON
NEW YORK • OXFORD • PARIS • SAN DIEGO
SAN FRANCISCO • SINGAPORE • SYDNEY • TOKYO
Butterworth-Heinemann is an imprint of Elsevier

Butterworth-Heinemann is an imprint of Elsevier
30 Corporate Drive, Suite 400, Burlington, MA 01803, USA
Linacre House, Jordan Hill, Oxford OX2 8DP, UK

Library of Congress Cataloging-in-Publication Data
Energy, sustainability, and the environment : technology, incentives, behavior / edited by Fereidoon P. Sioshansi.
p. cm.
Summary: "The complexity of carbon reduction and economic sustainability is significantly complicated by competing aspects of socioeconomic practices as well as legislative, regulatory, and scientific requirements and protocols. An easy to read and understand guide, Sioshansi, along with an international group of contributors, moves through the maze of carbon reduction methods and technologies, providing steps and insights to meet carbon reduction requirements and maintaining the health and welfare of the firm"– Provided by publisher.
Includes bibliographical references and index.
ISBN 978-0-12-810376-0 (paperback)
1. Renewable energy sources. 2. Energy conservation–Social aspects.
I. Sioshansi, Fereidoon P. (Fereidoon Perry)
TJ808.E588 2011
333.791'6--dc22

2010048926

British Library Cataloguing-in-Publication Data
A catalogue record for this book is available from the British Library.

ISBN: 978-0-12-810376-0

Printed and bound in the U.S.

11 12 13 10 9 8 7 6 5 4 3 2 1

Contents

Foreword

A *New Yorker* cartoon from an earlier era featured a man at a bar telling a companion, "It sure is a relief to meet somebody with no energy policy." If that is your view, you should run from this book. If you have at least a nagging sense that our energy affairs need immediate remedial attention, you have cause to appreciate what Fereidoon Sioshansi and his coauthors have achieved in this volume.

Sioshansi has challenged his small army of distinguished coauthors to bring him no small plans. An underlying theme in a very diverse collection is the imperative to imagine a wholly different future, and to apply tools far transcending those of the technologist. The book touches more than lightly on behavioral change, sociology, economics, anthropology, ethics, and philosophy.

Underlying every chapter is a well-justified sense of urgency, at a moment when the temptation for complacency is almost palpable. After all, U.S. energy consumption dropped 7% between 2007 and 2009, and our greenhouse gas emissions are down 10% since 2005. In 2008 and 2009, our electricity use dropped in consecutive years for the first time since World War II. Domestic oil use peaked in 2005, and by 2009 daily consumption was down by 10%. The trend of fossil fuel prices since mid-2008 is generally declining, and reports abound that plentiful natural gas supplies will persist for decades thanks largely to advances in drilling technology. Worldwide, despite the continued economic surges of giants like China, Brazil, and India, total energy use dropped by 1.2% in 2008 and another 2.2% in 2009. Can't we all just relax for a while?

Sioshansi and his coauthors don't think so, and May 2010 projections from the U.S. Energy Information Administration bear them out.[1] EIA sees global energy consumption growing by almost 50% over the next quarter century if business as usual is allowed to reassert itself. Greenhouse gas emissions and fossil fuel use would increase at comparable rates. That would make today's dangerous oil dependence much worse and all but eliminate any chance to suspend a uniquely dangerous global experiment with climate instability.

Our best hope for a different outcome lies, ironically, in a common misunderstanding that surfaces from time to time in this volume—as do forceful rebuttals. I refer to the contention that comfortable lifestyles, for those who have them worldwide, depend on continued access to cheap and plentiful

1. U.S. Energy Information Administration. (2010). *International Energy Outlook 2010*. *www.eia.doe.gov/oiaf/ieo/index.html*

energy. But as Sioshansi himself would be the first to remind us, the real issue is not the cost of energy but the cost of the services that it provides, and the affordability of humanity's energy services hinges now on a race between energy efficiency improvements and increases in the cost of energy commodities. Readers of this volume will emerge with a strong rooting interest in the energy efficiency improvements and a well-grounded optimism about their prospects. Those readers will also have a much fuller understanding of prospects for changing the human behaviors that matter most in determining the energy intensity of energy services, with profound implications for both economic and environmental welfare.

Thoughtful observers of the energy scene already had ample cause to appreciate the insights of Fereidoon Sioshansi and his redoubtable coauthors. Together they have achieved a formidable addition to an impressive legacy.

Ralph Cavanagh
Energy Program Co-Director
Natural Resources Defense Council

Preface

Pursuing a low-carbon energy system is arguably the single most important step that can be taken towards achieving a sustainable economy and lifestyle in the developed as well as in the developing world. Demand for energy is set to increase for the foreseeable future due to continuous population and economic growth.

The services from energy use affect most aspects of modern life and range from necessities, to essentials, and to leisure activities. Hence energy use is a function of derived demand and ultimately driven by our needs, wants, and desires. Meanwhile, the steady rise in our incomes has reduced the effect of energy prices on demand and incentives for exploiting the large energy efficiency and saving potential.

The search for a low-carbon energy system has now been a major economic and environmental priority for nearly four decades. Initially, the focus of this search was on finding supply-side solutions in renewable and low-carbon energy sources. While there has been substantial progress on this front, there remains much scope for technical progress and cost reductions. As we recognized the limits of supply-oriented solutions we began to explore the possibilities of demand-side-management options. Also here, while some progress has been made there was a need to further widen the scope of the search for ideas and solutions. More recently, much hope has been vested in the promise of smart electricity networks in integrating large amounts of renewable energy sources while at the same time allowing for an active and flexible demand-side to participate in the energy system and markets.

However, as part of this process, we have come to realize that technological and economic approaches are not sufficient given the scale of the challenges in question. The cost of most new energy production technologies are still high, smart energy networks are yet to be implemented and prove their merit, and market failure remains pervasive in various aspects of demand for and usage of energy. Faced with these challenges, some researchers and decision-makers have begun to widen the scope of their search for solutions that are beyond the conventional boundaries of the energy supply and demand systems and economic instruments to energy and environmental issues. Some of these efforts explore innovative approaches that focus on values, behaviour, and lifestyle aspects of individual consumers, communities, and the societies as a whole.

It is not sufficient to debate how the future energy systems may or should be formed. We are yet to translate and define sustainable energy future into better and viable technological, economic, social, and political terms. We also need to be conscious of the path and transition to the sustainable futures. The benefits of a low-carbon energy system are obvious and are widely recognized. At the same time, the importance of the critical issue of energy equity at national as well as international level for the sustainable energy future cannot be overemphasised. Failure to explicitly address the equity and distributional aspects of measures and policies can undermine the long-term continuity of the efforts towards achieving a low-carbon energy future and lifestyle.

This book can be seen within the context of these challenges as it rightly attempts to go beyond the traditional realm of technical and economic conceptions of energy systems where the users are regarded as consumers or customers. This volume also recognizes the role of the energy users as citizens and communities whose demand for energy services is also driven by their norms, values, and lifestyles.

This book brings together the latest ideas from an international and interdisciplinary group of authors as they explore the premises of a sustainable energy future and the path that can lead to it. In this regard, the book offers a timely and welcome contribution towards enhancing our understanding of some emerging aspects of a viable energy system and the role of sustainable lifestyle in achieving this.

Tooraj Jamasb
Heriot-Watt University

Contributors

Clinton J. Andrews is a professor of planning in the Edward J. Bloustein School of Planning and Public Policy at Rutgers University.

His expertise is in the substance and processes of energy and environmental planning and policy. Previous experience includes working in the private sector on energy issues, helping to launch an energy policy project at the Massachusetts Institute of Technology, and helping to found a science policy program at Princeton University. Andrews currently serves on the Board of Governors of the American Collegiate Schools of Planning, and is a past member of the Board of Directors of the Institute for Electrical and Electronics Engineers (IEEE) and the International Society for Industrial Ecology, and a winner of the IEEE's 3rd Millennium Medal. His books include *Industrial Ecology and Global Change; Regulating Regional Power Systems; and Humble Analysis: The Practice of Joint Fact Finding*.

He was educated at Brown and MIT as an engineer and planner. He is a member of the American Institute of Certified Planners, a LEED Accredited Professional, and a licensed Professional Engineer.

Doug Arent is Executive Director of the Joint Institute for Strategic Energy Analysis at the National Renewable Energy Laboratory (NREL) in Golden, Colorado. He specializes in strategic planning and financial analysis competencies; clean energy technologies and energy and water issues; and international and governmental policies. In addition to his NREL responsibilities, Arent is a member of the U.S. Government Review Panel for the IPCC Reports on climate change and is a senior visiting fellow at the Center for Strategic and International Studies.

In 2008 he was appointed to serve on the National Academy of Sciences panel on limiting the magnitude of future climate change. He is on the Executive Council of the U.S. Association of Energy Economists; a member of the Keystone Energy Board; on the Advisory Board of E+C; and serves on the Chancellor's Committee on Energy, Environment and Sustainability Carbon Neutrality Group, University of Colorado, among others.

He has a Ph.D. from Princeton University, an MBA from Regis University, and a Bachelor of Science degree from Harvey Mudd College in California.

Françoise Bartiaux is Senior Research Associate at the Belgian Fund for Scientific Research and Professor at the Université catholique de Louvain in Belgium, where she teaches environment sociology, family sociology, and research methodology.

Her research interests are focused on energy-related practices and social and family factors at the household level in European countries. She has conducted or participated to several multidisciplinary research projects including collecting and analyzing both quantitative and qualitative data.

She was educated at the Université catholique de Louvain in Belgium and holds a Master in Sociology and a Ph.D. in Demography.

Klaas Bauermann is a Ph.D. student and research associate to the Chair for Management Sciences and Energy Economics at the University of Duisburg-Essen. He has gained work experience in several national and European utilities.

His main areas of research are efficient and sustainable long-term solutions for space heating, distributed generation, and combined heat and power.

Bauermann has studied in Essen and Paris and holds an MBA from the University of Duisburg-Essen.

Carlo Andrea Bollino is Professor of Economics at University of Perugia, and Professor of Energy Economics at Luiss University of Rome, Italy. Previous experience includes working as an economist at the Italian Ministry of Treasury, the Bank of Italy, and in the private sector. He is a past president of the International Association for Energy Economics (IAEE).

His research interests include consumer behavior, electricity markets, renewable energy, and growth economics. He coordinated research projects on energy markets, growth, and efficiency of local economic systems, and has testified before the Italian Congress.

He received his M.A. in Economics at Bocconi University, Milan, Italy and Ph.D. in Economics from the University of Pennsylvania.

Marilyn Brown is Professor of Energy Policy in the School of Public Policy at the Georgia Institute of Technology. Her research has focused on the impacts of policies and programs aimed at accelerating the development and deployment of sustainable energy technologies. She has led several energy technology and policy scenario studies and is a national leader in the analysis and interpretation of energy futures in the U.S. In this capacity, she has testified before the Congress and state legislatures.

Prior to her present position she was at Oak Ridge National Laboratory where she led programs in energy efficiency, renewable energy, engineering science, and technology. She remains a Distinguished Visiting Scientist at ORNL. Dr. Brown serves on the boards of directors of the American Council for an Energy-Efficient Economy, the Alliance to Save Energy, is a member of the National Commission on Energy Policy and the National Academies Board of Engineering and Environmental Systems. Her latest edited book, *Energy and American Society: Thirteen Myths*, was published in 2007 by Springer.

Dr. Brown has a Ph.D. in Geography from the Ohio State University and a Masters Degree in Resource Planning from the University of Massachusetts.

Bram Buijs is a researcher at the Clingendael International Energy Programme (CIEP) in The Hague, the Netherlands.

His research interests are in energy and geopolitics and include Asian energy affairs, European energy policy, global energy security issues, and sustainability challenges. He has focused in particular on China's energy sector, emerging low-carbon industries, and China's role in the international climate treaty negotiations.

Buijs holds a Master of Science degree in Mathematics and in Contemporary Asian Studies, both from the University of Amsterdam. He lived and worked in China for two years while studying Mandarin Chinese at the Shanghai International Studies University.

Pierre Bull is an energy policy analyst at the Natural Resources Defense Council where he advocates for state and utility-level programs and policies that promote the rapid scale-up of energy efficiency and distributed renewable energy deployment.

Prior to NRDC he was a project manager in the energy efficiency market transformation program at the New York State Energy Research and Development Authority (NYSERDA) focusing on appliance standards and efficiency strategies for hospitals and IT data centers.

Mr. Bull has a Master of Science degree in Natural Resources Policy from the University of Michigan and a Bachelor of Science degree in Environmental Science from the University of Illinois at Urbana-Champaign.

Ralph Cavanagh is a senior attorney and co-director of NRDC's energy program, which he joined in 1979. In addition, Cavanagh has been a Visiting Professor of Law at Stanford and UC Berkeley and served as a member of the U.S. Secretary of Energy's Advisory Board from 1993—2003.

He has received the National Association of Regulatory Utility Commissioners' Mary Kilmarx Award, the Heinz Award for Public Policy, the Yale Law School's Preiskel-Silverman Fellowship, the Lifetime Achievement in Energy Efficiency Award from California's Flex Your Power Campaign, the Headwaters Award from the Northwest Energy Coalition, and the Bonneville Power Administration's Award for Exceptional Public Service.

He is a graduate of Yale College and the Yale Law School.

Jennifer Clymer is a senior associate with ICF International, a global consulting firm specializing in energy and the environment. Ms. Clymer supports national energy efficiency programs and projects, including the U.S. Environmental Protection Agency's ENERGY STAR® program. Other areas of expertise include: assessing the air quality and climate change impacts of multimodal transportation, energy use, and waste management.

Prior to returning to ICF International, Ms. Clymer was an Environmental Program Coordinator for the City of Austin, Texas, where she was responsible for the implementation of Austin's internal greenhouse gas reduction goals and strategies as part of the Austin Climate Protection Program. She has also served as an energy and environmental policy aide for a state legislator.

Ms. Clymer holds a Master of Public Affairs from the Lyndon Baines Johnson School of Public Affairs at the University of Texas at Austin and a Bachelor of Arts in Environmental Sciences and Environmental Management from the University of Virginia.

Rodrigo Cortes-Lobos is a graduate research assistant at the School of Public Policy at Georgia Institute of Technology. His research interests include science, technology, and energy policy. Currently he is conducting research into industrial energy efficiency.

He was Executive Director of the Masters program in Technology Management at the University of Talca-Chile and coordinator of the American Academy of Science and Technology in Santiago-Chile.

Cortes-Lobos holds an M.Sc. in International Agriculture from the University of Georg August, Göttingen, Germany and undergraduate degree in Agricultural Economics from the University of Talca-Chile.

Matthew Cox is a graduate research assistant and Ph.D. student at the Georgia Institute of Technology. His previous work has involved implementing and developing wide-ranging sustainability projects at the university and municipal level. His current research interests include industrial energy efficiency, renewable energy, and sustainable agriculture.

He has a Master of Science degree in Public Policy with a concentration in Energy and Environmental Policy from the Georgia Institute of Technology and a Bachelor of Science degree in Environmental Biology from the University of Dayton.

Kat A. Donnelly is the Founder and President of EMpower Devices and Associates, an organization that provides simple, cost-effective energy management solutions emphasizing the customer/user perspective. Previously, she managed numerous local and regional planning studies, as well as large transit and transportation infrastructure projects at the San Diego Association of Governments in California.

Donnelly is currently a Ph.D. student at the Massachusetts Institute of Technology studying the impact of technology and consumer behavior on energy consumption and efficiency.

She has dual Master of Science degrees in Sustainable Energy Technology & Policy, and Civil & Environmental Engineering from MIT with a Bachelor of Science degree in Civil Engineering from San Diego State University.

Paul Denholm is a senior energy analyst in the Strategic Energy Analysis Center at the National Renewable Energy Laboratory (NREL) in Golden, Colorado, where he examines system integration of renewable electricity generation sources such as wind and solar.

Denholm's main research interests include examining the technical, economic, and environmental benefits and impacts of large-scale deployment of renewable electricity generation. This includes the limits of variable generation sources and quantifying the need for enabling technologies such as energy storage, plug-in hybrid electric vehicles, and long-distance transmission.

He has degrees in Physics and a Ph.D. in Environmental Studies and Energy Analysis from the University of Wisconsin-Madison.

Easan Drury is an energy analyst at the National Renewable Energy Laboratory (NREL) in Golden, Colorado, where he examines the technical, economic, and environmental impacts and benefits of large-scale renewable energy deployment.

Drury's main research interests include modeling the U.S. photovoltaics (PV) market by quantifying the impacts of PV costs, electricity rate structures, net metering, carbon prices, and state and federal incentives on PV adoption.

He has a B.A. in physics from the University of California, Berkeley, and an M.S. and Ph.D. in Engineering Sciences from Harvard University.

Karen Ehrhardt-Martinez is a Senior Research Associate with the Renewable and Sustainable Energy Institute (RASEI) at the University of Colorado. Previously, she was a Senior Researcher with the American Council for an Energy-Efficient Economy (ACEEE) and as an Assistant Professor of Sociology and Environmental Studies at Denison University.

Ehrhardt-Martinez has studied the social and behavioral dimensions of energy consumption, efficiency initiatives, energy management strategies, and conservation. Of particular interest are efforts to identify how social and technological systems interact and change over time; understand the variation in individual perceptions, individual energy service demands, and household consumption patterns and management strategies; and reveal the importance of spatial and demographic factors.

Dr. Ehrhardt-Martinez earned her undergraduate degree in International Studies and her M.A. and Ph.D. in Sociology from The Ohio State University in Columbus, Ohio.

Frank A. Felder is Director of the Center for Energy, Economic and Environmental Policy and Associate Research Professor at the Bloustein School of Planning and Public Policy at Rutgers University.

Felder directs applied energy and environmental research. Ongoing and recent projects include energy efficiency evaluation studies, economic impact of renewable portfolio standards, and power system and economic modeling of state energy plans. He is also an expert on restructured electricity markets. He has published widely in professional and academic journals on market power and mitigation, wholesale market design, reliability, transmission planning, market power, and rate design issues. He was a nuclear engineer and submarine officer in the U.S. Navy.

Dr. Felder holds undergraduate degrees from Columbia College and the School of Engineering and Applied Sciences and a Masters and Doctorate from MIT in Technology, Management and Policy.

Scott Finlinson is manager of organizational efficiency at NORESCO, a leading energy services company. For the past 10 years, he has been developing, implementing, and assessing custom-tailored behavior change programs aimed at increasing and sustaining organizational energy conservation within ongoing energy efficiency building retrofit projects. He manages successful behavior change programs across the country with clients in higher education, K-12, and county and state governments.

Finlinson has a Ph.D. degree in Industrial/Organizational Psychology from Ohio University.

Nathalie Frogneux is Professor of Philosophy at the Université catholique de Louvain in Belgium, where she teaches philosophical anthropology.

Her research interests are focused on anthropological phenomenology with a particular interest on the principle of responsibility and on the collective responsibility toward the future generations.

She was educated at the Université catholique de Louvain in Belgium, and at the Sorbonne (Paris IV) and the Catholic Institute, both in Paris, France. She holds a Masters degree in Philosophy and a Ph.D. in Philosophy and History of Religions and Religious Anthropology.

Rachel Gelman is an energy and markets analyst at the National Renewable Energy Laboratory (NREL) in Golden, Colorado, where she focuses on various analyses relating to market acceleration. Gelman was previously an analyst in Merrill Lynch's Global Power and Energy investment banking group where she was involved in the modeling and valuation of power generating assets and oversaw the deal flow process for a wide range of financial solutions for utilities.

Her interests include corporate sustainability, energy efficiency, smart grid and the electric distribution infrastructure, and renewable energy technologies. She recently produced the *2008 Renewable Energy Data Book*, published by the U.S. Department of Energy's Office of Energy Efficiency and Renewable Energy.

She has Bachelors degrees in Statistics and Economics from Rice University and is pursuing a Master's degree in Mineral Economics from the Colorado School of Mines.

Meredith Gray currently works for the City of Austin Office of Homeland Security and Emergency Management as a grant coordinator. She is a Leadership in Energy and Environmental Design (LEED) accredited professional. She has worked on environmental remediation projects in Switzerland, China, and Texas.

Ms. Gray's interests include habitat conservation, reducing the environmental footprint of buildings, emergency mitigation, disaster-resilient buildings, integrating nature into public spaces, water use reduction, alternative energy, and the promotion of native plant species. She holds dual master's degrees in Community and Regional Planning from the School of Architecture and in Public Affairs from the LBJ School of Public Affairs at the University of Texas. Her master's professional report was recently produced and is titled Designing for Disasters: Incorporating Hazard Mitigation Methods into the LEED for New Construction and Major Renovations Framework. Her undergraduate degrees are from Rice University with a Bachelor of Science in Ecology and Evolutionary Biology and a Bachelor of Arts in English.

Seth D. Hulkower is President of Strategic Energy Advisory Services, a consulting firm focused on the regulatory, technology, and management challenges in the energy market. Prior to SEAS, he was Chief Operating Officer of the Long Island Power Authority (LIPA), responsible for the day-to-day operation of one of the largest U.S. public power utilities. At LIPA he led a team in the $7 billion takeover of the Long Island Lighting Company.

Before joining LIPA, Mr. Hulkower developed and sought investment opportunities in overseas power projects including development of wind, natural gas, and cogeneration projects overseas at a time when nonregulated and renewable energy project development and financing was in its infancy.

Mr. Hulkower holds an M.S. in Technology and Policy from the Massachusetts Institute of Technology and a B.A. in Economics and a BSME in Mechanical Engineering, both from Tufts University.

Frede Hvelplund is a professor in Energy Planning at the Department of Development and Planning, Aalborg University. He is a member of an interdisciplinary Research Group for Sustainable Energy with interest in public regulation, sustainable energy, political economy, and social anthropology.

His current research is focused on technologies that can facilitate transition from fossil fuel-based to sustainable energy systems. He is author and coauthor of several books and articles on the interrelationships between sustainable energy systems and public regulation.

Dr. Hvelplund's educational background is in Economy and Social Anthropology and he has a Dr. Techn. degree from Aalborg University.

Tooraj Jamasb holds the post of SIRE Chair in Energy Economics at Heriot-Watt University, Edinburgh. He is a research associate at Electricity Policy Research Group, University of Cambridge and MIT Center for Energy and Environmental Policy Research. His research interests include energy sector regulation, reform issues, electricity networks, energy policy, efficiency and productivity analysis, energy technology and innovation, and energy demand.

He is coeditor of the books *Future Electricity Technologies and Systems* and *Delivering a Low-Carbon Electricity System* and *The Future of Electricity Demand*. He has participated on projects for the Council of European Energy Regulators (CEER), several European energy regulators, energy companies, and the World Bank.

He holds a Ph.D. in Energy Economics from the University of Cambridge and has Masters' degrees in Energy Management and Policy from the University of Pennsylvania, the French Institute of Petroleum, and the Norwegian School of Management, and a BBA from Tehran Business School.

Chuck Kutscher is a principal engineer and manager at the Thermal Systems Group at the National Renewable Energy Laboratory (NREL) in Golden, Colorado, where he currently leads the research on parabolic trough solar collector systems. He is an adjunct professor at the University of Colorado at Boulder, a founding fellow of the UCB Renewable and Sustainable Energy Institute, and writes a monthly column for *SOLAR TODAY* magazine.

He is a past chair and a fellow of the American Solar Energy Society (ASES) and was General Chair of the SOLAR 2006 national solar energy conference held in Denver. He is editor of the ASES report, *Tackling Climate Change in the U.S.,* which details how energy efficiency and six renewable technologies can greatly reduce U.S. carbon emissions by 2030.

He has a B.S. in Physics from the State University of New York at Albany, an M.S. in Nuclear Engineering from the University of Illinois at Champaign-Urbana, and a Ph.D. in Mechanical Engineering from the University of Colorado at Boulder.

John A. "Skip" Laitner is Director of Economic and Social Analysis with the American Council for an Energy-Efficient Economy (ACEEE) in Washington, D.C. He previously served as a Senior Economist for Technology Policy for the U.S. Environmental Protection Agency (EPA).

In 1998 he was awarded EPA's Gold Medal for his evaluation of the impact of different greenhouse gas emissions reduction policies and his 2004 paper, "How Far Energy Efficiency?" catalyzed new research in the proper characterization of efficiency as a long-term resource. A prolific author, he has more than 39 years of involvement in the environmental and energy policy arenas and served as an adjunct faculty at the Virginia Polytechnic Institute and State University and the University of Oregon.

He has a Master's degree in Resource Economics from Antioch University in Yellow Springs, Ohio.

Anna LaRue is an independent consultant working on projects focused on reducing energy and water use and solid waste. Previously, she was a Senior Regulatory Analyst with the Pacific Gas & Electric Company (PG&E) in charge of the Zero Net Energy (ZNE) Pilot Program to support ZNE development in California. Prior to that she was an architectural program coordinator with the PG&E Pacific Energy Center. She has been an instructor for Boston Architectural College, a researcher on demand response technology through the UC Berkeley Center for the Built Environment, and has worked for Building Science Corporation and the Cleveland Green Building Coalition.

Ms. LaRue earned a Master of Science in Architecture with a concentration in Building Science from the University of California, Berkeley, and a B.A. in Physics from Smith College.

Noah Long is an energy attorney with the Natural Resources Defense Council in San Francisco, California. He works on energy policy in California and across the west with a focus on electricity and gas utility regulation.

Mr. Long's work and research focus on energy state regulation to align customer and utility incentives toward minimizing the environmental impacts of energy services by increasing investments in energy efficiency and renewable energy sources and preventing investment in dirty fuel sources.

Mr. Long holds a JD from Stanford University, M.Sc. from the London School of Economics, and a B.A. from Bowdoin College.

Margaret Mann is a senior chemical process engineer and group manager at the National Renewable Energy Laboratory (NREL) in Golden, Colorado, where she leads the Technology Systems and Sustainability Analysis Group in the Strategic Energy Analysis Center. She is on the executive board of the American Society of Life Cycle Assessment and is on the editorial board of the *Journal of Power and Energy.*

She has more than 15 years' experience in process design and simulation, process cost analysis, environmental life cycle assessment (LCA), and project management. In 2003, Mann worked with the Department of Energy to form the H2A (for Hydrogen Analysis) group, which established a consistent methodology for conducting analyses on hydrogen systems. She is an expert in analysis of the environmental consequences of various renewable and fossil-based energy conversion systems, including LCAs of coal, natural gas, several biomass power technologies, and hydrogen systems.

She has a B.S. in Chemical Engineering from the University of Colorado.

Mark Mehos is a principal program manager in the Concentrating Solar Power (CSP) Program at the National Renewable Energy Laboratory (NREL) in Golden, Colorado. Mehos has led the High Temperature Solar Thermal Team at NREL since 1998 and has managed the Concentrating Solar Power Program since 2001.

His emphasis with the CSP program has been the development of low-cost, high-performance, and high-reliability systems that use concentrated sunlight to generate power. He has participated on and conducted analysis for several task forces including New Mexico Governor Richardson's Concentrating Solar Power Task Force and the Solar Task Force for the WGA Clean and Diversified Energy Initiative. He is currently the leader for the International Energy Agency's SolarPACES "Solar Thermal Electric Power Systems" task.

He has an M.S. in Mechanical Engineering from the University of California at Berkeley, and a B.S. in Mechanical Engineering from the University of Colorado.

Niels I. Meyer is Emeritus Professor of Physics at the Technical University of Denmark.

Dr. Meyer's main research interests include policies for sustainable energy development with focus on renewable energy systems. Dr. Meyer is past President of the Danish Academy of Technical Sciences and past member of the Danish Energy Supervisory Commission.

Dr. Meyer has a Ph.D. and Dr.Sc. in Physics from the Technical University of Denmark.

William C. Miller works for Sentech, Inc. where he is assigned to the Energy Efficiency and Renewable Energy Program in the U.S. Department of Energy, advising on energy efficiency measurement and utility and legislative matters. Prior to his current position, he managed the Strategic Regulatory Issues group for the Customer Energy Efficiency department at Pacific Gas and Electric Company (PG&E) where he oversaw the company's Zero Net Energy Pilot

Program, a demand side management program that initiates research, development, and demonstration projects to support super-efficient buildings as the most significant challenge in achieving zero net-energy buildings. Prior to joining PG&E, he spent 10 years in academia, engaged in teaching and research.

He has worked for over 25 years in the areas of energy forecasting, strategic market analysis, energy efficiency planning, measurement, policy, and litigation. From 1990 to early 2010, he managed many of the evaluation, policy, and regulatory activities for the energy efficiency department at PG&E.

Dr. Miller received a Bachelor of Science degree in Economics from Stanford University and a Ph.D. in Economics from the University of Minnesota.

Alan Moran is Director of the Deregulation Unit at the Institute of Public Affairs (IPA). His work covers regulatory issues concerning energy, water, housing, and infrastructure.

He has published widely on energy and regulation including the Australian chapter in *Electricity Market Reform: An International Perspective* edited by Sioshansi and Pfaffanberger in 2006 as well as in *Generating Electricity in a Carbon Constrained World* edited by Sioshansi. In 2007 he published *Regulation of Infrastructure* coauthored with Warren Pengilley. In 2010 he was the editor and an author of a compendium including papers from both scientists and economists, *Climate Change: The Facts.*

He has degrees in Economics including a Ph.D. from the University of Liverpool in the U.K.

Jørgen S. Nørgård is Emeritus Associate Professor, Department of Civil Engineering at the Technical University of Denmark.

His research for more than three decades has been focused on combining environmental sustainability with a decent quality of life including improved technology, eco-efficiency, and adaptation of alternative lifestyles that would lead to steady state economics. He has served on the board of various government committees in Denmark.

Nørgård has an education as a Mechanic, followed by an M.S. in Engineering and a Ph.D. in Applied Physics, both from the Technical University of Denmark.

Paolo Polinori is Professor of Economics and Public Economics, Faculty of Political Science, at the University of Perugia, Italy.

His research interests include electricity markets, economic valuation of renewable energy sources, economics and transport policy, and the analysis of the agri-food business. He coordinated research projects on energy markets and efficiency in the public sector.

He holds a Master's degree in Agricultural Economics from the University of Naples and Ph.D. in Agricultural Economics from University of Molise.

William Prindle is a vice president with ICF International, a global energy-environment consultancy. He helps lead the firm's energy efficiency work for government and business clients, including major support work for the U.S.

EPA's ENERGY STAR® and related efficiency programs, as well as utility efficiency program development. He also supports the firm's corporate energy management, carbon management, and sustainability advisory services.

He was previously Policy Director at the American Council for an Energy-Efficient Economy (ACEEE), where he led research and advocacy work on energy and climate policy for federal and state governments. Prior to that, he directed buildings and utilities programs for the Alliance to Save Energy, and was a management consultant before that.

Mr. Prindle received his B.A. in Psychology from Swarthmore College and M.S. in Energy Management and Policy from the University of Pennsylvania.

Nicholas B. Rajkovich is a Ph.D. student in the Urban and Regional Planning Program at the University of Michigan. Prior to entering his Ph.D. program, he was responsible for coordinating a Zero Net Energy Pilot Program for Pacific Gas and Electric Company (PG&E) in San Francisco, California.

His current research focuses on coupling climate change mitigation and adaptation strategies in the built environment. Prior to working for PG&E, Rajkovich taught several courses on lighting, acoustics, and building systems in the Department of Architecture at Cornell University. He also was an associate at Einhorn Yaffee Prescott Architecture and Engineering, P.C. in Albany, New York.

Rajkovich received a Bachelor of Architecture from Cornell University and a Master of Architecture from the University of Oregon. He is a licensed architect and a LEED Accredited Professional.

Olivier Servais is Associate Professor at Université catholique de Louvain where he teaches Anthropology of Nature, Environment and Symbolic Systems.

His research interests are focused on values, religions, and their relation to the natural environment. He has conducted or participated in a number of multidisciplinary research projects including collecting and analyzing both quantitative and qualitative data in Belgium, Cabo-Verde, Canada, and Guatemala.

He was educated at Université catholique de Louvain in Belgium, University of Paris-Sorbonne in France, and holds a Masters in History and a Ph.D. in Anthropology.

René Sigg is a Principal at Intep, a consulting firm with offices in Germany, Switzerland and the U.S., reflecting 20 years of sustainable design and integrated planning experience. René Sigg has consulted in Europe and the U.S. for large companies and organizations in the industrial, finance, health care, and public sector.

Mr. Sigg received his Environmental Science degree from the University of Zurich.

Fereidoon P. Sioshansi is President of Menlo Energy Economics, a consulting firm based in San Francisco, California serving the energy sector. Dr. Sioshansi's professional experience includes working at Southern California Edison Company (SCE), the Electric Power Research Institute

(EPRI), National Economic Research Associates (NERA), and most recently, Global Energy Decisions (GED), now called Ventyx. He is the editor and publisher of *EEnergy Informer*, is on the Editorial Advisory Board of *The Electricity Journal*, and serves on the editorial board of *Utilities Policy*.

Dr Sioshansi's interests include climate change and sustainability, energy efficiency, renewable energy technologies, regulatory policy, corporate strategy, and integrated resource planning. His three recent edited books, *Electricity Market Reform: An International Perspective*, with W. Pfaffenberger, *Competitive Electricity Markets: Design, Implementation, Performance*, and *Generating Electricity in a Carbon Constrained World* were published in 2006, 2008, and 2009, respectively.

He has degrees in Engineering and Economics, including an M.S. and Ph.D. in Economics from Purdue University.

Benjamin K. Sovacool is Assistant Professor at the Lee Kuan Yew School of Public Policy, part of the National University of Singapore. He is also a Research Fellow in the Energy Governance Program at the Centre on Asia and Globalization.

Dr. Sovacool has worked as a researcher, professor, and consultant on issues pertaining to energy policy, the environment, and science and technology policy. He has served in advisory and research capacities at the U.S. National Science Foundation's Electric Power Networks Efficiency and Security Program, Virginia Tech Consortium on Energy Restructuring, Virginia Center for Coal and Energy Research, New York State Energy Research and Development Authority, Oak Ridge National Laboratory, Semiconductor Materials and Equipment International, U.S. Department of Energy's Climate Change Technology Program, and the International Institute for Applied Systems and Analysis near Vienna, Austria. Dr. Sovacool has published four books and more than 80 academic articles, and presented at more than 30 international conferences and symposia.

He holds a Ph.D. in Science & Technology Studies from the Virginia Polytechnic Institute & State University, an M.A. in Rhetoric from Wayne State University, and a B.A. in Philosophy from John Carroll University.

Roland Stulz is Executive Director of Novatlantis at the Swiss Federal Institute of Technology (ETH) in Zurich. This program has the mission to translate cutting-edge technology research relevant for sustainability into practical applications, especially in the fields of mobility, renewable energy, and building construction. Mr. Stulz has many years of experience as an architect, urban planner, and consultant for energy efficiency and ecology in building construction.

He has been involved in many leading pilot and demonstration buildings in the field of sustainability including issues of appropriate technologies and low-cost building constructions in Europe, the U.S. and in developing countries. Mr. Stulz is a member of the board of the Sustainability Forum Zurich, the steering committee of the Center of Competence for Energy and Mobility at ETH, and has been president of the Energy Commission of the Swiss

Association of Engineers and Architects (SIA). He is the initiator and cofounder of the International Sustainable Campus Network.

Mr. Stulz graduated in Architecture at the Swiss Federal Institute of Technology (ETH) Zurich in 1970.

Stephan Tanner is a Principal at Intep, a consulting firm with offices in Germany, Switzerland and the United States of America. He is experienced in project development, planning, and design, combining his professional interest with a passion for people and their diverse requirements for living space. He has led a number of large scale, high-performance development projects in the U.S., Europe, China, and Korea. In the U.S. he is known as the architect of the first certified Passive House Standard building the implementation energy goal of the 2000-Watt Societies in buildings.

Mr. Tanner is a licensed architect and started his career in Switzerland as an architect apprentice and graduated from the Minneapolis College of Art & Design as a licensed architect.

Ted Trainer is a Visiting Fellow in the School of Social Work, University of New South Wales, where he has taught on sustainability since the 1970s.

Trainer's main focus has been radical critiques of industrial-consumer society and alternatives focused on less energy-intensive ways. He has written numerous books and articles on these topics including *The Conserver Society: Alternatives for Sustainability*; *Saving the Environment: What It Will Take*, and *Renewable Energy Cannot Sustain a Consumer Society*. He is also developing Pigface Point, an alternative lifestyle educational site near Sydney, which hosts guided tours explaining The Simpler Way and a website for use by critical educators, *http://ssis.arts.unsw.edu.au/tsw/*

Trainer holds a Ph.D. in Sociology of Education from Sydney University.

Christoph Weber is the Chair for Management Sciences and Energy Economics at the University of Duisburg-Essen. He was head of the Department of Energy Use and Management at IER, University of Stuttgart.

His main research interests are in energy risk management, energy market liberalization, and application of operations research methods to energy issues. He has published in national and international journals on these issues.

He holds a Diploma in Mechanical Engineering from University of Stuttgart with a specialization in Energy Technology and obtained his Ph.D. in 1999.

Alison Wise is Director of Career Services at Ecotech Institute. She previously was a senior strategic analyst at the National Renewable Energy Laboratory (NREL) in Golden, Colorado. Her professional experience spans almost 20 years in strategies surrounding environmental sustainability issues including policy analysis and advocacy, socially responsible investing, stakeholder engagement, and market research analysis. She founded a trade association of environmentally sustainable businesses called Sea Change, was head of public policy for Future 500, and the head of market research for Clean Edge, a market research firm focused on the clean tech sector.

Her interests include the intersection between network dynamics and distributed renewable generation, carbon pricing, and digital energy infrastructure. Her most recent project is a web presence called the Clean Energy Economy Gateway, a virtual portal for aggregating and disseminating knowledge about this new economic system.

She has undergraduate degrees in History and Biology from Reed College, and an MBA from the University of Oregon.

Jay Zarnikau is president of Frontier Associates where he provides consulting assistance to utilities and government agencies associations in the design and evaluation of energy efficiency programs, retail market strategies, electricity pricing, demand forecasting, and energy policy. He is also a part-time visiting professor at The University of Texas LBJ School of Public Affairs and the College of Natural Sciences Division of Statistics.

Zarnikau formerly served as a vice president at Planergy, manager at the University of Texas at Austin Center for Energy Studies, and a division director at the Public Utility Commission of Texas. His publications include over 20 articles in academic journals as well as numerous articles in trade publications and conference proceedings. His research interests include the pricing of energy and water resources, electric utility planning, the measurement of national energy efficiency achievements, and the selection of functional forms in energy demand modeling.

Zarnikau has a Ph.D. in Economics from The University of Texas at Austin.

Nick Zigelbaum is a freelance consultant specializing in energy policy for buildings and appliances. Formerly he was an energy analyst with the Natural Resource Defense Council where he focused on legislative and regulatory advocacy for state and federal building energy codes in the U.S. and China. Prior to joining NRDC he worked at a construction company involved in the construction of Cape Cod's first wind turbine in Massachusetts and managed energy-efficient design projects including zero-carbon, zero-waste, and zero-energy developments in the Bay Area.

His current interests include advancing the energy efficiency and environmental synergy of the built environment through policy, market, and regulatory driven initiatives.

He earned his B.S. in Mechanical Engineering from Cornell University's College of Engineering. He is also a certified CHEERS rater in the state of California.

Introduction

Can We Have Our Cake and Eat It Too?

Fereidoon P. Sioshansi
Menlo Energy Economics

INTRODUCTION

There is a growing recognition among some experts that humankind may have gotten itself into an unsustainable path – *ecologically*, *economically* and *ethically*[1]. Those who subscribe to this belief point out that we have, perhaps unwittingly, adopted a socio-economic system that encourages resource extraction, industrial production, consumption and economic growth as if these were the sole means to an end, namely prosperity, high living standards, and human well-being[2]. Most economists *define* and *measure* the well-being of a country by looking at its gross domestic product (GDP), per capita income, or economic growth rate, metrics that are increasingly being questioned as inadequately measuring a country's overall welfare or status.[3] Today, there is

1. Trainer's chapter provides a rather bleak assessment of this predicament, supported by Meyer et al. But, of course, such views are not universally accepted.

2. Moran, for example, points out that we have evolved the system because – despite its shortcomings – it is the best there is (personal communications). Speaking about democracy, Winston Churchill reportedly said that despite its many problems, it is the best form of government anyone has come up with.

3. Referring to a report co-authored by the Nobel laureate Joseph Stiglitz and Jean-Paul Fitoussi, in a speech on 14 Sept 2009, French President Nocolas Sarkozy proposed that factors such as happiness, long holidays and a sense of well-being be included in an overhaul of the world's accounting systems. Mr. Stiglitz has long proposed a multi-attribute measure of a country's well-being not exclusively focused on GDP.

growing interest in measuring prosperity, happiness and human welfare by including non-economic attributes.[4]

Our adopted capitalistic, profit-motivated and consumption-oriented socio-economic system has resulted in highly unequal distribution of income, typically across but also within societies, some enjoying extremely lavish standards of living sustained at considerable ecological cost while others live in abject poverty. The current debate on how to address climate change and who should bear the costs – to a large extent – is a manifestation of these inequalities. It pits the rich, who have attained high living standards by emitting enormous quantities of carbon into the atmosphere in the past, against emerging economies, who aspire to reach the same living standards following a similar development path – namely reliance on fossil-fuels and other natural resources.

In this context, we examine the following four important questions:

- Are we in a bind?
- How did we end up in this bind?
- Where do we go from here?
- How can we get there?

ARE WE IN A BIND?

As we enter the 21st century, there are growing concerns about the long-term viability and *sustainability* of the business-as-usual paradigm.

Some experts, for example, debate how much of the remaining finite reserves of fossil fuels can be extracted at reasonable cost. More importantly, there are serious concerns about mankind's long-term impact on the environment. With growing population and rapidly rising aspirations for higher living standards in developing countries, we will not only need more energy, but more water, more food, more mobility, more of everything for more people at *affordable* prices.

Human ingenuity and technological advancements have historically gotten us out of the bind[5]. New York City was not buried under horse manure as was once feared, neither did the stone age end because man ran out of stones, as the

4. A recent survey by Legatum, for example, ranks Finland as the most prosperous nation in the world – not because it has the highest per-capita GDP but because of other attributes including income equality, good institutions and a decent education, health and welfare system. US, by comparison, was ranked #9. Further details at http://www.prosperity.com/rankings.aspx. Similarly, the Human Development Index (HDI), developed by the United Nations Development Program (UNPD) in the 1990s proposes to shift the focus of development economics from national income accounting to people centered policies and measures. HDI is a composite measure that evaluates development not only by economic advances but also improvements in human well-being.

5. There is a debate among those who believe in technological fixes vs. those who believe technology can only go so far in addressing longer-term scarcity issues. For an example, see R. Heinber's Searching for a Miracle: Net Energy Limits & the Fate of Industrial Society, Sept 2009.

famous saying goes. But there are growing concerns that with rising population and growing demand for higher living standards, we must seriously consider major adjustments in how we use energy – and other natural resources – and how the timing of these choices may affect the quality of life for our children and grandchildren.

As developing economies continue to grow, the ranks of the global middle class are rapidly increasing. With over 2 billion living in China and India alone, a new consumer society who can afford higher standards of living is emerging. Many citizens of oil-rich Persian Gulf states already have higher per capita energy and carbon footprints than those in the West. Many experts believe that feeding, housing and providing for the growing ranks of the global middle class may push us beyond what is practical with the natural resources and technologies at our disposal. In the words of Trainer, whose views are further described in the book,

"There is no possibility that the per capita rates of resource consumption the 1.5 billion rich (currently) have can be extended to the other 7.5 billion we will soon have on Earth. The Australian footprint of 8 ha of productive land per capita is about 10 times the area that will be available per capita in 2050."

Trainer is among those who are predicting that we will run into resource shortages, the ecological consequences of climate change, destruction of soils and forests, and loss of critical habitat for species long before 2050 unless we get on a path different than the *business-as-usual*.

While such views may be in the minority today, they cannot be ignored or dismissed. Aside from the issue of resource scarcity, there is the growing inequalities among the rich and the poor, not just in terms of income or energy per head but in terms of per capita carbon and ecological footprint. Many view these as unsustainable, unjust, and unethical.

Another issue receiving increased attention in recent years, of course, is climate change and how to allocate the significant costs needed to address the problem given the highly unequal contribution of the rich and the poor on a per capita basis – not to mention the unequal contribution of the rich in the past and the poor going forward.

The short answer to the question "are we in a bind?" may be that, at a minimum, we have to make significant adjustments in our business-as-usual economic, social and ecological paradigms to avoid impending scarcities including the worst consequences of climate change[6]. Technological fixes will no doubt help, and our economic system has plenty of built-in, self-correcting mechanisms to deal with emerging scarcities through powerful price signals that drive investments to address emerging problems.

---

6. A recent study compiled by the US National Academy of Science, released in May 2010, for example, concluded that "Climate change is occurring, is caused largely by human activities, and poses significant risks for – and in many cases is already affecting – a broad range of human and natural systems."

At least one contributing author to this book does *not* subscribe to the view that we are in fact in a serious bind while another believes that our existing socio-economic system – assisted by technological fixes and adequate investment – offers the means to get us out of the bind. Climate change, for example, is viewed as yet another difficult challenge that we shall, in time, overcome. Others are not so sure.

HOW DID WE END UP IN THIS BIND?

Since the dawn of the industrial revolution in 1880s, humankind has enjoyed a period of sustained – if geographically uneven – economic growth accompanied by significant population growth. Much of this can be attributed to our ability to exploit the earth's abundant natural resources, particularly but not exclusively fossil fuels, literally and figuratively, to feed the growing population and the economic engine.

While segments of the world's population remain in abject poverty and subsist on extremely low per capita food and energy diets, a growing portion of the world population enjoy high standards of living made possible, to a large extent, thanks to *plentiful and cheap energy*, and one might add, because we have been able to ignore the significant externality costs associated with our unsustainable habits up to now[7].

In developed countries, plentiful and cheap energy makes it possible for a small fraction of the population to feed the masses of urban dwellers[8]. Cheap and plentiful energy also provides affordable transportation and mobility, allowing people and goods to travel long distances to destinations and markets around the world. Moreover, cheap and plentiful energy offers a wide range of affordable *energy services*, allowing us to live in comfortable homes, work in functional offices and factories, and provides the means of transportation to and from home to work. Finally, cheap and plentiful energy defines our travel and leisure habits, allowing a growing number of people to travel on business or for pleasure to locations far from where they live.

More broadly, our entire economic system is predicated on cheap and plentiful energy as are the design of our infrastructure and layout of our urban centers. It is not an exaggeration to say that our comfortable lifestyles – for those who are privileged to enjoy it – is very much dependent on continued access to cheap and plentiful energy.

---

7. A 2009 study by the National Research Council puts the externality costs associated with energy production & consumption in the US at a staggering $120 billion per annum – and this figure does not include climate related costs nor the loss of habitat. Further details at Hidden Costs of Energy: Unpriced Consequences of Energy Production & Use, NAS, Oct 2009

8. Food Inc, a documentary movie nominated for 2010 Academy Awards, describes how multinational corporations produce large amounts of "food" at extremely low cost and significant ecological footprint.

In *$20 per gallon* (2009), Christopher Steiner asks what would happen should oil prices rise to levels far above what we have historically taken for granted. The short answer is that if gasoline prices were to rise to, say, $20 per gallon, as the book's title suggests, nothing will be the same. Habits and practices that are routine and make perfect sense today become exorbitantly expensive or outright impractical. Our world, in other words, will turn upside down under such a scenario.

The short answer to the question "how did we get into this bind?" may be that our current economic paradigm – which is based on ever increasing production, consumption and growth – has resulted in an arguably wasteful and potentially unsustainable consumer-oriented culture. Our economic system encourages growth and profit on ever increasing scale through globalization of manufacturing and trade with little regard for sustainability or negative externalities associated with use of natural resources. The intractable nature of climate change is a mere manifestation of an economic system that has not put an appropriate price on emissions of greenhouse gases up to now – and that may merely be the tip of the iceberg.

WHERE DO GO FROM HERE?

As the following chapters of this book explain, there is no universal consensus on where we want to – or should – go given where we are and how we have gotten here. Most, if not all, of the contributors to this book would probably subscribe to the notion that we need to change, modify or adjust our ways, our economic paradigm, and/or our lifestyles to get on a more sustainable path:

- Some come to this conclusion by looking at economic or price indicators while others are compelled by environmental, ecological or ethical issues;
- Some are in favor of gradual approaches while others propose more abrupt and radical ideas; and
- Some favor that the changes be introduced primarily or exclusively through price signals and financial incentives within our *current* economic systems while others believe that we need to go beyond existing mechanisms by imposing regulations and/or resorting to strict and mandatory command and control measures.

HOW CAN WE GET THERE[9]?

Given the fact that it has taken us a couple of centuries to get to where we are suggests that it may take an equally long while to get on a different path,

9. Book's Epilogue reflects on the same issues

assuming we can agree on which path we need to be on. The current disagreements on reaching a binding international treaty to combat climate change is an example of different beliefs on the gravity of the problem, the costs and benefits of inaction or delayed action, and serious disagreements on what needs to be done, by whom, at what cost, and when. There is, of course, drastic differences of opinion on how fast we need to change our ways – driven by differences of opinion on how much time we have before resource scarcities and/or ecological catastrophes may confront us[10].

The following chapters of this book provide a lively debate on the nature, scale and seriousness of the problems while offering suggestions, insights and solutions on how they may be addressed.

As the book's editor, I have allowed the contributing authors the maximum flexibility to express their views and perspectives while encouraging them to support their ideas and conclusions by rigorous analysis and reasoning. This means that there are disagreements among the contributors on how serious the problems are and what needs to be done to address them. I believe that the readers would benefit from such diversity of opinions and perspectives.

SETTING THE CONTEXT

A previous volume titled *Electricity Generation in a Carbon Constrained World* (2009) explored the various means of de-carbonizing electricity generation. That book's focus was on the electric power sector since, along with transportation, it is among the most significant emitters of GHGs.

The book examined a wide range of options to reduce the carbon footprint of the electric power sector, from increased reliance on renewable energy resources, to more nuclear energy, to carbon capture and sequestration (CCS) on a massive scale. It also examined energy efficiency – using less energy by using it wisely and sparingly.

The prior book's epilogue, which is particularly germane to the present volume, is repeated below in its entirety:

> When I started working on this project (the preceding volume), my genuine hope and expectation was that by capturing the inherent synergies in the options and solutions examined by the various contributing experts to this book, one would be able to claim that the problem of carbon emissions associated with electricity generation could be successfully addressed. Moreover, I was convinced that while the electricity sector faces a challenging transition to a lower carbon future, there were opportunities to become a part of the solution – rather than remain part of the problem.

---

10. Moran, for example, refers to studies that examine the relative merits of taking drastic measures today given what we know vs. waiting a while before deciding. There are advantages and disadvantages to these alternatives.

Along the way, I was confronted by two surprises and one insight – perhaps trivial to every one else. The first surprise was that as I learned more about the scale of the (climate change) problem, I became alarmed about the sheer immensity of the challenges ahead. The second surprise was that as I studied the various chapters describing the obstacles and limitations of each technology or solution (to de-carbonize electricity generation), it began to dawn on me that the task at hand is more daunting than I had originally imagined. My personal insight – and I must emphasize that this is *not* necessarily shared by the contributors to the book and should *not* be attributed to anyone other than myself – is that while a lot can, and should, be done on the electricity *generation side*, in all likelihood, these efforts will *not* be sufficient. If the goal is to limit concentration of carbon in the atmosphere to, say, 450-550 parts per million by 2050, as many scientists have suggested, I am now convinced that we must work equally hard on the *demand-side*, and that effort should not be limited to improving the efficiency with which electricity – or energy – is used but on altering the nature of the way we *work* and ultimately examining our *values* and *lifestyles*.
The excerpt below from the Preface to the (preceding) book by **Wolfgang Pfaffenberger** states the same, but more eloquently:

> "As the chapters of this book make clear, we cannot rely entirely on changes in the supply-side of the equation to reduce the industry's carbon footprint. Changes in the demand-side as well as changes in energy consumption habits – and perhaps more profoundly – lifestyles changes may ultimately be needed to address the carbon problem."

The present volume essentially starts where the prior volume ended.

SCOPE AND OUTLINE OF THE BOOK

The preceding discussion illustrates that man's ingenuity and technological prowess to deliver vast quantities of fossil fuels at affordable – and declining – prices over many decades has resulted in an economic system, infrastructure, lifestyle and habits that assumes the continuation of the same for an indefinite future. If you believe that the status quo can go on indefinitely with little or no modifications, no need to bother with the rest of this chapter or the book.

By all accounts, there are ample reserves of oil, natural gas, coal, and uranium to feed our energy needs for decades to come. Renewable energy resources are abundant and with improved technology will become more economic and prevalent. Moreover, energy and resource scarcities will inevitably lead to rising prices, which will in turn lead to adjustments in how and how much resources are produced and consumed. The system, given enough time, will self-regulate.

This book *is not* about doing without or sacrificing high living standards that many have come to enjoy and take for granted. On the contrary, this book *is* about how more of us, globally speaking, can enjoy adequate living standards while living in harmony with the environment. This entails using energy *wisely*, *efficiently* and *sparingly* – perhaps *frugally*. As further described in Chapter 1, it asks if we need a tank to kill a fly? Do we need to drag 2 tons of steel to pick up a carton of milk from the corner grocery store?

This book is about what it takes to keep the beer cold and the shower hot, using the famous words of Amory Lovins, the acknowledged energy efficiency guru. But as has been pointed out, keeping the beer cold and the shower hot can be accomplished by many means and in many ways. The hot shower, for example, may be provided by absorbing the heat of the sun in a solar hot water heater, by burning natural gas in a water heater, by absorbing the exhaust heat from a fossil-fueled power plant, or a variety of other ways. Each option has certain advantages and disadvantages and entails a different energy/carbon footprint.

This book explores, not only which option has the lowest energy/carbon footprint, but asks if we can also install a low-flow showerhead to economize the amount of water used for the hot shower and –moreover – what does it take to encourage taking a shorter shower[11]? If the intent is to enjoy a hot shower with minimum ecological footprint, a multitude of factors must be taken into account.

The book's chapters are organized into three complimentary parts briefly outlined below:

Part 1 of the book, ***challenge of sustainability***, examines the fundamental drivers of energy demand – economic growth, the need for basic energy services, and the interdependence of economic, political, environmental, social, equity, legacy and policy issues.

In chapter 1, "Why do we use so much energy, and what for?" **Fereidoon Sioshansi** examines what constitutes an *adequate standard of living* and how much energy – or more accurately *energy services* – is required to support such a lifestyle?

As it turns out, this is not a trivial question. There are many reasons why different societies and cultures use vastly different amounts of energy – not entirely explained by differing income levels, climate, population density, energy intensity or energy prices. Another complicating factor is that vast amounts of energy are wasted between the primary source and the final desired end use or application.

The chapter examines how, and how much, energy is currently used to support our modern lifestyles and suggests that dramatic savings, of the order of 90+%, may be possible in many applications if we can reorganize how we go about meeting the ultimate services delivered by energy – with important

11. The chapter by Prindle & Finlinson indicates that the average shower taken by university residence hall dwellers in the US is 14 minutes. How can we motivate these students to take shorter showers?

implications for sustainable energy consumption – the main theme of the book.

In chapter 2, "Which energy future?" **Frank Felder, Clinton Andrews**, and **Seth Hulkower** examine alternative future scenarios with different trajectories of population, economic, energy and GHG emissions growth to 2050 and beyond and describe their implications.

The authors point out that given the lasting environmental impacts of energy production and consumption and the long lead time needed to change the global energy infrastructure, envisioning future energy possibilities is critical, although fraught with challenges and uncertainties. The chapter examines two major countries – China and the US – with significant energy and carbon footprints – in detail.

The chapter's conclusions are that energy policy and planning can be directed at economic development, geopolitical security, environmental enhancement and public acceptability. Major forecasts of the long-term future global energy picture will fall short of achieving these four objectives. Any sizeable improvement over the business as usual scenario requires major technological advances, changes in attitudes towards the environment, and substantial world-wide cooperation.

In chapter 3, entitled "Energy 'need', desire and wish: anthropological insights and prospective views," **Françoise Bartiaux, Nathalie Frogneux** and **Olivier Servais** argue that human energy "needs" do not provide the right basis for projecting energy demand.

This chapter begins with a deconstruction of the notion of "needs" in general, and energy "needs," in particular by recasting energy consumption and production into sociopolitical components in the context of our common human condition, a call for social and cultural diversity, the paradigm of climate justice, and the crucial role of local energy policies.

The chapter's main conclusions are that individual and tradable carbon quotas could respect the relative necessity of "needs" as well as the paradox of the choices to be made for others in the context of global ecological constraints. A few prospective suggestions are made to make carbon quotas more efficient in reducing social inequalities about energy "need", desire and wish.

In chapter 4, "Equity, economic growth and lifestyle," **Niels I. Meyer, Frede Hvelplund** and **Jørgen S. Nørgård** point out that the current slow pace of political response to climate change is increasing the risk of passing a number of critical *tipping points* despite the fact that many of the required policy options are within reach.

The authors describe a number of driving forces for this risky development and analyze the barriers to efficient and timely action. These include addressing population growth, lack of equity between and among nations, vested interests and a socio-economic system with priorities other than sustainability.

The chapter presents a number of policy proposals that may support sustainable development before it is too late. This includes alternative concepts concerning *employment*, balance between *state* and *market,* and alternative priorities in our socio-economic systems and cultures exemplified by selected case studies.

In chapter 5, "Why we *can't* have our cake and eat it too", **Ted Trainer** argues that underlying the current debate on climate change is the fundamental assumption that the problem *can* be solved within our current socio-economic system while maintaining continued growth, affluent lifestyles and acquisitive consumer-oriented culture. The chapter casts doubt on this premise and explains why sustainability is fundamentally inconsistent with today's consumer-capitalistic system and values – a rather sobering message and a minority view, but one that cannot be ignored or dismissed.

The author points out that the true magnitude and seriousness of the global climate predicament facing humankind is not widely recognized and that levels of production and consumption in rich countries are far beyond those that are sustainable for long or could be extended to those in developing countries. Sustainability, it is argued, requires *significant reductions* in consumption of natural resources, which cannot be achieved by technical or legislative means alone.

The chapter concludes that renewable energy resources cannot indefinitely sustain the current affluent lifestyles and historical economic growth rates. The salvation has to be a transition to a simpler lifestyle based on frugality, mostly small and highly self-sufficient local economies and very different values.

In chapter 6, "Will there be the will and the means?" **Carlo Andrea Bollino** and **Paolo Polinori** explain how we can have our cake and eat it too, namely how we can achieve sustainability while maintaining high standards of living – not just for the rich but for the developing world – in stark contrast to the preceding chapter and the one that follows. The chapter argues that it is difficult to find solutions to address climate change, not because the costs are excessive, but because there is lack of international cooperation. From a macro-economic perspective, the size of required sacrifice is comparable with past periods of large-scale investment.

Pointing to historical precedents – e.g., the Marshall Plan after WWII – the authors argue that we can rely on existing technical, legal, regulatory – and most importantly – financial incentives and economic price signals – to make the necessary transition to a sustainable future. The transition may not be painless or cheap and it will take time, but is feasible, provided there is the political will to justify the means.

The chapter concludes that there is no need for measures which are more severe than what was done in the past. The political problem of coordination – as exemplified by the failure of Copenhagen – can be solved by a combination of standard tax and credible punishment measures.

In chapter 7, "Is it possible to have it both ways?" **Alan Moran** explores the implications of emission limits estimated to be necessary to limit global temperature rises while maintaining high living standards in rich countries and without impinging on rapid economic growth in developing countries. It takes a rather different perspective on the question of whether we can have our cake and eat it to, addressed in the two preceding chapters.

The chapter is based on the concept of resource scarcity. When resources become scarce, this results in higher prices, which lead to conservation of these resources as well as a search for more of them and for their substitutes. It examines the means of imposing a global carbon limit; the economic, social, and political implications of attempting to force marked reductions in emission levels in developed countries and preventing increases in the developing world; and what technological measures could emerge to deliver the reduction levels said to be required and how likely are these developments to be forthcoming.

The author concludes that fossil fuel energy has been basic to creating a world that supports more people than ever with a degree of affluence, even for the poorest, that far exceeds any previous levels. He points out that peer reviewed economic analyses of the costs of global warming range from a positive one to a negative 2.5 per cent in the context of a per capita growth of 66 percent even in highly developed countries. He argues that markedly reducing greenhouse gases on our present knowledge base, would require a very high price for energy with unknown, but almost certainly highly adverse, consequences for world income levels.

In chapter 8, "Efficiency first: Designing markets to save energy, and the planet," **Noah Long, Pierre Bull** and **Nick Zigelbaum** discuss successful policies to advance energy efficiency in buildings, appliances and utilities.

The policies are categorized into three areas: minimum performance standards, removing market barriers to efficiency and cost effective incentives to achieve high levels of efficiency in buildings, appliances and within the utility sector. The policies focus on realigning market incentives, removing barriers to promote innovation, reward successful technologies and practices that reduce energy use while meeting energy service demands.

The chapter concludes that appropriate policy drivers can and should be seen as an engine for improving energy efficiency and reducing the societal cost of energy services.

Part 2 of the book, ***technological solutions***, examines how energy can be used smartly, judiciously, sparingly and frugally to support the basic energy service needs of homes, commercial and industrial facilities and for other applications. As it turns out, enabling technology is often necessary but not sufficient to deliver the desired outcomes.

In chapter 9, "Getting to Zero: Green Building and Net Zero Energy Homes," **Meredith Gray** and **Jay Zarnikau** examine how energy is currently used within the household sector and how current uses can be altered/modified over time to achieve drastic reductions in energy consumption.

The authors examine the concept of the *net zero energy home* of the future and ways to reduce carbon footprint of typical household. Also included are an examination of behavioral issues and enabling technologies that provide the means and the motivations for improved energy use. Green building efforts, such as the LEED certification, are discussed to determine if an effective standard exists. The discussion ties in with the chapter that follows on smart grid technologies and a related chapter under the case studies.

The chapter concludes that energy usage in households can be reduced by one-quarter to one-half using existing commercially available technologies and practices. Green taxes, higher energy prices and consumer behavior modification through educational campaigns are considered viable options. Significant savings can be achieved in lighting, refrigeration and by integrating in-home sources and uses of heat. Key barriers include cost, longevity of the building stock, and lack of credible information. Ultimately, net zero energy homes are possible with upgrades, regulation, and consumer behavioral shifts.

In chapter 10, "Beyond the meter: Enabling better home energy management," **Karen Ehrhardt-Martinez, John A. "Skip" Laitner** and **Kat A. Donnelly** look beyond the utility-based benefits of new smart grid technologies and smart meters to identify how households, individuals and communities can empower themselves, reshape current energy consumption patterns, and reconfigure energy production systems.

The authors examine existing patterns of household energy consumption as well as the potential role of advanced metering technologies, in-home display devices, and behavior-savvy feedback programs for informing and motivating better energy practices and reducing energy consumption. The chapter illustrates the range of new feedback technologies from web-based initiatives to real-time in home displays that can provide consumers with appliance-specific energy consumption data and documents past feedback studies that have achieved savings of 4 to 12%.

Based on an assessment of the specific energy savings attributable to distinct feedback mechanisms and important behavioral changes that people have reported making in their habits, lifestyles and choices, the authors conclude with estimates of the potential economy-wide energy savings that well-designed, behavior-savvy feedback programs could achieve.

In chapter 11, "How Organizations Can Drive Behavior-Based Energy Efficiency," **William Prindle** and **Scott Finlinson** examine current energy performance trends, benchmarks in commercial and organizational settings and how behavior-based energy efficiency strategies can change performance in the commercial sector.

The chapter examines best-practice energy management programs in corporations, universities and other organizations and shows how organizational-level behavior is necessary to create the framework for a future of zero-net-energy performance.

The authors conclude that by combining information systems, motivation systems, organizational culture, and technology solutions into efficiency programs leading organizations can go far beyond conventional technology-based results. The outcome is substantial reductions in energy consumption, leading to wider and deeper innovations in energy and other resource use.

In chapter 12, "Reinventing industrial energy use in a resource-constrained world," **Marilyn Brown, Rodrigo Cortes** and **Matthew Cox** point out that in an increasingly competitive and carbon-constrained world, improving the energy efficiency of industry is essential for maintaining its viability, especially as energy-intensive manufacturing shifts to developing countries.

The chapter examines the energy productivity of US industry and shortfalls in adopting energy-efficient and low-carbon technologies and fuels. It surveys the advanced engineering, materials, and information technology concepts that could transform industrial processes. The authors also focus on the key role of industry as a technology innovator, supplying the economy with next-generation energy technologies.

The chapter concludes that we need a new vision of industry – factories-of-the-future with minimal resource requirements, that clean up our ecosystems, contribute to human health, produce valuable goods, and improve standards of living.

In chapter 13, "Prospects for renewable energy," **Douglas Arent, Paul Denholm, Eason Drury, Rachel Gelman, Chuck Kutscher, Margaret Mann, Mark Mehos** and **Alison Wise** cover the broad range of renewable energy resources, their technical potential, long-term prospects, and challenges, in meeting a growing percentage of world's energy needs.

The chapter examines the technical, economic and policy issues affecting future investment in renewable energy resources and barriers that may impede their penetration. With double-digit growth rates experienced in the recent past due to rapid technology advancement and mass production, some renewable technologies are cost competitive and rapidly gaining market share.

The authors conclude that continued technological progress, economies of scale and strong policy support will make renewable energy an increasingly attractive alternative to traditional fossil fuels. In this context, it is not unrealistic to envision a scenario for a distant future where renewables may provide as much as 80% of US energy needs – assuming that a number of significant obstacles can be overcome.

In chapter 14, "Future of heating when little heating is needed," **Christoph Weber** and **Klaas Bauermann** examine energy demand for space heating in a future where buildings are much better insulated and heating technologies are significantly more efficient – requiring little heating – including an assessment of the implications for traditional providers of district heating, popular in parts of Europe.

The chapter provides an overview of building energy use, primarily focusing on heating, and why future developments in this area are crucial to meeting energy reduction goals, especially in colder climates. Using Germany as a case study, options to reduce buildings' heating load are examined.

The chapter concludes that the energy saving potential of the built environment is huge. High standard for new construction and regulations mandating improved insulation of the existing building stock, respectively, are the logical first step. Additionally, substitution of the current heating systems with more efficient varieties is far more effective. As heating systems become more efficient and buildings become better insulated, providers of district heating must rethink their core business strategies.

Part 3 of the book, ***case studies***, covers a number of innovative projects, initiatives, concepts or self-imposed targets in different parts of the world with the aim of significantly reducing energy use and carbon footprint of a company, a community, a city or an entire country. Following the failure of the UN-sponsored conference in Copenhagen in Dec 2009, many experts believe that such self-imposed initiatives, supported by local communities or through corporate leadership or philanthropies may pave the way for meeting more concrete and binding global targets.

In chapter 15, "Why China matters," **Bram Buijs** points out that given the size and rapid development of China, its future and the future of the globe are increasingly intertwined, especially when it comes to the fate of climate change and future global sustainability.

The chapter examines China's current coal-based energy system and energy policy and outlines the key obstacles that the country faces in attempting to decouple its economic development from growth in energy demand and carbon emissions.

The author concludes that China offers huge mitigation opportunities, as it is still in the midst of massive infrastructure development. However, an analysis of alternative growth scenarios suggests that even with more ambitious policies it will be extremely challenging to reconcile China's growing energy demand with desired global emission reduction objectives.

In chapter 16, "Swiss 2,000-Watt Society: A sustainable energy vision for the future," **Roland Stulz, Stephan Tanner** and **René Sigg** provide a vision of future where sustainability *and* high standard of living may co-exist on a meager energy diet and carbon footprint.

The authors describe the concept of Swiss 2000-Watt Society from its inception to adoption by cities of Basel and Zurich and provide an update. The chapter highlights the lessons learned so far and the remaining challenges while speculating on the implications of wide-scale adoption of similar concepts elsewhere and its potential impact on quality of life and sustainability on a global scale.

The chapter concludes that sharing and communicating a common long-term sustainable energy vision, such as the Swiss 2,000 Watt concept, helps focus research and implementation for the challenges we face and can gain public and political buy in, as demonstrated by the referendum in the City of Zurich in the fall of 2008.

In chapter 17, "Zeroing in on Zero Net Energy," **Nicholas Rajkovich, William Miller** and **Anna M. LaRue** discuss a goal in California to achieve

ZNE in residential buildings by the year 2020, commercial buildings by 2030. The chapter ties with the discussion by Gray and Zarnikau as well as the Clymer on the same topic.

Under the auspices of the California Public Utility Commission, the California investor owned utilities have proposed a series of energy efficiency programs to put the state on a path to ZNE. The authors provide a brief history of energy efficiency in California, describe how the ZNE goal was adopted by the regulators and the utilities, and discuss how one utility, Pacific Gas & Electric Company, is positioning itself to achieve it.

The authors point out that as the utilities and regulators have reviewed the program plans, it has become apparent that existing regulatory frameworks may actually impede the state's progress toward reaching the ZNE goal. Understanding some of these issues are important as other jurisdictions nationally and internationally consider similar attempts to achieve ZNE.

In chapter 18, "Towards carbon neutrality: The case of the City of Austin, Texas," **Jennifer Clymer** explains why and, more importantly, how the city of Austin, Texas, is striving to drastically reduce its carbon footprint.

The author describes Austin's Climate Protection Plan, a voluntary program to make Austin's municipal operations carbon-neutral by 2020; de-carbonize the energy provided by the municipally owned electric utility; increase the energy efficiency of the city's buildings through code changes and expanded voluntary conservation programs; and lower the community's collective carbon footprint through stakeholder involvement and education. The chapter provides a summary of Austin's progress to date, a comparison to other U.S. cities' climate action plans, and an assessment of key lessons learned.

The chapter concludes with a forward-looking view of a world in which all cities achieve, or beat, Austin's goal of carbon neutrality.

In chapter 19, "Rising to the challenge of sustainability," **Benjamin Sovacool** examines three case studies where communities and corporations are planning to or have successfully reduced their energy usage *and* greenhouse gas emissions. The Clinton Climate Initiative illustrates how NGOs can bring the industry, the public sector and financiers together to help cities improve energy efficiency. The Motorola case study shows how the telecommunications giant has aggressively curtailed its GHG emissions while reducing its energy costs. Masdar City is a carbon-neutral, zero-waste, 100% renewable energy economic zone being built in UAE.

The author points out that while each case study has confronted its own unique challenges, adequate living standards can be attained with the right mix of policies, technologies, and commitment.

In the book's epilogue, "How can we get there from here," **Fereidoon Sioshansi** summarizes the collective insights gained from the preceding chapters.

Part I

Challenge of Sustainability

Chapter 1

Why Do We Use So Much Energy, and What For?

Fereidoon P. Sioshansi
Menlo Energy Economics

Chapter Outline

1. INTRODUCTION

When asked if he wanted India to attain a living standard similar to that enjoyed by Britain, Gandhi reportedly replied—I am not sure about his exact words—that Britain had attained its high standard of living by exploiting the resources of half the world, a reference to its colonies. He then asked, "How many worlds do you think India will need to reach a similar living standard?"[1]

To be sure, a significant portion of India's population enjoys much higher living standards today and we have not run out of natural resources yet. One can also excuse Gandhi's politically loaded response in the context of British colonial policies during India's long struggle for independence. But his observation about our ever-increasing reliance on finite natural resources to provide a decent standard of living for people of India—or for that matter, the world—is as relevant today as when he made the statement. Moreover, Gandhi's remarkable remark goes to the core of the central question for this book, namely can we maintain an adequate standard of living in a sustainable fashion for the world's growing population—not just for the rich and the privileged?

1. Another colorful quote attributed to Gandhi, mentioned by Bartiaux et al. in this book's Chapter 3, is "Live simply so others can simply live."

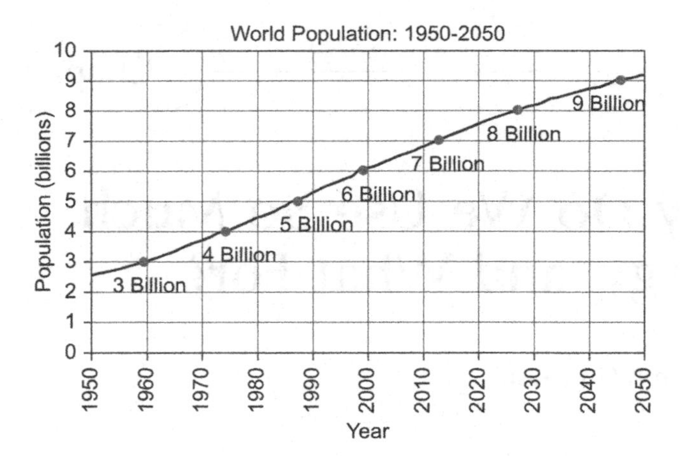

FIGURE 1 World population projection to 2050. *Source: US Census Bureau*

As further explored in the following chapters, different scholars come up with different answers to Gandhi's question. Some believe that continued technological advances will allow us to meet the ever-increasing demands of a growing population (Figure 1), no problem. Others, while acknowledging the significant role of technological fixes, are not so sure.

Aside from long-term sustainability is the concern for the lack of equity and fairness in the current distribution and use of resources. Today, we live in a world where an estimated one-third of the global grain harvest is used to feed the animals that produce the meat that is consumed by the rich, while millions of people in less fortunate circumstances struggle to feed themselves, lack clean water, electricity, or other basic necessities.[2]

It is in this context that this chapter, and those that follow, explore a number of critical issues affecting the future course of human evolution on planet Earth—although this book does not claim to be the only one with such a grand agenda, nor does it claim to have the answers.

This chapter, benefiting from the insights of the others and referring to them selectively, is organized as follows:

- Section 2 asks *why* we use so much energy.[3]
- Section 3 briefly examines *what* we use energy for.
- Section 4 asks what constitutes an adequate standard of living and how much energy we need to maintain it.

---

2. For further discussion of inequities, see Chapter 5. The current head of the Inter-governmental Panel on Climate Change (IPCC), Dr. Ravindra Pachauri, a vegetarian, has proposed that rich countries should consider going meatless for one day per week as a way to reduce human impact on the environment.

3. This, of course, applies to rich countries where a lot of energy is used on a per capita basis.

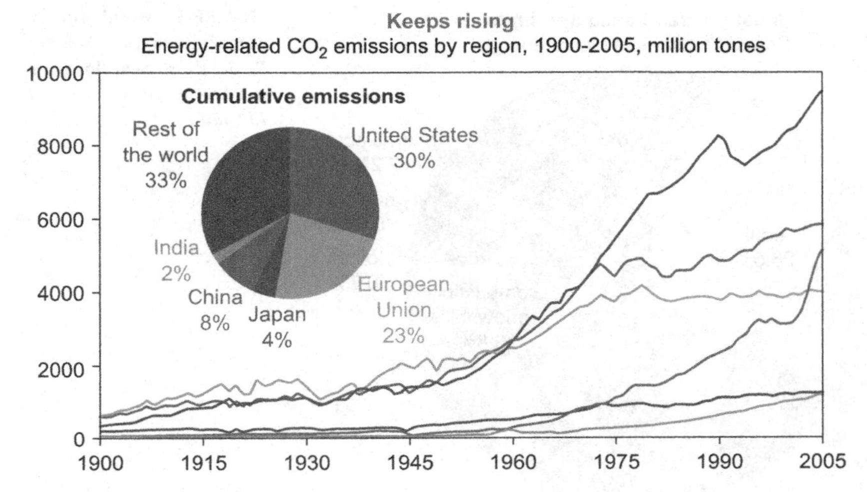

FIGURE 2 Energy-related CO_2 emissions. *Source: International Energy Agency, World Energy Outlook, 2007*

- Section 5 examines alternative lifestyles, habits, and socio-economic systems that might provide equally satisfying—or potentially superior—standards of living on a fraction of our current levels of energy consumption—a recurring theme in the book.
- Section 6 asks what would it take—in terms of technological improvements, changes in policies, socio-economic systems, regulations, education, cultural, and behavioral adjustments—to lead us toward a more sustainable future path.

This chapter, and the book, is primarily focused on energy use and its environmental consequences, but one can easily substitute the term *natural resources* for *energy*. By the same token, many chapters in the book address greenhouse gas (GHG) emissions associated with burning of fossil fuels (Figure 2) and resource depletion issues—visible symptoms of our current unsustainable practices.

The chapter's main contribution is to encourage a fundamental search for alternatives to the *status quo* by asking questions such as, why can't things be radically different in the future than they have been in the past? Or, why can't we accomplish far more with far less?

2. WHY DO WE USE SO MUCH ENERGY?

The short answer is that *up to now* energy has been relatively cheap and seemingly plentiful and few were concerned about climate change, sustainability, or equity issues (Figure 3).

Economic theory predicts that, all else being equal, one would use more of one critical input in the production process if it were inexpensive relative to

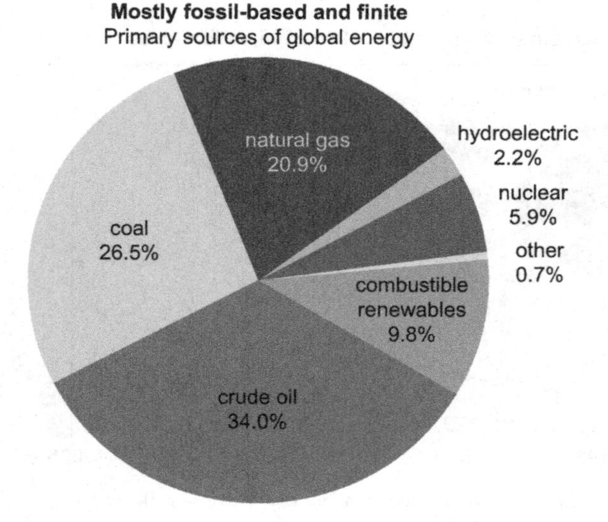

FIGURE 3 World primary energy sources. *Source: R. Heinberg,* **Searching for a Miracle***, 2009, based on EIA data*

others. If energy were cheap relative to labor and capital—typical variables in the production process—the rational user would substitute more energy while cutting back on labor and/or capital. This has been a recurring pattern since the beginning of the industrial revolution, as illustrated in the following anecdotes:

- The longevity of the notoriously inefficient incandescent light bulb, virtually unchanged since Thomas Edison invented it a century ago, provides a manifestation of this phenomenon. These bulbs typically *waste* over 90% of the energy in the form of unwanted heat,[4] converting less than 10% of the energy into useful light. Yet the incandescent light bulb is still in widespread use because electricity has traditionally been relatively cheap, subsidized, or inappropriately priced.[5]
- Until recently, computer makers paid scant attention to energy use, focusing exclusively on other features such as speed, capacity, and other product attributes. All that was needed to get rid of the generated heat was a bigger fan—or in the case of large main-frame computers, more air-conditioning—at the expense of the users, not the manufacturers.[6] It was not until portable computers became popular that computer makers began to pay serious attention to power consumption, primarily driven by limited capacity of batteries, especially in early years.

4. The wasted heat, of course, is appreciated in colder climates as a useful by-product of light.

5. As explained by Long et al. in Chapter 8 of this book, even today, private utilities in many parts of the world have perverse incentives to sell more, rather than less—which means they have little or no interest in encouraging energy conservation.

6. This is among a number of the market *failures* or *barriers* to energy efficiency often mentioned in the literature, including Long et al. (Chapter 8).

- Detroit's big three carmakers, all suffering financially today, for decades paid virtually no attention to fuel economy, focusing entirely on other features—including making progressively bigger, heavier, and more gas-guzzling models.
- Planes, trains, buses, trucks, ships—and virtually all other major energy consuming devices—were and still are designed and marketed on features other than energy efficiency, be it speed, capacity, range, performance, longevity, maintenance, and most important, initial price tag. If a plane's initial price is less than a rival model or if has a longer range, *that* becomes the main selling point—not its fuel consumption, which often dwarfs the initial price tag. Fuel consumption has become a selling point only in recent times due to rising fuel prices and concerns about pending regulations (see "What Happens When Fuel Is Cheap and Emissions Are Free" sidebar).

What Happens When Fuel Is Cheap and Emissions Are Free?

It was recently reported that commercial transatlantic flights can reduce their fuel consumption by roughly 2% by making trivial modifications in operating procedures such as flying at optimal altitudes for maximum efficiency rather than arbitrary altitudes set by air traffic controllers. This requires *no* technological improvements and virtually *no* investment, in other words *zero pain, 2% gain.*

Given the current traffic levels in the key transatlantic market—100,000 one-way flights annually, each one on average using 25 metric tons of fuel—a 2% saving adds up quickly. Similar savings can be had on other routes and other markets. Less fuel burned means less GHG emissions, which are especially damaging when released at high altitudes. The question is why such trivial measures were not explored until now?

The answer is that airlines began to pay serious attention to their fuel consumption only after recent rises in the cost of jet fuel, beginning in 2008 when oil prices approached nearly $150 per barrel for the first time. They became even more interested in fuel consumption only after the European Commission began talking about putting restrictions on GHG emissions associated with air transportation, believed to represent roughly 3% of global GHG emissions.

Two observations:

- First, if energy costs are a trivial part of the overall cost of a business, they get scant attention.
- Second, regulations—or the mere threat of regulations—are usually needed to prompt the industry to take corrective measures.

As further explored in chapters of this book, the implications of this example are twofold.

- First, prices must reflect the true and full costs. To the extent that they do not capture externality costs—in this example GHG emissions associated with planes at high altitudes—they lead to suboptimal outcomes.

(Continued)

What Happens When Fuel Is Cheap and Emissions Are Free?—cont'd

- Second, regulations, standards, and policy imperatives are often needed to correct market imperfections, which are typically driven by short-term profit motives. The question of how much regulation and how intrusive it should be, however, is not as trivial as it may seem.

Source: *Atlantic Interoperability Initiative to Reduce Emissions (AIRE), reported in* The Wall Street Journal, *10 March 2010.*

As these examples illustrate, the pattern is pervasive.[7] It is not limited to a few products, few industries, or a few countries. Even today, manufacturers of most consumer products focus almost exclusively on minimizing the initial price of the finished product—using material and designs that reduce manufacturing costs even when this increases life cycle energy consumption since these costs are borne by consumers.[8] This raises important questions about w*hether* and *how* to protect consumers who may not necessarily be well-informed and/or sufficiently motivated to compare initial purchase price of a product to its lifelong energy consumption costs.[9]

Over time, the *illusion* of cheap and seemingly abundant energy[10] has instilled wasteful habits in many of us—and these habits have permeated throughout our lives, lifestyles, livelihood, infrastructure, and institutions to the point were we are no longer even aware of them. People in rich countries have acquired energy-thirsty lifestyles including long commutes to work, for example. In many cases, prevailing customs, social norms, and even our tax codes actually *favor* increased consumption[11] (see "Do You Want to Start a Revolution" sidebar).

And because energy has historically been cheap and seemingly plentiful, no one thinks twice about traveling to a distant city for a sales call or a weekend skiing holiday. Dubai boasts a giant indoor ski resort, which is kept at freezing temperatures in a hot climate, not to mention the glittering lights of Las Vegas and numerous other examples of gluttonous and wasteful energy use. How can

7. In Chapter 4, Meyer et al. illustrate a few examples of tax codes and prevailing laws that encourage more consumption, rather than conservation.

8. As several chapters in this book point out, there are a number of persistent obstacles that prevent the optimal choice of investments in favor of lowest up-front costs as opposed to life-cycle costs, including energy costs over the long life of appliances, buildings, cars, and other devices.

9. This leads to philosophical and ideological arguments on whether consumers in fact need protection in the form of minimum energy efficiency standards and/or energy efficiency labeling or should we leave it to buyers to rely on their own wits and motivations.

10. Chapter 10, by Ehrhardt-Martinez et al., refers to this phenomenon.

11. Meyer et al. (Chapter 4) provide examples of tax codes and pricing tariffs that encourage consumption.

Do You Want to Start a Revolution?

During a recent trip to Germany, I was picked up by a friend in a Mercedes Benz. I could not resist congratulating him on owning such a fine car. He confessed that it was a *company car.* Further questioning revealed that he commuted a long distance from home to work, and beyond, on a daily basis. When asked about the cost of gas, parking, and maintenance, he said everything was covered by his employer. Why bother with public transportation, he implied?

Noticing how puzzled I was, he explained that his case was not unique, that thousands of German mid- to top-level managers enjoy similar corporate benefits, the higher the rank, the more prestigious the car. In fact, he said many of his colleagues would prefer a fancier car to higher pay. More than your salary or title, the brand of the car and the size of the engine determines your ranking and status within the company, he explained. And, of course, the same goes for the location of the assigned parking space in the corporate garage.

When I pointed out that this encourages more driving, more traffic congestion, more gas consumption, more pollution—all *negative* externalities associated with driving cars—he looked at me as if I had come from Mars. He asked, "What are you trying to do, start a revolution?" More company cars, he explained, means more jobs for the likes of Mercedes Benz and the entire value chain that produces the parts and components. Driving cars generates additional jobs for the petrol industry, for road construction and maintenance, not to mention generating tax revenues to pay for other services. Plus all the other *positive* benefits—jobs for the tire industry, car service industry, insurance industry, and so on.

This, and numerous examples like this, shows the pervasive consumption and production culture we live in. My conversation with my friend made me aware of the myriad customs and conventions that lead to *more* consumption, because it creates demand for *more* production, and that is what keeps the economic engine moving—but to what end?

anyone seriously begin to discuss energy efficiency in Las Vegas, Abu Dhabi, or Houston?

The rapidly growing ranks of the well-off of the world can afford their huge carbon-footprint because energy is a small percentage of their disposable income. The popularity of SUVs, second holiday homes, and remote resorts in exotic places are symptoms of the energy glut supported through hidden subsidies, convoluted tax laws, and other perverse incentives that encourage consumption as opposed to conservation.

Making matters worse, in many countries, energy continues to be heavily subsidized, resulting in excessive consumption (see "Promoting Wasteful Consumption Through Energy Subsidies" sidebar). Is it any surprise that citizens of many oil-rich countries in the Persian Gulf, who enjoy subsidized petrol at below market prices aimlessly roam around in gas-guzzling SUVs without a second thought (Figure 4)? The same is true of other precious necessities,

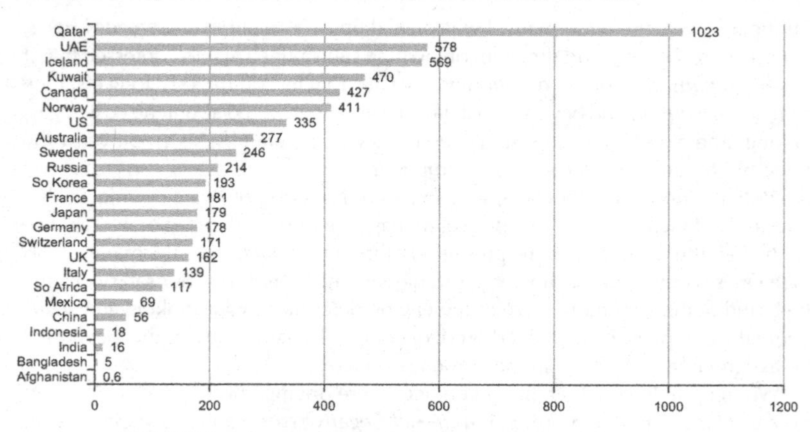

FIGURE 4 *Source: Searching for a Miracle based on EIA data*

notably water and food, which are subsidized in many parts of the world, resulting in excessive waste and inappropriate use.[12]

Surprisingly, energy subsidies are not limited to developing countries with vast domestic energy resources. Many net importers of energy (e.g., India and China) subsidize energy prices, in some cases literally bankrupting their economies in the process (e.g., Iran and Venezuela) while others are projected to become net importers of oil at current growth rates (e.g., Malaysia and Egypt). The total amount of subsidies outside the Organisation for Economic Co-operation and Development (OECD) countries is estimated at a staggering $310 billion a year.

Nor is the practice limited to developing countries. The U.S. fossil fuel industry—oil, gas, and coal—reportedly received $72 billion in subsidies between 2002 and 2008, according to the Environmental Law Institute. The nuclear industry gets significant subsidies also—no one can be sure exactly how much. More recently, renewable energy resources have been receiving subsidies. Their proponents consider these more deserving than subsidies given to fossil fuels on environmental grounds as well as for energy security and fuel diversity reasons.

We not only *use* a lot of energy but we also *waste* a lot of it unnecessarily. According to one estimate, as much as a third of energy used in the U.S. commercial sector may be *unnecessarily* consumed or *wasted.*[13] And this explains why much of our infrastructure is poorly designed, our cities and urban

12. For a discussion of what is an appropriate price, refer to Sioshansi, F. (2010). What is the right price? In D. Reeves (Ed.), *Current Affairs – Perspectives on Electricity Policy for Ontario*. University of Toronto Press.

13. Estimate from U.S. Environmental Protection Agency (EPA) reported in *The Wall Street Journal*, "Encouraging Business to Turn Off the Lights," 27 April 2010.

Promoting Wasteful Consumption Through Energy Subsidies

Energy subsidies, if they ever made sense, were justified on the grounds that they help the poor. And some forms of subsidy, say for bread or mass transit, could conceivably be justified in some cases and they may in fact help the poor. But energy price subsidies hardly ever qualify.

Since the middle class and the rich use far more energy than the poor—they drive bigger cars, own bigger homes, have bigger air conditioners, fly more often, and so on—they benefit disproportionately from energy price subsidies, be it petrol prices, electricity, or natural gas. It is hard to imagine the opposite. In this sense, energy price subsidies are among the most distasteful forms of regressive tax imaginable.

But that is not the end of it. Artificially subsidized energy prices encourage increased consumption, which runs counter to the goal of encouraging energy conservation and more efficient energy utilization. Subsidies also empty government coffers, money that could be put to better use elsewhere.

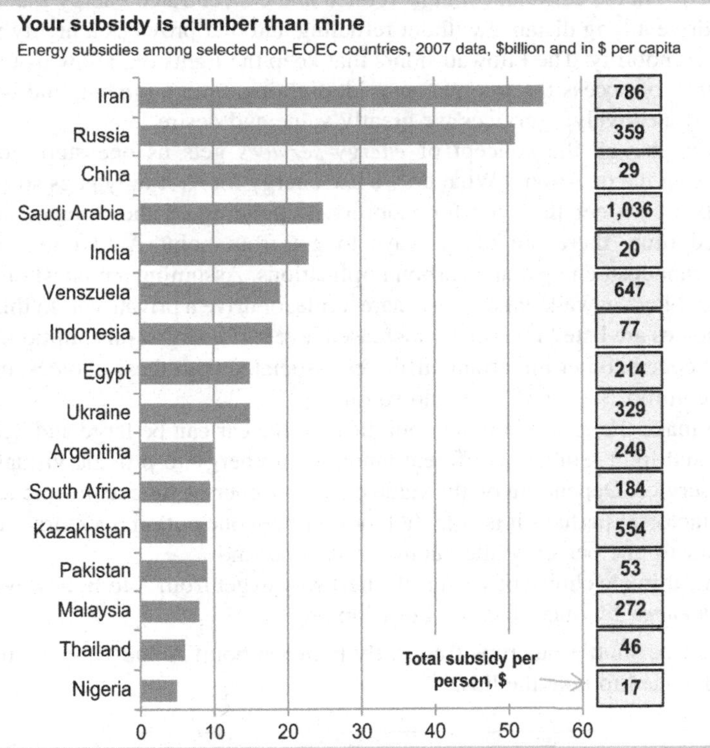

Source: The Economist 3 Oct 09 based on data from IEA & IMF

And of course, more consumption results in more carbon emissions. At the G20 Summit in Pittsburgh in 2009, leaders of the 19 biggest economies plus the European Union agreed to *phase out energy price subsidies* in *medium term*, at least in principle. This is expected to result in a 10% drop in global GHG emissions by 2050.

centers have turned into endless sprawls, and our personal habits and lifestyles have become energy intensive.

Another issue, of course, is not just *how much* energy but *what kind* of energy? As several chapters in the book explain, for a long list of reasons, we have become overly dependent on fossil fuels, with significant implications for the environment.[14]

3. WHAT DO WE USE ENERGY FOR?

The short answer is that we use energy to *derive* needed services.

A lump of coal, a barrel of oil, a kilowatt-hour of electricity, or a cubic feet of natural gas are only valued because they can heat a furnace, run a car, allow us to watch TV, or fry an egg, respectively. This is referred to as *energy services*; we need the *services* provided by energy, not the energy source itself. The petrol in the car tank contains refined and concentrated energy that allows us to drive a long distance without refueling and that provides a highly valued service: mobility. The kilowatt-hours that keep the lights on, allow us to listen to music, or access the Internet, provide lighting, entertainment, and connectivity, respectively—services we greatly value and desire.

Even though the concept of *energy services* gets us one step closer to answering the question, "What do we use energy for?", it only goes so far. For example, consider the need for mobility. Depending on the distance and the allotted time, there are many ways to get from point A to point B with significant cost, energy, and carbon implications. Assuming a modest distance, one can bicycle, walk, take a bus, share a ride, or drive a private car. In this case, the choices are listed in terms of increased levels of energy consumption. There are, of course, other important attributes associated with these choices, notably time, comfort, safety, style,[15] and so on.

To make the example more complicated, the car can be large and heavy or small and light requiring different amounts of energy to provide virtually the same service. Depending on the value placed on energy, time, convenience, and other factors—perhaps it is cold, hot, or raining—one option may turn out to be superior to one person while another suits a second.

The ultimate choice of what is the best way to get from A to B, however, can be *influenced* by many factors. For example:

- If a free shuttle bus runs frequently between point A and B, many may be persuaded to take the bus.

---

14. Arent et al. describe the prospects for low-carbon renewable energy resources in Chapter 13 of this book.

15. In Chapter 6 of this book, Bollino et al. refer to the significance of positional goods and the element of style and prestige associated with many personal consumption decisions. For example, if driving a private car is perceived as superior to riding a bike or taking public transit, this may influence the ultimate choice.

- Having safe designated bicycle lanes between A and B and parking racks at both ends may persuade some people to ride the bike.[16]
- Availability and price of parking is often a significant determinant of choice. If parking is scarce and expensive, other options become more attractive.[17]
- Road congestion is often an important factor. Some countries offer multiple-occupancy lanes, allowing cars with two or more riders to use specially designated lanes as a way to promote high occupancy in private cars.[18]
- A pleasant pedestrian walkway between A and B may encourage more people to consider walking relatively short distances, especially in congested urban areas.
- Prices, including fuel tax, bus fare, and parking fees, and also road access restrictions will influence the choice.[19]

Why dwell on such a trivial example? The point is that urban planners and governments can *influence* the final choice of travel mode, say, away from private cars to other options with multiple benefits including less congestion, less energy consumption, less pollution, and less space allocated to roads and car parking, leaving more to city parks and open space. Many cities are promoting improved mobility and access, and less congestion and pollution, by increasing urban density through better planning.[20]

Private cars, considered by many as the ultimate epitome of comfort, personal convenience, and flexibility will be hard to replace by buses, bikes, or other means of transportation, but much can be done to reduce their energy consumption *and* emissions—the most troubling source of pollution in urban areas.[21] The good news is that today's cars, like incandescent light bulbs, are energy guzzling monsters compared to what they can become.

Depending on many variables, today roughly 90% of the energy content of the fuel in a typical internal combustion (IC) engine car may be wasted as

16. Some cities are experimenting with offering free public bicycles that people can take from one point to another and leave at designated racks for others to use. Others have devised inexpensive bike rental systems where users can pay by the hour using a credit or debit card.

17. In Chapter 19 of this book, Sovacool provides an example where city planners can essentially ration use of private cars in city centers by making parking scarce and expensive. Residents of big congested cities such as New York, London, or Tokyo are better off without a private car for lack of parking and congestion of inner-city traffic.

18. The city of Bogota in Colombia has specially designated lanes for express buses that whisk passengers around while traffic snarls force private cars to sit and wait in endless traffic jams.

19. A few cities, including London, Stockholm, and Singapore, have introduced access charges within the central business district (CBD) while others ban private vehicles from certain areas altogether.

20. Car-obsessed California has recently introduced measures to meet its ambitious climate change target that requires statewide emissions to be reduced to 1990 levels by 2020.

21. In Chapter 7 of this book, Moran provides counter-arguments in describing evolving traffic patterns in urban centers.

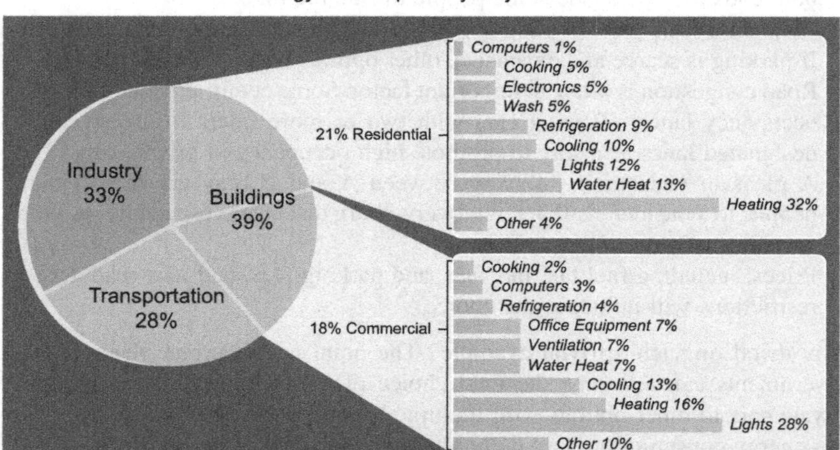

FIGURE 5 *Source: Lawrence Berkeley National Laboratory*

heat—not delivered as mobility. One cursory look at an IC engine reveals the ingenious engineering that goes into capturing and disposing of enormous quantities of heat into the atmosphere via the radiator, the hoses, and the fans. By comparison, an electrical engine uses roughly 90% of the energy to provide mobility with no emissions at the point of use—which explains the current interest in electric and hybrid vehicles.[22]

The most energy frugal, zero congestion, zero emission mobility option, of course, is to avoid the need for *physically* going from point A to B altogether. Perhaps we can conduct the required business on the phone or by accessing the Internet or by sending a text message. Examples include remote banking, remote shopping, and telecommuting. An increasing number of people now routinely work from a home-based office or telecommute,[23] with considerable implications for energy use and GHG emissions while saving time and avoiding the stress of daily commuting.[24]

Figure 5 provides some clues to the question of what we use so much energy for. The mobility example above can be repeated in countless other contexts. Take the case of lighting, a significant contributor to energy use in both

22. Of course, one must consider the energy and pollution to generate the electricity that is stored in car batteries and their associated conversion losses.

23. According to one estimate, the percentage may be as high as 20% in the U.S.

24. Of course, as usual, there is no free lunch. In this case, Internet service providers and data centers consume energy to maintain the infrastructure, but the net energy savings is still substantial relative to the alternatives.

residential and commercial sectors.[25] In many cases, artificial lighting can be significantly reduced by increased reliance on natural lighting during daylight hours. This, of course, requires better design, building orientation, windows, shades, and other means to adjust interior lighting without increasing the buildings' heating load.

Artificial lighting can be provided by more or less efficient lights and lighting fixtures. For example compact fluorescent lights (CFLs) use a quarter of energy per lumen than traditional incandescent light bulbs. And light emitting diodes (LEDs) offer even more drastic energy savings—albeit at a cost.

But the cost is not fixed. It can be *influenced* by supportive policy measures, standards, and regulations. CFLs used to be pricey and their performance and light quality was poor when first introduced.[26] CFL prices have plummeted in the recent past as a mass market has evolved, and prices continue to decline while quality and performance improves.[27] The same can hopefully be said about LEDs, and the next generation of super-efficient lighting technologies and fixtures.

Moreover, there is more attention devoted to how the artificial light is directed to where it is needed. Directional lighting and more efficient lighting fixtures deliver a large percentage of the light to where it is needed, which means little energy is wasted to illuminate space that does not need the lumens. The savings from such simple applications, aggregated over a large population of users, can be enormous.

There is, of course, an extremely low-tech, zero-cost option to cut down energy used for lighting to zero, a parallel to the case of telecommuting—the option that literally negates the need for mobility in the prior example. It is turning off the lights when not needed, or not turning them on in the first place.

Other examples of "What do we use so much energy for?" can come from the commercial and industrial sector.[28] In many commercial establishments, huge amounts of energy are wasted; for example, the air conditioners work overtime to keep the premises cool while doors and windows remain open. In many restaurants, the oven and various heating devices used to cook food generate enormous amounts of heat that the air conditioner and the circulation system have to get rid of. Better, more integrated designs can overcome many of these types of wasteful energy consumption or recycle the wasted heat into a useful application.

---

25. Refer to this book's Chapter 9 by Gray and Zarnikau and Chapter 11 by Prindle and Finlinson for further discussion of residential and commercial energy use.

26. California's three investor-owned utilities, with the support of state regulators, have sold over 95 million CFLs at subsidized prices to consumers, roughly three per capita.

27. A number of countries in Europe as well as Australia have banned the use of incandescent lights altogether, further increasing the demand for CFLs.

28. Refer to Brown et al. (Chapter 12) for a discussion of energy efficiency opportunities in the industrial sector.

The commercial sector, which captures everything that is not residential, industrial, or agricultural, poses special challenges when it comes to energy use because the typical users/occupants do not generally pay for the energy consumed directly, nor do they have direct incentives to conserve. Moreover, energy costs tend to be a relatively small percentage of overall costs of products and services in the commercial sector, making energy efficiency a difficult proposition in many organizations.[29]

Typical thermal power plants generate enormous heat that must be exhausted, at considerable cost and harm to the environment. Combined heat-and power, cogeneration, and district heating provide useful applications for the wasted heat.[30]

The average homes in many countries are poorly insulated, allowing too much heat to get in during the summer, when it is not needed, and too much heat to escape from the interior during the winter, when it is. Making matters worse, generations of architects and engineers were taught to treat buildings as isolated spaces to be artificially heated, cooled, lighted, humidified, and ventilated with little consideration of the ambient environment. This thinking is now being challenged, and changed, in favor of designs that *integrate* the building into the surrounding environment, taking advantage of the orientation, natural lighting, and the ambient air to meet most of the comfort needs of the occupants.[31]

In sum, we typically *waste* a lot of energy in the process of getting to what we really want and need. It is perhaps a slight exaggeration to ask, "Do we need a tank to kill a fly?"

4. WHAT CONSTITUTES AN ADEQUATE STANDARD OF LIVING?

This is an important question and highly relevant if we agree, as the prior discussion suggests, that when it comes to energy consumption, we, in fact, often use a tank to kill a fly, rather than a fly swatter.

If we can define an *adequate* standard of living, and identify how much energy services are needed to support that level of comfort, then we can focus on ways to provide the services with the least amount of energy.[32]

By most measures, people living in rich countries, on average, enjoy high standards of living made possible by prodigious use of energy and other resources (Figure 6).[33] Referring to Gandhi's observation at the beginning of

29. In Chapter 11 of this book, Prindle and Finlinson address ways of overcoming these barriers.

30. This book's Chapter 14 by Bauermann et al. discusses district heating.

31. In Chapter 19, Sovacool describes how the planners for the carbon-neutral city of Masdar in UAE are employing many design elements used in traditional buildings in hot, arid areas.

32. In practice, of course, one must strive to minimize overall resource use, not *just* energy.

33. In Chapter 15, Buijs describes the significance of rapidly developing countries such as China.

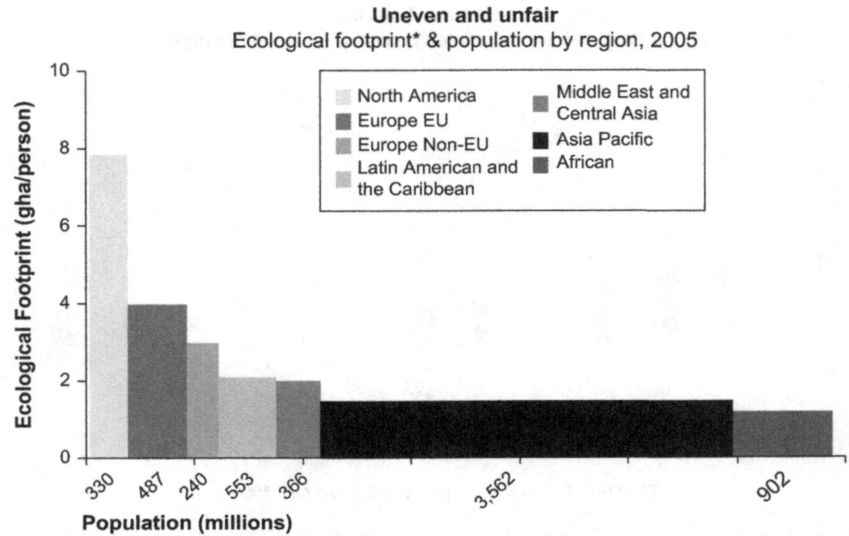

FIGURE 6 Ecological footprint. *Source: Living Planet 2008*

the chapter, these high living standards are made possible by reliance on cheap imports of natural resources and, increasingly, manufactured products, from developing countries.[34]

In contrast, many people in developing countries subsist on the mere necessities on a fraction of energy and natural resources used by citizens of developed countries. The average per capita energy use and carbon emissions of the former are miniscule compared to the latter group (Figure 7).[35]

Defining what constitutes an adequate standard of living, however, is fraught with difficulties:

- First, standard of living for whom?
- Second, even for a given population, for some of the reasons described in preceding sections, it is not trivial to define how much energy and natural resources are *needed* to sustain a given standard of living.

The first problem goes back to the immense inequalities that have arisen among various societies and cultures over the course of human evolution and economic history.[36] As further explained by Bartiaux et al. (Chapter 3), different societies

34. Refer to Trainer (Chapter 5) for a discussion of this.

35. In Chapter 7, Moran compares and contrasts these differences.

36. For an evolutionary perspective on how various societies have evolved through history, refer to Jared Diamond's *Guns, Germs and Steel* (1999).

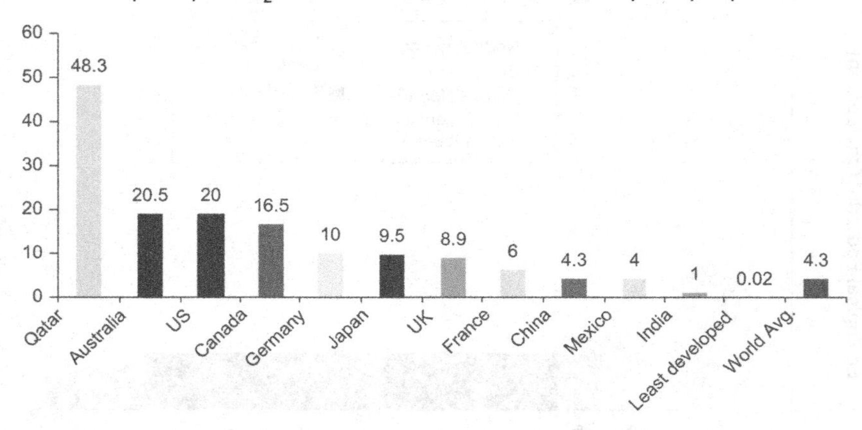

FIGURE 7 *Source:* Energy Informer, *Oct 2009*

have different definitions for what constitutes basic needs.[37] In fact, these authors suggest that *energy needs* may not provide a useful basis for defining *energy demand* or arriving at future energy projections. More fundamentally, we are faced with the reality that the definition of human need is problematic on so many dimensions as to be virtually indefinable (see "How Many Gallons Does It Take for a Shower?" sidebar).

How Many Gallons Does It Take for a Shower?

Water consumption is energy-intensive. It must be collected, pumped, stored, purified, disinfected, and piped. After usage, it must be collected, pumped, treated, and discharged—all of which requires energy. Not to mention that it is becoming increasingly scarce in many parts of the world, requiring highly energy intensive desalination of sea water.

For affluent consumers, none of this matters. In the last few years, there has been a trend toward multiple nozzles and massive showerheads that discharge enormous quantities of water. Moreover, consumers in rich countries take more frequent and longer showers. As described in Chapter 11, "How Organizations Can Drive Behavior-Based Energy Efficiency" by Prindle and Finlinson, students residing in U.S. college dormitories average 14-minute showers, and counting. Many luxury spas now advertise the fact that you can enjoy an 18-showerhead bath with incredible water flow. Obviously no one is going to time how long a shower takes.

Given water scarcity and its energy intensity, water flow, pressure, and the definition of showerhead becomes critical. As it happens, a 1992 U.S. federal law

37. Refer to Bartiaux et al. (Chapter 3) for an anthropological perspective on the concept of needs, wants, and desires.

How Many Gallons Does It Take for a Shower?—cont'd

specifies that a *showerhead* cannot deliver more than 2.5 gallons per minute (GPM) at a pressure of 80 pounds per square inch (PSI). Some current showerheads in the U.S. deliver as much as five times this limit. Making matters worse, an increasing number of upscale new homes—industry estimates put the number in the 1%–4% range and growing—now feature showers with multiple showerheads, as many as 18, spraying water from all directions.

The U.S. Department of Energy (DOE) has decided that enough is enough. In May 2010, according to an article in *The Wall Street Journal* (21 July 2010), Scott Harris, DOE's General Counsel, fined four shower manufacturers for failure to abide by the law, putting others on notice. Moreover, the DOE has decided that the definition of a showerhead needs further elaboration.

Watch out for water police

Water usage for a 3-minute shower with a showerhead compliant with 1992 U.S. federal law and a 15-minute shower using 18-nozzle showerheads, in gallons:

Current U.S. law	Luxury 18-nozzle showerhead
3-minute shower 7.5 gallons	15-minute shower 675 gallons

Manufacturers' interpretation of the existing law is that as long as *each* showerhead complies with the existing mandate, there is no limit to how many *nozzles* there can be in a shower. Hence, a bathroom with 18 nozzles each delivering 2.5 GPM, or 45 GPM in total, would be legitimate. In this case, a 15-minute shower will use 675 gallons of water, more than people in some impoverished countries use in a whole year. DOE's interpretation, however, is that *all* nozzles would count as a single showerhead and subject to the 2.5 GPM limit. That would mean that if you choose to have 18 nozzles, each would deliver a trickle, not a flood of water.

This has caused quite a fury among the manufacturers. Barbara Higgens, Executive Director of the Plumbing Manufacturers Institute—yes, apparently there is such an institute—complained to the *Wall Street Journal* that DOE's Mr. Harris is making a "value judgement," adding that, "One person's waste is another person's therapeutic use of water." Pedro Mier, Vice President of Grupo Helvex in Mexico, agreed, pointing out that his firm's customers "just like to feel they are getting a lot of water." He confessed that until he received the DOE's letter recently, he was not even aware of the federal law. Referring to Mr. Harris' letter, he said, "At first, I thought it was a scam."

"Did Congress limit consumer choice?" DOE's Mr. Harris asks rhetorically, adding, "Absolutely. When you waste *water*, you waste *energy*," (emphasis added) and that's where DOE comes into the picture. Mr. Harris estimates that each multihead shower fixture uses the equivalent of roughly one barrel of oil per year. It may not sound like a lot, but multiplied over many homes built each year, it quickly adds up.

So where does one draw the line between personal choice and societal costs—in this case waste of energy and water in a finite planet? Where does basic human *needs*, such as cleanliness, end and where does frivilous consumption begin? Clearly, the question is not limited to use of water and goes beyond the definition of a showerhead.

The point of this anecdote is to illustrate the elastic nature of need, in this case water for taking a shower—considered a rather basic human need—discussed more eloquently by Bartiaux et al. in Chapter 3.

Swiss citizens might define an adequate standard of living based on what they have come to expect as reasonable and customary.[38] The same question posed to citizens of Swaziland may produce a rather different answer. The difference in the two answers can be attributed to different standards of living in the two countries, as well as other variables including population density, climate, culture, lifestyles, energy prices, income levels, and so on.

But it gets even more complicated than that (see "How to Measure Prosperity or Happiness" sidebar). Currently, a typical Swiss citizen uses roughly half of the energy of a typical American on a per capita basis. Most people would agree that we cannot therefore conclude that a typical American enjoys a standard of living twice that of a typical Swiss. In fact, by some measures, standards of living may be higher in Switzerland than in the U.S.[39] Clearly other factors play a role. Switzerland is a small, mountainous, landlocked country with an extensive train and mass-transit system that allows many citizens to enjoy comfortable lives without the need for an automobile, let alone a big SUV. The typical Swiss lives in a smaller house, with smaller appliances, drives a smaller car fewer miles—if at all—and far less often than the typical American.

How to Measure Prosperity or Happiness?

Economists traditionally measured a nation's prosperity based on its economic output, per capita GDP, or similar monetary indexes. But as everyone knows, money alone doesn't buy happiness. The debate about how to measure prosperity, and happiness, has received more attention in recent years with suggestions that nonmonetary factors, such as quality of a nation's institutions, the state of its democracy, environment, its health, education, public services, income distribution, freedom, security, and leisure time play important roles.

French President Nicolas Sarkozy recently commissioned a study led by prominent economist Joseph Stiglitz to come up with alternative measures, pointing out that a nation's GDP is an "insufficient measure of its wellbeing." Few would disagree.

38. Refer to Chapter 16 by Stulz et al. for a discussion of energy use in Switzerland and efforts to reduce it.

39. According to 2009 Legatum Prosperity Index, Switzerland is ranked as the second "most prosperous" country; the U.S. is ranked #9.

How to Measure Prosperity or Happiness?—cont'd

Legatum, a London-based think tank has come up with a multidimensional prosperity index, which includes a number of nonmonetary measures. It ranks Finland on top, followed by Switzerland, Sweden, Denmark, and Norway—all reasonably wealthy countries but also enjoying other positive attributes such as stable democratic institutions, good social welfare systems, reasonable distribution of income, among other reasons.

Overall Rank	Country	Economic Fundamentals	Entrepreneurship & Innovation	Democratic Institutions	Education	Health	Safety & Security	Governance	Personal Freedom	Social Capital
1	Finland	10	9	9	3	7	2	2	7	6
2	Switzerland	2	13	1	22	3	6	3	11	2
3	Sweden	16	3	7	4	15	7	5	5	3
4	Denmark	15	6	12	2	12	4	1	2	13
5	Norway	18	17	8	1	10	1	7	1	10
6	Australia	7	15	5	6	21	14	10	4	4
7	Canada	6	4	6	16	22	9	9	3	9
8	Netherlands	3	5	19	14	8	15	8	10	8
9	United States	14	1	2	7	27	19	16	8	7
10	New Zealand	27	18	4	10	19	13	11	6	1

Source: 2009 Legatum Prosperity Index, http://www.prosperity.com/rankings.aspx

The U.S. is ranked ninth, ahead of Britain, Germany, and France. Zimbabwe, Sudan, and Yemen did not do well—not only because they are poor but, more importantly, because they suffer from unstable and undemocratic governments, if one can dignify their ruling system as a government.

But the issue of how much energy is needed to sustain an adequate lifestyle is even more complicated than comparing Swiss to American lifestyles. Looking at per capita energy consumption and carbon emissions among selected U.S. states, what explains the vast difference between Wyoming and California (Figure 8)? Retail electricity prices, population density, the composition of the economy, climate, home-building codes, and appliance standards explain some of the difference but not all. In the case of Wyoming, a mining and coal producing state with a small population, the per capita

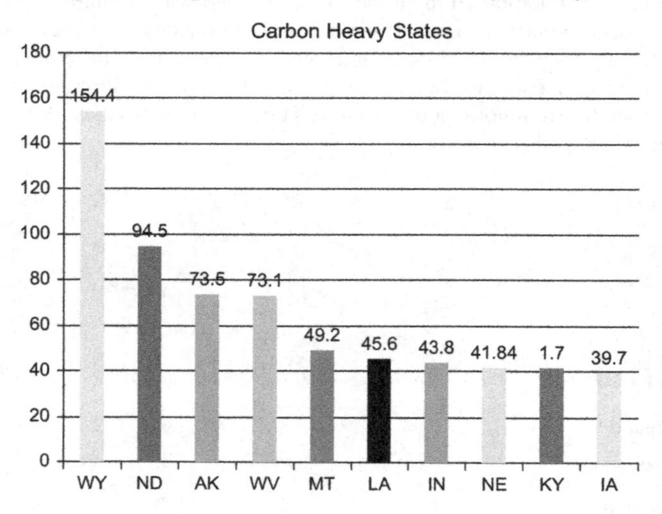

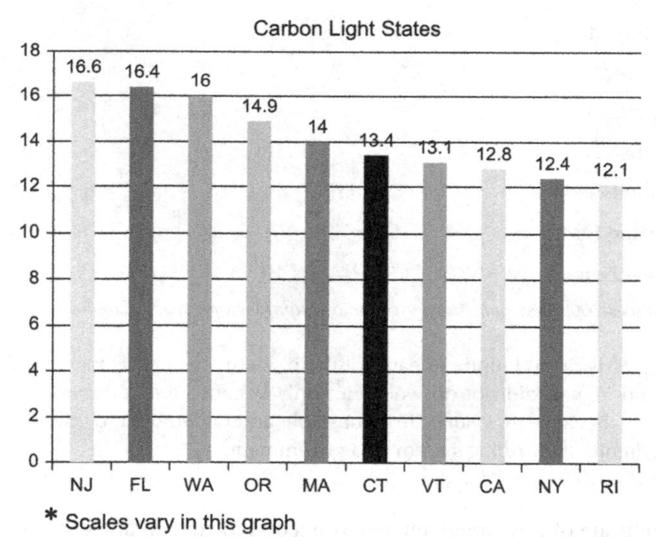

FIGURE 8 *Source: Climate Analysis Indicators Database, World Resources Institute*

number looks excessively high. California, on the other hand, is a state with a relatively mild climate, little heavy industry, and high retail electricity prices.

Even if one adjusts for these variables, defining how much energy—and other natural resources—is needed to sustain an *adequate* standard of living is still fraught with challenges. The reason, as already explained, is that our basic *energy service needs* can be met through a combination of energy, capital, and human factors. By adjusting these variables, one can arrive at vastly different answers. Pushing the concept to its logical limits, one can approach *zero net energy use* in buildings if costs were not an obstacle and one could invest not in the *leading edge*, but as Amory Lovins likes to say, *bleeding edge* technologies.[40]

To demonstrate the point, let's examine the energy use of a building. A highly insulated building with state-of-the-art windows, lighting, and space conditioning equipment and controlled with a sophisticated energy monitoring and management system would use very little energy, offering significant cost savings over its extended life.[41] But such a building comes at a relatively high up-front cost.[42] A *zero net energy home*, will—at least in theory—have a *zero* net energy consumption over time but to get there would require a significant investment in design and construction, not to mention highly efficient energy using and generating devices and behavioral adjustments by its occupants.[43]

Since there are obvious tradeoffs between how much is invested in the capital stock versus their energy usage over time, this can quickly turn into a circular argument.[44] From an economic point of view, energy efficiency investments should be pursued to the point where they are deemed economic in terms of reduced operating costs. This entails putting a value on future energy savings versus present investments in better equipment, better design and construction, more insulation, and so on. Since future savings are worth less than present investments, consumers must make tradeoffs, explicitly or implicitly, when buying energy-intensive appliances, cars, or homes. A number of chapters in this book examine the implications of these intertwined issues and how appropriate policies, incentives, information, standards, and codes can *influence* the ultimate decision.

This section must end inconclusively on both questions posed, namely,

- There is *no* universally accepted definition of what constitutes an adequate standard of living,
- There is *no* definite answer to how much energy—and other resources—it takes to sustain it even if a universal definition existed.

---

40. Chapter 9, by Gray and Zarnikau, discusses the cost implications of moving to zero net energy buildings.

41. Chapter 10, by Ehrhardt et al., covers the so-called "beyond the meter" applications that enable reduced energy consumption through improved utilization and management.

42. Chapter 18, by Clymer, describes some of the measures pursued in the case of city of Austin, Texas.

43. Chapter 9, by Gray and Zarnikau, and Chapter 17, by Rajkovich et al., describe the concept of zero net energy.

44. Energy efficiency guru Amory Lovins lives in a zero net energy home using the most advanced appliances and devices on the "bleeding edge" of technology. But these concepts are still beyond the means *and* reach of average homeowners.

For pragmatic purposes, however, one can assume that living standards enjoyed by Western European countries today are adequate, and average Europeans appear to get by on half as much energy per capita than their American, Canadian, or Australian counterparts. In the case of Switzerland, a country with enviable living standards, there is an effort to cut the current per capita energy use to a third of current levels as described by Stulz et al. in Chapter 16.

Switzerland's ambitious target may provide a useful goalpost for this book, say, the equivalent of 2000 Watts, or even less, per person. To put this number in perspective, the corresponding number for the U.S. is currently around 12,000 Watts per person—implying that the U.S. would have to cut its per capita energy consumption six times as much to achieve what the Swiss have established as a goal. This, however, is a gross simplification given the vast differences in the composition of the economies, population density, and numerous other factors.

5. LIVES, LIFESTYLES, AND SOCIO-ECONOMIC SYSTEMS

Our lives and lifestyles are defined by habits, culture, conventions, prices, and income levels in the context of a socio-economic capitalistic system, which encourages production and consumption and is predicated on continuous growth.[45]

Moreover, a consumer-oriented culture, incessantly promoted through advertising and marketing, pervades all aspects of life in rich countries.[46] Powerful multinational corporations are the beneficiaries of this system and encourage further propagation of this culture globally. With powerful brands and massive marketing budgets, they do their best to influence not only our tastes and our choices, but also broadly encourage more consumption.[47] Money may not buy happiness, but in our materialistic culture, it comes close.

One might argue that multinational corporations even try to define our values and our needs.[48] How else can one explain the demand for nonessential luxury items, expensive branded products, mega-sized homes, yachts, and private planes? Why else would anyone market cigarettes, sugar-loaded fizzy drinks, or unhealthy fast food to consumers?[49]

---

45. A number of economists are questioning the necessity of continuous economic growth including Tim Jackson in his book *Prosperity Without Growth: Economics for a Finite Planet* (Earthscan, 2009) and Peter Victor in his book *Managing Without Growth: Slower by Design, Not Disaster* (Edward Elgar, 2008).

46. In Chapter 5 of this book, Trainer offers further discussion of the evils of what he considers excessive consumerism.

47. But as Trainer explains in Chapter 5, our capitalistic system not only affords the rich to live extravagant lifestyles but encourages more of us to aspire to reach similar status.

48. This, of course, is a circular argument, since profit-seeking companies can only sell what consumers will willingly buy.

49. In Chapter 7, Moran questions consumer "paternalism" pointing out that consumers are smarter than many consumer advocates give them credit for.

Such discussions, while intellectually interesting, can sidetrack us from our main focus.[50] Returning to *energy services* that support the basic necessities of life, it is clear that culture and convention play critical roles. The British, to use a stereotype example, are accustomed to drinking their beer at room temperature while Americans like it ice cold. Americans like their showers hot while the British may put up with lukewarm water. Americans, on average, are used to keeping their homes warmer in the winter than the British.[51]

Conditioning is another important factor. Older generations that survived through the Great Depression and the rationing of the Second World War tend to be frugal, using less of everything even when they can afford it. The new generation grown during an age of plenty, rising standards, and relative affluence tend to be frivolous spenders and energy hogs by comparison.

Attitudes, however, continue to change, particularly among the educated and upwardly mobile, who exhibit increased awareness about the environment, are fond of organic food, recycling, and back-to-nature lifestyles. These powerful demographic and attitude shifts have not been lost on profit-seeking companies who are increasingly catering to these affluent consumers with a growing array of green and organic products.[52]

Social norms and income levels are other important determinants of energy use. For example, Americans, on average, live in much larger homes, own bigger cars, and drive longer distances while paying relatively lower gasoline prices.[53] Energy prices and their relation to average household disposable income play an important role in energy consumption patterns. Everything else being equal, people in Kentucky, West Virginia, or Wyoming—low-cost electricity states—can be expected to use significantly more electricity on a per capita basis than those in California or New York (Table 1). For the same reason, people in California and New York can be expected to be far more receptive to invest in energy efficiency measures than those in Kentucky, West Virginia, or Wyoming.

Similarly, price differentials for gasoline including taxes, explain why people in some countries own bigger and more cars and drive them farther—even accounting for population density and income levels.[54] If energy

50. Another prominent feature of our current system, disturbing to many, is the vast money spent on arms, defense, and wars. World military spending doubled during the past decade, reaching $1.53 trillion in 2009, according to a Swedish study reported in *The Wall Street Journal*, 2 June 2010.

51. Refer to Chapter 9, by Gray and Zarnikau.

52. There are numerous examples of this trend, including General Electric's Ecomagination campaign, Wal-Mart's efforts to cater to green consumers, and Intel buying all its electricity from renewable resources.

53. Gray and Zarnikau provide comparisons in Chapter 9.

54. Refer to Chapter 6 by Bollino and Polinori on the effect of energy taxes on energy consumption.

TABLE 1 Average Retail Electricity Prices Among US States

Want cheap juice?					
Average retail and residential rates*, cents/kWh, 12 month average ending June 2006					
Region/state	**Avg. retail rate**	**Avg. residential rate**	**Region/state**	**Avg. retail rate**	**Avg. residential rate**
New England	13.29	14.93	MD	8.58	8.60
CT	13.20	15.02	NC	7.42	8.94
ME	10.63	13.98	SC	6.89	9.75
MA	14.04	15.44	VA	6.74	8.31
NH	13.65	14.44	WV	5.05	6.22
RI	13.32	14.49	East South Central	6.55	7.82
VT	11.22	13.27	AL	6.85	8.41
Mid Atlantic	11.20	13.00	KY	5.18	6.74
NJ	11.33	12.11	MI	8.27	9.44
NY	13.81	16.49	TN	6.68	7.35
PA	8.48	10.19	West South Central	9.18	10.82
East North Central	7.23	8.81	AK	6.54	8.29
IL	7.04	8.43	LA	8.61	9.38
IN	6.20	7.89	OK	7.36	8.47

MI	7.92	9.20	TX	9.96	11.88
OH	7.38	8.94	Mountain	7.40	8.84
WI	7.81	10.07	AZ	7.94	9.04
West North Central	6.51	7.83	CO	7.84	9.18
IA	6.88	9.51	ID	5.14	6.35
KS	6.79	8.11	Mont	6.81	8.20
MN	6.81	8.51	NV	9.26	10.57
MO	6.18	7.25	NM	7.64	9.25
ND	5.99	7.12	UT	6.02	7.63
SD	6.66	7.41	WY	5.16	7.52
South Atlantic	7.98	9.25	Pacific	9.93	10.72
DL	8.14	9.68	CA	11.89	13.00
DC	9.57	9.33	OR	6.38	7.37
FL	9.56	10.39	WA	5.90	6.65
GA	7.73	8.96	Hawaii	20.06	22.55
			US Avg.	8.51	9.92

** These are averages for all applicable retail and residential rates, respectively, for each state including IOUs, munis and co-ops. Average rates are weighted by taking the total revenues and total kWh sales for each company. Non-IOU data is from EIA, No data is provided for Alaska, nor Nebraska, which has no IOUs*

Source: Typical Bills and Average Rates, Oct 07, Edison Electric Institute

becomes scarce and more expensive—for example through the introduction of a carbon tax—energy consumption can be expected to drop, which is why many economists favor economy-wide, technology-neutral carbon taxes as a means of combating climate change.[55]

Difference in income levels is another factor that explains differences in energy consumption levels. For high-income earners, energy constitutes a relatively small fraction of the disposable income while the opposite is true for low-income consumers. If energy prices continue to rise, as some experts predict, the percentage of average disposable income devoted to energy expenses are likely to become more pronounced. If carbon taxes are introduced, either directly or indirectly, this will add to energy budgets for petrol, electricity, and heating. Sooner or later, consumers—and politicians—will notice.

OFGEM, the United Kingdom's energy regulator, for example, estimates that by 2016, the average annual electricity and household gas bill in Britain could rise by 60% to £2,000 (approx. $3,100), or roughly 10% of average household disposable income. How would consumers react to higher energy bills? Utility executives believe that they will have to get used to the idea. But will they? *Fuel poverty*, already a major political issue in Britain, is likely to become more pronounced if OFGEM's predictions materialize.

In his book *$20 Per Gallon*, Christopher Steiner (2009) envisions a future where petrol may cost the consumer $20 per gallon and examines how peoples' lives, lifestyles, and livelihoods may be affected. It is hard to imagine anything with a more dramatic impact on life and lifestyles than the price of energy—because it affects the price of everything else, from the cost of transportation to food to nearly all other products and services we need and use.

The book's most interesting contribution, however, is to point out the many *benefits* of expensive oil once we make the necessary adjustments. Steiner portrays a future where we *could* enjoy better lifestyles on a fraction of energy we currently use. One may not agree with his vision or conclusions but it is an intriguing and promising perspective.

The point of his arguments, which are echoed in a number of chapters of this book, is that lifestyles, habits, and norms are adjustable and can be *influenced* over time. Plastic water bottles, once fashionable, have become the curse of environmentalists. More shoppers bring reusable bags to carry groceries home. Some stores actually charge for them or give token rewards if you bring your own bags. More aluminum cans are recycled. As these examples illustrate, humans are adaptable to change, and given the means and the motivations, can be influenced to do the right things.

---

55. It must, however, be noted that elasticity of demand for energy tends to be low, meaning that significant increases in prices are required to produce rather small reductions in energy use.

6. TOWARD A MORE SUSTAINABLE LIFESTYLE

The examples and anecdotes offered thus far lead to three important observations:

- The amount of energy required to support basic human needs is *not* a preordained or set number.
- The amount of energy required to sustain a given lifestyle can be *adjusted* through substitution, investment, by adjusting prices, influencing habits, social and cultural norms, policies, and standards.
- Markets, while powerful and efficient, need occasional fine-tuning, prodding, and appropriate incentives—and such meddling can be supported so long as the end justifies the means.[56]

The main instruments of change, illustrated through examples, include:

- Investment and substitution—more capital and superior technology, all else being equal, typically results in lower energy usage with substantial savings over extended life of the energy using stock.
- Higher prices will result in reduced energy consumption and encourage substitution of other input variables such as labor and capital.[57] Fuel and carbon taxes,[58] for example, are among well-known means of discouraging energy use in general and carbon-heavy forms of energy, specifically. Moreover, rising price signals offer powerful incentives in directing investment and resources to address scarcities as they are encountered.
- Norms and habits are amenable to gradual change through education, conditioning, and with the emergence of superior alternatives and substitutes.[59] Recent popularity of CFLs, recycling aluminum cans, and reusable grocery bags are good examples.
- Energy policies can induce gradual change,[60] for example, by switching to low-carbon fuels and more efficient utilization of resources.
- Standards are among the most potent means of encouraging efficient utilization of energy and other scarce resources.

---

56. Needless to say, there are considerable differences in views on how best to regulate markets and to what extent. For example, refer to Meyer et al. (Chapter 4), who prefer a strong hand for government in controlling the excesses of free markets.

57. The elasticity of demand for energy appears to be low, however, suggesting that relatively large price increases may be necessary to promote energy efficiency or substitution. This suggests that higher prices, combined with other incentives, may be needed to affect consumption and behavior.

58. Refer to Chapter 6, by Bollino and Polinori, for more on petrol taxes.

59. In Chapter 11, Prindle and Finlinson, for example, examine how college students living in dormitories may be persuaded to take 3-minute showers instead of the current 14-minute variety.

60. Chapter 2, by Felder et al., describes alternative energy futures.

- There is increased recognition of the role of governments to control or modify private sector investments in ways that increases human welfare while reducing wasteful energy consumption.

Viewed in this context, the energy and sustainability problem, which appears insurmountable at first glance, may in fact be amenable to *adjustments* at *multiple* levels. Granted, the energy infrastructure is long-lasting and formed habits are hard to change, but persistent application of available instruments over time will deliver substantive results.

Take the case of zero net energy buildings, mandated in the state of California starting in 2020 for *new* homes and 2030 for *new* commercial buildings. Admittedly, the impact of the new regulation will be modest at first since it only applies to new construction, but will become pronounced over time. The same applies to the 33% renewable mandate in California by 2020. Opponents of such measures complain about the extra costs and the fact that it will drive more business away from the state. Proponents believe that these requirements will eventually create more jobs than they destroy, will encourage technological innovation, and benefit the environment in the long run. California's climate bill, which requires state-level GHG emissions to be reduced to 1990 levels by 2020, have also come under attack by the same critics but are supported by the proponents on its positive merits.

These topics are repeated in the balance of the book from different perspectives. While the problem of sustainability is not trivial and there are no simple single silver bullets, there is enormous *flexibility* in how we can meet our basic human needs, and in this sense, I prefer to view the glass as half full rather than half empty.

Chapter 2

Which Energy Future?

Frank A. Felder,* Clinton J. Andrews,† and Seth D. Hulkower**
*Rutgers University, †Rutgers University, **Strategic Energy Advisory Services

Chapter Outline

1. INTRODUCTION

Energy is such a critical contributor to prosperity and national strength that we regularly worry about where it will come from in the future. No government leaves its energy supply solely to the marketplace, and stakeholders have strong opinions about both the ends and means of energy policy. By asking "Which energy future?" this chapter identifies those ends and means, and highlights which aspects of our energy future are controllable (via our choices) and uncontrollable (because they are exogenous or uncertain).

A "buyer beware" notice belongs right here at the beginning of the chapter. Thinking systematically about the future is something of an act of faith, a belief that some of today's knowledge will be relevant tomorrow. This rational act has practical value when we get our *determinisms* right, that is, when the technological, environmental, demographic, economic, or political patterns that we extrapolate do in fact hold true in the future. Part of the challenge is to balance the appropriate extrapolation of historical trends against our imagined possibilities for the future. There is also an under-appreciated communicative challenge, because plans, forecasts, and projections are stylized forms of storytelling that fail when they are incomprehensible to or do not engage their intended audience.

The global energy system faces immense economic, geopolitical, environmental, and public-acceptability challenges. Energy decisions that are being made today to address these multiple challenges will affect all corners of the globe for decades, if not centuries. Despite the large uncertainties in energy demand growth (such as the dip caused by the recent financial crisis), energy supplies, climate impacts, technological advances, and changing economic relationships, it is important for energy policymakers and planners to envision and articulate possible energy futures. In order for this envisioning exercise to move beyond wishful thinking, it must be acknowledged that different countries, and even different policymakers within a country, have different priorities with respect to economic, geopolitical, environmental, and public acceptability challenges.

The remainder of this chapter discusses future scenarios and associated tradeoffs, business-as-usual energy trends, the cases of China and the United States, the multiplicity of solutions to global energy problems, and the linkage between scenarios and policy and planning objectives.

2. PROJECTING ENERGY FUTURES

We approach projecting global energy futures in several different ways. In this section, we review internally consistent qualitative projections that are based on four different worldviews. In the next section, we review two business-as-usual forecasts to try to understand what the future holds if there is no major departure from existing trends. In Appendix A, we summarize and compare multiple scenarios' analyses conducted by a wide range of organizations.

One approach to thinking about our energy future is to make projections instead of forecasts. That is, to consider the logical implications of several very different worldviews. While still grounded in historical data, this approach investigates what would happen if different *determinisms* govern future global relations. It produces a wide range of outcomes and illuminates how political and cultural factors might derail or pervert technological and economic planning.

In considering possible energy futures, it is important to balance utopian visions with pragmatic expectations. Realism regarding expected human behavior is especially helpful, because, as psychologists like to say, "technology changes rapidly, people change slowly" (Norman, 2002). In economics, the slowness with which human responses change forms the basis for income and price elasticities of demand. These elasticities are stable enough to find predictive usage, although sometimes the usage is inappropriate because elasticities are very situation-specific. People are also more socially embedded than their economic caricature as self-interested, atomistic, rational maximizers typically assume, but there are limits to their altruism. Over time, people can be educated to perform a modest range of environmentally friendly behaviors such as recycling and energy conservation. The point here is to think about future energy scenarios that change human behavior at a plausible pace and with a significant level of effort, and not overnight and without cost. Organizations, institutions, and national governments are subject to similar social frictions, as the ongoing negotiations to craft a worldwide response to global warming demonstrate.

The International Energy Agency (IEA) has conducted a thorough review of energy scenarios and developed its own scenarios (IEA, 2003). It recognizes that a sustainable and secure future is not likely to occur unless policy interventions are made (ibid., p. 17). Furthermore, it finds that those scenarios that envision a sustainable world or stabilization of greenhouse gases (GHGs) require either technological breakthroughs or gigantic efforts to change either political will or consumer preferences (ibid., p. 52). The IEA then develops three exploratory scenarios: *Dynamic but Careless* (fast technological change and an unconcerned attitude toward the global environment), *Clean but Not Sparkling* (slow technological change but concerned attitude toward the global environment), and *Bright Skies* (fast technological change and concern toward the global environment).

- Article I. These three scenarios have six items in common. Population is expected to continue to grow but at a slower rate.
- Article II. Growth will be concentrated in developing countries along with increasing urbanization and population aging.
- Article III. Overall income will also grow and at a slower rate than in the past but with developing countries growing faster than developed ones.
- Article IV. Energy supplies will be sufficient in general to meet growing energy demand, but regional imbalances may occur.
- Article V. As affluence increases, concern for the environment will as well.
- Article VI. The world will be increasingly interdependent and interconnected and market liberalization will continue.

The IEA 2003's multiscenario analysis has three major conclusions. First, it is not necessarily the case that pro-environment policies always lead to rapid development of environmentally friendly technologies. In the Clean but Not Sparkling scenario, there are pessimistic perceptions about technology and overzealous policy interventions restrict technological development to both the detriment of economic and environmental outcomes (IEA, 2003, p. 65). Second, the spatial distribution of oil and natural gas has the potential to disrupt energy markets even when there are adequate, extractable resources (ibid., p. 104). Third, technological advancement is necessary for sustainability but unless it is accompanied by a fundamental change in values and attitudes toward the global environment, technological change will not be sufficient (ibid., p. 104).

The Netherlands Environmental Assessment Agency (*www.pbl.nl/en/*) offers another approach to evaluate different energy scenarios along two dimensions: globalization versus regionalization and efficiency versus solidarity. The four resulting quadrants, or scenarios, are Global Market (globalization/efficiency), Caring Region (regionalization/solidarity), Safe Region (regional/efficiency), and Global Solidarity (globalization/solidarity).

The Global Market scenario (globalization/efficiency) consists of a liberal, individualizing, efficiency-seeking environment with high economic growth, energy consumption, and GHG emissions unless a global cap-and-trade regime can be implemented. The outcomes of this scenario are technological innovation and cost savings resulting in energy prices growing at or less than the rate of inflation, a +100% increase in energy consumption in 2030, an energy supply mix dominated by fossil fuels, and associated GHG emissions, so that renewable energy resources are pushed out to the longer term and the global population reaches 9 billion at mid-century.

The Global Market scenario places a great deal of weight on economic outcomes, with some emphasis on GHG emissions, assuming that a combination of economic incentives and technological advances can reduce the direct and social cost of energy production and consumption. It places little emphasis on energy security; in fact, it may increase security vulnerabilities by relying on global trade. It may be the case that economic progress and development accomplished through globalization reduces global tension. On the other hand, the race for low-cost fossil fuels, particularly oil, may create substantial energy security concerns, as has happened historically. To the extent that this scenario is accompanied by democratization, the public process would be improved, although it depends in part on the willingness of different nations and cultures to accept the homogenization that accompanies globalism.

This scenario is similar to the Intergovernmental Panel on Climate Change's (IPCC) B1 scenario described in Riahi, Grübler, and Nakicenovic (2007). In 2100, population is 7 billion and global carbon emissions are 5 Gt. This is achieved through global integration and free flow of knowledge and technologies. Per capita gross domestic product (GDP) is high and there is a large emphasis on equity. Economic growth results in eventual negative population growth and more equity. In addition, urbanization is reduced since growth is information-intensive and less materially oriented. This outcome starts looking like the Global Solidarity one described below, but it is economic growth that leads to global solidarity, not the other way around.

The Caring Region scenario (regionalization/solidarity) assumes the presence of community spirit, civic duty, cultural diversity, and value placed on immaterial goods such as free time and community. Its outcomes include consumer behaviors that change dramatically toward clean energy, energy consumption that increases +50% in 2030, energy prices that increase by +80% in 2030, renewable energy deployment, and reductions in energy consumption to stabilize GHG emissions.

This scenario promotes the public process; in fact, it assumes that such a public process leading to widespread public acceptance results in this scenario. This scenario does poorly regarding the economic objective, particularly with respect to short-term economic growth, and with respect to efficiency. That being said, it does very well on the environmental objective, assuming that regions that adopt this approach do not "export" the costs associated with energy consumption to other regions, that is, leakage. Thus, the environmental benefits depend on whether this caring region approach is adopted globally. This scenario scores well on security because regions reduce energy consumption and use renewable resources located within their region. One open question is whether regions are seen as competing with each other, perhaps undermining some of the security benefits, or viewed as nonthreatening, thereby enhancing security.

The Safe Region scenario (regionalization/efficiency) consists of protection of economic and cultural interests, sharply opposed cultural blocks, and emphases on national safety, law and order, and limiting free trade. The outcomes are energy consumption that increases by +75% in 2030, energy prices that are up by +100% in 2030, energy consumption and GHG emissions that continue to rise, limited development of renewable energy based on regional resources, and a global population of 11 billion at mid-century.

This scenario scores poorly on all three substantive objectives, although it might score well on the public process objective within each region. It limits economic growth due to regional isolationism, has poor environmental outcomes due to emissions and large global population, and aggravates interregional frictions and security concerns.

This Safe Region scenario has a lot of similarities with the IPCC's A2r scenario in Riahi, Grübler, and Nakicenovic (2007). Fertility patterns converge slowly resulting in 12 billion people in 2100, emitting 30 gigatonnes (Gt) of carbon dioxide (CO_2) equivalent that year. Regions are fragmented and urbanization within regions is high resulting in high income disparities. The supply focus is on regionally available fuels such as coal.

It also is reminiscent of a Global Mercantilism scenario developed by Kahane (1992). This scenario consists of regionalization and fragmentation, a widening gap between rich and poor countries, large oil price swings, and low levels of political support for environmental concerns.

The Global Solidarity scenario (globalization/solidarity) asserts that social justice is a priority and in the energy context this is manifested in sustainable development in order to resolve the tensions between ecology and economics; worldwide government coordination is achieved. The outcomes are effective global climate policy, energy consumption that is up by +50% in 2030, renewable energy and energy-saving technology are prevalent, and use of fossil fuels and emissions of GHG emissions decline.

TABLE 1 Comparison of Outcomes of the Netherlands Environmental Assessment Agency

Outcome	Global Market	Caring Region	Safe Region	Global Solidarity
Energy Prices	Grow with inflation	+80%	+100%	Not available
Energy Consumption in 2030 (relative to today)	+100%	+50%	+75%	+50%
GHG Emissions	Increase	Stabilized	Not available	Decline
Population 2050	9 billion	Not available	11 billion	Not available

GHG means greenhouse gas.

This scenario assumes an effective public process that can garner global acceptance of a fundamental change in energy, and has much in common with the Sustainable World scenario developed by Kahane (1992), which emphasizes the focus on the resolution of common problems, continued and successful development of institutional structures to deal with these problems, and aid and technology transfer from rich to poor countries.

If such an outcome could be achieved, then it would score well on global security, environment, and economics, particularly long-term environmentally conscious economic growth. The question is whether this ideal outcome is something that could reasonably be achieved, albeit with great effort and uncertainty, or whether the pursuit of such an outcome is not realistic and therefore takes away from more modest but doable gains. In short, does this scenario risk making the perfect enemy of the good?

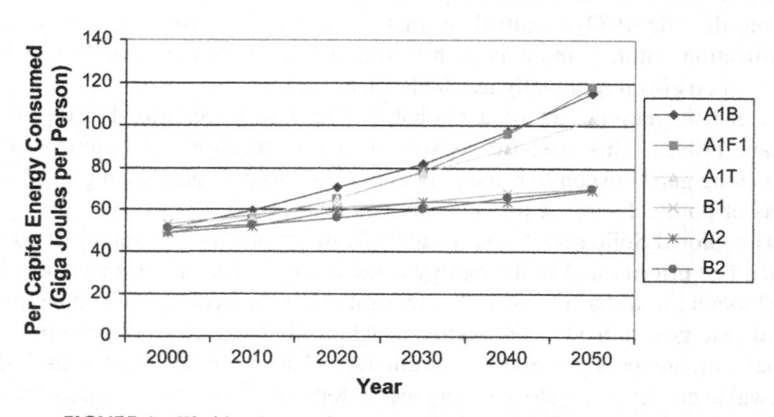

FIGURE 1 World average primary energy consumption (IPCC scenarios)

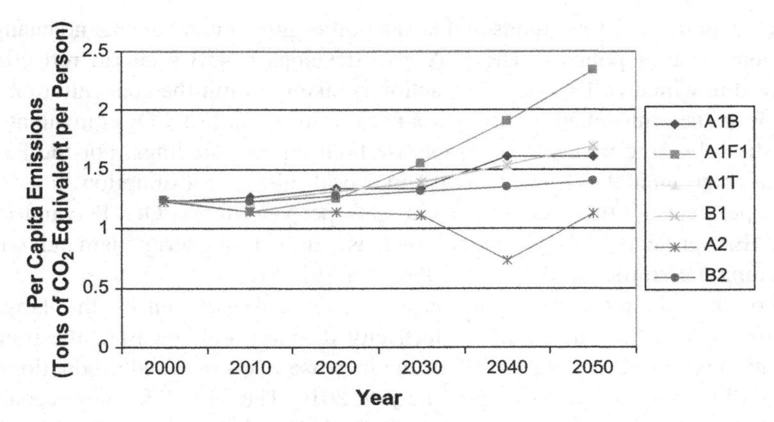

FIGURE 2 World annual CO_2 equivalent emissions (IPCC scenarios)

Table 1 summarizes the four Netherlands Environmental Assessment Agency's scenarios.

Given the overlap of some of the Netherlands Environmental Assessment Agencies and some of the IPCC scenarios, Figures 1 and 2 compare different IPCC scenarios for per capita energy consumption and per capita CO_2 equivalent emissions. Family A scenarios are various business-as-usual scenarios, and Family B scenarios provide a range of different policy intervention scenarios. Appendix A provides more information on these and other future energy scenarios.

3. BUSINESS-AS-USUAL GLOBAL ENERGY TRENDS ARE DOMINATED BY INCREASING FOSSIL FUEL USAGE

In contrast to what *could* happen, in this section we compare two long-term forecasts of the global energy future under business-as-usual conditions. Both rely heavily on extrapolation from historical trends, informed by engineering-economic analysis of future alternatives. They are forecasts, meaning that they attempt to predict the most probable trajectory for the future energy economy assuming no major changes in worldwide energy policy. These two studies result in remarkably similar forecasts of population, economic, energy, and GHG growth rates.

The forecast from the U.S. Energy Information Administration (EIA) at the time of writing is the *International Energy Outlook*, published in May 2009. EIA's forecast assumes no major action on climate change, extends through the year 2030, and divides nations into two groups, within and outside of the Organization for Economic Cooperation and Development (OECD). In November 2009, the World Energy Outlook released by the IEA provides a similar scenario to EIA's reference case. It develops a Reference scenario, which captures existing trends and also assumes governments make no change to their existing policies. The IEA also develops a 450 scenario reflecting a world in which collective global action is taken to limit the concentration of GHGs in the atmosphere to 450 parts per million (ppm) of CO_2-equivalent.

The following paragraphs paraphrase both reports' findings. For the EIA, the average annual percent increase in world energy consumption is 1.5%, 0.6% per year in OECD countries and 2.5% per year in non-OECD countries. The IEA projects a 1.5% annual increase in world energy demand with developing countries in Asia being the main drivers.

Fossil fuels retain their dominance, and coal experiences the largest increase due to the increases in electricity demand. For the EIA, the transportation sector accounts for 97% of the increase in oil use, and production in non-OPEC countries peaks around the year 2010. The IEA Reference scenario highlights energy security concerns. It forecasts that China overtakes the United States around 2025 to become the world's largest spender on oil and gas imports, with India becoming third, ahead of Japan. According to the EIA, liquid fuels, due to their importance in transportation and industrial use, are anticipated to be the dominant energy source, rising from 85 million barrels per day in 2006 to 107 million barrels per day in 2030. The EIA anticipates OPEC retaining an approximate 40% share of global liquids production; the IEA believes that OPEC's share grows from 44% today to 52% in 2030. Even with large uncertainties in the price of oil, varying between $50 to $200 per barrel in both forecasts, oil demand is high due to transportation. Eighty percent of the worlds proven oil reserves are located in eight countries and only two of them, Canada and Russia, are not in OPEC (*Oil & Gas Journal*, 2008).

The EIA forecasts that 40% of the world's natural gas supply will be used for industrial purposes, and electricity generation will account for 35% of gas consumption in 2030. The IEA forecasts that 45% of the increase in the demand for natural gas through 2030 will be due to power stations, primarily using combined-cycle gas turbine technology. World coal consumption is expected to increase from 127 quadrillion BTUs in 2006 to 190 quadrillion BTUs in 2030, with much of this increase due to non-OECD countries in Asia, particularly China (EIA, 2009). China is forecasted to triple its coal use between 2006 and 2030, according to the EIA, but only to double according to the IEA. The IEA projects that 97% of the increase in global coal demand occurs in non-OECD countries, mainly Asia. According to the EIA, renewable energy sources increase their share, primarily from hydroelectric facilities in non-OECD countries and from wind and biomass in OECD ones. Two-thirds of the new nuclear power plant additions are expected to occur in China, India, and Russia (EIA, 2009).

The EIA also forecasts global economic growth is anticipated to be 3.5% per year with 2.2% in OECD countries and 4.9% in non-OECD countries. The IEA has a slightly lower rate of global economic growth of 3.1%. According to the EIA, world carbon dioxide emissions rise from 29 billion metric tons in 2009 to 40.4 billion metric tons in 2030 due to economic growth and use of

TABLE 2 Comparison of U.S. EIA to IEA Reference Scenario

Parameter (2007-2030)	U.S. EIA	IEA Reference scenario
Annual World Population Growth Rate	1.0%	1.0%
Annual Energy Growth Rate	1.5%	1.5%
Real GDP Annual Growth Rates	3.5%	3.1%
Annual CO_2 Emissions Growth Rate	1.7%	1.5%

Note: IEA reports compound average annual growth rates.

fossil fuels. Although emissions from coal exceed those of liquids and natural gas, each fuel source's contribution is substantial. Coal is both the most carbon-intensive fuel and is the fastest growing carbon-emitting energy source. China alone counts for 74% of the expected total increase in the world's coal-related CO_2 emissions. The world CO_2 emission per person is anticipated to be 4.9 metric tons per person, with OECD nations at 11.2 and non-OECD nations at 3.7 metric tons per person.

Table 2 compares some of the major forecasted parameters of these two forecasts, assuming current trends.

The 450 scenario envisions a 14% reduction in total energy demand, approximately a 50% reduction in coal, smaller reductions in oil and gas demand, and increases in nuclear hydro-electric, biomass, and renewables compared to the IEA Reference scenarios. Figure 3 provides fuel-by-fuel comparison for 2030.

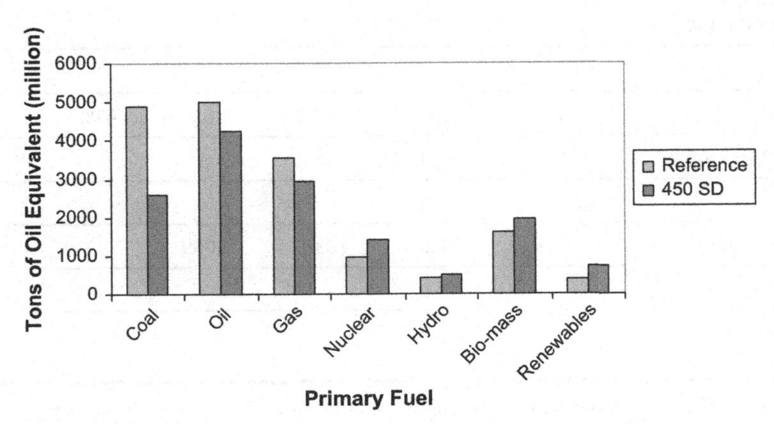

FIGURE 3 Comparison of world primary energy demand in 2030 by fuel type between the reference case and 450 ppm scenarios (IEA-WEO, 2009)

Each of these approaches for developing future energy scenarios highlights different challenges. Forecasts strive to predict, while projections strive to explain. We suspect that the real future will not match any of these scenarios, but hopefully by studying them we will be better prepared for whatever unfolds.

4. COOPERATION BETWEEN CHINA AND THE UNITED STATES IS A NECESSARY CONDITION TO IMPROVE THE WORLD'S ENERGY FUTURE

Since much but not all of the world energy's future depends on the individual and collective actions taken by China and the United States, each country's energy outlook is examined in more detail. (For readers who are further interested in China, Bram Buijs devotes an entire chapter in this volume on the subject, Chapter 15, "Why China Matters.")

Table 3 compares key parameters of the IEA forecast for China and the U.S. in the year 2030 and makes clear how critical the combination of China and the U.S. are to global energy issues. China and the U.S. combined are 37%

TABLE 3 Comparison of Key Forecasted Parameters between China and U.S. in 2030 (IEA Reference Case)

Parameter	China	U.S.	China + U.S.	World	(China + U.S.)/World
GDP ($2008 Trillion, PPP)	$28.5	$22.4	$50.9	$137	37%
Population (Billion)	1.461	0.367	1.828	8.2	22%
GDP/person ($2008, PPP)	$19,507	$61,035		$16,707	
CO_2 (Mt)	11,615	5,535	17,150	40,226	43%
CO_2/person (t)	8	15.1		4.9	
Oil (Mtoe)	758	772	1530	5009	31%
450 ppm CO_2/ person (t)	4.8	8.6		3.2	

CO_2 means carbon dioxide; GDP means gross domestic product; PPP means purchasing power parity; Mtoe means million tonnes of oil equivalent.

of the global economy and 43% of energy-related CO_2 emissions. To illustrate the importance of China and the U.S., if both countries were to reduce their CO_2 emissions to zero in 2030, the rest of the world would not have to make any reductions to achieve the IEA's 450 scenario. Similarly, these two countries combined are forecasted to consume 31% of the world's oil in 2030. In line with the IEA figures presented in Table 3, the EIA forecasts that in 2030, China and the U.S. are expected to account for 32% of world oil demand, 61% of world coal demand, and 43% of the energy-related CO_2 emissions (EIA, 2009).

4.1. China

China's most recent long-term National Energy Strategy is available online in the format of a multivolume report (China National Energy Strategy, undated). China has a growing oil security problem. In 2020, almost 60% of its oil will be imported (China National Energy Strategy, undated). The report acknowledges the obvious fact that countries have oil as part of their core energy strategies and that China, like many other oil-importing countries, is concerned with a temporary and abrupt cutoff or shortage in oil. Not surprisingly, it is imperative for China to maintain energy security, and it has been stockpiling oil in recent years, further adding to its oil imports (EIA, 2009b). According to the EIA, "The Chinese government's energy policies are dominated by the country's growing demand for oil and its reliance on oil imports" (EIA, 2009b). Although oil will be China's major energy security issue, it is also expected to continue importing natural gas via LNG and is considering imports via pipelines from neighboring countries (EIA, 2009b).

China, as its energy strategy recognizes, is extremely dependent on coal. In 2003, 68% of China's primary energy comes from coal and that percentage is expected to be above 60% in 2020 despite China's push for increasing fuel diversity with natural gas, nuclear, and renewable resources. In addition, despite China having the third largest coal reserves in the world, it is expected to be a net importer of coal in the next 5 to 10 years (EIA, 2009b). China's increasing demand for coal—expected to almost double in 2030—will account for 65% of the world's increase in coal demand and is a major reason that worldwide demand for coal grows more than any other energy source except nonhydro renewable (IEA, 2009). China is also pursuing a large coal-to-liquids industry (EIA, 2009b).

Economic growth for China is still the top priority but more emphasis is being placed on environmental issues. In the past, over-emphasis on growth has resulted in substantial pollution, including but not limited to air emissions. The energy strategy acknowledges the tension between the environment and the economy. China's current position in the international Conference of the Parties (COP) is that the developed countries must contribute 1% of their GDP to developing countries to address global climate change. China's economic situation is partially dependent on the U.S. economy. China is a major exporter to the U.S. and holds approximately $800 billion of U.S. federal government debt. This interdependency, fully recognized by China's energy strategy, is a two-edged sword in furthering cooperation between these two countries.

China's National Energy Strategy also pays substantial homage to market oriented reform of its energy sector. In the overview portion of the report, it refers to such reforms approximately a dozen times. It also specifically acknowledges that energy prices are currently subsidized and do not include the cost of environmental externalities. The report contains contradictory statements, referring to relaxing economic control while noting that increased social and regulatory control of the energy sector is also necessary.

The report refers to the need to have public involvement in energy policy and planning. It also raises concerns about equity, particularly in the context of raising energy prices, whether by removing subsidies or internalizing negative externalities. In December 2008, the government adjusted its oil pricing mechanism to more closely align internal prices to the international crude oil market (EIA, 2009b).

Finally, the document supports the diversification of fuel for China away from coal and oil. Currently, hydroelectric provides 6%, natural gas 3%, and nuclear power 1% of China's energy consumption mix (EIA, 2009b). China is vigorously pursuing nuclear power, planning for a fivefold increase in nuclear capacity by 2020 and a further threefold to fourfold increase of 120–160 GWe by 2030 (Guang and Wenjie, 2010). China is also one of the world's largest wind producers with over 25 gigawatts. China also may have substantial reserves of unconventional natural gas, although there are major potential obstacles to their development (IEA, 2009). Given the large amount of energy China gets from coal, the dominance of oil in its transportation sector, and the high expected energy demand growth rates, even with very aggressive fuel diversification policies and energy efficiency policies, little progress on GHG reductions is expected to be made by the year 2020.

4.2. United States

The U.S. energy situation is evolving under the influences of its historical legacy as a major energy exporter, its strong market orientation, and its fragmented, decentralized political structure. These factors frame the drivers of energy demand and supply, and energy plans and policies.

Energy demand in the U.S. will continue to be tied to economic growth, although the increasing penetration of efficient consumption devices and technologies will slow the rate of growth as compared to the growth of U.S. GDP. The U.S. economy has been on a long path toward increasing electrification of all energy-consuming sectors and this is expected to continue—especially in transportation with the trend toward electric vehicles (EV) and plug-in hybrid electric vehicles (PHEV). Electricity will continue to be the most convenient energy transport medium over the next 50 years with the ability to provide high levels of power at the point of consumption when needed.

Over the next 50 years, economic activity and growth, energy intensity of the economic growth, environmental concerns and the resulting policy, efficiency improvements impelled by cost and policy, and domestic and global supply will be the factors driving U.S. energy consumption. Each of these factors is interrelated.

The U.S. economy continues to move away from heavy industry material production (aluminum and steel) and heavy manufacturing (white goods and vehicles) and toward service and high technology. This trend moves energy consumption from the use of primary fuels in the production and manufacturing process and again toward electricity to be used in light manufacturing and lower energy density production. The one area of high-energy density that is uniquely suited to electricity and not to the use of primary fuels is computer data and server facilities, although progress is being made to make those facilities more energy efficient.

Transportation for both private use and shipping will continue to be dominated by the direct consumption of primary fuels in terms of personal vehicles and trucking to support industrial and commercial activity. Rail transportation for commuting is largely powered by electricity while intercity and interstate transportation for shipping is generally fossil fueled. Rail transportation for shipping is highly energy efficient but the prospects for growth are dependent on the development of intramodal transportation hubs.

Major demographic trends will also influence the U.S. energy picture. As more of the population moves to urban and semi-urban environments, the per capita energy consumption will go down. This is driven by population density as multifamily dwellings are generally smaller and more energy efficient and people travel more by public transportation in urban settings. Population growth will largely be driven by immigration policies over the next 50 years. The U.S. has displayed a cyclic pattern of anti-immigrant sentiment and more open borders and seems to currently be moving toward tighter limitations on immigration. A youthful population is needed to fuel economic growth and provide entry-level labor, which suggests that immigration policies will need to be revisited and relaxed at some point in the next 10 years.

The U.S. coal, oil, natural gas, and electricity industries all started as highly local, private enterprises. Over a century, as they grew in scale, they encountered challenges associated with financing capital-intensive facilities and became viewed as critical infrastructures. Government involvement first occurred at the local level, it then moved to the state level, and later to the federal level, but many jurisdictional overlaps remain. One result is the persistence of a regulatory patchwork that continues to hinder the development of a truly continental marketplace for electricity and natural gas. Although China's unitary governmental system is having trouble delegating the effective implementation of energy policies to its provinces, the U.S. federal system of government arguably has demonstrated an even greater degree of incoherence in its energy policymaking.

The U.S. and Canada currently operate in tandem on many energy supply policies. U.S. investment is routinely made in Canadian energy projects such as large-scale hydroelectric, shale oil recovery, and natural gas pipelines (through long-term take or pay contracts). Remarkably, while Mexico continues to be the second largest exporter of petroleum products to the U.S. after Canada, there is little long-term investment or contractual coordination. There are numerous historical and political reasons for this pattern with Mexico, dating back to the nationalization of the Mexican oil industry in 1938. The question now is what changes will take place in this relationship over the next 50 years. A closer relationship isn't expected to result in advantageous pricing but could be anticipated to move Mexican production toward greater efficiency and higher production levels, which would have an influence on world oil prices. The major policy question facing the U.S., Canada, and Mexico rests on willingness to extract petroleum from nonconventional sources (such as shale) and from more challenging and (in some cases) environmentally pristine regions. Offshore gas and oil reserves are being discovered around the globe, but to date the U.S. has prohibited coastal exploration outside the Gulf of Mexico. These resources are expensive to extract and require high world prices to make them economical, so there has been little interest in changing the policy. Energy security concerns may alter that policy over the long-term.

Renewable energy can be anticipated to play an increasing role in electric power supply for the U.S. (and China) primarily through the growth of wind and solar power. The limiting technical factors will be the ability to economically store the power and the ability of the grid to manage an intermittent supply. Energy storage technologies are still expensive and each comes with an energy transfer penalty; there are losses putting the energy into storage, whether charging batteries, compressing air, or pumping water up a hill. There are additional energy losses when the stored energy is converted to electricity. Each of these losses must be added to the cost, as well as the capital cost and operating cost of the storage medium to calculate the full cost of renewable energy. The intermittent nature of renewable energy poses a different set of technical challenges. Solar power is reasonably predictable from day to day but is not available every day and every hour of the day and so it must rely on energy storage systems. Wind power presents the additional challenge coming on and off rapidly as the wind surges. Wind power can result in the need to dump low-cost power from other sources, which can further impact the operating efficiency of base-load power plants. The wind energy must be absorbed rapidly to prevent instability in the transmission grid.

Coal represents half the primary energy for electric production in the U.S. and is a proven domestic resource. Electricity derived from coal enjoys a significant cost advantage against all competing sources and while cap-and-trade policy or explicit carbon taxes will drive up the cost of operation of coal, the incremental costs will not eliminate coal from the energy supply mix. Policy initiatives may also limit the cost impact on coal to prevent adverse impacts on the economy.

U.S. energy and environmental policy has been and is a mixture of markets, market-based mechanisms such as cap-and-trade policies for sulfur dioxide and nitrogen oxide, and command-and control-policies. In some areas it has been successful, such as the implementation of cap-and-trade policies for air emissions (Ellerman et al., 2000). In other areas, the policy has been mixed, for example the introduction of wholesale electricity markets. The one area that has been a clear failure is the dependence on imported oil. The percentage of imported oil into the U.S. has increased from 8% in 1950 to 68% in 2009.

The U.S. enjoyed a long period as the dominant global energy producer—it was a leading oil exporter until the end of World War II—and it remains the dominant global energy consumer. The U.S. has already become a marginal energy producer and in the next 50 years it will be one of many large energy consumers. The U.S. will increasingly be a price-taker with less ability to influence world prices by altering its consumption levels. The ample supplies of the past have forestalled a national consensus on an energy security policy that goes beyond the use of domestic resources.

China and the U.S. have substantial overlap in their energy resources, needs, and objectives. On a technological front, overlaps include addressing the intermittent nature of wind and solar energy, finding ways of using coal more efficiently and with a lower environmental impact, particularly in carbon capture and sequestration, reducing dependence on imported oil and associated risks, improving energy efficiency, and reducing energy intensity. From a policy and planning perspective, finding the right combination of international arrangements and agreements, domestic legislation, energy plans, mandates, and market-based mechanisms to address the multifarious energy objectives is also critical.

4.3. U.S. and China Energy Interactions Over the Next Several Decades

In many ways the two countries are two sides of the same coin, competing for resources and leadership, but looking at the world from completely different perspectives. On climate change matters, neither country wants to commit to reduction and management programs without the other. They do agree that they don't want to be subject to international mandates. Where they diverge is the internal approaches that are being taken.

Whereas China is refusing to accept any fixed targets, it has been moving forward aggressively with wind power projects and is a leading manufacturer of solar panels. It is also funding carbon capture and storage projects and integrated gasification combined cycle (IGCC) plants to develop the technologies. In contrast, the U.S. is moving more tentatively on both these technologies and part of what is holding the U.S. back is the weakness in the economy, while China is moving ahead with its comparatively robust economy.

In the area of new resources, China is aggressively seeking and securing access to oil and LNG in overseas markets, especially in Africa through the use of foreign aid and direct investment in energy and infrastructure projects. The U.S. is primarily relying on multinational oil companies to develop new supplies. The U.S. continues to argue internally regarding the further development of domestic resources—especially in Alaska and in the coastal waters along the Eastern Seaboard. The U.S. appears poised to develop new natural gas reserves from shale deposits but has not begun to address the water demands that this new resource will require. China and the U.S. will continue to exploit their extensive coal resources. Over time the coal burning fleet will become more energy efficient and will produce less CO_2 per kilowatt-hour as new technology is developed and applied. China is developing a domestic wind turbine manufacturing industry. Utility scale wind turbines in the U.S. are all imported. China is the leading manufacturer of photovoltaic systems.

5. NO SINGLE SOLUTION WILL ADDRESS GLOBAL ENERGY OBJECTIVES

Focusing on only one objective comes at the expense of one or more of the others. If national security is the priority, that would point to using coal, with its huge environmental costs along with substantial economic costs to wean the transportation sector off oil. If economic development is the priority, then national security suffers due to dependence on oil and so do environmental objectives due to the emissions associated with coal and oil. If environmental objectives have priority, then economic development is limited, although security may be enhanced depending on the use of nuclear power.

There is no single technology, either existing or emerging, that alone will address society's energy objectives; all technologies involve substantial tradeoffs between objectives. Solar is too expensive, intermittent, and large-scale and inexpensive storage does not exist. Wind is expensive and suffers from the same intermittency issues as solar. Nuclear has safety, proliferation, and long-term storage concerns and cannot be ramped up quickly. Hydroelectric is not available everywhere, takes a long time to build, is extremely capital intensive, requires extensive transmission, and has negative environmental implications. Biomass has very low energy density and negative environmental implications. Many of these options are not scalable to meet substantial global energy needs.

Thus, a combination of existing and new technologies is needed, but the optimal mix both among technologies and between investing in existing technologies versus research and development to improve future technologies is not at all clear. It becomes important to compare how alternative energy technology portfolios are likely to perform as the future unfolds.

The projection of current global trends in energy usage, technologies, and outcomes will fall short of achieving the above four major categories of objectives. If one were to infer the goals behind the current trends, they would be short-term economic growth even at the risk of energy price shocks and increasing levels of GHGs.

Past energy policies and plans have rarely been able to overcome the large inertia of the current energy technologies. Some examples include the U.S. response to the oil crisis starting in 1973 and the Kyoto Treaty. This inertia has several causes. First, existing energy assets are capital intensive and have long lives. Once they are built they continue to be maintained and operated so long as their going-forward costs, which do not include their sunk capital costs, remain low enough. As a result, any new technology's total cost must be less expensive than an existing technology's going-forward costs to replace it prior to the existing technology's end of life. If a new technology's total cost is lower than an existing technology's total cost, but not its going forward cost, then the replacement time of the existing capital structure is on the order of several decades, if not longer. Second, existing technology attracts political interest groups that have a large incentive to maintain the dominance of their technology backed by substantial resources compared to nascent technologies. Energy policymakers and planners should articulate the tradeoffs among objectives of various proposals so that the policymaking process is as informed as possible as opposed to advocating particular solutions that inherently reflect the advocates preferred tradeoffs. One way to do this is with scenario planning. Scenario planning, organized around internally consistent narratives and visions of the future, can help explore the above tradeoffs.

Before looking ahead, it is worthwhile to look at past energy forecasts. One key lesson that can be learned from energy forecasts made about U.S. energy use is that forecasters have often underestimated the importance of uncertainty (Craig et al., 2002). For example, in the 1970s, forecasters underestimated the ability of the U.S. economy to respond to higher energy prices, particularly oil, by increasing efficiency. Forecasters must not only forecast new technologies but must also forecast changes in human behavior and social networks.

A substantial amount of uncertainty exists in technologies, fossil fuel energy reserves, world economic and population growth, and so on, which must be addressed in scenario analysis. Within a given scenario, sensitivity analysis can be conducted to further explore uncertainty with the parameters of a particular vision of the future. The difference between scenario and sensitivity analysis is that sensitivity analysis occurs within a given scenario. Under the current trends scenario, global population is expected to peak around the year 2050. A sensitivity analysis on population may assume that the peak occurs in 2075, but within the internally consistent narrative that existing trends dominate the global energy picture.

6. LINKING SCENARIOS TO ENERGY POLICY AND PLANNING OBJECTIVES

One approach to energy policy and planning is considering it as a multi-objective problem under uncertainty (Hobbs and Meier, 2000). Although there are many objectives associated with energy planning and policy, one can group them into four categories: economic, geopolitical, environmental, and public acceptability. The first three categories are the "three Es" of sound energy policy: economic development, energy security, and environmental protection (IEA, 2009).

Economic objectives include economic development and macroeconomic improvements. A comparison between developed and developing countries is instructive. The developed countries achieved their current status in some measure due to access to inexpensive energy, although part of the "low" cost of energy has been, and is likely to continue to be, at the expense of the environment, along with pronounced security concerns. In addition, there is increasing recognition that sustained economic progress also requires maintaining and improving natural assets (Kahane, 1992). In contrast, developing countries and their populations are pursuing similar economic status at growth rates at twice or more than the growth rates of developed countries. An important feedback exists between economic development and population growth rates. History has shown that as the former increases, the latter decreases. Moreover, the reduction in fertility rates associated with economic growth rates is occurring over a much shorter time period now for developing countries than when it occurred for now developed countries during the industrial revolution.

Geopolitical energy challenges include national and international energy security, nuclear proliferation, and terrorism. Yergin states that "the objective of energy security is to assure adequate, reliable supplies of energy at reasonable prices and in ways that do not jeopardize major national values and objectives" (1988, p. 111). Consistent with Yergin is the International Energy Agency, which defines energy security as "access to adequate, affordable, and reliable supplies of energy" (2009, p. 115). Andrews introduces the notion of energy vulnerability as the "state of energy insecurity, typically due to insufficient or inadequately protected domestic supplies" (2006, p. 17). Energy security is not limited to oil and geopolitics, although that has been the dominant focus, but instead has multiple meanings that depend on an evolving political, economic, and social context (Chester, 2010).

Energy security is dictated in part by the location of fossil fuels or large scale hydroelectric facilities. In the case of oil, it is relatively concentrated in a few countries, enabling suppliers to control, in part, the world oil market. Although definitions of energy security and vulnerability cover multiple fuels and even energy carriers such as electricity, oil security is the concern that is most acute. The IEA identifies the specific vulnerabilities in the oil-supply chain and its emergency response mechanism (2009), and a review of the academic literature concludes that the U.S. macroeconomic costs of disruptions and adjustments—combining economic objectives with security—range from $2 to $8 per barrel (NRC, 2009).

Energy security concerns are not limited to oil. Natural gas resources are more dispersed than oil, but also have important security risks. Due to the cost of transporting it long distances, particularly over water, natural gas is susceptible to threats of interruption. In sharp contrast, coal is much more evenly dispersed throughout the globe and large resources exist near large populations. With electricity, notions of energy security are perhaps better related to reliability instead of security (IEA, 2009). Both concepts of *energy security* and *energy vulnerability* need to be expanded to include proliferation and terrorism that may not be related to supplies of energy but the use of the energy infrastructure to achieve violent outcomes. Nuclear proliferation, which can occur either through the commercial production of electricity with nuclear energy or directly via nuclear weapons, is another major security concern (MIT, 2003). In the Middle East and Africa, pipelines and export facilities have been targets in local and regional political disputes. Finally, energy systems are valued targets for terrorists either to interrupt supplies of energy or to use the targets themselves as means of spreading terror, such as attacking a nuclear power plant or liquefied natural gas facility.

Energy security and economic development overlap in several ways. One way is the *resource curse*, which has three variations (Kolstad and Wiig, 2009). The Dutch Disease involves a loss of productivity due to the appreciation of the resource-rich country's currency resulting in a contraction in manufacturing. Another is that patronage leads to inefficient employment and investment allocation. The third is that the economic rents associated with the resource result in unproductive rent-seeking activities to control the natural resource. For example, in the U.S. states of Texas and Louisiana, natural resource dependence contributes to slower economic growth, poorer developmental performance, and less competitive politics (Goldberg et al., 2008). Whoever can control the resource, for example an oil field, thereby controls enormous wealth potentially leading to endless fighting between competing military forces and instability for that country and region.

A third major objective of energy policy and planning is to address environmental concerns associated with both ecological and human-health impacts, such as GHG emissions, other air emissions, water impacts, toxic materials, and safety and longer term sustainability. Environmental impacts due to the production and consumption of energy have a range of implications over geography and time. Some are local, such as strip mining of coal, others are regional, such as acid rain, and others are global, such as the emission of GHGs. This category also includes damages associated with energy accidents (Felder, 2009).

A fourth objective, public acceptability, is both substantive and procedural (Andrews, 2002). It is not sufficient, particularly in democracies, for major energy policies and plans to be formulated and adopted without public input. The democratic process itself requires such input, at least indirectly through elections; it is common to add other direct stakeholder processes. Nondemocratic regimes also depend on the support of the people, not through formal means such as elections, but in acceptance of energy policies and plans, which is why they, along with democracies, heavily subsidize energy. As an illustration, when oil prices in 2008 were at $140 per barrel, the Chinese government decided to reduce gasoline subsidies, which sparked demonstrations and minor riots. Those subsidies were quickly restored. China is not the only country to subsidize energy prices (IEA, 2009; Pearce and von Finckenstein, undated). Another illustration of the importance of public acceptability is nuclear power, for which support is growing internationally (Adamantiades and Kessides, 2009).

Within each of these four broad categories of objectives—economic, geopolitical, environmental, and public acceptability—there are multiple subobjectives. In some cases these subobjectives are not ends in themselves, but serve to achieve other ends. This same categorization can apply to the four major objectives. Reducing GHG is a means to achieving end objectives of human health and economic wellbeing, itself a means to achieving human health and happiness. To further complicate the issue, there may be multiple indicators for various objectives and subobjectives. In the case of oil security, one reference found 12 indicators of oil import diversification and argues that diversification is not enough for energy security (Vivoda, 2009). Cabalu uses four indicators for the security of natural gas supplies in the context of Asia (2010).

Two problems arise with this multitude of objectives. First, all objectives, subobjectives, and their relationships need to be identified and constructed. This is a huge semantics problem as different people, groups, and nations use different terminology to refer to the same objectives, as was discussed above for energy security.

Furthermore, how objectives are characterized is not independent of how different people value that objective, so the process of establishing a common taxonomy of objectives is not merely a question of obtaining common understanding. It quickly bleeds into the second problem, which is that different people, groups, and nations have vastly different preferences for the myriad objectives. For example, Chester notes that different stakeholders will have different definitions of energy security (2010). Thus, developing this taxonomy is not a simple matter of imposing clarity and consistency on the objectives and subobjectives. It is itself a dispute between different groups that are trying to establish their preferences for different objectives and subobjectives, and any such dispute risks power trumping rationality (Flyvbjerg, 1998).

Underlying these objectives and their many supporting subobjectives are fundamental debates about how humans should live their lives, captured in part by the term *sustainability*, the organizing theme of this volume. Thus, to many the energy future is not just optimizing a set of objectives, but living in such a manner as to achieve more encompassing outcomes that are social, cultural, and political in nature that cannot easily be reduced to a vector of objectives. Despite the multitude of definitions of sustainability and their associated ambiguities (Bent et al., 2002), the term does capture a broader notion that the sum of the individual objectives, no matter how precisely laid out, does not. Table 4 summarizes these four objectives and some of their subobjectives.

TABLE 4 Energy Policy and Planning Objectives and Some Important Sub-Objectives

Objectives			
Economic	Security/ Geopolitical	Environment	Public Acceptability
GDP	Attacks on energy installations designed to cause loss of life, property damage, and widespread panic	GHG emissions	Equity
Economic growth rate	Interruption of key energy supplies	Other air emissions (mercury, sulfur dioxide, oxides of nitrogen, particulate matter, diesel particles, etc.)	Public input
Reduction of poverty	Energy blackmail and associated threats	Solid waste (nuclear waste, coal ash, heavy metals)	Accountability of policymakers
Volatility of energy prices and associated economic shocks	Nuclear proliferation	Liquid waste (slug, waste water, thermal waste, oil spills)	Transparent and above-board process
Affordable energy		Aesthetics /noise	Social justice
		Water quality and water availability	Local impacts vs. local benefits

GDP means gross domestic product; GHG means greenhouse gas. References: IEA, 2009; NAS, 2009; authors.

Incorporating objectives other than global climate change mitigation into the consideration of energy policies and plans may facilitate achieving greater GHG emission reductions than would solely focusing on climate change. This is an example of the old adage that in order to solve a big problem, make it bigger. The focus in this chapter is to envision energy futures across the three major categories of substantive objectives (geopolitical, economic, and environmental) while keeping in mind that at all times, broad public support is necessary if our collective future is to be shaped by these objectives as opposed to the inertia of the past.

CONCLUSIONS

On their current course, in 2050 China and the U.S. will be in an intense competition for oil. Even if both countries were able to emit no carbon except for the consumption of oil, the amount of GHG emissions would exceed GHG reduction goals. Thus, in terms of energy security, alternatives to oil must be developed and implemented relatively rapidly. A necessary but insufficient condition to address global warming is that the alternative fuels to oil be carbon neutral. This interdependency clarifies why a global GHG management regime is so necessary.

This chapter starts with a question: "Which energy future?" The energy economy has great inertia, so that many historical trends will in fact carry forward far into the future. Patience on a generational time scale is probably a prerequisite for a satisfying career in energy policy. However, the future can and will eventually diverge from the past. Powerful dynamics of change include demographic processes; the discovery, trading, and depletion of varied energy resources; numerous technological and institutional innovations; and shifting geopolitical roles for actors on the world stage. Some of these dynamics are in fact choices that people, organizations, and nations can make about the future. Much of the value of forecasts and scenarios for the future lies in encouraging thoughtful discussion about which energy future we really want. A key part of that discussion is whether to accept a probable but undesirable energy future, or to pursue a more desirable, but also more challenging energy future.

ACKNOWLEDGMENTS

The authors would like to thank Shankar Chandramowli, Erin Coughlin, and Haiyan Zhang for their editorial and research assistance.

BIBLIOGRAPHY

Adamantiades, A., & Kessides, I. (2009). Nuclear power for sustainable development: Current status and future prospects. *Energy Policy, 37*, 5149–5166.

Andrews, C. (2002). *Humble analysis: The practice of joint fact-finding*. Praeger.

Andrews, C. (2005). Energy security as a rationale for governmental intervention. *IEEE Technology and Society Magazine, 24*(2), 10–25.

Andrews, C. (2006). National responses to energy vulnerability. *IEEE Technology and Society Magazine, 25*(3), 16–25.

Bent, R., Orr, L., & Baker, R. (2002). *Energy: Science, policy, and the pursuit of sustainability.* Island Press.

Cabalu, H. (2010). Indicators of security of natural gas supply in Asia. *Energy Policy, 38*, 218–225.

Chester, L. (2010). Conceptualizing energy security and making explicit its polysemic nature. *Energy Policy, 38*, 887–895.

China's National Comprehensive Energy Strategy and Policy. (undated). Retrieved from *www.efchina.org/FReports.do?act=detail&id=155*

Craig, P. P., Gadgil, A., & Koomey, J. G. (2002). What can history teach us? A retrospective examination of long-term energy forecasts for the United States. *Annual Review of Energy and the Environment, 27*, 83–118.

Ellerman, A., Joskow, P., Schmalensee, R., Montero, J., & Bailey, E. M. (2000). *Markets for clean air: The U.S. acid rain program.* Cambridge University Press.

Felder, F. (2009). A critical assessment of energy accident studies. *Energy Policy, 37*, 5744–5751.

Felder, F., & Haut, R. (2008). Balancing alternatives and avoiding false dichotomies to make informed U.S. electricity policy. *Policy Sciences, 41*, 165–180.

Flyvbjerg, B. (1998). *Rationality and power: Democracy in practice.* University of Chicago Press.

Goldberg, E., Wibbels, E., & Mvukiyehe, E. (2008). Lessons from strange cases: Democracy, development, and the resource curse in the United States. *Comparative Political Studies, 41*, 477–514.

Greenberg, M., Mantell, N., Lahr, M., Zimmerman, R., & Felder, F. (2007). Short and intermediate economic impacts of a terrorist-initiated loss of electric power: Case study of New Jersey. *Energy Policy, 35*, 722–733.

Guang, Y., & Wenjie, H. (2010). The status quo of China's nuclear power and the uranium gap solution. *Energy Policy, 38*, 968–975.

Hobbs, B., & Meier, P. (2000). *Energy decisions and the environment: A guide to the use of multicriteria methods.* Kluwer Academic Publishers.

International Energy Agency. (2003). *Energy to 2050: Scenarios for a sustainable future.*

International Energy Agency. (2009). *World economic outlook.*

Jefferson, M. (2000). Long-term energy scenarios: The approach of the World Energy Council. *International Journal of Global Energy Issues, 13*, 1–3.

Kahane, A. (1992). Scenarios for energy: Sustainable world vs. global mercantilism. *Long Range Planning, 25*(4), 38–46.

Kolstad, I., & Wiig, A. (2009). It's the rents, stupid! The political economy of the resource curse. *Energy Policy, 37*, 5317–5325.

MIT. (2003). *The Future of Nuclear Power.*

Norman, D. (2002). *The design of everyday things.* Basic Books.

NERC. *Understanding the grid: reliability concepts.* Retrieved from *www.nerc.com/page.php?cid=1%7C15%7C123*

Oil & Gas Journal. (December 22, 2008). (2008). Worldwide look at reserves and production. *Oil & Gas Journal,* 106(48), 23–24.

Pearce, D., & von Finckenstein, D. (undated). Advancing subsidy reform: Towards a viable policy package. Unpublished.

Riahi, K., Grübler, A., & Nakicenovic, N. (2007). Scenarios of long-term socio-economic and environmental development under climate stabilization. *Technological Forecasting & Social Change, 74*, 887–935.

Sachs, J., & Andrew, M. (1995). NBER working paper 5398: *Natural resource abundance and economic growth.*

United States Energy Information Agency. (May 2009a). *International Energy Outlook 2010*. Retrieved from *www.eia.doe.gov/oiaf/ieo/index.html*.

United States Energy Information Agency. (July 2009b). *Country Analysis Briefs, China*. Retrieved from *http://www.eia.doe.gov/emeu/cabs/China/Background.html*.

United States National Research Council. (2009). *Hidden costs of energy: Unpriced consequences of energy production and use*.

Vivoda, V. (2009). Diversification of oil import sources and energy security: A key strategy or an elusive objective. *Energy Policy, 37*, 4615–4623.

Yergin, D. (1988). Energy Security in the 1990s. *Foreign Affairs*, p.111.

APPENDIX A: SUMMARY OF ENERGY SCENARIOS

This appendix summarizes numerous energy scenarios conducted by a wide range of organizations both qualitatively and quantitatively.

Summary of Scenarios:

Scenario Developer	Target Year	Scenario Name	Scenario Description	Type	Main Drivers/ Normative Goals
Inter-governmental Panel on Climate Change					
Special Report on Emissions Scenario - 1996-2001	2100	**A1 Family**	High economic growth, technology absorption		Population changes, economic growth, environmental quality, equity, technology and globalisation
		A1F1 Family	Emphasis on fossil fuel	Explorative	
		A1B Family	Balanced technology mix	Explorative	
		A1T Family	Non-fossil fuel technologies	Explorative	
		A2 Fmaily	Slow economic growth, self reliance	Explorative	
		B1 Family	Gloabalised economy, sustained economic growth	Explorative	
		B2 Family	Economic growth slower, localised solutions	Explorative	
International Energy Agency (IEA)					
Sustainable Development Vision (2003)	2050	SD Vision	Refer Normative goals.	Normative	[1] 60% share of "zero carbon" sources in total world primary energy supply, by the year 2050. [2] Reducing dependence of oil in transportation sector by less than 40% by 2050. [3] Supplying electricity to at least 95% of the World population by 2050.

Three Exploratory Scenario (2003)	2050	Clean not sparkling Dynamic but careless Bright Skies	Goal of global sustainability vision is missed, lack of appropriate technologies Increasing pressure on fossil resources, environmental threat, and technology options Long term sustainability and security of energy supply	Explorative	Population, Income, Energy Supply and Energy Demand in two phases 2000-30 and 2030-50.
World Energy Outlook-2009	2030	Reference Scenario	Business As Usual Scenario	Explorative	Population, Economic Growth, Energy Prices and Technology
World Energy Outlook-2009	2030	450 Scenario	Refer Normative Goals	Normative	450 ppm CO2 eqv. by 2030
Stockholm Environment Institute					
Global Scenario Group	2050	**Conventional Worlds** Reference Policy Reform	Strong economic growth, mid-range population& development projections and gradual technological change	Explorative Normative	Values, desires, knowledge, population, economic growth, governance, technology, etc.
		Barbarization Breakdown Fortes World	Global Political and economic changes, growing populations, income inequity and persistent poverty, environmental degradation and technological innovation	Explorative Explorative	
		Great Transisition Eco-communalism New Sustainability	Sustainablility vision achievable, powerful cooperation among state/community level actors	Explorative Normative	

Shell Scenario					
	2050	**Scramble**	Mandate driven abatement measures, externalities excluded.	Explorative	Choices, prices, efficiency, climate, water, innovation, and implementation, etc.
		Blueprints	Market driven abatement measures, externalities included.	Explorative	
World Business Council on Sustainable Development					
		FROG	Economic growth accorded highest importance	Explorative	Human response, values, and beliefs, technology, innovation, population increase, etc.
	2050	GEOpolity	International cooperation, lowered economic growth	Explorative	
		JAZZ	Social and technological innovations, global markets	Explorative	
American Council- United Nations University*					
Millennium Project	2050	Cybertopia	Open trade and increasing globalization, low government involvement, intensely developed communications and high security.	Explorative	Major "dimensions": degree of globalization, communications technology, threats to global security, quality of life, government participation in society
		The rich get better	Open trade, low government involvement, intense communications and low security.	Explorative	
		A passive mean world	Isolation for global trade, stagnant communications but high government involvement and high security.	Explorative	
		Trading places	Open trade, low government involvement, intense communications but low security.	Explorative	
		Normative world		Normative	

US Energy Information Administration					
International Energy Outlook-2009	2030	Reference	Business as usual scenario	Explorative	Economic growth, energy price (low, high, and reference)
Canada- Energy Technology Futures					
	2050	Life goes on	Slow paced innovation, closed global markets, and grey environmental etiquette.	Explorative	Market conditions, rate of economic growth, pace of innovation, degree of environmental etiquette
		Grasping at straws	Slow paced innovation, green environmental etiquette, reasonable economic growth (about 2% annually), and open global markets.		
		Taking Care of Business	Rapid innovation, open markets robust economic growth (about 4% annually), and grey environmental etiquette		
		Come Together	Open global markets, rapid innovation, and high levels of environmental etiquette.		
Netherlands: Long Term Energy Outlook					
		Free Trade	Thriving economy, competitive markets, close economic cooperation.	Explorative	Level of cooperation between states, economic growth and environmental awareness
		Ecology on a small scale	Non-material values and environment accorded greater priority, economic growth slow.		
		Isolation	Policy driven by short term monetary gain, no international cooperation.		
		Great Solidarity	International cooperation, environmental awareness, rapid technological development.		

United Kingdom Foresight Program				
Energy Futures	World Markets	High levels of consumption and integrated world trading systems and lack of sustainability vision.	Explorative	Sustainability vision, consumerist values, degree of trading cooperation
	Provincial Enterprise	High levels of consumption, regional trading systems, lack of sustainability vision.		
	Global Sustainability	Sustainability vision, high degree of international cooperation.		
	Local Stewardship	Sustainability vision permeates into all tiers of economic and social system.		

Source: International Energy Agency (2003). *Energy to 2050: Scenarios for a Sustainable Future,* Paris: IEA, OECD . 1:19-38, 53-55, 190-196.
International Energy Agency (2009). *World Energy Outlook 2009*, Paris: IEA, OECD. 1: 73-126
Shell International BV, (2009). *Shell Energy Scenarios to 2050.*
US Energy Information Administration (2009). *International Energy Outlook-2009.*

Summary Predictions for 2050

Model	Scenario	Energy Mix (in EJ)							Primary Energy Mix (in percentage)						
		Coal	Oil	Gas	Nuclear	Biomass	Other	Total	Coal	Oil	Gas	Nuclear	Biomass	Other	Total
IPCC	**A1F1**	**186**	**214**	**465**	**123**	**193**	**167**	**1347**	**13.81**	**15.89**	**34.52**	**9.13**	**14.33**	**12.40**	**100**
	A1B	**475**	**283**	**398**	**137**	**52**	**86**	**1431**	**33.19**	**19.78**	**27.81**	**9.57**	**3.63**	**6.01**	**100**
	A1T	**119**	**250**	**324**	**115**	**183**	**222**	**1213**	**9.81**	**20.61**	**26.71**	**9.48**	**15.09**	**18.30**	**100**
	B1	**167**	**228**	**173**	**105**	**95**	**46**	**813**	**20.54**	**28.04**	**21.28**	**12.92**	**11.69**	**5.66**	**100**
	A2	**294**	**228**	**275**	**62**	**71**	**42**	**971**	**30.28**	**23.48**	**28.32**	**6.39**	**7.31**	**4.33**	**100**
	B2	**86**	**227**	**297**	**48**	**105**	**107**	**869**	**9.90**	**26.12**	**34.18**	**5.52**	**12.08**	**12.31**	**100**
IEA	**SD Vision**	**99.3**	**181.3**	**267.1**	**114.5**	**159**	**191.8**	**1013**	**9.802567**	**17.90**	**26.37**	**11.30**	**15.70**	**18.93**	**100**
Shell Scenario	**Scramble**	**141**	**108**	**263**	**43**	**131**	**199**	**881**	**16.00454**	**12.26**	**29.85**	**4.88**	**14.87**	**22.59**	**100**
	Blueprints	**157**	**122**	**208**	**50**	**57**	**175**	**769**	**20.41612**	**15.86**	**27.05**	**6.50**	**7.41**	**22.76**	**100**

Summary Predictions for 2030

Model	Scenario	Primary Energy Mix (in MTOe)							Primary Energy Mix (in percentage)						
		Coal	Oil	Gas	Nuclear	Biomass	Other	Total	Coal	Oil	Gas	Nuclear	Biomass	Other	Total
IEA-WEO	**Reference**	**4887**	**5009**	**3561**	**956**	**2376**		**16790**	**29.11**	**29.83**	**21.21**	**5.69**	**14.15**	**0.00**	**100.00**
IEA-WEO	**450 Scenario**	**2614**	**4250**	**2941**	**1426**	**3159**		**14389**	**18.17**	**29.54**	**20.44**	**9.91**	**21.95**	**0.00**	**100.00**
		Energy Mix (in Quadrillion BTU)													
US EIA- IEO	**Reference**	**190.2**	**106.2**	**152.2**	**155.5***	**74.1**		**678.3**	**28.04**	**15.66**	**22.44**	**22.93**	**10.92**	**0.00**	**100.00**

* *Actual figures quoted as 3844 Billion kWh. EJ: Exa (10^18) Joule; MToE: Million Ton of Oil Equivalent; BTU: British Thermal Units.*

Source: International Energy Agency. (2003). *Energy to 2050: Scenarios for a sustainable future.* Paris: IEA, OECD. 1:19–38, 53–55, 190–196.

International Energy Agency. (2009). *World Energy Outlook 2009.* Paris: IEA, OECD. 1: 73–126

Shell International BV. (2009). *Shell Energy Scenarios to 2050.*

U.S. Energy Information Administration. (2009). *International Energy Outlook 2009.*

Summary Predictions for 2050

Model	Scenario	Population (in millions)	GDP/GNP trillion $ (2008 PPP basis)	GDP/GNP trillion $ (1990 PPP basis)	World Average Energy Use per capita per annum (GJ per capita)	World Energy Use per capita per annum (Toe per capita)	Emmissions level per capita (Tons per capita)	Emissions level (GTC eqv.)
IPCC	**A1F1**	**8704**	**254.4**	**181.3**	**154.76**	**3.68**	**1.61**	**730.60**
	A1B	**8703**	**230.5**	**164**	**164.43**	**3.91**	**2.34**	**820.90**
	A1T	**8704**	**262.8**	**187.1**	**139.36**	**3.32**	**1.69**	**730.60**
	B1	**8708**	**189.7**	**135.6**	**93.36**	**2.22**	**1.70**	**730.60**
	A2	**11296**	**113.8**	**81.6**	**85.96**	**2.05**	**1.13**	**730.60**
	B2	**9367**	**153.9**	**109.5**	**92.77**	**2.21**	**1.40**	**730.60**
IEA SD Vision	**SD Vision**	**8704**	**243.4**	**173.2**	**116.38**	**2.77**	**1.15**	**9.99****
Shell Scenario	**Scramble**	**9100**			**96.81**	**2.31**		
	Blueprints	**9100**			**84.51**	**2.01**		

Summary Predictions for 2030

Model	Scenario	Population (in millions)	GDP trillion $ (2008 PPP Basis)	GDP/GNP trillion $ (1990 PPP basis)	World Average Energy Use per capita per annum (GJ per capita)	World Energy Use per capita per annum (Toe per capita)	Emmissions level per capita (Tons per capita)	Emissions level (GtC eqv.)
IEA-WEO	**Reference**	**8286**	**137**	**97.48**	**85.10**	**2.03**	**1.51**	**12.49****
IEA-WEO	**450**	**8286**	**137**	**97.48**	**72.93**	**1.74**		
US EIA- IEO	**Reference**	**8327**	**137.48**	**97.82**	**85.94**	**2.05**	**4.85**	**40.385****

** *Indicates noncumulative annual figure. Toe: Ton of Oil Equivalent; GJ: Giga Joule; 1 Toe = 42 GJ.*

Source: International Energy Agency. (2003). *Energy to 2050: Scenarios for a sustainable future.* Paris: IEA, OECD. 1:19–38, 53–55, 190–196.

International Energy Agency. (2009). *World Energy Outlook 2009.* Paris: IEA, OECD. 1: 73–126

Shell International BV. (2009). *Shell Energy Scenarios to 2050.*

U.S. Energy Information Administration. (2009). *International Energy Outlook 2009.*

Chapter 3

Energy "Needs", Desires, and Wishes: Anthropological Insights and Prospective Views

Françoise Bartiaux,[*,†] Nathalie Frogneux,[†] and Olivier Servais[†]
[*]*National Fund for Scientific Research,* [†]*Université Catholique de Louvain, Belgium.*

Chapter Outline

1. INTRODUCTION

This chapter[1] begins by addressing the question: "Can we provide a decent standard of living for 9+ billion people by 2050 on a sustainable basis?" One could ask: "What is a 'decent' standard of living?" and, "How much energy is 'needed' to sustain a decent standard of living?"

Let's be clear from the outset: sociologists and philosophers have difficulty with such questions. Indeed, from what cultural perspective will these questions be answered? Each culture defines in its own way what a decent standard of living is or what the energy requirements for it are. Perhaps the most appropriate solution would be to focus first on the origin of the notion of *need* in our civilization, which is the path we have chosen to follow.

1. We wish to thank Dr Mithra Moezzi (Portland State University), whom we asked to correct the language of this chapter. Her remarkable competence and availability led to additional discussion of the substance of this chapter, bringing valuable improvements to its final version.

To reduce energy consumption, one could indeed be tempted to define the notion of "needs", to hierarchize and prioritize these needs in order to arrive at energy demand, but such an approach leads to many difficulties. This chapter addresses some of these difficulties from four perspectives, namely: philosophy, psychology, anthropology, and sociology. In so doing, this chapter attempts to continue the discussion initiated by Douglas et al. (1998) who observed that "the present social science conceptualization of human needs and wants sits awkwardly in the global climate change debate." (pp. 259–260).

Section 2 of this chapter is devoted to a deconstruction of the notion of "needs" in general and "energy needs" in particular. In Section 3, we recast energy consumption and production into sociopolitical stakes and argue that a local perspective is the best way to examine the links between energy consumption and its environmental consequences. We will also discuss the notion of a "decent" lifestyle.

Overall, we suggest that instead of looking only at energy needs and demand from a traditional economic point of view—which leads to significant investments in the supply side of the equation—it is preferable to focus on the energy sociopolitics directly involving the concerned actors. The participants of a decision should be everyone who is implicated in this decision, for she or he will have to support its consequences. This criterion is not from economics, and can be endorsed only by a voluntary politics.

2. CRITIQUE OF THE NOTION OF "NEEDS" IN THE CONTEXT OF "ENERGY NEEDS"

This section is devoted to a critical examination of the notion of "needs" in our occidental tradition including the supposedly common and physiological nature of "needs", the misleading hierarchy between physiological and other human "needs", the so-called "needs" in production-oriented societies, and the question of society's roles in defining and answering human "needs". We conclude this section with a critique of the neoliberal consumerist society.

2.1. The Relativity of Physiological "Needs"

The notion of "needs" often implies that there exists a "human nature" resting on a common biological reality that is universally shared. However, cultural diversity and the variety of conceptions of the "good life" show the relativity of "needs" and by so doing, raise questions about the very notion of "needs", understood as what life requires absolutely and necessarily. The misunderstanding originates in the assumption that "needs" are objective and universal whereas, as illustrated below, they are *relative* to cultural frameworks and individual subjectivities.

In the field of direct energy consumption, the importance of individual variation has been demonstrated in a study of Lutzenhiser (1993), who notes that in nearly identical buildings occupied by families with similar demographic characteristics, 200%–300% variations in energy use have been reported.

Food and water consumption seem, at first glance, to be answers to physiological "needs"—answers requiring a lot of energy in the production stage of food as well as during transportation. Regarding water, however, Cohen (1995) demonstrates that as with other natural constraints, limitations associated with water strongly depend on human choices and time limits. These human choices—made by those living today and their descendants—point to the fact that there is no universal definition of the water quantity each human should enjoy.

Similarly, the UNFPA (1991, p. 72) as well as Bartiaux and van Ypersele (2006) cite the study by the World Hunger Program (Chen et al., 1990). According to this study, the planetary ecosystem could, in the present state of agricultural techniques and with the share of foodstuffs equalized, accommodate 5.5 billion individuals under good conditions, but only if they were satisfied with a vegetarian diet. If these individuals obtained 15% of their calories from animal products, as is generally the case in South America, the tolerable effective total would fall to 3.7 billion. The Earth could only accommodate 2.8 billion human beings if they derived 25% of their calories from animal products, as is currently the case with the majority of inhabitants in North America. The variations in these figures clearly show that capacity depends on the definition given to an acceptable diet.

Food production necessitates energy, and, furthermore, food items differ substantially with respect to greenhouse gas (GHG) emissions when these emissions are calculated from farm to table, taking into account anthropogenic warming caused mainly by emissions of GHGs, such as carbon dioxide (CO_2), methane, and nitrous oxide. Agriculture is the main contributor of the last two gases. Other parts of the food system contribute CO_2 emissions that emanate from the use of fossil fuels in transportation, processing, retailing, storage, and preparation, setting aside additional emissions as well as carbon capture deficit caused by deforestation to grow animal's food such as soy. In a recent study of 20 items sold in Sweden, Carlsson-Kanyama and González (2009) showed a range of 0.4 to 30 kg CO_2 equivalents/kg edible product. For protein-rich food, such as legumes, meat, fish, cheese, and eggs, the difference in emissions is a factor of 30, with the lowest emissions per kilogram for legumes, poultry, and eggs and the highest for beef, cheese, and pork. Large emissions for ruminants are explained mainly by methane emissions from enteric fermentation.

2.2. Deconstructing the Hierarchy of "Needs"

Were "needs" to be related only to the physical and metabolic nature of our body—which is not the case, as just shown—it could be tempting to distinguish between "primary needs" (to drink, eat, sleep, breath, take shelter, and cure oneself) and "secondary needs" (to dress, to speak...), which would be more superficial and have a lesser priority.

The psychologist Abraham Maslow has been a major contributor in the development of the hierarchical approach to survival in anthropology and Western thought in general. In his seminal work on the theory of human motivation (Maslow, 1943), Maslow developed an approach to needs now famous for its conceptualization in the form of a pyramid (see Figure 1).[2] In this perspective, he introduces a key distinction between physiological needs and other human needs: security, socialization, esteem, and achievement. Doing so helps build a narrow conception of "needs", relegating survival to the solely physiological level.

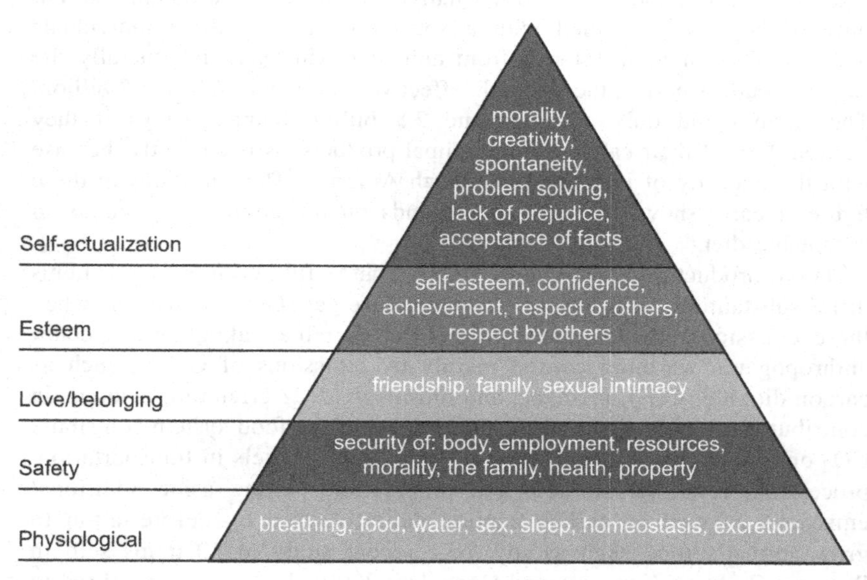

FIGURE 1 Maslow's pyramid of needs. *Source: http://upload.wikimedia.org/wikipedia/commons/thumb/6/60/Maslow%27s_Hierarchy_of_Needs.svg/500px-Maslow%27s_Hierarchy_of_Needs.svg.png*

2. Maslow's hierarchy from the base to the top of the pyramid is: 1) physiological needs (to eat, drink, breath, sleep); 2) body security; 3) social needs (communication, expression, and affectivity); 4) self-esteem within respectful relations; 5) implication; and 6) accomplishment and personal evolution.

Maslow's seminal needs pyramid raises many problems: satisfying the security "need" is necessary for satisfying "physiological needs" so both types of "needs" are intermingled. Indeed, it has been proven that babies raised with food but without love and words do not develop properly (Strivay, 2004 and 2006). Does this mean that language is a more fundamental "need" than food? Are affective bonds a condition for physical survival?

Maslow thought that the physiological needs came first. Applying these ideas in the energy field, one can justify the importance of energy production and consumption as a primary "need" of our modern civilizations. This gives a logical justification for saving energy and finding new means of energy production, but fails to allow the energy-production system to be thoroughly questioned and limits ability to think from another point of view.

An illustration of the inadequacy of such a hierarchization may be found among the Cree Amerindians of the far north of Canada. Traditional territories often radically change or even disappear because of a hydroelectric project (James, 2001; Niezen, 1993, 1998; Roue, 1999; Scott, 1984). "Green" energy production is favored by Canadian society to the detriment of transmission (to the next Cree generations) of Cree hunting habitats, routines, and rituals. These profound changes to territory are understandably experienced as traumatic events. They affect the very foundation of history and thus the identity of the Cree. An emblematic case that illustrates this situation is the Cree of James Bay during the impoundment of the hydroelectric dam EM-1 (Bréda et al., 2008). To construct this dam, and thus meet supposed energy "needs", ancestral territory is being submerged. In this case, the "energy needs" of one people, urban dwellers of Montreal and New York, supersede the identity marker (the ancestral territory) of another people, namely the native Cree. It implicitly endorses a hierarchy of "needs" that are paramount: "energy" is more necessary than "identity". But which arguments could justify this hierarchy? Needs of the others always seem weaker than the needs of the more powerful.

These choices should be made by democratic procedures—we later come to this point in Section 3. Without democratic procedures or moral considerations, the economic logic or the search for profit and particular interest may result in conflicts, as first hypothesized by Marx, whose thoughts on "needs" are presented below.

2.3. "Needs" in Production Societies

Even if the Marxist concept of survival societies as opposed to production-oriented societies lacks empirical basis, this concept is of great interest for this chapter. Indeed, it highlights an important distinction concerning the opposition of usage value and exchange value, as well as introduces "needs" as social constructions aimed at supplying exchanges imposed by the necessity of producing increase in value (Marx, 1845). So two notions are important to remember: on the one hand, some "needs" are related to our physical survival and thus are not created, and on the other hand, production-oriented societies manufacture an ideology of the "human rich in needs" to justify a logic of overproduction of which consumption society is the most radical version (Heller, 1978).

In his essay on the economics of primitive societies, Sahlins is the first truly systematic critic of Marx on empirical grounds (Sahlins, 1972). For Sahlins, if the hunter-gatherers did not produce and accumulate, it is primarily because they did not feel the need to do so. It is therefore unnecessary to adopt a production-oriented economy to have abundance; rather, simply by desiring little, one can create abundance. In other words, survival does not *necessarily* depend on production capacity, but on concepts that underlie the desire of nonproduction and accumulation. The priority is given to the symbolic rather than to the material dimension. Sahlins' hierarchy then reverses Marx's hierarchy, at least for the societies Sahlins studies.

In so doing, Sahlins remains within the logic of Marx's argument, because survival remains rooted in a concept of physical reality. The prominence given to the hierarchy of "needs" is symptomatic of an ethnocentric and western concept of survival.

We later return to this consumption society and its implication for energy.

2.4. Society as a Response to Human "Needs"

According to Malinowski, a Polish anthropologist of the first half of the twentieth century, institutions are the organizational means that human groups create to meet their biological "basic needs" and their cultural "derivative needs". The objective of Malinowski's theory (1944) is to study the function of culture by examining how any culture defines and answers the "basic needs" of humankind. Ethnology is, for him, the science of culture, which seeks to discover the laws of social organization. Culture is thus the way humans adapt to the natural and environmental conditions. Its function—an important concept for Malinowski—is to satisfy biological and social "needs". A culture is a coherent totality, an assemblage of institutions that are all, without exception, attached to the satisfaction of a "need". In his view, everything has a function in culture and everything is linked seamlessly to best meet human "needs".

For Malinowski, everything thus rests on the notion of "need", and in particular "physiological needs". And these "needs" determine requirements and create institutions and social practices. Institutions have functions that can be isolated, and, according to Malinowski, these functions would be pretty much the same in all societies—from which he infers that all human societies are reducible to a number of specific models related to these functions. Malinowski's functionalism assumes that a practice has to function to meet the "needs" of individuals. But at the same time, it is always the entire society, not its separate components, that responds to individual "needs". For Malinowski, culture is an undivided whole whose various parts are interdependent.

In energy policy and research, this way of thinking is still quite alive: states, and/or international institutions such as the United Nations' organizations, are seen as responsible for providing their citizens or members with adequate answers to energy needs that are conceptualized as being individual. This focus on individuals fits well with the way of thinking in economics.

However, some time ago, the social psychologist Paul C. Stern (1986), uncovered "what economics doesn't say about energy use", in particular because it "reduc[es] analysis of energy use to an application of economic theory" by focusing almost exclusively on a limited set of determinants of consumer behavior: prices and attention, the role of information, and the investments in energy efficiency (Stern 1986). This economics paradigm causes "blind spots in policy analysis" by ignoring "nonfinancial motives", such as consumers' commitment to save energy, attitudinal factors and beliefs, and "personal values concerning 'voluntary simplicity', beliefs about the effects of temperature levels in the home on comfort and health, and a sense of moral obligation to use energy efficiently." (ibid., p. 207)

This economics' focus on prices and decisions of end-users not only ignores psychological variables, as criticized by Stern (1986), it also "has the effect of excluding other questions about the social organization of energy consumption. More than that, it has the effect of sustaining a view of the policy process as pulling switches and juggling incentives so as to influence individual action", according to Shove et al. (1998, p. 301). These authors have been pioneers in opening energy issues to sociological investigation, as shown with several examples on energy below.

The British anthropologist Radcliffe Brown proposed an alternative to Malinowski's analysis, by comparing the different functions of culture as they relate not to individual "needs" but to the ones of society taken as a whole: the function of a particular social usage is the contribution it makes to society conceived of as the operation of the entire social system.

Therefore, both Malinowski and Radcliffe Brown limit every society to a response to human "needs". Such a concept of society determines a sole model of society whose unique function is to answer to human "needs"—a society of production and consumption.

A consequence of this systemic concept of society is that—because of its orientation to answering "needs"—it has crept into public politics as an ideology that precludes fully questioning conventional habits of thinking. So, in environmental matters, this analytical approach to the contribution ("function") of each component to the broader social system may limit a more comprehensive approach of the whole social system and a reappraisal of its objectives and habitual ways of thinking or doing. For example, in the energy research field, the dominance of technological research in changing schemes of energy production and consumption is not questioned, as in the European Union's Seventh Framework Programme (Moezzi and Bartiaux, forthcoming). The focus thus remains on technical fixes mostly on the supply side of the energy equation, as opposed to the end-use and behavioral aspects.

Similarly, on even more global questions of sustainability, the Agenda 21 (UNCED, 1992), states that, in order to "[develop] national policies and strategies to encourage changes in unsustainable consumption patterns" (4.1) "... governments, working with appropriate organizations, should strive to meet the following broad objectives:

(a) To promote efficiency in production processes and reduce wasteful consumption in the process of economic growth, taking into account the development needs of developing countries.
(b) To develop a domestic policy framework that will encourage a shift to more sustainable patterns of production and consumption.
(c) To reinforce both values that encourage sustainable production and consumption patterns and policies that encourage the transfer of environmentally sound technologies to developing countries." (4.17).

So, the paradigm of economic growth is not questioned in these Agenda 21 objectives.

2.5. "Need", Desire, and Wish

"Needs" are not static but instead dynamic. The "needs" dynamic is like a spiral, because it is able to integrate new objects of desire as new necessities, for instance new media of communication and quick transportation, both requiring more and more energy per person. In other words, new possibilities become new desirable objects and further necessities. Peculiar wishes and "needs" seem to work within a logic of a more fundamental desire that is usually translated into singular and various "needs" and wishes, except when, being conscious of this structure of desire, one applies oneself to break it, namely by spiritual means (such as in Buddhist groups).

So "needs" related to our human condition are inseparable of the logic of desire. Aristotle and Spinoza call it "perseverance" and Freud, "libido". Arendt (1958) underscores that the characteristics of "needs" that are related to our corporeal condition makes them complex and underlies their evolving character: "needs" are not given at once but are related to our human condition and to the conditions of our existence. Thus if a possibility appears, whether natural or artificial—as in the case of Internet use or air-conditioning, further described below—and if it corresponds to a desire, this possibility is turned into a "need" or even an addiction. For example, as new services or products are introduced, such as air-conditioners, the Internet, mobile phones, or flat-screen TVs, consumers take them for granted and demand them. Therefore, it is impossible to completely separate desire and "need", since desire focuses on new objects that it transforms into "objects of needs".

Like the notion of "need", the one of a "decent" lifestyle is worth thinking over. Indeed, it is impossible to objectively establish what a decent life would be as the answers and conceptions of a good life vary according to culture, time, and even individuals within a period or society. Many parts of a societal system may interact to *produce* needs that become less and less negotiable.

In the energy field, Shove (2003b, p. 399) illustrates this very well with the example of air-conditioning and goes one step further by showing how the new possibilities are invented and diffused by social processes. By doing so, she clearly demonstrates that "energy needs" are socially defined and embedded in a specific sociotechnical system made of building technologies, codes and standards, social practices, such as siesta, and shared expectations about a "normal" temperature that is adequate to answer to what is perceived as a physical "need" of coolness: "The conclusions of scientific research are embedded in codes and standards that are in turn reproduced in the built environment and in peoples' expectations of what it should be like. By redesigning homes and offices *for* air-conditioning, designers have condemned homeowners and workers to an air-conditioned way of life from which there appears to be no way back." Shove compares this one-directional process to a ratchet[3] to which dressing practices contribute: "Sure enough, the suit (or its thermal equivalent) has indeed become 'normal' wear all over the world and all year round. Conventions of this kind further restrict the range of actions people can take in making themselves comfortable, so increasing their reliance on the uniform provision of standard conditions at home, at work, in the car, on the train, and all points in-between. In this way, mechanisms of path dependent ratcheting also foster standardization within and between societies." (Shove, 2003b, p. 400.)

This sociological approach is thus far from "individual energy needs" because it shows the sociotechnical construction of this "need for coolness". This construction is made through a convergence of technologies (here, air-conditioning), social practices (building techniques, other type of windows, e.g., Wilhite (2008), new ways of adequate dressing, and so on) and norms (codes and standards as well as expectations on adequate temperature).

Another example of the social construction of "needs" is the following. When American astronauts arrived on the moon in 1969, one place on Earth was particularly visible at night: Belgium and its highways. This road network, especially dense at the time, was the most illuminated in the world. This collective habit of lighting the highways, as practiced in Belgium, is another example of "energy needs" that were artificially created and collectively integrated. This practice was developed in the 1960s to use excess nuclear capacity during the nights. Consequently, this practice equalizes what would otherwise have been low electricity consumption at night to be closer to daily

3. "I track the history of thermal comfort, using this to illustrate a path-dependent process involving a ratcheting of energy intensity from which there is no obvious way back." (Shove, 2003b, p. 397.)

consumption. In doing so, it legitimizes a particular logic of power consumption. Today, though, Belgium is no longer an exception, as highway and nighttime illumination have dramatically increased in many countries.

Both examples indicate that if energy consumption is to be diminished, one, two, or all three components of the sociotechnical system have to be redefined—techniques, social norms, and social practices—in order to socially recast normality in a way that reduces energy consumption. For example, for building cooling/heating, what social practices (dress, schedule, and so on…) and/or building techniques would save energy and how could they become normal? This recasting may be compared to processes of social diffusion of (new) practices or norms: the diffusion is either vertical, from the upper social classes to the lower classes as studied by French sociologist Bourdieu in his famous book on "distinction" (1979), or the diffusion is horizontal, and comparable to contagion, with all social classes changing at the same time. In environmental matters, such a horizontal diffusion occurred in Belgium in the late 1990s for new practices of sorting household waste because all segments of the population changed their routine at the same time (Bartiaux, 2007).

Bauman (2001) further discusses the notion of "need" within the framework of what he calls a *liquid society* that he defines: "I chose the metaphor of 'liquidity' mostly because of one trait all liquids share: the feebleness, weakness, brevity, and frailty of bonds and thus inability to keep shape for long." (Rojek, 2004, p. 301). In a liquid society, "The future—the realistic future and the desirable future—can be grasped only as a succession of 'nows'." (Bauman, 2001, p. 22).

Historically, during the nineteenth century, Bauman shows that "need" was "the very epitome of 'solidity'—inflexible, permanently circumscribed and finite." Later, during the twentieth century, consumption, and thus production, was driven by "desire, much more 'fluid' and therefore expandable than need" (Bauman, 2001, p. 14). Now, in our liquid society, "to keep the acceleration of consumer demand on a level with the rising volume of consumer offer", desire is replaced by wish as a motivating force of consumption (ibid.). With a wish, "a want, a whim, an impulse (…) satisfaction is instantaneous (…) it is the blissful instant of *acquisition* which is the contemporary consumer's prime mover." (Rojek, 2004, p. 299).

Bauman identifies "a 'mutual fit' between consumer culture and the task posed to individuals under conditions of modernity: to produce for themselves the continuity no longer provided by society." He therefore explores the new forms of consumption formed by a shift from the functionality of needs to the diffuse plasticity and volatility of desire, arguing that this principle of instability has become functional to a modernity that seems to conjure stability out of an entire lack of solidity. These anxieties are "born of and perpetuated by institutional erosion coupled with enforced individualization" (Bauman, 2001, pp. 9 and 28). Put otherwise, "Happiness-named-consumption is a *private* utopia (…) 'deregulated' and 'depolitized' (…) and ceded to individual enterprise." (Rojek, 2004, p. 309).

He concludes that "To avoid confusion, it would be better to follow that fateful change in the nature of consumption and get rid of the notion of 'need' altogether, accepting that consumer society and consumerism are *not about satisfying needs*—not even the more sublime needs of identification or self-assurance as to the degree of 'adequacy'." (ibid., p. 13).

To sum up, according to Bauman's conclusion, any initiative to save energy should match both individualization and institutional erosion. Thus on the one hand, it must be related to individual consumption behaviors, and on the other hand, it must instantly provide consumers with an answer to their anxiety with whimsical satisfaction. This is quite different from actual European awareness-rising energy policies, which deal with comparison of energy labels on appliances, or analysis of payback times of energy renovations, as indicated on the new European energy certificate, which is issued for each private dwelling to let or sell. As a matter of fact, neither environmental information nor customized energy-saving recommendations provided after energy assessments seem to themselves bring change in consumers' practices or energy-saving renovations works, as a comparative study realized in Denmark and Belgium has shown; indeed, if information or advice to save energy are not embedded in social life by being supported by the social networks of the receiver, they are not applied (Gram-Hanssen et al., 2007; Bartiaux, 2008).

2.6. Summary

In the section above, we criticized the notions of "needs", energy "needs", and "decent life". The argument began with the fact that physiological "needs", even for basics such as food and water, are relative to a society's way of life, as illustrated by the number of persons the Earth could accommodate according to the type of their diet, vegetarian or not. This relativity leads us to a wider deconstruction of the hierarchy of "needs" that was first proposed by psychologist Maslow. To define priority, "energy needs" has political stakes, as illustrated by the hydroelectric projects in Amerindian territories. There is thus neither a universal definition of "needs" that defines a "decent" lifestyle, nor consequently a universal definition of "energy needs": each society implicitly or explicitly defines them and will probably have to do so in an explicit way to meet post-Kyoto agreements.

To contextualize the notion of "needs" in production societies, we summarized Marx's concept, which dominated the twentieth century. For example, functionalist anthropologists establish that the primary function of society is to meet the "needs" of individuals or of the entire society itself. This way of thinking is close to the one in economics with its focus on individual agents, and it leads to ignoring nonfinancial motives as well as sociopolitical factors.

Finally, "needs" are dynamic. They integrate new objects of desire as new necessities supported by social processes such as building techniques, social practices related to work schedules, adequate dress, codes, standards, and norms on adequate work environment, as illustrated by air-conditioning. Furthermore, according to Bauman (2001), in our ever-changing societies, desire is replaced by wish as a motivating force of consumption. With a wish, satisfaction is instantaneous at the time of acquisition. He concludes, and so do we, that "To avoid confusion, it would be better to (...) get rid of the notion of 'need' altogether, accepting that consumer society and consumerism are *not about satisfying needs*".

3. RELEVANT ELEMENTS FOR A REAPPRAISAL OF CONSUMPTION POLITICS

This section brings together several concepts that we find useful in answering the criticisms of "needs" and "energy needs" offered in the previous section. We focus on a few building blocks that enable a reappraisal of energy policies without being grounded in the notion of "energy needs". They include a reference to our common human condition, a call for social and cultural diversity, the paradigm of climate justice, the crucial role of local energy policies, and an inverse scale of permissibility.

3.1. Our Common Human Condition

Lifestyle diversity and the variety of concepts of the good life lead to the acknowledgment that we share a common human condition—rather than a common human nature—whose conditions are peculiar to each society and define the social-historic (Castoriadis, 1987) or specific societies. The human condition enables us to define "needs" through a minimal set of what cannot be absent: no human being could stop eating, sleeping, desiring, or loving. But it seems impossible to positively define these "needs" such as the necessity to absorb this quantity of calories per day and per person, to sleep that number of hours... society institutes these criteria itself and in so doing, society institutes itself. Therefore, it is impossible to define *a priori* what it means for the society in question to satisfy human "needs". This introduces a policy dimension of the notion of "needs" via its normative dimension—a requisite and "an obligation to provide"—and thus "needs" correspond to rights, to which society should answer (Soper, 2006, p. 355 et seq.).

In the same way, the notion of "decency" or adequacy that underlies the present book requires further precision because, as discussed above, this notion does not make it possible to establish one lifestyle that would be acceptable by everyone. Choices and priorities may vary according to individuals, cultures, and periods. Evidently, individuals, cultures, and periods define for themselves their own lifestyles in very different manners, and characterize them as decent according to various criteria. An adequate or decent life might be variously defined as enjoyable, interesting, respectful, dignified, moral, entertaining, and so on. So, in terms of energy consumption, some will define a decent life as a life with low consumption of energy whereas others would not accept as decent this definition of a good life; it would seem austere to them. This definition of a lifestyle is valid only for those who choose their lifestyle *for themselves*; it cannot be decided for others. These lifestyles must be defined while respecting global and local ecological constraints. We need to move away from extravagant wasting as well as frivolous and myopic choices attached to a particular era, from the industrial revolution to the current green revolution, that ignore what precedes or follows it.

The question is how to have this constraint accepted by those who want to remain in a selfish and expensive lifestyle. (See the failure of Copenhagen summit.) A frugal but decent lifestyle must be accepted by all segments of population and this lifestyle must be sustainable according to the given context, for example, ecological and social constraints when it comes to energy costs. Current ecological conditions must enter into the determination of whether our choices on lifestyles are decent for ourselves, our contemporaries, and for future generations.

On the contrary, and when it comes to establishing *for others* (and not for oneself) what a decent life is, it is clear that this decent life has to take into account certain physiological functions as priorities: food, hygiene, security, and so on. In other words, a decent life for those others could be one that would allow them to feel free enough from the satisfactions related to the "naked life"—that life on which rests the possibility of giving and choosing meanings, and that could be characterized as secure for the maintenance and reproduction of life.

As an example, energy policy in Wallonia, the Southern Region of Belgium, includes several social measures, among which the gas and electricity utilities are prohibited from interrupting delivery of gas or electricity to their debtor consumers during winter because access to warmth is seen as a condition of a decent life (the details of the procedures are explained in Énergie Wallonie, 2008). If it is impossible to know which priorities individuals would specify for their lifestyle, the alternative is to guarantee a minimum threshold.

For example, if there were personal electricity quotas, a person could choose for herself to use her quota to play games on the Internet rather than to cook a meal. But this person can make this choice only for himself/herself. When the choice is to be made for others, the question of meanings and significance must remain open in such a way that every group and every individual has different possibilities from which to choose.

3.2. Individual and Collective Changes

So claims, expectations, or "needs" should be understood as having an objective dimension on the one hand (for they focus on specific objects) and a subjective dimension on the other hand—since the human condition is characterized by desire. This is because what one person judges as superfluous another may see as necessary, for example, in the case of saving on food to buy electronic devices. It seems misguided to reduce "needs" in an authoritarian or paternalist way, even while paternalistic measures might be applied transiently.

In environmental matters, the creation of a feeling of obligation as well as public infrastructure to enable fulfilling this obligation made behavioral change possible for sorting domestic waste during the 1990s in Belgium (Bartiaux, 2007). This obligation also relieved the consumers of making individual choices that would conflict with what they perceived as social normality. This positive aspect of many mandatory and relatively environmentally-friendly measures is often overlooked by policymakers.

Another solution for relieving consumers from making individual choice is to have the green option by default, as demonstrated by Pichert and Katsikopoulos (2008) in four experiments that show that people use the kind of electricity that is offered to them as the default. These authors conclude that changing defaults can be used to promote pro-environmental behavior.

But these mandatory containment measures may also awaken the desire for transgressing them, as seen, for example, in the difficulties in respecting the obligation to wear safety belts in private vehicles.

Other than mandatory measures, only voluntary limitation of "needs" and desires can have much effect when it comes to changing energy-related practices. Of course, authoritarian and paternalistic measures for saving energy may be considered on a temporary or even very short-term basis and still have an educational effect. However, to last, these changed practices should have a convincing justification in terms of what individuals can identify with, such as to be a good citizen, a respected professional, an adequate parent, and so on, which may be developed very quickly when people are obliged to modify their habits (Bartiaux, 2002). But given the structure of desire, only individuals or collectives themselves can trigger a sustained decrease in the hunger for energy, especially as society calls into question any fundamental heteronomy.

One solution would be to provide access to certain uses under certain conditions, such as using the seat belt when flying. Nobody obliges us to fly, but if we do choose to fly, then the conditions established by a collective can be imposed on travelers. In this respect, energy prices could include additional taxes for dealing with energy poverty, installing new infrastructures for producing renewable energy, and so on; this tax could be progressive, increasing with the quantity of energy used.

It would be better to readjust our "needs" on an acceptable scale for both the present generations—which must accept a diminution of current consumption—and future generations, who will have fewer resources. The choices of present generations should not mortgage the choices of future generations: for example, they should not choose irreversible uses of nonrenewable energy resources.

But how can we reduce levels of desires that are very expensive in terms of energy, either during production or through disposal of their unwanted effects, such as nonrecyclable waste? Provided that frugality and austerity are not wanted for their own sakes but rather to constrain frivolous behavior, several methods can be implemented. Quotas have the great benefit of quelling the illusory sense of infinite availability of energy. They allow the imposition of choices that are mutually exclusive; for example, either use a household appliance or water the lawn. Furthermore, they highlight the futility of certain behaviors where only the enjoyment of the finished product is beneficial and the other aspects of the life cycle are costly. For example do we want to enjoy a mobile phone, but not suffer the fumes resulting from reprocessing?

3.3. Diversity of Social and Cultural Ways of Life

Although there are neither unique nor worldwide definitions of "need", globalization of the economy and extensions of western lifestyle have made "needs", and how they are met, converge worldwide. For example, Shove (2003a,b) shows how air-conditioning is increasingly replacing traditional building techniques and traditional social practices such as the siesta: "Likewise, those who work in uniformly controlled climate conditions have no 'need' to pause for a siesta during the heat of the day. The fact that the siesta is in decline, even being officially banned in Mexican offices in 1999, is thus a compelling illustration of the extent to which whole societies have come to take a year-round pattern of a nine-to-five working day, and mechanical cooling, more or less for granted." (2003b, p. 399).

A suggestion to thwart worldwide trends leading to escalating energy consumption is made by Shove (2003a, p. 199): "[E]nvironmentalists should argue for social and cultural diversity. They should do all that can be done to engender multiple meanings of comfort, diverse conventions of cleanliness, and forms of social order less reliant on individual modes of coordination."[4]

Traditional building techniques are indeed culturally diverse. "If one begins to think green in a locally appropriate way, one will realize that traditional architecture was green in many ways. Every part of India had its unique stamp of buildings. This is because creative and architectural diversity was built on biological diversity. So buildings in hot regions would ensure corridors directed the wind so that it naturally cooled the interiors. (…) Today, Indians have forgotten how to build for their environment. Instead, modern buildings are examples of monocultures—lifted from the building books of cold countries where glass facades are good to look at and appropriate for their climate. The same building in India is a nightmare; the glass traps the heat." (Narain, 2010).

---

4. These individual modes of coordination often require car use and appliances such as freezers and dryers to save time. For example, day care centers at workplaces would reduce car use as an individual mode of coordination.

With the example of Kerala in Southern India, Wilhite (2008) underscores that imported technologies such as refrigerators bring along "scripts"—their way to be used—and so potentially reframe practices and social representations (on good food, on women's paid work, and so on) to the extent that they are consistent with other social changes. So Shove's call for more social and cultural diversity around the world faces another obstacle in the agentive capacity of bringing about change of some technologies themselves.

3.4. Climate Policies and Social Justice

Making policy links between social cohesion and sustainability objectives such as climate mitigation policies, is a new challenge, both in the academic world and in policy arenas of all levels, including European Union institutions. Pye et al. (2008) did pioneering research in "addressing the social dimensions of environmental policy" and studying "the linkages between environmental and social sustainability in Europe". They showed that "environmental policy interventions are likely to be regressive unless designed to mitigate such effects", for example by "increased social benefit payments to vulnerable groups, targeted subsidies for improved home insulation or energy-efficient products (e.g., the U.K.'s Warm Front Scheme) or general subsidies for public transport." (ibid., p. 6). Several policy instruments to alleviate and fix energy poverty are detailed by Boardman (2009) and many policy recommendations to find "the right balance" between climate change mitigation policies and social justice policies in Europe have been defined in a recent conference (King Baudouin Foundation, 2010).

The following paragraphs present two different concepts on how climate mitigation policies or environmentally-friendlier practices could be linked with social justice objectives. The first concept is voluntary and deals with more affluent consumers, whereas the second is mandatory and includes all consumers of the given political unit (i.e., country).

Soper (2007) theorizes a concept of "alternative hedonism", where new practices of affluent consumers, such as biking instead of driving a car, or eating organic food, arise as self-interested forms of disaffection with "consumerist" consumption. These affluent consumers themselves revise thinking about the "good life" and what is conducive to human flourishing and personal fulfillment. Soper argues that this "alternative hedonist" framework might "help to set off this relay of political pressure for a fairer global distribution of resources." (p. 223.)

Soper is probably overestimating the political impact of these affluent consumers around the world and their individual actions within this "alternative hedonism" framework. Voluntary actions of affluent consumers—such as occasionally biking—will certainly be insufficient to substantially and rapidly reduce their carbon footprints.

Individual carbon rations are a much more binding arrangement and therefore would probably be more effective if important energy savings had to be obtained. The main features of carbon rations or quotas would be the following: equal rations for all individuals, tradable rations, progression reduction of the annual ration, signaled well in advance, personal transport and household energy use included, and being a mandatory, rather than voluntary arrangement (Fawcett, 2005; see also this book's Chapter 4, by Meyer et al.).

Should such carbon quotas be implemented, their amount and their evolution should be debated in democratic ways in different fora such as the Parliament, perhaps in representative councils on sustainable development, or suchlike. For example, with the support of empirical data, Pett (2009) raises the following question: "Is an equal carbon allowance equitable for those with chronic diseases, who are always home and need high levels of warmth?" Before getting to that stage, many issues on the procedure remain to be tested (Fawcett et al., 2007), for example, should the allowance be per household or per individual? If by household, which proportion of the allowance should be granted per child?

3.5. Policy at a Local Scale with Visible Consequences

Furthermore, this notion of "needs" is like a mirror. We also identify ourselves as having such a "need" by telling what we need something for. The definition of needs is specular in the sense that it reflects both the subject that defines the object of "need" and the defined object of "need" itself. The question of needs cannot end at: What do we want? It must also involve the topic: Who do we want to be? It should thus be possible to regulate the pretentiousness of our claims about needs by raising the question of what we want to be, or preferably, who we want to be, whether as individuals or societies. These questions are to be locally answered. Indeed, the definition of identity and common goals for a good life cannot be conceived on a large scale, but only for small and medium-sized groups. The envisioned dialogue should not be about the set of values shared, but rather (only) about the environmental costs of particular standards of living and foreseeable consequences of those options, even recognizing that not all consequences of our actions can be predicted.

Couvin, a small town in Belgium, illustrates this point. One windmill with three specially-profiled blades produces an average of 450 kW, enough to power half the local households. The project is novel in its local and social characteristics, and is a result of briefings and field visits as well as innovative technical design conceived to eliminate any mechanical noise. The total set up cost is €2.8 million, according to the citizen-owned company Greenelec Europe. This one time amount includes €370,000 funded by the European Union and the Walloon Region. The balance is funded by bank loans and citizens from Couvin. It is an interesting example of citizens of a small town aspiring to become major producers of energy.

"Being both a producer and consumer is the best way to regulate the market" says Bernard Delville a citizen-sponsor of the project. "The problem is the same as for food: it is the problem of intermediaries. Relocating the production of renewable energy offers many opportunities. It is an argument that we have been defending for a long time."

This approach is a quite different strategy from the current binary approach to energy: either macro-production settlements (nuclear, hydroelectric, etc.), or individual devices such as photovoltaic panels in the U.K. (Keirstead, 2005), or small windmills in Sweden (Tengvard and Palm, 2009). "Today, two windmills per village would be sufficient to produce all residential electricity for Wallonia or 15% of overall consumption", says Jean-François Mitsch, a citizen involved in the Couvin project.

Local authorities seem to be a relevant locus for these debates, which could be framed by national governments with the following principles: local production of renewable energy, without causing troubles for neighboring municipalities, and with democratic procedures to guarantee that if energy production did increase, the local community would have to assume its potential harmful consequences. Furthermore, local authorities should become key actors in energy production (Cose, 1983; Tatum and Bradshaw, 1986; Hoff, 2000; Mesbah et al., 2007; Farrell and Morris, 2008).

These debates would have the advantage of recasting energy consumption and production as matters with clear political stakes (Morris, 2001). Indeed, energy production is typically made invisible and seemingly infinite. In Europe, when an energy shortage occurs, compensation is immediate as grids are interconnected. In addition, parts of energy-production costs are not quantified: for example, what is the cost of nuclear waste that will remain substantively radioactive for several thousand years? People do not see a stockpile of oil or coal dwindle as they turn on their appliances, and are typically unaware of how much energy they use, how much carbon this usage emits, and so on.

Therefore, we argue that what's at stake with energy use should be made visible and practical again, by situating energy production at the local level. That would mean, for example, that if citizens want more electricity, they would have to accept additional windmills or a depository of a proportional quantity of nuclear waste within the municipal territory. Awareness of links between energy claims or demands and their visible consequences would then be immediate. Bringing new knowledge from practical (and hidden) to discursive (and explicit) consciousness has been identified as a necessary condition for bringing about change in energy consumption (Hobson, 2003, following Giddens, 1984).

Linking energy consumption to the visible and potentially harmful consequences of energy production is in line with the "inverse scale of permissibility" (Jonas, 1980; Frogneux, forthcoming). This scale argues that an innovation should be legitimate only if its originators—those who are able to understand its stakes, who are motivated by this innovation, and who may benefit from it either directly or indirectly—apply this innovation for their own selves and their own children. In other words, no innovation or technology can be defended if the ones who understand it, benefit from it, and are motivated to develop it do not assume its consequences. By so checking the originators' integrity, it would be possible to better protect ourselves from immoral politics.

This is a remarkably practical way to check that the enlarged mentality for an equitable justice is at work when a technological option is considered or when a costly option in terms of energy development is taken. Several philosophers, from Montesquieu (1748) to Rawls (1971), have argued that to use a "thought experiment" in which one adopts another point of view than one's own is an effective way of broadening the terms of debate and of simulating having a collective decision based on the highest number of parties or persons. This argument suggests that one's judgment is only valid if it receives the agreement of someone else who is situated in another socio-economic position, in another place, in another culture, in another time, or in another health condition. Without pretending to have a universal point of view, it is possible to proceed from one point of view to the next one to develop a broadened judgment and to aim at a universal position. In such a way, decision-makers would no longer decide based on their own interests. If it is expected that a decision considered today would meet the disagreement of future generations, this decision should not be taken, even if the temptation is strong and there is low risk of disapprobation during the decision-makers' lifetimes. Of course, this method to enlarge the decisions to all implied persons does suppose a moral concern.

The inverse order of permissibility is another method for evaluating the distributive justice of the various options under consideration. If those who have the greatest economic interest in an option (for example nuclear) are not willing to suffer the effects of this process from its beginning to its end (to live near power plants, to live near nuclear waste, accept that their children live near nuclear waste, and so on), then this option should be either dismantled or not developed any further. Indeed, to assume a responsible energy choice is to accept to take in all its aspects and its ultimate consequences. If a technology is acceptable only if it is to be exported and if it is chosen by those who do not experience its worst effects, then this technology should not be developed.

3.6. Summary

In this section, we have tried to answer our own criticisms of the notions of "needs" and "energy needs" by presenting a few building blocks for reframing energy policies other than with this notion of "energy needs".

The first and perhaps most important point calls out our common human condition—rather than a common human nature—which is peculiar to each society and should be defined accordingly, possibly with an international mechanism to avoid free-rider societies or states. Defining the minimal conditions in reference to our human condition also means that a given society must clarify what a "decent lifestyle" is, along with its consequences in terms of energy policy; for example, a right for enough warmth during winter, even for consumers who are unable to pay for it. When defining a "decent lifestyle" for others, the priority physiological functions must be taken into account and a minimal quantity of energy must be granted. In addition, meanings and significance must remain open and allow different possibilities among which to choose.

We then turned to the question of how to reduce our "needs" to an acceptably low but sufficient level for present as well as future generations—knowing that no exact estimation is possible given the uncertainty of accumulated consequences—and compared self-selected restrictions to imposed and mandatory measures. We note that mandatory measures may offer relief to consumers by ensuring that individual choices would not conflict with social normality. Other policy instruments mentioned are offering the greenest options as defaults, quotas, and access to certain uses on certain conditions.

Another building block noted is the call for social and cultural diversity in the definitions of normal lifestyles that respect ecological constraints. Longstanding traditional building techniques, for example, are often less energy demanding in terms of regulating indoor temperature than are newer ones.

Furthermore, energy policies should, in our opinion, engage with a paradigm of climate justice, which links social cohesion policies and sustainability issues, namely climate mitigation policies. Indeed, environmental policies are likely to have adverse effects on social cohesion unless they are designed to mitigate such effects.

We then argued for local energy policies discussed and partly designed by citizens in democratic debates to reframe energy consumption and production in terms of political stakes and to make visible the consequences of energy production. Local energy policies should favor and implement local production of renewable energy, without causing trouble to neighbor municipalities, and with democratic procedures to guarantee that energy production will not increase. Two methods to guarantee democratic procedures were presented: the inverse scale of permissibility and the broadened point of view aiming at a universal position.

CONCLUSIONS

In this chapter, we investigated the question of "energy needs" and of "decent lifestyle" and their social and cultural preconceptions. We have acknowledged on one hand that there are "biologic needs" related to every human existence and, on the other hand, that these are radically relative, historically and culturally. Therefore, it seems misguided to try to define "needs" as a particular set of objective conditions, which means neutrally and validly for others; this impossibility explains why we prefer to speak about human condition rather than human nature. A consequence of this observed relativity is the impossibility of defining as frivolous certain choices that a person or collective might make for themselves and consider an absolute necessity (e.g., to play games on the Internet rather than to prepare and eat a meal). The logic of "need" is never far from the logic of desire and wishes, and resists external constraints, even if they are collectively imposed and accepted, as illustrated by infringements of traffic codes. What makes the difference between survival defined as satisfying physiological necessities versus a life that is human (possibly in an austere way) and respects the environment and available resources may be the possibility of choosing for oneself diminished "needs" that are assumed as meaningful and not only as constraining. The autonomy principle defines the difference between physical survival and austere life.

In negotiating this relative necessity of "needs" with the paradox of the choices to be made for others *sensu stricto*, should only the concerned persons make this choice for themselves while respecting the global ecological constraints? This seems to be the challenge of the contemporary world in dealing with the management of energy demand. Below we try to provide a few elements of solutions that could be both efficient and acceptable, in order to progress toward the dual aims of reducing nonrenewable energy consumption and in reducing social and economic inequalities at a global level and within countries. To be realistic and therefore more readily acceptable, energy policies should also take into account two contexts. First, they should take into account the points of view of those choosing measures of common restriction, who would strive to make these chosen measures respected. And second, they should take into account the individualistic contexts in which these collective measures would be imposed, which in turn points to the relevance of including individual measures in energy policies.

Would individual and tradable carbon quotas or rations allow both these dimensions to be respected?

Following Fawcett (2005), carbon quotas are interesting in that the points of departure to calculate them refer both to global energy demand and ecological criteria. Carbon quotas would be attributed to individuals (children's quotas would be managed by the responsible adults). They could be annual and initially calculated at the scale of a given country and according to various elements of context (transportation means, climate, etc.).

To be acceptable, restrictions of carbon quotas should be progressive, as proposed by Fawcett (2005). This characteristic should enable both a sustainable change of behaviors (and maybe of representations) and a progressive acceptability and motivation toward measures initially imposed or, in other terms, a transition from heteronymous restrictions to autonomous restrictions (even if they make sense to some people for other reasons than the ones initially motivating this policy measure).

Although these carbon quotas would have to be imposed, they could become accepted by the citizenry, since they allow an adjustment of one's "needs", both on their "quality" and their "quantity". From a qualitative point of view, everyone would be allowed to establish their own preferences on how to use the quota and whether to use it completely or not; on a quantitative point of view, everyone would be entitled both to sell or buy carbon rations on a carbon market and to produce renewable energy on a autarkic way—perhaps adding this production into the quotas, or instead using two distinct accounting systems (one for the carbon quotas, another one for the energy produced).

But the rules of the market risk favoring both overconsumption by most affluent people and impoverishment of poor people if carbon prices are low. Therefore, it could be necessary to correct this functioning by foreseeing quotas that would be managed by public authorities for other energy policies such as energy-saving infrastructures and research on these matters, as well as for others (such as people at risk who would sell their entire ration to survive, or for those who could not manage this system on their own). An additional mechanism could be taxation, potentially progressive, on transactions when buying additional carbon rations. Furthermore, these two mechanisms could be extended to reduce social and energy-related inequalities between countries.

Of course, the persons who impose these carbon quotas (and who are able to understand the stakes and the interest of these quotas) would have to be the first ones to respect them according to the inverse scale of permissibility and the most vulnerable people will be the last ones to be hit.

These are only a few ideas that should be investigated and specified with actual and simulated estimates on energy consumption. If they are supplemented by adequate collective infrastructure (public transportation network, possible district heating, etc.), individual carbon quotas would indeed be a way of implementing the phrase often attributed to Mahatma Ghandi "Live simply so that others may simply live".

BIBLIOGRAPHY

Arendt, H. (1958). *The human condition*. University of Chicago Press.

Bartiaux, F. (2002). Relégation et identité: les déchets domestiques et la sphère privée. In M. Pierre (Coord.) (Ed.), *Les déchets ménagers, entre privé et public. Approches sociologiques* (pp. 123–146). Paris: L'Harmattan (Dossiers Sciences Humaines et Sociales).

Bartiaux, F. (2007). Greening some consumption behaviours: Do new routines require agency and reflexivity? In E. Zaccaï (Ed.), *Sustainable consumption, ecology and fair trade* (pp. 91–108) London: Routledge.

Bartiaux, F. (2008). Does environmental information overcome practice compartmentalisation and change consumers' behaviours? *Journal of Cleaner Production, 16*(11), 1170–1180.

Bartiaux, F., & van Ypersele, J.-P. (2006). The relationships between population and environment. In G. Caselli, J. Vallin & G. Wunsch (Eds.), *Demography: Analysis and synthesis, Vol. 1* (pp. 383–396). London: Academic Press.

Bauman, Z. (2001). Consuming life. *Journal of Consumer Culture, 1*(1), 9–29.

Boardman, B. (2009). *Fixing fuel poverty: Challenges and solutions*. Earthscan Ltd.

Bourdieu, P. (1979). *La distinction: critique sociale du jugement*. Paris: Minuit.

Bréda, C., Chaplier, M., & Servais, O. (2008). De la survie traditionnelle aux débats identitaires contemporains: 3 études de cas algonquiennes. *Recherches Amérindiennes au Québec, 38* (2–3), 13–18.

Carlsson-Kanyama, A., & González, A. D. (2009). Potential contributions of food consumption patterns to climate change. *American Journal of Clinical Nutrition, 89*(5), 1704S–1709S.

Castoriadis, C. (1987). *The imaginary institution of society.* Cambridge: Polity Press and Oxford: Blackwell.

Chen, R. S., Bender, W. H., Kates, R. W., Messer, E., & Millman, S. R. (1990). *The hunger report.* Providence: Brown University World Hunger Program.

Cohen, J. E. (1995). *How many people can the Earth support?* New York: Norton.

Cose, E. (1983). *Decentralizing energy decisions: The rebirth of community power.* Boulder: Westview Press.

Douglas, M., Gasper, D., Ney, S., & Thompson, M. (1998). Human needs and wants. In R. Rayner & E. Malone (Eds.), *Human choice & climate change. Vol. 1. The societal framework* (pp. 195–263). Ohio: Battelle Press.

Energie Wallonie. (2008). Des mesures sociales pour les ménages en difficulté. http://energie.wallonie.be/fr/des-mesures-sociales-pour-les-menages-en-difficulte.html?IDC=6272&IDD=11805&highlighttext=clac Accessed 03.04.10.

Farrell, J., & Morris, D. (2008). *Rural power: Community-scaled renewable energy and rural economic development.* Minneapolis: New Rules Project.

Fawcett, T. (2005). Making the case for personal carbon rations. In *ECEEE 2005 Summer study: What works and who delivers?* (pp. 1483–1493). *www.eceee.org*

Fawcett, T., Bottrill, C., Boardman, B., & Lye, G. (2007). *Trialling personal carbon allowances. UKERC Research Report* (2007). http://www.eci.ox.ac.uk/research/energy/downloads/fawcett-pca07.pdf.

Frogneux, N. (forthcoming). Paradoxes et apories de la peur dans la sur-civilisation. In C. Bréda & M. Deridder (Eds.), *Modernité insécurisée.* Louvain-la-Neuve and Brussels, Academia—L'Harmattan.

Giddens, A. (1984). *The constitution of society.* Cambridge: Polity Press.

Gram-Hanssen, K., Bartiaux, F., Jensen, O. M., & Cantaert, M. (2007). Do homeowners use energy labels? A comparison between Denmark and Belgium. *Energy Policy, 35*, 2879–2888.

Heller, A. (1978). *La théorie des besoins chez Marx.* Paris: 10/18.

Hobson, K. (2003). Thinking habits into action: The role of knowledge and process in questioning household consumption practices. *Local Environment, 8*(1), 95–112.

Hoff, T. E. (2000). *The benefits of distributed resources to local governments: An introduction.* Napa: Clean Power Research.

James, C. (2001). Cultural change in Mistissini: Implications for self-determination and cultural survival. In C. Scott (Ed.), *Aboriginal autonomy and development in northern Quebec and Labrador* (pp. 316–331). Vancouver: University of British Columbia Press.

Jonas, H. (1980). Philosophical reflections on experimenting with human subjects. In H. Jonas (Ed.), *Philosophical Essays* (pp. 217–242). Midway Reprints: University of Chicago Press.

Keirstead, K. (2005). Photovoltaics in the UK domestic sector; a double-dividend? *Energy savings: what works and who delivers? ECEEE 2005 Summer Study* 1249–1258.

King Baudouin Foundation. (2010). *Climate change mitigation and social justice in Europe: Striking the right balance. Ideas for actions.* Brussels. *www.kbs-frb.be/uploadedFiles/KBS-FRB/05)_Pictures,_documents_and_external_sites/09)_Publications/PUB2009_1951_ClimateChange.pdf*

Lutzenhiser, L. (1993). A cultural model of household energy consumption. *Annual Review of Energy and the Environment, 18*. 247-89.

Malinowski, B. (1944). *The scientific theory of culture*. University of North Carolina Press.

Marx, K. (1845, 1932). *German Ideology*. Paris: Marx-Engels Institute.

Maslow, A. H. (1943). A theory of human motivation. *Psychological Review, 50*, 370–396.

Mesbah, T., Bounaya, K., & Chetate, B. (2007). Localisation of production placement and reduction of energy transfer. *Asian Journal of Information Technology, 6*(9), 948–951. *http://works.bepress.com/boukhemis_chetate/62*.

Moezzi, M., & Bartiaux, F. (forthcoming). An anthropological look at the energy efficiency field. *Energy Efficiency*.

Montesquieu. (1748). De l'esprit des Lois (The spirit of the laws). Cambridge University Press.

Morris, D. (2001). *Seeing the light regaining control of our electricity system*. Minneapolis: New Rules Project.

Narain, S. (2010). Green buildings: How to redesign. *www.eceee.org/columnists/Sunita_Narain/Green_buildings/*.

Niezen, R. (1993). Power and dignity: The social consequences of hydro-electric development for the James Bay Cree. *The Canadian Review of Sociology and Anthropology, 30*(4), 510–529.

Niezen, R. (1998). Defending the land. Sovereignty and forest life in James Bay Cree society. Needham Heights: Allyn and Bacon.

Pett, J. (2009). Carbon footprints of low income households; does addressing fuel poverty conflict with carbon saving? In ECEEE (Eds.), *Act! Innovate! Deliver!* (pp. 1675–1686). ECEEE 2009 Summer Study.

Pichert, D., & Katsikopoulos, K. V. (2008). Green defaults: Information presentation and pro-environmental behaviour. *Journal of Environmental Psychology, 28*, 63–73.

Pye, S., Skinner, I., Meyer-Ohlendorf, N., Leipprand, A., Lucas, K., & Salmons, R. (2008). Addressing the social dimensions of environmental policy: A study on the linkages between environmental and social sustainability in Europe. In *Social and Demographic Analysis* (p. 54). European Commission, Directorate-General "Employment, Social Affairs and Equal Opportunities," Unit E1.

Rawls, J. (1971). *Theory of justice*. Cambridge, Massachusetts: Belknap Press of Harvard University Press.

Rojek, C. (2004). The consumerist syndrome in contemporary societies. An interview with Zygmunt Bauman. *Journal of Consumer Culture, 4*(3), 291–312.

Roue, M. (1999). Les Indiens Cris de la baie James: Chronique d'une dépossession. In S. Bobbé (Ed.), *Taïga-Toundra: Au nord, la démesure* (pp. 59–96). Paris: Autrement,.

Sahlins, M. (1972). *Stone age economics*. New York: Aldine.

Scott, C. (1984). Between "original affluence" and consumer affluence: Domestic production and guaranteed income for James Bay Cree hunters. In R. F. Salisbury & E. Tooker (Eds.), *Affluence and cultural survival* (pp. 74–86). Washington: American Ethnological Society.

Shove, E. (2003a). *Comfort, cleanliness and convenience*: The social organization of normality. Oxford and New York: Berg.

Shove, E. (2003b). Converging conventions of comfort, cleanliness and convenience. *Journal of Consumer Policy, 26*, 395–418.

Shove, E., Lutzenhiser, L., Guy, S., Hackett, B., & Wilhite, H. (1998). Energy and social systems. In Rayner., & Malone. (Eds.), *Human choice & climate change, Vol. 2, resources and technology* (pp. 291–327). Ohio: Battelle Press.

Soper, K. (2006). Conceptualizing needs in the context of consumer politics. *Journal of Consumer Policy, 29*, 355–372.

Soper, K. (2007). Rethinking the "Good Life": The citizenship dimension of consumer disaffection with consumerism. *Journal of Consumer Culture, 7*(2), 205–229.

Stern, P. C. (1986). Blind spots in policy analysis: What economics doesn't say about energy use. *Journal of Policy Analysis and Management, 5*(2), 200–227.

Strivay, L. (2004). Enfants-loups, enfant-mouton, enfants-ours, enfants seuls..... *Communications, 76*, 41–57.

Strivay, L. (2006). *Enfants sauvages: Approches anthropologiques*. Paris: Gallimard (Bibliothèque des sciences humaines).

Tatum, J. S., & Bradshaw, T. K. (1986). Energy production by local governments: An expanding role. *Annual review of energy, 11*, 471–512.

Tengvard, M., & Palm, J. (2009). Adopting small-scale production of electricity. In ECEEE (Ed (s).), *Act! Innovate! Deliver*! (pp. 1705–1713). ECEEE 2009 Summer Study.

UNCED. (1992). United Nations Conference on Environment and Development. *Agenda, 21*, UNCED.

UNFPA. (1991). *United Nations Fund for Population Activities*. In: *Population, resources, and the environment: the critical challenges*. New York: UNFPA.

Wilhite, H. (2008). New thinking on the agentive relationship between end-use technologies and energy-using practices. *Energy Efficiency, 1*(2), 121–130.

Chapter 4

Equity, Economic Growth, and Lifestyle

Niels I. Meyer,* Frede Hvelplund,** and Jørgen S. Nørgård*
*The Technical University of Denmark, **Aalborg University

Chapter Outline

1. INTRODUCTION

Consequences of global warming are appearing much faster than assumed just a few years ago and irreversible "tipping points" are a few years ahead (IPCC, 2007; Hansen et al., 2008; Kopp et al., 2009). So far, strategies for mitigation of global warming have mostly focused on technological solutions, for example, renewable energy sources (RES) on the supply side and energy efficiency on the demand side. Much less attention has been given to potential contributions from changes in lifestyles and alternative economic, institutional, and social systems.

The progress of international negotiations on mitigation of climate change due to global warming has been slow, insufficient, and hampered by controversies between industrial and developing countries. The latter demand that the industrial countries commit themselves to much more ambitious reductions of greenhouse gas (GHG) emission in the short term than those proposed at the United Nations (UN) Conference of Parties in Copenhagen in December 2009 (COP15). They are supported by scientific analyses that point to the need for a reduction by 2020 of 40% in the industrial world compared to the 1990 level (Hansen et al., 2008), while the offers from industrial countries in total amounts to less than 17%. This underlines the need to make lifestyle changes in the rich countries part of the solution and implies the need for changes in economic, institutional, and social systems.

Despite long and tedious preparations for COP15, the final result (Copenhagen Accord, 2009) lacked sufficient concrete commitments for reduction of GHGs after 2012 when the Kyoto Protocol expires. Moreover, the "Copenhagen Accord" is not a binding international treaty, but an offer of cooperation that individual nations can choose to support. In addition, only loose intentions and goals without concrete policy means were included concerning the financing for a Green Fund and other means to support mitigation of climate change and adaptation to the unavoidable consequences in developing countries. A specified goal of a maximum of 2° C increase in global mean temperature was included in the Copenhagen Consensus but concrete policy means were missing. Recent analyses indicate that this goal will require that about half of present fossil reserves, especially coal, should remain underground (Meinshausen et al., 2009). No reference to such a requirement was even mentioned in the Copenhagen Accord.

Human activities in their present form are strongly dependent on the supply of energy for heating and cooling, electric equipment, transport, food and industrial products, and other services. A dominant part of the global energy supply is based on fossil fuels and a dominant part of the climate change is due to emission of carbon dioxide (CO_2) from the use of fossil fuels. For simplicity, this chapter focuses on CO_2 emissions from fossil fuels as a dominant contributor to climate change, but CO_2 from deforestation as well as methane (CH4), laughing gas (N_2O), and a number of industrial GHGs should be included in a more comprehensive analysis.

If the global energy supply system including fuels for the transport sector was completely based on RES most of the problems in relation to climate change could be eliminated. However, even the most optimistic energy experts do not expect this to be the case before the second half of this century, if then. This may well be too late to avoid serious disruptions of the global climate system, likely to be accelerated by exceeding the so-called "tipping points" with positive feedback effects and irreversible consequences. One example is the release of large quantities of the potent GHG CH4 now frozen in tundras in arctic areas. New data from 2010 point to an accelerating increase in the CH4 concentration in the atmosphere since 2007 (Dlugokencky and Nisbet, 2010).

It should be added that in practice the global potential of RES for energy supply is not unlimited and not without environmental impact. With a growing world population and a growing demand for energy services, the practical potential for supply systems based on RES will eventually meet its limits (Trainer, 2001).[1] Furthermore, even if a clean and infinite energy source

1. See also this book's Chapter 5 by Ted Trainer.

became available, this might lead to an unrestricted exploitation and pollution of nature with significant losses of natural habitat and biodiversity.

This chapter will focus mainly on future *nontechnological* strategies for mitigation of climate change including addressing such questions as national and international equity, limits to growth, population policies, alternative employment policies, and alternatives to the traditional GDP concept as the dominant indicator of welfare and of sustainable development.

It should be stressed that this chapter is focusing on developments in affluent countries, such as those in OECD, with the aim of leaving more *environmental space* for the less developed countries, where growth in material living standard is often a more pressing goal. The general concept of sustainable development includes a number of other elements besides energy. However, this chapter focuses on the policy means for sustainable *energy* development.

Section 2 describes the driving forces behind the climate change and Section 3 analyses how to counteract these driving forces by a number of different policy means. Special attention is given in Section 4 to the concept of *limits to growth* and its consequences for employment policy. Section 5 describes the shortcomings of the present economic system in relation to the promotion of a sustainable energy development. The chapter's conclusions and recommendations are lastly presented.

2. DRIVING FORCES FOR CLIMATE CHANGE

The main factors behind the present climate change may be described by the following simple equation:

$$I = P * A * T$$

where I denotes the impact on the environment, P is population, A is affluence in the sense of general consumption per person, and T is a technological factor representing the ecoimpact per consumption provided (Ehrlich and Holdren, 1971). These factors enable one to evaluate solutions for mitigation of the climate change.

The debate about this equation has been dominated by disputes about which one of the factors P, A, and T is mainly responsible for the environmental problems and hence holds the key to a potential solution (Chertow, 2001). As a matter of fact, growth in any one of these three factors will tend to push upwards the total impact, while a decrease will have the opposite effect. In affluent countries, solutions are almost as a rule sought in the T-factor, while the two other factors are ignored or even encouraged to grow.[2] However, with the acute challenge of climate change and the recognition of overall limits to

2. See also this book's Chapter 5 by Ted Trainer.

growth, it is necessary to consider all three factors as summarized below while more details are given in subsequent sections.

- *Population growth*: Over the last five decades world population has grown from about 3 billion to around 7 billion. Most of this growth has occurred—and still occurs—in countries with a low CO_2-emission per person. However, in a future where people in these countries are expected to improve their material welfare, the number of people will play a significant role in global environmental problems.

Political and religious taboos prevent debate on how to handle the population issue, which fortunately is characterized by a large flexibility in options over the long term. This should be taken into account in political strategies for mitigation of climate change, as described in Section 3.

- *Lack of equity*: Lack of economic equity nationally and between nations is an important driving force for economic growth in spite of the fact that in many countries economic growth has actually resulted in less equity. At the same time, economic growth in its present form creates a number of problems for sustainability. This complex coupling between inequity, economic growth, and sustainability is often overlooked in international negotiations concerning sustainability.[3]
- *Vested interests*: Energy production sectors engaged in oil, gas, and coal exploration, and traditional electrical utilities have invested large sums in infrastructure including drilling, mining, extraction, refining, and distribution of fossil fuels and in energy production plants primarily based on fossil fuels. As a consequence they have no interest in reducing the demand for their products or any change in the status quo when mitigation of global warming requires exactly this sort of intervention.
- *Liberalized markets*: Commercial markets typically have relatively short time horizons, for example, demanding less than five years payback time for investments. In contrast to this, desired radical changes of the energy supply systems require planning horizons of 50 years or more.[4] If this is not taken into account, shortsighted investments based on market competition may block necessary long-term solutions. Investments in new coal plants without carbon capture and storage (CCS) and in oil production from tar sand are examples of this. New systems thinking is needed, in some cases requiring that planning and promotion of investments in vital sectors are transferred from the commercial market to government institutions. This applies in particular to an energy sector that has a goal of sustainable development.

---

3. See also this book's Chapter 5 by Ted Trainer.

4. For example, see this book's Chapter 16 on Swiss 2000 Watt Society with a future vision extending to 2150 .

- *Perception of unlimited resources*: Over the last century, development of economics as a discipline has increasingly, with few exceptions, been dominated by a perception of living in an unlimited world with unlimited resources and pollution drains. Resource and pollution problems in one area were supposed to be solved by moving production or people to cleaner and more resource rich parts of the world. The very hint of an overall global limitation as suggested in the report *The Limits to Growth* (Meadows et al., 1972) has generally been met with disbelief and rejection by businesses and most economists. However, this conclusion was mostly based on false premises, as further discussed in Section 4.
- *Fear of unemployment*: One of the main arguments for continued economic growth is based on the experience that in OECD countries the productivity in the production sector, and to a certain degree also in the service sector, generally increases by about 2% or more per year. The assumption is that without economic growth this would create more unemployment. This assumption overlooks the flexibility of the employment concept, as further discussed in Section 4.
- *Exclusive focus on GDP*: Gross Domestic Product, GDP, is an indicator of the monetary economic activities in a society, but was never meant as a measure of how well-off its citizens really are. Several attempts have been made to define a better indicator, by incorporating, for example, the cost of environmental degradation as well as the benefits of leisure time. In this chapter, we consider the ultimate goal of the economy and lifestyle to be wellbeing, not just in the sense of income and social welfare, but a more general satisfaction or happiness, although such an indicator is not easily quantifiable (Jackson, 2009). Two Nobel laureates in economics, Joseph Stiglitz and Amartya Sen, stated in their key message in a commission report to the French President Nicolas Sarkozy that "the time is ripe for our measurement system to shift emphasis from measuring economic production to measuring people's well-being" (Stiglitz et al., 2009).

In recent times, a number of studies by economists and other researchers have revealed that in societies where citizens' basic needs are met, further income and consumption seems *not* to increase people's satisfaction and happiness (Jackson, 2009; Layard, 2005; Jackson, 2005; Nørgård, 2006). In other words, in affluent parts of the world, continued growth in GDP is not necessary for a good life.

3. POLICIES FOR MITIGATION OF GLOBAL WARMING

A wide range of policies and strategies are being applied and more are suggested to mitigate climate change with special attention to energy production and consumption. The policies may be divided into *hard* means such as binding standards, carbon quotas, taxes, and subsidies, and *soft* means such as information campaigns to influence behavior and lifestyle. In the case of energy, the goal is to promote energy conservation and encourage low carbon alternatives to fossil fuels, like RES and proposals for new technologies like CCS. This section discusses the merits and shortcomings of different policies by focusing on their effectiveness in relation to change of lifestyle.

3.1. Population Policies

According to official UN estimates, world population with no new measures is projected to grow to around 9 billion by 2050. It turns out that with relatively small changes in number of births per woman, large changes are possible in the long term. For example, with 1.6 or 2.6 births per woman, global population in 2050 could be 8 billion or 12 billion respectively (Figure 1). The corresponding numbers with these birth rates by 2150 would be 3.6 billion—around half the present population—or 27 billion, respectively. Maintaining 1998 average birth rates in different world regions would, taking into account the different regional age structures, result in 296 billion global citizens by 2150 (The Population Council, 1998). Sustainability, including mitigation of climate change, would be a lot easier in a world of material equity and welfare with the numbers in the lower range of these population scenarios.

In many parts of the Western world population is slowly declining. Government policies in these densely populated, high CO_2 emitting countries often encourage higher birth rates rather than lower, however. Most of the global population growth will, nevertheless, take place in the developing world where current energy consumption per capita is much lower than in industrial countries. In recent decades a number of developing countries, especially in the

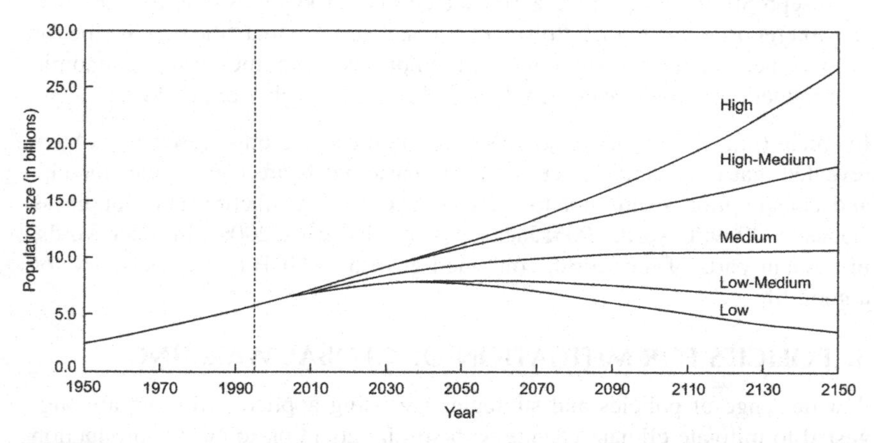

FIGURE 1 Five scenarios for world population development to the year 2150. 'Low' corresponds to 1.6 and 'High' to 2.6 average births per woman. As shown, the relatively small differences in these anticipated birth rates make a huge difference in the long term. *Source: [The Population Council, 1998].*

Asian region, have successfully reduced birth rates to around 2.0 or below (UN Population Division, 2008; Mason, 2001). In general this has been associated with a better material standard of living.

Based on concerns for the consequences for human welfare and the environment, policies encouraging birth rates below 2.0 should be promoted in poor as well as affluent countries. In the latter, a population policy that avoids economic and social incentives to *increase* birth rates will often be sufficient. In poorer countries, soft strategies like information campaigns on birth control often seem to suffice, while in other cases stronger means like economic incentives may be used. Population control is a sensitive issue, and coercive policies often conflict with basic principles of personal freedom. Similarly, birth control may conflict with traditions and religious beliefs. Nevertheless, as indicated above, population policy should be part of any climate change policy.

3.2. Equity

The goal of more equity plays an important role in the quest for sustainable development as it tends to acknowledge and promote economic satiation in affluent societies. Recognition of a world with limited natural resources will tend to make demands for equal right to the use of these resources more morally and politically legitimate.

More affluence is often strongly correlated with more energy and CO_2-intensive lifestyle, as illustrated in the case of Sweden in Figure 2 (Larson and Wadeskog, 2003). Furthermore, powerful decision-makers are often found in the upper crust of society and tend to be biased toward the desires of the rich. This constitutes a political barrier against broad changes to less energy intensive lifestyles and should be taken into account when framing environmental and energy policy.

With present mainstream (neoclassical) economic principles, changes of lifestyle are expected to occur too slowly in relation to mitigation of climate change. This is reflected in the proposals for policy changes in later sections in this chapter.

3.3. Mandatory Energy Standards

The efficacy of mandatory energy standards established and enforced by government is widely acknowledged from experiences in the building sector using building codes. This tool has also been applied to energy consuming appliances such as refrigerators and washing machines and for private cars. Mandatory standards are an efficient means for reducing CO_2 emission but are restricted by political and ideological barriers and inflexible commercial considerations. In the European Union (EU), as an example, high priority was given in the creation of the Single Market to promote commercial competition,

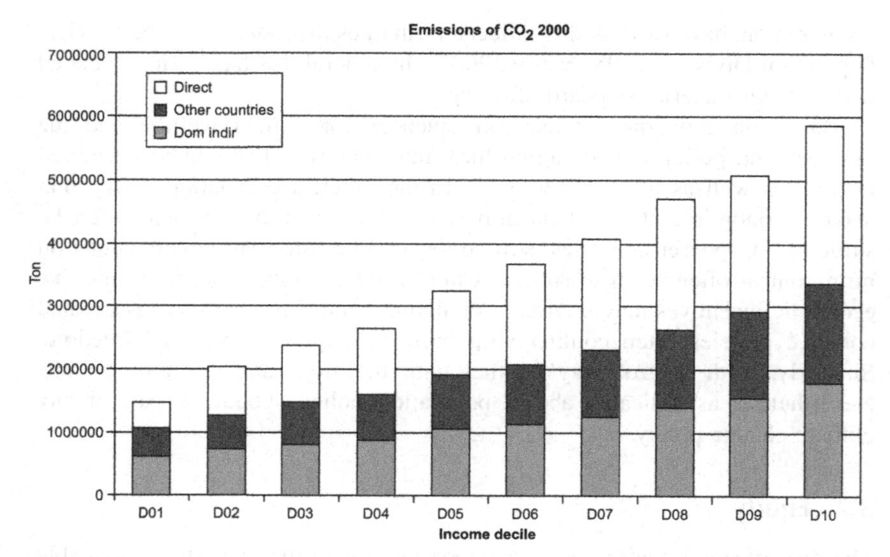

FIGURE 2 Sweden's total emission of CO_2 is here distributed between the households according to their disposable income deciles. Each decile represents one-tenth of the total income of all households, the poorest at the left. The CO_2 emissions is divided into: 1) direct emission from household's use of fuel for transport and heating, 2) indirect emission in consumer goods and services made in the country, and 3) indirect emission in imported products consumed. *Source: [Larson and Wadeskog, 2003].*

and the EU Court is taking measures against standards that, in the opinion of the Court, may give relative advantages to national production. In the U.S., individual states have more freedom to proceed with stricter codes and standards. However, the full potential of legally binding standards cannot be reached as long as politicians are primarily engaged in promoting GDP growth through free competition rather than addressing environmental problems.[5]

Experience has shown that in large unions like the EU and U.S. it is a long and tedious process to agree on common standards for either labeling or for minimum energy efficiency. Instead, the use of standards has been pioneered by states and nations like California in the U.S. and the Netherlands and Denmark in the EU. It should be more broadly recognized that the system of standards and labeling is an efficient and low-cost strategy for promoting consumers' choice of energy efficient appliances, as well as houses, cars, and so on.

The standards cannot, however, prevent people from buying more or larger energy-consuming equipment like refrigerators, air-conditioners, lamps, TVs, and so on, and thereby eating up the energy savings achieved from the more efficient technologies. On the contrary, through the so-called rebound effect,

5. In this book's Chapter 2, Felder et al., also address the issue of often conflicting objectives of economic growth, energy security, and the environment

the money saved on energy bills can make more equipment affordable, as discussed below.

3.4. Industrial Carbon Quota

The EU's Emission Trading System (ETS) where energy-intensive industries including electric utilities are given carbon emission quotas specified by the EU Commission and national governments has been in effect for some time. The experience with this cap-and-trade scheme has revealed a number of shortcomings, resulting in unimpressive carbon savings within the EU to date. It remains to be seen whether stricter caps and other reforms will produce the desired results during the second period of the scheme after 2012 (EU Commission, 2008).

The topic of the industrial ETS will not be further elaborated in this chapter since the scheme does not have a *direct* influence on people's choice of life-style. EU's ETS scheme has, however, been an *indirect* psychological barrier for energy savings in households that want to support mitigation of global warming because their individual savings in electricity consumption, for example, means that utilities can increase their sale to other customers without exceeding their quota. Thus the net effect of the efforts of idealistic households is likely to be zero.

3.5. Carbon Taxes

Carbon taxes on fossil fuels improve the competitiveness of RES, nuclear energy, and other low-carbon sources. The concept is also consistent with the *polluter pays principle*. By increasing the price of carbon-loaded fuels, it promotes changes in consumer lifestyles, encourages energy conservation, and fosters transition to energy efficient appliances and processes. The main barriers for this system are social and political. In addition, the energy intensive production sector typically opposes such taxes for competitive reasons if the taxes are not universally applied or at least cover a large region like the EU or the U.S.

The social barrier is due to the fact that in most cases an increase in energy price will be a greater economic burden for low-income than for high-income households (Ekins and Dresner, 2004). This may, to some extent, be compensated by changes in the general tax system at the cost of increased administration. Another energy tax system, which can avoid the high burden to low-income households, is a progressive tax on energy or carbon, meaning that a certain tax-free "lifeline" consumption is granted to each person or household, while higher taxes are applied to more "luxury" levels of consumption.

It is a political problem that the carbon tax has to be quite high to yield sufficient emission reductions. In addition, even within the EU it has not been possible to introduce harmonized carbon taxes for all member states. As a consequence, a high carbon tax will meet strong opposition from the various production sectors for competitive reasons.

From an administrative point of view the simplest system would be to introduce a global carbon tax on the basic production sources, that is, at oil wells, coal mines, and so on. The owners of these sources are a combination of states and private investors. A direct global tax at the production site of fossil sources will thus require a global agreement between all nations that are owners of such sources and a global enforcement of the tax on private owners of production sources. Alternatively, it should be legally accepted to adding import tax on all products from countries refusing to join a global carbon tax system. Judging from the experiences of COP15 in Copenhagen, this type of global agreement with interference in national policies will not be high on the present international agenda.

3.6. Personal Carbon Quota

Another scheme for reducing GHG emission is based on Personal Carbon Allowances (PCA) where every adult is allotted an equal, tradable ration of CO_2 emission per year related to their consumption of selected energy services for private households. Compared to the cap-and-trade system used in the industrial sector, PCAs apply to individuals and directly influence their behavior. PCA establishes a direct mechanism with negligible uncertainty for emission reduction in the household sector. For simplicity, it is proposed that PCA should be related only to "direct" energy consumption, that is, energy used for personal travel and within the household (Fawcett et al., 2009).

So far no country has introduced a PCA scheme and only a few industrial countries are giving serious consideration to the scheme. The most extensive discussion of PCA has taken place in the U.K. where the U.K.'s Department for Environment, Food, and Rural Affairs (Defra) published a report on the scheme in 2008 (Defra, 2008) with a positive evaluation of its potential. A similar conclusion was expressed by the chairman of the U.K.'s Environment Agency, Lord Smith of Finsbury, in a speech at the agency's annual conference in London in November 2009. In both cases it was stated that the scheme needs a relatively long time to develop and that the cost of administration should be analyzed more closely.

The efficiency of a PCA system is dependent on the particular energy system in the country and on the institutional arrangements. Analysis of the potential of PCA in the U.K. and Denmark (Fawcett et al., 2009) indicates that for these two countries the scheme could cover 30%–50% of the total national emissions. The PCA requires systematic government support to the households in order that they may benefit from the scheme and accept it as a positive challenge. Considering the slow progress in mitigation of climate change, PCA deserves further attention as a supplement to existing schemes.

3.7. Subsidies for Energy Supply and Conservation

Government subsidies for energy supply in industrial countries have mainly been aimed at fossil fuels and nuclear power. Change of economic balance in favor of RES and energy conservation is a relatively recent phenomenon. An early Danish example is investment subsidies for wind turbines introduced in 1979. This policy was quite successful but the scheme was phased out in 1989 as wind power became more competitive.

In the United Nations Environment Programme the global subsidies to conventional energy are estimated to U.S.$250–300 billion a year, based on 2005 data (UNEP, 2008), while subsidies for RES were an order of magnitude lower. A detailed analysis by the European Environment Agency (EEA) has estimated energy subsidies in the EU-15 to be in the order of €29 billion in 2001 (approximately U.S.$40 billion) with almost three quarters oriented toward the support of fossil fuels (EEA, 2008). Only about 19% of the subsidy was used for RES in that year. The figures are only indicative, due to the lack of consistency of data across countries and of assumptions made.

It is now increasingly recognized that fossil-fuel subsidies can drain government budgets and furthermore increase GHG emissions. In September 2009, the Leaders of the Group of Twenty (G-20) at a meeting in Pittsburgh, U.S.A. agreed to phase out fossil fuel subsidies estimated to be on the order of U.S.$500 billion per year. The actual amount of subsidies is rather uncertain because there is no international framework for regularly monitoring such subsidies (Global Subsidies Initiative, 2009).

In many industrial countries the energy demand for heat and electricity in buildings accounts for 30%–40% of the total national energy consumption.[6] There is a large potential for energy conservation in this sector but experience has shown that this potential will not be fully exploited without state subsidies and/or regular mandates.

It is technically possible to build so-called Passive Houses with maximum heating consumption of 15 kWh/m2 per year at a competitive price, as demonstrated by thousands of houses in Austria and Germany (Klingenberg et al., 2008; EEnergy Informer, 2010). In 2009, California policymakers set a mandate that requires all *new* residential dwellings built in the state starting in 2020 to be *zero net energy* houses. The same would apply to new commercial buildings starting in 2030 (EEnergy Informer, 2010).[7] It should be noted that with the significantly reduced heat losses in these new buildings, the consumption of primary energy is mostly related to electricity for lighting, electric appliances, and other forms of electric equipment. This is taken into account in the EU directive on energy performance of new buildings

6. See also this book's Chapter 9 by Gray and Zarnikau.

7. See also this book's Chapter 9 by Gray and Zarnikau, Chapter 17 by Rajkovich, Miller, and LaRue, and Chapter 18 by Clymer.

(EU, 2003), which includes energy used for space heating, hot water, space cooling, ventilation, and fixed lighting but excludes movable equipment.[8]

A major problem in the building sector is that most buildings have a life of 50 years or more, hence the transition to low-energy buildings takes a long time. This problem may be overcome by complete renovation of the old building stock, which would often require government subsidies to be feasible. Some proposals even go so far as to subsidize owners of buildings who tear down the old house and replace it by a low-energy house (Jensen, 2009).

3.8. Energy-Saving Campaigns

Technological energy efficiency improvements can contribute substantially to reducing energy consumption, but only if they are implemented. It turns out to be a complicated task to convince individual consumers and businesses to choose the most efficient technologies available.[9] Convincing energy consumers about the options available is best achieved through campaigns that combine different measures, as illustrated below. An example is labeling of appliances and cars with clear information about energy efficiency, combined with public information programs in the media and on websites. An additional effect may be obtained by subsidizing the purchase of the more efficient technologies.

Such campaigns have been conducted in a number of countries, but often with only moderate results. The actors chosen to implement the campaigns may partly be responsible for the lack of success. Often governments have left it to the energy suppliers to run such campaigns despite the fact that this will reduce their own sales. This clearly involves a "letting the wolves watch the sheep" dilemma.[10] To eliminate this dilemma, the Danish government in 1996 established a Danish Electricity Saving Trust (DEST) independent of government and business, and financed by a small extra electricity tax of €0.001 per kWh, or around €2.5 annually per person (Nørgård et al., 2007). DEST has been successful, for instance, in getting thousands of buildings with electric space heating converted to more appropriate heating sources like natural gas or, even better, district heating from combined heat and power (CHP). Another successful DEST campaign included improved labeling, temporary subsidies, and website information aimed at moving consumers to the most efficient electric appliances. Within a few years, the Danish sale of the most efficient refrigerating appliances soared from only a few percent to about 90% (Nørgård et al., 2007).

8. In the Danish version, energy in the form of electricity is multiplied by a factor of 2.5 when calculating the total energy consumption of the building (Danish Ministry of Economy and Business Affairs, 2007).

9. See also this book's Chapter 8 by Long et al.

10. In a few states in the U.S., such as California, regulators have decoupled utility kWh sales from profits to eliminate the disincentives to promote energy efficiency.

Government policies are, however, usually aimed only at reducing the consumers' *direct* energy use, while their *indirect* energy consumption is not included. Direct energy is defined as energy bought by consumers as electricity, gasoline, natural gas, fuel oil, district heat, and so on. Indirect energy refers to the embedded energy used to manufacture, transport, and provide nonenergy goods and services for the consumers. Direct and indirect energy consumption are of roughly equal magnitude in industrial countries. In EU countries the indirect energy share varies from 34%–64% (Reinders et al., 2003). In Figure 2, the *direct* energy consumption by private households in Sweden includes only fuel for transportation and for heating houses, but not electricity. The absence of indirect energy in governments' energy saving campaigns reflects a fundamental built-in conflict in the present economic system. On the one hand, most governments want households to reduce energy consumption, but at the same time they want the general consumption of goods and services to increase—neglecting its embedded energy consumption—to support economic growth and employment. This illustrates a basic dilemma in solving the climate problems while maintaining lifestyles with growing consumerism as the target.

3.9. Rebound Effect

Technical energy efficiency improvements are usually cost effective, which leads to an extra "income" for those investing in energy savings. In an unsatiable economy, this will typically result in extra consumption or investment implying some extra energy consumption and hence modifying, sometimes substantially, the energy savings achieved directly from the technical improvement. One example of this so-called *rebound effect* is the replacement of an old car by a car running more kilometers per liter of gasoline, where the saved fuel cost is spent on driving more kilometers. Another common case is when electricity savings from replacing incandescent lamps with CFL or LED energy saving lamps are partly turned into leaving more lights on. In the production sector an energy saving technology may lead to cheaper products and hence higher demand. With continued economic growth, the climate benefits from higher energy efficiency might eventually be cancelled out (Nørgård, 2009). This would not happen to the same extent in affluent societies with demand satiation.

3.10. International Negotiations

After the UN Climate Conference in Rio in 1992, it took until COP5 in Kyoto in 1997 to reach an agreement among the industrial countries on commitments for reductions of GHG emissions between 2008 and 2012. Subsequently, it took until 2005 before the Kyoto Protocol went into effect after a sufficient number of countries had ratified the agreement. The same slow pace has manifested itself in connection with the international negotiations concerning the second commitment period after 2012. Most industrial countries are primarily concerned with their own economic problems and they often try to postpone concrete offers for economic support for developing countries in relation to climate protection.

It is highly questionable whether it will be possible to bridge the gap between different demands for emission reductions. COP15 has clearly illustrated how far the large CO_2-emitters are from consensus on the required commitments and strategies.[11] The bottom-up activities initiated by mayors of large cities, local and regional politicians, "green" commercial enterprises, and the NGOs may in the long run provide more efficient contributions to mitigation of global warming than the top-down international schemes.

4. LIMITS TO GROWTH REVISITED

During the history of economics as a discipline, several pioneers have warned that pursuing growth forever was not desirable (Mill, 1900; Keynes, 1931). Instead, growth should be considered as a temporary phase, until the global population has achieved satisfactory material welfare, including the preservation of a satisfactory natural environment. After the establishment of the GDP in 1930s as a measure of a nation's activities, and especially after World War II, the Western world experienced a quarter century of rather stable growth in GDP. This was fueled by an abundance of cheap and seemingly unlimited fossil fuels as well as other resources from the less developed parts of the world. During this period the level of GDP gradually became effectively synonymous with human welfare.

4.1. The Limits to Growth Report

Through the 1960s, however, a number of studies revealed the hitherto ignored human and environmental costs of the GDP growth, and in the early 1970s several reports questioned the net blessings of continuing this path (e.g., Schumacher, 1973; Goldsmith et al., 1972). One of the more extensive studies was initiated by The Club of Rome, resulting in the 1972 report *The Limits to Growth* (LtG) (Meadows et al., 1972). In the following years about 30 million copies of this report were sold in about 30 languages. The study's results are based on an aggregated computer model, simulating the global development in population, food production, use of nonrenewable resources like fossil fuels, industrial output, and in pollution such as the emission of CO_2. A key feature of the model is the interactions between these parameters.

The basic conclusion of the report was that continuation of the growth policies in population, industrialization, pollution, food production, and

11. See also this book's Chapter 2 by Felder et al., and Chapter 15 by Buijs.

consumption of nonrenewable resources would most likely lead to some kind of collapse during the twenty-first century, due to resource scarcity, over-pollution, over-population, and so on, as shown in the left of Figure 3. Although this catastrophic growth scenario got most attention, alternative scenarios were also presented in the report, including the one shown in the right of Figure 3, which illustrates that it is possible to change course and reach an environmentally sustainable development path, one that is able to satisfy all people's physical needs. Finally, the report stressed that due to delays in natural and manmade systems, it is essential for achievement of sustainability that global society acts before the environment undergoes irreversible changes and forces undesired changes on us.

The very title of the report was a provocation to many politicians, economists, and others having devoted their work and life to the goal of measuring progress in terms of increased income, profits, and GDP. Consequently a cohort of critics reacted intuitively against the report, often using groundless arguments. For instance, many critics focused exclusively on the growth and collapse scenario, and stated in the 1990s that the collapse had failed to materialize. However, as illustrated in Figure 3, the analysis in that scenario finds the collapse to occur decades later (Nørgård et al., 2010). Some critics do not seem to have read the report at all. In this way the critics managed for a couple of decades to derail the debate by strong, but not well-documented statements such as the report has "been damned as foolishness or fraud by every serious economic critic" (Simon and Kahn, 1984). The recent recognition of global warming from CO_2 concentration in the atmosphere and an approaching peak in oil supply combined with the current financial crises fits rather well with the growth scenario in LtG. As shown in Figure 4, real developments in the main parameters in LtG have followed quite closely the main trends in the report's standard scenario, which leads to collapse around 2040 (van Vuuren and Faber, 2009; Turner, 2009). This underlines the fact that the basic structure and drivers of the global economy have not changed. Not surprisingly, recent environmental reports have presented concepts and warnings similar to those in LtG (Rockström et al., 2009). Thus, in today's debates on strategies for sustainable development it would be wise to revisit the analysis of the original 1972 version of LtG and its two later revisions (Meadows et al., 1992; Meadows et al., 2004).

4.2. The Illusion of Decoupling

Since all economic activities accounted for in GDP involve some energy consumption, the so-called *decoupling* between GDP and energy consumption is a statistical myth or illusion based on macroeconomic abstractions (Nørgård, 2009). While it is a historical fact that GDP and energy consumption have often grown at different rates, this does not imply that there is no coupling, no connection, between them in the real world as further described below.

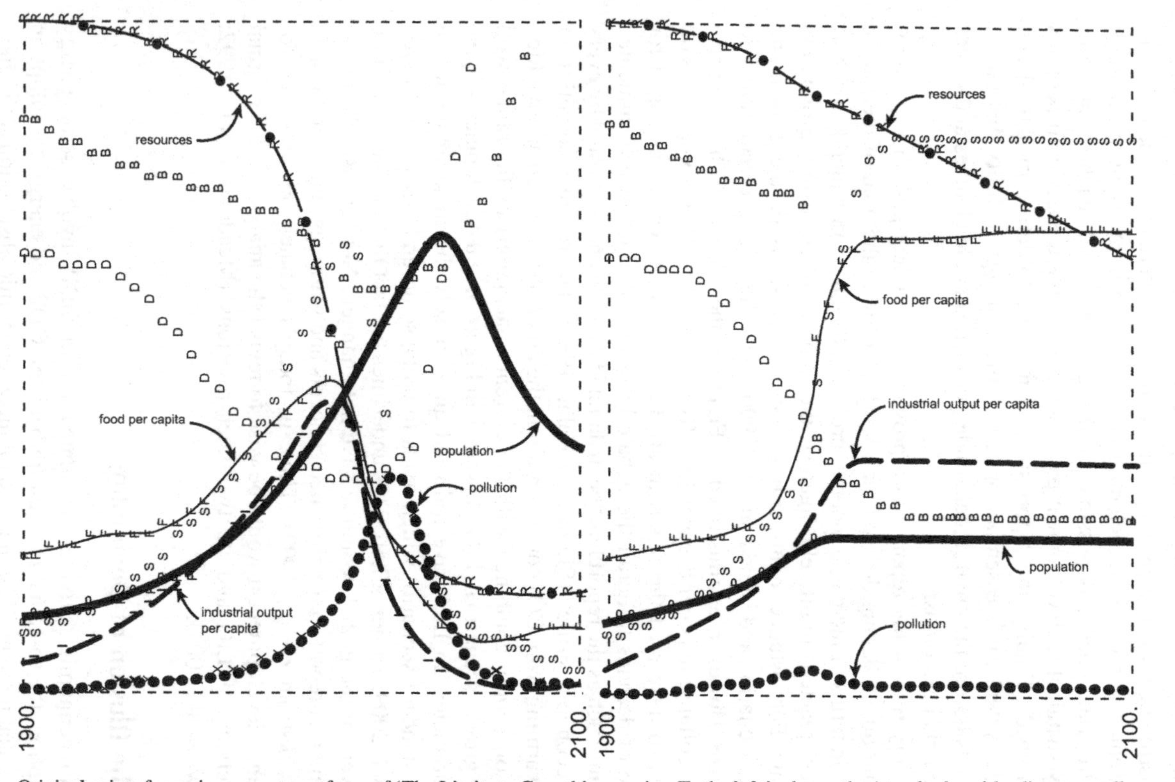

FIGURE 3 Original prints for main parameters of two of 'The Limits to Growth' scenarios. To the left is shown the 'standard run' leading to a collapse some time in the 21st century. To the right is shown another option, a stabilized development. *Source [Meadows et al., 1972].*

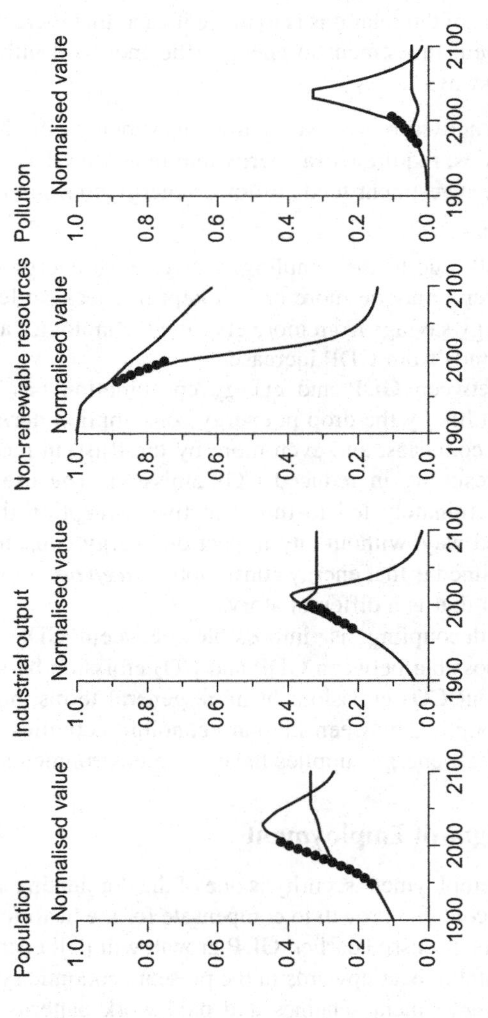

FIGURE 4 The observed global development of four essential parameters, here shown as black dots, are shown to roughly follow the outcome in the Standard run with growth and collapse, see Fig. 3, rather than the Stabilized run from 'The Limits to Growth' model. *[van Vuuren and Faber, 2009]*.

Denmark, for example, has experienced more than 30 years with no increase in energy consumption, while GDP has increased by 70% during the same period. This is often used as an example of decoupling between the two. But the technological improvements in energy efficiency over this period would actually have resulted in a marked decline in energy consumption, had there not been any coupling to the growing GDP.

A fruitful analysis of the relations is to make it clear that there are more than one coupling involved. Investment in energy efficiency will influence energy consumption in two ways:

- The production activities related to the investment will, like all other additional activities, require extra energy and increase GDP.
- The effect of the investment tends to lower energy consumption for operating the system.

A growing GDP will, due to the coupling, always push energy consumption upwards. But this trend may be more or less kept in check, at least for some time, as long as energy savings from more efficient technologies and structures offset the growth trends from GDP increase.

The coupling between GDP and energy consumption has clearly been demonstrated in real life by the drop in energy consumption during the recent recession in OECD countries, and even more by the Russian recession in the early 1990s, both resulting in reduced CO_2 emission. The use of the term *decoupling* has unfortunately led to the seductive perception that GDP can grow in an unlimited way, without any impact on energy consumption. What can rightfully be claimed is that energy consumption *need not grow at the same rate* as the GDP, but that is a different story.

While a true decoupling is impossible between GDP and energy consumption, it is possible between GDP and CO_2 emission by shifting to an energy supply without CO_2 emission. In more general terms, however, there will always be a coupling between human economic activities and *environmental impact* since all energy supplies have some environmental impact.

4.3. New Paradigm of Employment

In affluent societies employment security is one of the dominating arguments for continued push for economic growth to compensate for the historical increase in labor productivity. As discussed earlier, GDP growth will pull energy consumption and environmental impact upwards in the present economic system.

Consequently, employment schemes and paid work patterns are essential factors in the strategy for mitigating climate change. Rather than considering a lower demand for the workforce due to increased labor productivity as a *problem*, it should rather be considered as a *blessing* due to its associated increase in general welfare. A constructive solution to the unemployment problem may be a paid work-sharing program with fewer working hours per week or more

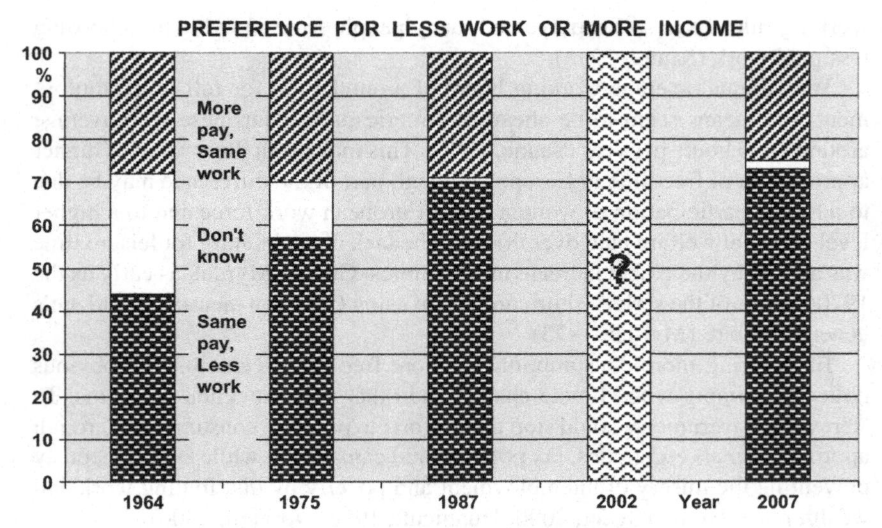

FIGURE 5 Development in Danish people's preference between more income and less work time. The survey in 2002 Danish National Social Research Institute left out exactly this question. *[Nørgård, 2009].*

vacations. Incidentally, in the 1920s the general perception in the U.S.A. was that basic needs were or would soon be satisfied and that further productivity increase should mainly be turned into more freedom and leisure time.[12]

Other studies focusing on how little labor is needed to provide a decent material standard of living, typically conclude that one or two days of work per week is sufficient (e.g., Gorz, 1983).[13] Flexible parameters in this connection are the weekly working hours, the length of vacations, the system of sabbatical leave, and the concept of *basic income* or *citizens salary* (Meyer et al., 1981).

In Europe, the gradual reduction of working hours has proceeded further than in the U.S., especially with the achievement of several weeks vacation per year. In many European countries, people increasingly prefer shorter paid work time over more income and consumption. In Denmark, for example, the proportion reached 73% in 2007 as shown in Figure 5 (Nørgård, 2009). This preference does not seem to be based on environmental concerns, but rather expresses a wish for a lifestyle with more freedom to enjoy time with family and friends, and leisure activities in general. However, social patterns and rigid

12. In 1933, U.S. president F.D. Roosevelt was only weeks from passing a bill to solve unemployment through paid work-sharing, by reducing the paid work week to 30 hours instead of the normal 40 hours, when the industrial lobby blocked the bill (Beder, 2000; Hunnicutt, 1988). Instead of a work-sharing policy, a work-creating policy aimed at increasing consumption and production was implemented. This has been the standard solution to employment problems ever since in Western industrialized countries.

13. Trainer reaches similar conclusions in this book's Chapter 5.

working rules have so far prevented many people from individually choosing less paid work (Sanne, 1995).

With annual average working hours of around 1700 for full-time employment, Europeans seem to be ahead of Americans and Japanese who average around 2000 hours per year (Sanne, 1995). This may be an illustration of higher appreciation of free time in Europe, although part of the difference may be due to a higher participation of women in the European work force and to a higher level of social welfare paid over the tax. The lack of accounting for leisure time was argued by the Nobel laureate in economics, Gunnar Myrdal, as early as the 1970s as one of the serious shortcomings in using GDP as a measure of citizen's general welfare (Myrdal, 1973).

Exchanging more consumption for more free time seems like an obvious policy for coping with climate change and other environmental problems. To start with, governments could stop their efforts to promote consumption through appropriate marketing rules, tax policies, and campaigns, while simultaneously preventing the misery of unemployment and poverty by distributing work and wealth more evenly (Beder, 2000; Hunnicutt, 1988; Nørgård, 2006).

4.4. Equity as an Environmental Policy

Global and national equity become a rational and legitimate moral goal in the context of a finite planet with limited resources. Globally, the lack of understanding the importance of equity or perhaps rather the lack of courage and political will to act has been demonstrated by the COP15 failure in Copenhagen. Governments in the affluent Western world were unwilling to give up their future GDP growth to make *environmental space* for the emerging and poor economies.

The ultimate goal of an economy ought to be human wellbeing in the sense of satisfaction and happiness. Due to the general observation of diminishing returns of increased income and consumption, equity tends to increase total human wellbeing (Daly, 2007). This is illustrated by the fact that the preference for more leisure over more income is typically more prevalent in countries with a high degree of equity and social security.

5. SHORTCOMINGS OF THE PRESENT ECONOMIC SYSTEM

The dominant world trade is supposed to be[14] based on neoclassical free market ideology. The basic assumption is that commercial competition in a free market leads to more efficient production and lower consumer prices. This is an oversimplified description of a complex economic system, where the market in many cases is actually not free, but is embedded in a set of historically constructed

14. We write "supposed to be" as the rules and institutions in which the market is embedded in many cases result in market outcomes that systematically deviates from what would happen under textbook free market conditions. The cases in this section are a few examples of this.

institutions. This has been analyzed in a large number of critical papers and books (e.g., Hvelplund, 2005a; Lund, 2010). The global financial crisis that started in 2007 has amplified the critical view on the present profit making adventures of the financial institutions. It has also emphasized the need for a new economic paradigm supporting sustainable development. The report in 2009 from the British Sustainable Development Commission chaired by Professor T. Jackson (British Sustainable Development Commission, 2009) and the subsequent book (Jackson, 2009) are constructive examples of this field of research. So far, however, the proposed adjustments and regulations have not fundamentally changed the basic paradigm of the present economic market system.

In most official economic reports and in applied neoclassical texts, economic development is regarded as a result of an optimization process which results in a GDP growth, and which optimizes societal welfare. This basic assumption is built into most econometric models used by ministries of finance in the industrial world. The reality of the assumptions and the shortcomings of the GDP concept is, however, increasingly being criticized in the professional economic discussions (Stiglitz et al., 2009; Jespersen, 2009; Hvelplund, 2005a; Lund, 2010). This general criticism of neoclassical economics will not be pursued further in this chapter except to note that the externality costs from energy production are not adequately included - if at all - in most econometric models. This applies especially to the externalities from global warming due to burning of fossil fuels. Examples below illustrate a number of cases where the optimization assumption is not fulfilled in practice in relation to a sustainable energy development.

5.1. Nonoptimal Welfare

Middle class families in rich countries normally have one or two cars and drive 20,000–30,000 kilometers every year. They travel on holidays by air one or more times per year, have a rather big house with fossil fuel based heating and air-conditioning, many appliances, eat around 300 grams of meat a day, and buy meat, fish, fruit, vegetables, and other consumer goods from all over the world.

As a result people in affluent societies have a yearly CO_2 emission in the range from 10–20 tons CO_2 per person.[15] In applied neoclassical economics this lifestyle is regarded as the result of an optimization process where free, well-informed, and rational consumers are deciding what should be produced in society by voting with their banknotes in the housing and transport sectors and in supermarkets and shopping malls. According to this economic theory, the resulting growth and distribution of production is therefore the result of the multitude of choices made by these free and rational people pursuing their individual wishes. In reality, more than six billion people in the world do not have the opportunities implied by this model, but a large part of these people

15. See also this book's Chapter 7 by Moran for per capita emission levels.

appear to aim at the same rich middle class lifestyle. This is a basic problem for sustainable development and in relation to the externality problem.

5.2. The Externality Problem

Some external costs, for instance arising from CO_2 emission, are often not or only partly included in the market prices. These costs have been calculated in the EU ExternE Project, and for coal-based electricity production they range between 3 and 8 euro cents per kWh in different European countries (European Commission, 2003). The same applies for a number of other externalities. A number of examples where the present market construction does not lead to a sustainable welfare optimum are given in the following, partly based on Danish case studies.

In many countries there is no CO_2 taxation on production and transportation. This is partly based on a wish to protect national production from unfair competition as the rules of WTO accepts trade where neither the external costs of production nor of transportation are internalized in the market prices. The WTO rules do not allow a country with CO_2 taxation to put a CO_2 import levy on goods produced in a country without a similar tax. Furthermore, countries are allowed in practice to subsidize the use of fossil fuels. For instance the U.S. subsidized technologically mature fossil fuel technologies by U.S.$70.2 billion during the period 2002–2008 while the nascent renewable energy technologies were subsidized at U.S.$29 billion during the same period (Environmental Law Institute, 2009). Many countries including Iran, Russia, and China also give heavy subsidies to fossil fuel consumption.

Consequently, international trade rules favor an international competition process, where countries that do not internalize external environmental costs in their prices have a better competitive position than countries that do. In addition, the rules accept a competition process where mature fossil fuel technologies are heavily subsidized in some countries.

A special problem within the transportation sector is the lack of CO_2 tax on air and sea transport for goods and passengers. Thus, airlines can buy their fuel without paying energy tax, despite the fact that CO_2 pollution at an altitude of 10 km gives rise to a two to three times stronger greenhouse effect compared to the same emission at ground level.

The present market conditions thus do not represent a process that optimizes sustainable welfare, and the resulting GDP therefore does not represent a welfare optimum.

5.3. Historically Accumulated Market Institutions Often do not Lead to Sustainable Growth in Welfare

All over the world market rules are embedded in legal, fiscal, and cultural institutions that have accumulated as a result of an array of historical decisions.

As a consequence, these institutions typically support the technologies of the past. This is especially true in the energy sector, where coal, uranium, and oil technologies receive much higher state subsidies than technologies based on renewables, and where structural "support systems" for the old technologies are built into market mechanisms and tax rules.

In many countries, car taxation, car insurance, and costs of using the roads are fixed and thus do not depend on the number of kilometers driven. As a consequence it is relatively expensive to buy a car, but cheap to use it once you have bought one. The consequences of this system are illustrated by the following Danish case study where the typical costs and taxation of an average diesel car are shown in Table 1.

As can be seen from Table 1, the marginal cost of driving an extra km is 9.3 euro cents, and the total cost is 40 euro cents. This division of costs is strongly influenced by the present tax system, where 15.2 euro cents per km is a tax that is independent of the number of kilometers driven while 3.1 euro cents per km is a function of the driven kilometers. This means that with present institutionalized tax rules the system actually *encourages* people to drive more once they have purchased a car. This is even further promoted by the Danish system with tax deductions for commuting expenses between home and workplace. Thus, in some cases the marginal costs of driving 1 km is reduced to *minus* 1.3 euro cents, so that people are actually *earning* money for each extra kilometer they drive. For these people it does not pay either to buy a house close to work or to use public transportation.

The costs for society of driving one kilometer in an average Danish car, including pollution costs, are estimated to be between 27 and 53 euro cents, and Denmark has designed an incentive system where the car driver encounters a marginal cost of between −1.3 and 9.3 euro cents per km. This means that if Danish car owners act in an economically rational way they will choose to drive

TABLE 1 Cost Elements of a Private Car in €cent per km

Cost elements	Costs in €cent per km	Tax share of cost in €cent per km
Fixed costs/fixed tax	30.7	15.2
Variable costs /variable tax	9.3	3.1
Total cost per km	40	
CO_2 cost per km	0.27	

Assumptions: Efficiency of diesel car: 20 km/liter; price of car including tax: € 32,000; technical lifetime of the car: 10 years; discount rate 6% p.a.; diesel cost per litre including tax: €1.33; CO_2 price: 20 €/ton.

one extra kilometer if it increases their welfare by at least 9.3 euro cents. But if the car is driven one extra kilometer, society loses 27 to 53 euro cents as described above. So the total welfare *loss* of this market system will be at least 40 minus 9.3 euro cents, which equals 30.7 euro cents per km driven. At the same time this kilometer will result in a CO_2 emission for an average family car of around 150 grams. This example shows that the present Danish price and tax scheme for cars encourages increased car traffic while at the same time reducing sustainable welfare. Many other countries have systems with similar irrational consequences in the transport sector.

The above discussion implies that a cap-and-trade system extended to car transportation would have very limited effects compared to fixing the existing institutionalized tax system in place. At a CO_2 price of €20 per tonnes, this will only be 0.7% of total car expenses per kilometer (see Table 1). Instead, the system should be changed to a tax per kilometer where the cost of driving one extra kilometer equals the total societal costs including all environmental costs.

5.4. Fixed Energy Tariffs in the Danish District Heating System

In the Danish cogeneration energy system, the heat is distributed via a water-based pipe distribution system. The tariffs linked to this system include a fixed share and a share that depends on the consumption of heat. The fixed share varies from company to company from around 25% to 60% of the annual heat bill. Table 2 illustrates the data for a typical 140 m^2 house in the Aalborg district heating system.

In this case, the motivation for energy conservation is linked to the variable tariff, which is heavily reduced by the structure of the tariff. A CO_2 price of €20 per tons CO_2 only increases the economic conservation motivation from €700 to €760, in other words, by less than 9%. If the tariff structure is changed to a 100% variable tariff, the economic motivation is increased from €700 to €1150, or by 64%. This example illustrates that the introduction of

TABLE 2 Tariff Distribution and CO_2 Cost for a 140 m^2 House in Aalborg District Heating System [Calculations by Hvelplund]

	€ per year
Fixed tariff	450
Variable tariff	700
Annual heat bill	1.150
Annual CO_2 emission: 3 tons	60

a CO_2 tax has very little effect on consumption, unless it is combined with a radical change in tariff principles. Thus the change of market rules—in this case prevailing tariffs—matters far more than the introduction of a CO_2 tax on top of an inefficient tariff structure that actually discourages energy conservation.

5.5. Financing of Housing and Time Budgets of People

Adjustments in the number of annual working hours were discussed earlier in this chapter. But this is not an easy task, unless it is combined with changes in the cost structure of housing, transportation, and other necessities. Today, an average Danish family has to allocate around 800 to 1200 working hours per year to pay for housing costs. The number may be even higher in other high-cost countries.

Houses can technically last for at least 100 years, but are typically financed with a mixture of 20- and 30-year loans in Denmark. Thus the loans have a much shorter running time than the lifetime of typical houses. If the housing cost instead had a 70% share financed with 60-year low interest loans, the cost pressure from housing would be reduced by around 40% compared to the present financing system. This would make housing affordable with a lower number of working hours.

The above examples illustrate that it is possible to redesign society toward a direction where the consumption of fossil fuels and emission of CO_2 would be reduced while welfare is improved at the same time. This can be done without removing the market system, but rather by changing the institutions and tax structure in which the system is embedded.

5.6. Promotion of Intermittent RES

As already mentioned, the time horizon of commercial markets is too short to govern the necessary long range planning of a sustainable energy development. Thus, alternative economic schemes are needed.

The success of wind power in Denmark, Germany, and Spain is mainly due to the so-called Feed-In Tariffs (FITs) where owners of wind turbines are guaranteed a favorable tariff on a long-term basis, often combined with priority access to the grid. The same type of scheme may be used for other RES like PV and wave power. Flexibility is introduced by yearly reductions in the tariffs for new plants as the technology matures (Bechberger and Reiche, 2007).

A competing scheme in Europe is called Trading of Green Certificates (TGC) with certificates issued to individual wind power producers in accordance with a specified government target for the total national wind power production and with sanctions to utilities if the target is not fulfilled. In the U.S. this scheme is called Renewable Portfolio Standard (RPS). A comparison of the

European and the U.S. experiences may be found in (Rickerson and Grace, 2007).

Neither the FIT nor the TGC operates in accordance with traditional market principles. In the case of FIT, the price of the wind electricity is fixed while the volume is decided by the market; the opposite applies to TGC. A number of variations of these two schemes may be found in different countries, for example, in the U.S. (Schreurs, 2007). The advantages and problems of the FIT and TGC schemes have been discussed by Hvelplund and Meyer (Hvelplund, 2001a, 2001b, 2006; Meyer, 2003). It is concluded that the FIT so far has been most efficient for promotion of wind power as illustrated by the wind power development in Denmark, Germany, and Spain (Meyer, 2003).

It is not relevant to force intermittent RES like wind power into the traditional market system, when the main goal is to accelerate its penetration into the electricity supply system. After the investment has been made in a wind turbine, there is no way that the investor can compete more efficiently by a different production strategy. It all depends on the wind. This also supports the priority for the FIT scheme.

CONCLUSIONS

The process of international climate negotiations from Bali to COP15 in Copenhagen has illustrated the need for new and supplementary schemes for mitigation of climate change. This chapter presents proposals for new strategic thinking to overcome present barriers and promote efficient mitigation schemes. The chapter's main conclusions are summarized as follows:

- A new economic paradigm with less attention to GDP and more attention to sustainability and welfare is needed. This involves, among other things, a shift in the present balance between societal planning and commercial market principles to the advantage of long-range planning. Economic science should give high priority to the development of market constructs that lead to sustainable development.
- The level of the global population is an important factor in relation to global warming. Thus, regulation of birth rates should not be a taboo subject.
- Lack of economic and social equity is a serious barrier for sustainable development. More equity globally and within nations is needed.
- Limits to growth on a finite planet should be recognized. It is necessary to change the institutional market conditions in which households are embedded. One of the potential policy means could be to introduce Personal Carbon Allowances.
- Present employment policies promote material growth. The alternative is sharing of paid work and more free time. Introduction of a general citizens salary (basic income) may be worth pursuing.
- Stronger promotion of renewable energy and energy conservation is strongly needed.

ACKNOWLEDGMENTS

The contributions of Frede Hvelplund and Niels I. Meyer have been supported economically by the research project Coherent Energy and Environmental System Analysis (CEESA), partly financed by the Danish Council for Strategic Research.

BIBLIOGRAPHY

Bechberger, M., & Reiche, D. (2007). Diffusion of renewable feed-in tariffs in the EU-28. In L. Mez (Ed.), *Green power markets* (pp. 31–50). U.K.: Multiscience Publishing.

Beder, S. (2000). *Selling the work ethic*. Melbourne: Scribe.

British Sustainable Development Commission. (2009). *Prosperity without growth? The transition to a sustainable economy*. London, U.K.

Chertow, M. R. (2001). The IPAT equation and its variants – changing view of technology and environmental impact. *Journal of Industrial Ecology, 4*(4), 13–29.

Copenhagen Accord. (2009). *www.un.org/climatechange*

Daly, H. E. (2007). *Ecological economics and sustainable development*. Massachusetts: Edward Elgar Publishers.

Danish Ministry of Economic and Business Affairs. (2007). In: *Building regulations*. Copenhagen: Denmark.

Defra. (2008). *Synthesis report on the findings from Defra's pre-feasibility study into personal carbon trading*. London, U.K.: Department for Environment, Food, and Rural Affairs.

Dlugokencky, E., & Nisbet, E. (2010). *Global atmospheric methane in 2010: Budget, changes, and dangers*. London, U.K.: Presented at the Royal Society. February 2010.

EEA. (2008). *EN34 energy subsidies*. Copenhagen, Denmark: European Environment Agency.

EEnergy Informer. (2010). *Zeroing in on zero net energy buildings* (pp. 4–6). California. www.eenergyinformer.com

Ehrlich, P., & Holdren, J. P. (1971). Impact of population growth. *Science (New Series), 171*(3977), 1212–1217.

Ekins, P., & Dresner, S. (2004). *Green taxes and charges: Reducing their impact on low-income households*. York, U.K.: Joseph Rowntree Foundation.

Environmental Law Institute. (2009). *Estimating U.S. government subsidies to energy sources: 2002–2008*. Washington D.C.: Environmental Law Institute.

EU. (2003). *Directive 2002/91/EC of the European Parliament and of the Council of 16 December 2002 on the energy performance of buildings*. Brussels: Belgium. *http://eur-lex.europa.eu/LexUriServ/LexUriServ.do?uri=OJ:* L2003:001:0065:0071:EN: PDF.

EU Commission. (2008). Questions and answers on the commission's proposal to revise the EU Emissions Trading System. *MEMO/08/35*, January 2008, Brussels, Belgium.

European Commission. (2003). *External costs – research results on socio-environmental damages due to electricity and transport*. Belgium Brussels: European Commission, Directorate General for Research.

Fawcett, T., Hvelplund, F., & Meyer, N. I. (2009). Making it personal: Per capita carbon allowances? In F. P. Sioshansi (Ed.), *Generating electricity in a carbon constrained world*. Holland: Elsevier.

Global Subsidies Initiative. (2009). *Fossil-fuel subsidies*. Geneva, Switzerland. *www.globalsubsidies.org*

Goldsmith, E., Allen, R., Allaby, M., Davoll, J., & Lawrence, S. (1972). A blueprint for survival. *The Ecologist, 2*(1).

Gorz, A. (1983). *Les chemins du paradis, l'agonie du capital*. Paris: Edition Galilée.

Hansen, J., Sato, M., Karecha, P., Beerling, D., Berner, R., Masson-Delmotte, V., et al. (2008). Target atmospheric CO_2: Where should humanity aim? *Open Atmospheric Science Journal*, Vol.2, (pp. 217–231), www.bentham.org/open/toascj/openaccess2.htm.

Hunnicutt, B. K. (1988). *Work without end – abandoning shorter hours for the right to work*. Philadelphia: Temple University Press.

Hvelplund, F. (1980). *The necessity of energy policy*, (Vol. 1). Aalborg, Denmark: Aalborg University Publishing Company.

Hvelplund, F. (2001a). Renewable energy governance systems. *Department of Development and Planning*. Denmark: Aalborg University. *www.wind-orks.org/FeedLaws/ARTsGovernance.html*.

Hvelplund, F. (2001b). Political prices or political quantities? A comparison of renewable energy support systems. *New Energy, 5*, pp.18–23.

Hvelplund, F. (2005a). Cognition and change. *Doctoral thesis (in Danish), Department of Development and Change*. Denmark: Aalborg University.

Hvelplund, F. (2005b). Renewable energy: Political prices or political quantities. In V. Lauber (Ed.), *Switching to renewable power: A framework for the 21st century/red* (pp. 228–245). London: Earthscan.

Hvelplund, F. (2006). Renewable Energy and the Need for Local Energy Markets. In *Energy, the International Journal* (pp. 1957–1966).

IPCC. (2007). Climate change 2007: The physical science basis. In S. Solomon, D. Qin, M. Manning, Z. Chen, M. Marquis & K. B. Averyt (Eds.), *Contribution of Working Group I to the fourth assessment report of the Intergovernmental Panel on Climate Change*. Cambridge, U.K. and New York, U.S.A.: Cambridge University Press.

Jackson, T (2005). Live better by consuming less. Is there a "double dividend" in sustainable consumption? *Journal of Industrial Ecology, 9*(1–2), 19–36.

Jackson, T. (2009). *Prosperity without growth – economics for a finite planet*. London: Earthscan.

Jensen, O.M. (2009). *Policies and means for promotion of energy savings in buildings*. Hørsholm, Denmark: SBI.

Jespersen, J. (2009). *Macroeconomic methodology – a post-Keynesian perspective*. London: Edward Elgar.

Keynes, J. M. (1931). *Essays in persuasion*. London: Milland and Co.

Klingenberg, K., Kernagis, M., & James, M. (2008). *Homes for a changing climate: Passive houses in the U.S. www.passivehouse.us*

Kopp, R. E., Simons, F. J., Mitrovica, K. X., Maloof, A. C., & Oppenheimer, M. (2009). Probabilistic assessment of sea level during the last interglacial state. *Nature, 462*, 786–867.

Larson, M., & Wadeskog, A. (2003). Households in the environmental accounts. *Statistics Sweden*. Prepared for DG Environment and Eurostat, Sweden. *www.scb.se/statistik/MI/MI1202/2004A01/MI1202_2004A01_BR_MIFT0408.pdf*.

Layard, R. (2005). *Happiness – lessons from a new science*. London: Penguin.

Lund, H. (2010). *Renewable energy systems*. Amsterdam: Elsevier.

Mason, A. (2001). Population policies and programs in East Asia. *East West Center, Occasional Papers. Population and Health Series, 123*, pp. 5–22. *www.eastwestcenter.org/fileadmin/stored/pdfs/POPop123.pdf*.

Meadows, D. H., Meadows, D. L., Randers, J., & Behrens, III, W. W. (1972). *The limits to growth*. New York: Universe Books.

Meadows, D. H., Meadows, D. L., & Randers, J. (1992). *Beyond the limits*. London: Earthscan.

Meadows, D. H., Randers, J., & Meadows, D. L. (2004). *Limits to growth. The 30-year update*. White River Junction, U.S.A: Chelsea Green.

Meinshausen, M., Meinshausen, N., Hare, W., Raper, S.C.B., Frieler, K., Knitti, R., et al. (2009). Greenhouse-gas emission targets for limiting global warming to 2°C. *Nature, 458*(7242), 1158–1162.

Meyer, N. I. (2003). European schemes for promoting renewables in liberalised markets. *Energy Policy, 31*, 665–676.

Meyer, N. I., Petersen, K. H., & Sørensen, V. (1981). *Revolt from the center*. London: Marion Boyars Publishing.

Mill, J. S. (1900). *Principles of political economy (revised edition, Vol. II; originally published in 1848)*. New York: Colonial Press.

Myrdal, G. (1973). *Against the stream – critical essays on economics*. New York: Pantheon Books.

Nørgård, J. S. (2006). Consumer efficiency in conflict with GDP growth. *Ecological Economics, 57*, 15–29.

Nørgård, J. S. (2009). Avoiding rebound through a steady state economy. In H. Herring & S. Sorell (Eds.), *Energy efficiency and sustainable consumption. The rebound effect* (pp. 204–223). Palgrave Macmillan.

Nørgård, J. S., Brange, B., Guldbrandsen, T., & Karbo, P. (2007). *Turning the appliance market around towards A++. Proceedings of the European Council for an Energy Efficient Economy, Summer Study 2007*. Stockholm: ECEEE. pp. 155–164.

Nørgård, J. S., Peet, J., & Ragnarsdóttir, K. V. (2010). The history of the limits to growth. *Solutions, 1*(2), 59–63. *www.thesolutionsjournal.com/node/569*.

Population Council. (1998). UN world population projections to 2150. The Population Division of the United Nations Population Secretariat. *Population and Development Review, 24*(1), 183–189. *www.jstor.org/stable/2808146*.

Reinders, A. H. M. E., Vringer, K., & Blok, K. (2003). The direct and indirect energy requirement of households in the European Union. *Energy Policy, 31*(2), 139–153.

Rickerson, W., & Grace, R. C. (2007). *The debate over fixed price incentives for renewable electricity in Europe and the United States: Fallout and future directions. A white paper prepared for the Heinrich Böll Foundation*. Berlin: Germany.

Rockström, J., et al. (2009). A safe operating space for humanity. *Nature, 461*(472), 24.

Sanne, C. (1995). *Working time. About working time reforms and consumption in the welfare state*. Stockholm: Carlsson Publishers (Bokforlag).

Schreurs, M. A. (2007). Renewable energy politics in the United States. In L. Mez (Ed.), *Green power markets: Support schemes, case studies and perspectives*. (pp. 227–250). Multiscience Publishing.

Schumacher, E. F. (1973). *Small is beautiful – a study of economics as if people mattered*. London: Blond & Briggs.

Simon, J. L., & Kahn, H. (1984). *The resourceful earth – a response to Global 2000*. New York: Basil Blackwell.

Stiglitz, J.E., Sen, A., & Fitoussi, J.-P. (2009). *Report by the Commission on the Measurement of Economic Performance and Social Progress*. Paris, France. *www.stiglitz-sen-fitoussi.fr*.

Trainer, F.E. (2001). *Natural capitalism cannot overcome resource limits. www.mnforsustain.org/trainer_fe_simon_lovins_critique.htm*

Turner, G. M. (2008). A comparison of the limits to growth with thirty years of reality. In *Global Environmental Change, 18*, 397–411. *www.csiro.au/files/files/plje.pdf*.

UNEP. (2008). United Nations Environment Programme, Division of Technology. *Reforming energy subsidies*. New York: Industry and Economics.

UN Population Division. (2008). World population policies 2007. *Department of Economic and Social Affairs*. New York: United Nations. *www.un.org/esa/population/publications/wpp2007/Publication_index.htm*

van Vuuren, D. P., & Faber, A. (2009). *Growing within limits – A report to the global assembly 2009 of the club of Rome*. The Netherlands Environmental Assessment Agency, Bilthoven, The Netherlands. www.rivm.nl/bibliotheek/rapporten/500201001.pdf

Chapter 5

We Can't Have Our Cake and Eat it Too: Why the Energy and Climate Problems Cannot Be Solved in Consumer-Capitalist Society

Ted Trainer
University of New South Wales

Chapter Outline

1. INTRODUCTION

There has been rapid increase in awareness of the climate change, energy, and environment situation. However three crucial assumptions have been evident underlying the responses. The first is that the problems can be solved. The second is that they can be solved without serious disruption to the economy and at negligible cost. The third is that, therefore, there is no need to question the universal belief in the possibility and desirability of pursuing ever more affluent "living standards" and economic growth, the central theme of this volume.

This chapter argues that all three of these assumptions are incorrect. It offers a quantitative analysis of the global energy and greenhouse gas (GHG) problems and of the technical possibilities for solving the problems, concluding that the difficulties involved in making a successful transition from the current carbon-based economic system to alternative energy sources are insurmountable without dramatic reduction in levels of production *and* consumption.

There has been very little discussion of these issues in the literature to date and it is taken for granted that the technologies required for a relatively smooth transition to a post-carbon energy era already exist or are on the drawing board and the problems are essentially only political and organizational. Many analyses claim to show how alternative energy technologies can meet future demand, but these typically do not deal satisfactorily, or at all, with limits and difficulties. *Renewable Energy Cannot Sustain a Consumer Society* (Trainer, 2007) seems to have been only the second attempt to carry out a critical analysis of the limits and problems in alternative energy sources.[1] A more recent account is available in Trainer 2008a.

The major reports influencing the discussion of the climate change problem and the global energy situation have failed to deal with these issues, notably the Intergovernmental Panel on Climate Change (IPCC) Third (2001) and Fourth (2007) Assessment Reports, the Stern Review (2006) and the Australian Garnaut report (2008). All essentially concluded that the problem can be solved at negligible economic cost, for instance 1% of GDP by 2050 according to Stern.[2] This optimistic conclusion seems to have been universally accepted, although some have argued that the cost will be much *lower* (e.g., Toll, 2006; Nordhaus, 2007).

Trainer (2009a) provides a detailed critique of this line of reasoning, pointing out that none of these reports consider possible limits to the application of alternative energy sources. The main two areas of difficulty are firstly the immense extent to which the alternative technologies would have to be scaled up to meet future demand, and secondly the variability and intermittency of renewable energy resources.

This chapter argues that these highly influential studies assert invalid and mistaken conclusions, and that it is not possible to explain how the required energy supplies can be provided within safe greenhouse limits. If this argument is accepted, it would be difficult to exaggerate the importance of implications for policy and indeed for the future of consumer society. It will be claimed that such a society is not viable and cannot be made so, and that the major global problems now accelerating cannot be solved without a radical and historically unprecedented transition to what the author calls "The Simpler Way." It will be argued, however, that such a transition is unlikely.

The perspective presented in this chapter is likely to be regarded as bleak and extreme. It departs from most current analyses, including many other

---

1. In this book's Chapter 4, Meyer et al. address the issue of limits to growth and similar constraints.

2. In this book's Chapter 6, Bollino et al. also discuss the cost issue, reaching similar conclusions.

chapters in the present volume, in concluding that the current global problems cannot be solved within or by the present society.[3]

The chapter is organized as follows: Section 2 outlines an analysis of the global sustainability and equity situation; Section 3 describes the form that a sustainable and just future socio-economic system must take if the author's main argument is accepted; and Section 4 offers some thoughts on the transition issues, followed by the chapter's conclusions.

2. WHY CONSUMER SOCIETY IS NOT SUSTAINABLE

Before looking in some detail at the greenhouse and energy problems, it is important to clarify the nature and especially the magnitude of the global predicament we find ourselves in. It is not commonly understood how grossly unsustainable today's consumer-capitalist society is. If we were just a little over sustainable levels of resource use and environmental impact, the problems might be tractable without radical social restructuring. But a glance at some basic indicators shows that we are far beyond such levels.

Brief examination of the way the global economy works reveals that our present system is also grossly unjust and far beyond moral acceptability. The argument in the first part of this chapter suggests that these two issues constitute an overwhelming case that an ecologically sustainable and a morally acceptable society cannot be achieved unless we undertake a radical transition to a different society of the kind sketched below.

2.1. Sustainability

The following points indicate the magnitude of some of the core sustainability problems set by the material living standards taken for granted in today's consumer-capitalist society.

- If all the estimated 9+ billion people likely to be living on Earth in 2050 were to consume resources at the present per capita rate in rich countries, world annual resource production rates would have to be about eight times as great as they are now (Trainer, 1985, 2006a). At that rate all estimated potentially recoverable resources of fossil fuels—even assuming 2 billion tones of coal—would be exhausted in about 18 years.
- If all 9+ billion were to have the present U.S. timber use per person, the forest area harvested would have to be three to four times all the forest area on the planet.
- It is now widely believed that global petroleum supply will peak within a decade, and be down to half the present level by about 2030 (Campbell 1997).

3. This is in stark contrast to this book's Chapter 6, by Bollino et al., who claim that we have both the means and the resolve to address the problems and doing so will not ruin us economically.

- *Footprint analysis* indicates that the amount of productive land required to provide one person in Australia with food, water, energy, and settlement area is about 8 hectares (ha), the U.S. figure is closer to 12 ha (World Wildlife Fund, 2009). If 9+ billion people in 2050 were to live as Australians do today, more than 70 billion ha of productive land would be required. However, the total amount available on the planet is only in the region of 8 billion ha. In other words our rich world footprint is about ten times as big as it will ever be possible for all people to have.
- The climate change problem is probably the most confronting testimony to the unsustainability issue. As discussed below, we probably should totally eliminate carbon dioxide (CO_2) emissions by 2050. It will be argued that this cannot be done in the context of our current consumer-capitalist society.[4]

But why should we think in terms of 9+ billion people aspiring to live as we do today in rich countries? Even if we ignore the moral issues, the Third World is determined to live as affluently as we do, whether we like it or not, and that is what conventional economics holds out as the goal of "development." So we have to consider whether it is possible and what is likely to happen if the goal is pursued.

These facts and figures make clear that rich world ways, systems, and living standards are grossly unsustainable, and it is impossible for all to rise to anywhere near the living standards we take for granted.

2.2. Economic Growth

The main worry is not the *present* levels of resource use and ecological impact discussed above but rather the levels we will rise to given the obsession with economic growth. The supreme goal is to raise incomes, living standards, and the GDP as much as possible, constantly, and without any limit. If we assume a 3% per annum economic growth, a population of 9+ billion, and all the world's people rising to the living standards we in the rich world would have in 2080 given a 3% growth until then, the total volume of world economic output would have to be 60 times as great as it is now.

So even though the present levels of production and consumption are grossly unsustainable, the fundamental goal in consumer-capitalist society is to continually increase incomes and economic output, multiplying them toward what would appear to be absurdly impossible levels in coming decades.[5]

---

4. In this book's Chapter 7, Moran also discusses the physical difficulties in meeting carbon reduction targets that are believed to be necessary by the scientific community by 2050.

5. Meyer et al. in this book's Chapter 4 are among numerous authors making similar observations as do the likes of Meadows, Meadows, and Randers, 1972; Ehrlich, 1972; Heinberg, 2003; Speth, 2001; and Kunstler, 2005.

2.3. Technical Fixes

The common response to these types of *limits to growth* arguments is to claim that the problems can be solved by technical advances that do not require fundamental change away from consumer-capitalist society. However, this position is highly implausible, given the magnitude of the problems and the multiples sketched above. Greater energy efficiency, recycling efforts, pollution control, and so on, are not likely to deal with present resource and ecological impacts, let alone those a growth economy will generate in future years.

Amory Lovins, possibly the best known "tech-fix" optimist, believes we could cut the resource and ecological costs per unit of economic output to one quarter of the present levels (von Weizacker and Lovins, 1997). But even if we accept those arguments, this would be far from sufficient. Let us assume that present resource and ecological impacts must be halved, although some of the above figures indicate the need for a much greater reduction. Again, if we had 9 billion people on the living standards Australians would have by 2080 given a 3% growth rate, then total world economic output would have to be 60 times as great as it is now. How likely is it that we could have 60 times as much production and consumption while we cut resource and ecological impacts to half their present levels, that is, achieve a factor 120 reduction? Hueseman and Hueseman (2008) document the way significant advances in efficiency improvement and impact reduction in recent decades have been outweighed by increasing consumption, resulting in considerable net increases in resource demands and ecological impacts within industrial societies.

2.4. The Unjust Global Economy

We in rich countries could not have anywhere near our present living standards if we were not getting far more than our fair share of world's resources. Our per-capita consumption of items such as petroleum is around 17 times that of the poorest half of the world's population. The rich one-fifth of the world's population are consuming around three-quarters of the resources produced globally. Many people receive such a minute share that 850 million are hungry and more than that number have dangerously dirty water to drink. Half the world's people live on an income of $2 per day or less (Fotopoulos, 1997; Speth, 2001; Chossudovsky, 1997).

This grotesque injustice is primarily due to the fact that the global economy operates on market principles. In a market-based system, need is irrelevant and is ignored, while goods go mostly to those who are richer, because they can offer to pay more for them. Thus, we in rich countries get almost all of the planet's scarce resources while billions of people in desperate need get little or none. This explains why one-third of the world's grain is fed to animals in rich countries while around 30,000 children die every day because they have insufficient food and clean water.

Even more importantly, the market system explains why Third World development is *inappropriate* to the needs of Third World people. What is *developed* is not what is most *needed*, it is what will make most profit for the few with capital to invest. Thus, there is development of export plantations and cosmetics factories but there is no development of farms and firms in which poor people can produce for themselves the basic items they need. In many countries there is almost no development because it does not suit corporations to develop anything there, even though those countries have the land, water, skills, and labor to produce most of the things they need for a good quality of life.[6]

Even when transnational corporations do invest in poor countries, wages offered in their factories can be as low as 15–20 cents an hour. Compare the miniscule benefit such workers get from conventional trickle-down development with what they could be getting from an approach to development that enabled them to receive all the benefits from their own labor, applied via mostly cooperative local firms producing the goods and services they most need. But development of this kind is explicitly prevented by the conditions written into the Structural Adjustment Packages inflicted on poor countries by the World Bank and IMF.

"Assistance" is given to indebted countries on the condition that they deregulate and eliminate protection and subsidies for their people, cut government spending on welfare and publicly beneficial infrastructure, open their economies to more foreign investment, devalue their currencies (making their exports cheaper for us and increasing what they must pay us for their imports), sell off their public enterprises, and increase the freedom for market forces to determine development. The beneficiaries of such policies are the transnational corporations, banks, and people who shop in rich world supermarkets. The corporations can buy up firms cheaply and have greater access to cheap labor, markets, forests, and land. The produce of the Third World's soils, labor, fisheries. and forests flows more readily to our supermarkets, not to Third World people.

In addition to the normal functioning of the global economy, rich countries go to considerable clandestine effort to keep Third World countries to the policies that suit us. Rich countries support repressive regimes willing to keep their economies to the policies that enable corporations to extract their resources, use Third World land for export crops, exploit cheap labor, and so on. They invade and control countries threatening to follow policies contrary to rich world interests. Rich countries could not have such high "living standards" if a great deal of repression and violence was not carried out to keep unjust systems in place, and rich countries contribute significantly to this (Chomsky, 1992; Goldsmith, 1997; Klein, 2007; Chossudovsky, 1997).

---

6. On the concept of "appropriate development" see Trainer, 2006c.

These are the reasons why conventional development can be regarded as a form of *plunder*. The Third World has been developed into a state whereby its land and labor benefit the rich few in the world, not the majority of Third World people. Rich world living standards could not be anywhere near as high as they are if the global economy was just and we had to live on something like our fair share of world resources.

These considerations of sustainability and economic justice show not only that our predicament is extreme, but more importantly that the problems *cannot* be solved in consumer-capitalist society. The problems are caused by the fundamental structure of our socio-economic system. The argument has been that there is no possibility of having an ecologically sustainable, just, peaceful, and morally satisfactory society if we continue to allow market forces and the profit motive to be the major determinant of what happens and if we seek economic growth and ever-higher "living standards" without limit.[7]

2.5. Can Alternative Energy Technologies Solve the Problems?

The faith in the capacity of alternative technologies to solve the problems and save affluence and growth society depends heavily on assumptions about the potential of nuclear energy, geo-sequestration, and renewable sources. It is therefore important here to briefly indicate the grounds for concluding that these alternatives cannot meet the energy demands set by consumer-capitalist society within safe GHG emission limits.[8]

2.5.1. Emission Limits

The IPCC Fourth Assessment Report concluded that a 50–80% reduction in emissions would be necessary by 2050 (IPCC, 2007; SPM 7). However these targets are now increasingly regarded as too low because, in the last few years, observed global warming effects have been more rapid than were anticipated by the IPCC Fourth Assessment Report of 2007 (Hansen et al., 2008). Other studies conclude in effect that emissions must be reduced to zero by 2050 (e.g., Meinschausen et al., 2009).

2.5.2. Geo-Sequestration

According to the IPCC, geo-sequestration is not likely to extract more than 80%–90% of the CO_2 generated at the power station and Hazledyne (2009) points out that when all elements in the process are taken into account, including for instance fugitive emissions from mining, the figure is 75%. If the 2050 emission limit is 5.7 Gt/y, as the IPCC says might be required, and if

7. This book's Chapter 16 by Stulz et al. proposes a system that attempts to share global resources more equitably among world's inhabitants without impinging on high quality of life.

8. Further details are given in Trainer (2006g, 2007, 2008a).

geo-sequestration captures 80% of CO_2 generated, then 28.5 Gt/y could be generated, corresponding to about 280 EJ of primary energy or 98 EJ of electricity. This would provide 9 billion people with 11 GJ per capita p.a.—about 33% of the present Australian per capita electricity consumption and only 4% of total Australian per capita energy consumption. Also 280 EJ of primary energy from coal would be more than twice the present rate of use and, according to some estimates, at the present consumption rate, global coal supply will plateau within two decades (Energy Watch Group, 2007).

Another major difficulty is that geo-sequestration is only applicable to stationary generating sources and would therefore not apply to perhaps 50% of carbon fuel use. At best, only about 60% of transport could be converted to electrical drive. Aircraft, ships, and heavy trucks cannot be run on batteries. This would still leave unaccounted the possibly 60% of present final energy demand that is in nonelectrical form, after electricity and that 60% of transport have been accounted for. To convert from electricity to these forms of energy will involve large losses, further discussed below. Fuel cells can power vehicles other than cars, but these involve the significant losses, costs, and other difficulties associated with hydrogen, referred to below in the discussion of energy conversion.

Other significant difficulties for geo-sequestration concern the availability of storage sites and the energy cost of the process. It is therefore not likely that that geo-sequestration of CO_2 can make a significant difference to the problems set by providing energy affluence to all people.

2.5.3. Nuclear Energy

Lenzen's (2009) review concludes that Uranium resources can sustain the present nuclear contribution for about 85 years, indicating a cumulative contribution of *ca.* 650 EJ. World energy demand by 2050 is likely to be 1000 EJ/y, meaning that unless fusion or fourth-generation breeders are introduced, nuclear energy cannot make a significant long term contribution to the global energy supply problem.[9]

2.5.4. Wind[10]

If wind was to supply one-third of the 1000 EJ/y required by 2050 then installed capacity would have to be around 600 times the early–mid 2000s capacity (derived from Coppin, 2008). It is not plausible that enough sites could be found, onshore or offshore, within tolerable distances of population centers, due to the density of settlement. Trieb, a strong believer in renewable energy

9. On the questions regarding the fourth-generation breeder that would have to be settled, see Trainer (2008a).

10. This book's Chapter 13 by Arent et al. covers future prospects for renewable energy technologies, including a discussion of biofuels.

resources, estimates European wind potential as about 2 EJ/y from onshore sites and another 2 EJ/y from offshore sites (Trieb, undated). Some European regions are probably close to their limits now. The main difficulty regarding wind, however, has to do with intermittency, which Lenzen's review concludes will limit its contribution to about 20% of *electricity* demand, meaning *ca.* 5% of total *energy* demand (Lenzen, 2009).

2.5.5. Biomass

Biomass is not likely to make a major contribution to global energy supply, in view of likely yields, conversion efficiencies, and land areas. Chapter 5 of Trainer (2007) reports what seem to be the most plausible estimates, that is, that the yield for very large scale cellulosic biomass production is not likely to exceed 7 tonnes per ha, and that ethanol or methanol production is likely to be around 7 GJ net per tonne of biomass. Mackay (2008) and Foran and Crane (2002) arrive at similar estimates, while Foran believes future output might rise to *ca.* 90 GJ/ha.

If all people were to consume the quantity of oil plus gas Australians now consume per annum—more than 128 GJ/y—via ethanol from cellulosic inputs, a plantation area of around 23 billion ha would be needed——on a planet with only 13 billion ha of land.

It is not likely that 1 billion ha could be put into biomass fuel production, given that the amount of productive land is in the region of 8 billion ha and cropland takes 1.4 billion ha. The conditions for biomass energy production will probably deteriorate in future. Land, forest, and water resources are already stretched. Fertilizer availability, especially phosphorus, and price are likely to set increasing difficulties, population is going to increase by 35%, and food demand by a greater amount, and the greenhouse problem is likely to reduce yields markedly.

One billion ha would yield 50 EJ/y of liquid fuel, an average of 5.5 GJ/y per capita for 9 billion compared with the present Australian per capita consumption of 128 GJ/y of liquid fuel (oil plus gas). Field, Campbell, and Lobell (2007) believe that the global biomass production limit is 27 EJ/y, corresponding to only about 9 EJ/y of ethanol.

2.5.6. Solar PV

The difficulties with solar PV are not primarily to do with quantity and cost but with their variability and their integration into supply systems. For instance if PV was to provide 20% of total supply, on a sunny day all this capacity would go off stream within perhaps two hours at the end of the day and would then need to be substituted for by some other source. Even on a good day PV provides no energy for about 15 hours, and an entire national capacity might provide almost no energy for weeks on end in a cloudy winter. A supply system would need enough plant of some other kind to substitute for the entire contribution its PV component makes in full sunshine—unless large-scale storage becomes possible, further discussed below.

2.5.7. Solar Thermal Systems

Solar thermal systems are likely to be the most effective renewable energy technologies, in view of their capacity to store energy as heat. However their capacity to deliver in higher latitudes, and in winter, is problematic. Unfortunately crucial data are not made available by generating companies. Trainer (2008b) attempts to estimate the electricity these systems might deliver to distant consumers from deserts in winter, net of all energy costs. The figure is likely to correspond to a flow of 20 W/m2 of collection area. If this is so, a power station capable of delivering 1000 MW at a distance would require some 50 million square meters of collectors, or 125,000 ANU Big Dishes.[11] At the anticipated future cost for the Big Dish (Luzzi, 2000) the power station would cost $18.4 billion, some five times the cost of a coal-fired station plus coal for lifetime operation. Not included in this estimate are the energy costs of constructing the dishes, the ammonia facility, the transmission lines, or the operations and management energy costs of all components.

The dollar cost is not the main concern. The frequency of occurrence of cloudy days would limit the capacity of solar thermal systems to bridge gaps in supply from the wind and PV components of a system. Equipping solar thermal plants with the capacity to supply over a four-day cloudy period would require 8 times the 12-hour storage capacity currently envisaged.

It would seem to be fairly clear, therefore, that in winter, solar thermal systems could not be the means whereby the gaps in supply from other components of a wholly renewable electricity supply system could be filled.

2.5.8. The Conversion Problem

Discussions of the potential of renewable energy sources usually fail to take into account the need to convert energy from forms that are available to forms that are needed. Conversion is typically quite energy-inefficient, meaning that much more primary energy needs to be generated than might appear to be the case. For instance, fuelling transport by hydrogen generated from electricity could require generation of about four times the amount of energy that is to drive wheels (Bossel, 2004).

Electricity accounts for only about 25% of Australian final energy consumption. If we assume that 60% of transport is to be run via electric vehicles, this would leave the question of where the perhaps 45% of energy other than direct electricity and transport electrical energy is to come from, and what the losses of energy might be in conversion of electricity into the needed forms.

11. The Big Dish refers to a 400-square-metre solar thermal parabolic concentrator driving a steam turbine that is being developed by the Australian National University.

The variability between summer and winter would more or less double the magnitude of the conversion and dumping problems for solar sources, given that in good solar regions, winter insolation is about half the summer value; the multiple is much greater in the higher latitudes.

2.5.9. Renewable Energy Conclusions

Trainer (2010) attempts to estimate an annual investment figure for a global energy system capable of meeting average demand through a mid-winter month, based on geo-sequestration, nuclear energy, and renewables. The key assumptions made in the paper are summarized in the appendix to this chapter. The sum is in the region of 30 times the present world annual energy investment total. It should be noted that this amount of plant would not solve the problem of intermittency; that is, there would still be times when long periods of calm and cloudy weather would cause supply to fall below demand.

The foregoing discussion points to a number of significant difficulties regarding the capacity of alternative energy sources to substitute for fossil fuels while maintaining consumer-capitalist society's energy requirements. The highly influential reports by Stern (2006), the IPCC (2007) and by Garnaut (2008) to the Australian government all fail to deal with these critical issues. The above analysis seems to seriously challenge or clearly invalidate major conclusions arrived at in these reports, as well as the general assumption held by governments, economists, and the general public that renewable energy resources can substitute for fossil fuels.

It should be stressed that these have not been arguments *against* the adoption of renewable energy technologies. On the contrary, as described in Trainer (2007, Chapter 11), we should move entirely to these sources and we could live well on them, but not in a society committed to high rates of energy consumption, affluence, and growth.

3. THE ALTERNATIVE PATH

If the foregoing analysis of the global situation is accepted, we must attempt to transition to an alternative path that would allow us to live well on a small fraction of our present per capita resource consumption and ecological impact. This, I believe, cannot be done without abandoning some of the basic systems and values of our current consumer-capitalist society. The basic features of a sustainable and just future socio-economic system would have to include:

- Far simpler material living standards. This does *not* mean hardship or deprivation but rather being content with what is sufficient.[12]

---

12. For a discussion of human needs, wants, and desires, refer to this book's Chapter 3 by Bartiaux et al.

- High levels of self-sufficiency at household, neighborhood, local, and national levels with relatively little travel, transport, or trade. This means mostly small local economies using local resources to meet local needs, and a relatively minor role for national governments.
- Control over these communities by *participatory* systems, such as citizen assemblies.[13]
- An economic system that is not driven by market forces and the profit motive, although it might have a significant role for private enterprise and markets, and without any growth.
- Most problematic, for such a socio-economic system to work there must be a radically different culture, in which competitive and acquisitive individualism is replaced by frugal, self-sufficient collectivism.

This vision does not threaten desirable high-tech or the training of highly skilled professionals. In an economy greatly reduced by the elimination of unnecessary and wasteful production, a greater share of resources than present could go into these purposes. Many could opt to continue full-time work in normal professions and trades, and it is assumed that significant changes do not need to be made to property relations or personal wealth and income differentials.

Advocates of "The Simpler Way" firmly believe that it would provide a much higher quality of life than most people experience in today's consumer society. Consider for example having to work for money only one day a week, living in a supportive community, and not having to worry about unemployment or insecurity in old age. These are among the benefits experienced by people presently living in alternative communities.[14]

4. HOW MIGHT THE TRANSITION BE MADE?

As has been noted, the chances of achieving such a radical transition would seem to be remote, given that the mainstream denies, ignores, and rejects the themes discussed in this chapter,[15] and given that we probably have no more than 30 years left to make the transition.

The task is essentially one of gradually altering the dominant ideology; in other words, working to change the worldview and values of ordinary people in existing towns and suburbs, so that there will eventually be a willingness to

---

13. Participatory democracy is distinguished from representative democracy by having all decisions made by all citizens. It is most easily practiced in small local contexts, but is applicable to larger communities via delegations to federated assemblies, from which recommendations are sent back to local assemblies for decisions.

14. For example, refer to the Global Ecovillage Network, 2009. For a more detailed account of the proposed vision, refer to The Simpler Way website, *http://ssis.arts.unsw.edu.au/tsw/*.

15. For example, this book's Chapter 6 by Bollino et al. reaches starkly different conclusions.

pursue "The Simpler Way," firstly, because it is understood as the solution to the global predicament, and secondly, because it is seen as enabling a higher quality of life.

There is powerful reluctance, indeed refusal, to think about the possibility that consumer-capitalist society with its determination to pursue limitless affluence and growth cannot be made sustainable or just. Above all, there is presently strong antipathy to suggestions that the salvation is not possible unless simple lifestyles and a zero growth economy are embraced.

These themes can be especially unpalatable for those who work in fields such as energy technology where experience has typically been of success at meeting demands for more and better supply within existing social system—at declining costs.

The goal of a thriving local economy under the control of conscientious and contented citizens cannot be achieved by the use of force, state power, or a centralized authority. It can only be learned, developed by, discovered by the people in the towns and suburbs where they live, as they find their way to the systems and procedures that work best for them in their unique social and economic conditions.

Therefore the most important action strategy is to join in the efforts to establish the new ways and systems in the towns and suburbs where we live. This is the strategy that the Anarchist philosophy terms "prefiguring," in other words, starting to build the new society here and now within the old one that is to be replaced. The main reason to do this is not to have established the new ways, important though that is, it is *to be in the best position to influence the thinking* of people in those neighborhoods, to try to ensure that their actions are driven by a critical global consciousness of the need for vast and radical change, as distinct from merely striving for reforms within our present growth and affluence society.

For 30 years now the Global Ecovillage Network (2009) has pioneered the building of settlements more or less of the kind proposed in this chapter. However since 2005 when the Transition Towns, now Transition Initiative (2009), burst onto the scene, it has taken the leadership of what might be termed the Global Alternative Society movement. Hundreds of towns are now seeking to develop more locally self-sufficient communities, motivated by the prospect of "peak oil" and similar concerns, including the threat of climate change (see, for example, Hopkins, 2009). If we make it to a sustainable and just world within the next few decades, it will have been through some kind of Transition Initiative process.

CONCLUSIONS

Most, if not all, of the recent discussions of the global situation have been based on the assumption—usually implicit—that consumer-capitalist society can be reformed to be sustainable, just, and socially acceptable. However the argument in this chapter has been that this assumption is seriously mistaken. The alarming problems now threatening our survival are being caused by some of the fundamental, defining characteristics of our society, and the main purpose of the discussion in Section 1 of this chapter has been to show that these must be replaced before the problems can be solved.

Most resistance to this view comes from the faith that technical advance, in general, and alternative energy sources, in particular, can head off any need for fundamental change in systems and values. Reasons for doubting these beliefs have been sketched, that is, for concluding that we "cannot have our cake and eat it too."

If this case is sound, the solution to our problems has to involve the abandonment of the quest for affluence and growth and a transition to a more sustainable path; but the prospects for achieving this would not seem to be encouraging.

APPENDIX

The key assumptions for the discussion in the body of chapter follow:

- Total primary supply required 1000 EJ/y
- Final energy supply 690 EJ/y
- Conservation effort and energy saving reduces this to 455 EJ/y
- Direct electricity 25%, transport 33% of final energy
- 60% of transport electrified
- Nuclear 8 EJ/y
- Hydro almost double present contribution to 19 EJ/y
- Geo-sequestration provides 51 EJ/y of electricity, corresponding to 23 Gt/y of CO_2 sequestered
- Biomass from 1 billion ha, yielding 50 EJ/y of ethanol
- Wind capacity .38 in winter
- Windmills of 1.5 MW(p) capacity assumed, costing $2.25 million
- PV solar to electricity efficiency at .13, and $6.5/W fully installed, 2.8 kWh/m/d winter insolation (compared with under 1 kWh/m/d in mid-European regions)
- Solar thermal 400 square meter dishes at .19 solar to electricity efficiency, 5.5 kWh/m/d DNI (the best U.S. locations but somewhat lower than for the Sahara)
- Ammonia heat storage at .7 efficiency
- $146,000 per dish, one-third present estimated commercial cost
- 10% of final demand as low-temperature heat supplied by solar panels
- Conversion of electricity to nonelectrical energy forms needed at .5 efficiency
- Wind and PV each contributing 25% of final energy supply, solar thermal 50%

BIBLIOGRAPHY

Anderson, K., & Bows, A. (2009). *Radical reframing of climate change agenda.* Tyndall Centre, Manchester University. *http://mikhulme.org/wp-content/uploads/2010/01/2010-Module-M594-syllabus.pdf.*

Baer, P., & Mastrandrea, M. (2006). High stakes; designing emissions pathways to reduce the risk of dangerous climate change. Institute of Public Policy Research, Nov. *http://www.stanford.edu/~mikemas/cv.htm.*

Bairol, F. (2003). World energy investment outlook to 2030. *IEA, Exploration and Production: The Oil & Gas Review, 2.*

Bossel, U. (2004). 'The hydrogen illusion; why electrons are a better energy carrier'. *Cogeneration and On-Site Power Production.* March – April, pp. 55–59.

Campbell, J. (1997). *The coming oil crisis. Brentwood.* England: Multiscience and Petroconsultants.

Chossudovsky, M. (1997). *The Globalisation of Poverty.* London: Zed Books.

Chomsky, N. (1992). *The prosperous few and the restless many.* New York: Odonian Press.

Climate Action Summit, Canberra. (2009). Target should be zero emissions by 2050, and 300 ppm. *Arena Journal, 99*(Feb–Mar), 2.

Coelingh, J. P. (1999). *Geographical dispersion of wind power output in Ireland.* The Netherlands: Ecofys. *www.ecofys.com.*

Coppin, P. (2008). Wind energy. In P. Newman (Ed.), *Transitions.* Canberra: CSIRO Publishing.

Davenport, R., et al. (undated). *Operation of second-generation dish/Stirling power systems.* San Diego: Science Applications International, Corp.

Davy, R., & Coppin, P. (2003). *South East Australian wind power study.* Canberra, Australia: Wind Energy Research Unit, CSIRO.

Ehrlich, P. (1972). *Population, resources and environment; issues in human ecology.* San Francisco: Freeman.

Energy Watch Group. (2007). *Coal resources and future production. www.energywatchgroup.org/fileadmin/global/pdf/EWG-Report_Coal_10-07-2007ms.pdf.*

On Netz, E. (2004). *Wind Report 2004. www.eon-netz.com. www.nowhinashwindfarm.co.uk/EON_Netz_Windreport_e_eng.pdf.*

Field, C. B., Campbell, J. E., & Lobell, D. B. (2007). Biomass energy; the scale of the potential resource. *Trends in Ecology and Evolution, 13*(2), 65–72.

Foran, B., & Crane, D. (2002). Testing the feasibility of biomass based transport fuels an electricity generation. *Australian Journal of Environmental Management, June(9)* 44–55.

Fotopoulos, T. (1997). *Towards an inclusive democracy.* London: Cassell.

Garnaut, R. (2008). The Garnaut climate change review: Final report. *www.garnautreview.org.au/index.htm.*

Global Ecovillage Network. (2009). *http://gen.ecovillage.org/*

Goldsmith, E. (1997). Development as colonialism. In J. Mander & E. Goldsmith (Eds.), *The case against the global economy.* San Francisco: Sierra.

Hansen, J., et al. (2008). Target atmospheric CO_2: Where should humanity aim? *Open Atmospheric Science Journal, 2,* 217–231.

Hazledyne, S. (Professor of Sedimentary Geology, Edinburgh University). (2009). Interviewed on *ABC Science Show,* 19 Sept.

Heinberg, R. (2003). *The party's over.* Gabriola Island: New Society.

Hendricks, C., Graus, W., & van Bergen, F. (2004). *Global carbon dioxide storage potential and costs*. Utrecht: Ecofys. *www.ecofys.nl*.

Hopkins, R. (2009). *The transition handbook*. London: Chelsea Green. (See also the Transition Towns blog, *www.transitionculture.org*).

Huesemann, M. H., & Huesemann, J. A. (2008). Will progress in science and technology avert or accelerate global collapse? A critical analysis and policy recommendations. *Environmental Development and Sustainability, 10*. 7687-825.

Intergovernmental Panel on Climate Change (IPCC). (1991). Climate change: The IPCC response strategies. Washington: Island Press.

Intergovernmental Panel on Climate Change (IPCC). (2007). Mitigation of climate change. Fourth assessment report, working group III report. *www.ipcc.ch/ipccreports/ar4-wg3.htm*.

Klare, M. T. (2001). Resource wars: The new landscape of global conflict. New York: Metropolitan Books.

Klein, N. (2007). *The Shock Doctrine: The Rise of Disaster Capitalism*. New York: Metropolitan Books/Hendry Holt.

Kunstler, J. (2005). *The long emergency: Surviving the converging catastrophes of the twenty-first century*. New York: Grove/Atlantic.

Leeuwin, J. W., & Smith, P. (2003). *Can nuclear power provide energy for the future; would it solve the CO_2 emission problem?* http://www.stormsmith.nl/

Lenzen, M. (2009). *Current state of development of electricity-generating technologies — a literature review*. Integrated Life Cycle Analysis. *Dept. of Physics*. University of Sydney.

Lovegrove, K., Luzzi, A., Solidiani, I., & Kreetz, H. (2004). Developing ammonia based thermochemical energy storage for dish power plants. *Solar Energy, 76*(1—3), 331—337.

Luzzi, A. C. (2000). Showcase project: 2 MWe solar thermal demonstration power plant. *Proceedings of the 10th Solar PACES Int. Symposium on Solar Thermal Concentrating Technologies*. Sydney.

Mackay, D. (2008). Sustainable energy — without the hot air. *Cavendish Laboratory*. www.withouthotair.com/download.html.

Mason, C. (2003). *The 2030 spike: Countdown to catastrophe*. Earthscan.

Meadows, D. H., Meadows, D., & Randers, J. (1972). *The limits to growth*. New York: Universe.

Meinschausen, M., Meinschausen, N., Hare, W., Raper, S. C. B., Frieler, K., Knuitti, R., et al. (2009). Greenhouse gas emission targets for limiting global warming to 2 degrees C. *Nature, 458*(April 30), 1158—1162, *http://www.stormsmith.nl/*.

Metz, B., Davidson, O., de Connick, H., Loos, M., & Meyer, L. (Undated). *Carbon dioxide capture and storage*. IPCC Special Report, ISBN 92-9169-119-4.

Nordhaus, W. (2007). *The Stern Review on the economics of climate change*. *http://ideas.repec.org/p/nbr/nberwo/12741.html*

Odeh, S. D., Behnia, M., & Morrison, G. L. (2003). Performance evaluation of solar thermal electric generation systems. *Energy Conversion and Management, V44*, 2425—2443.

Oswald Consulting. (2006). *25GW of distributed wind on the U.K. electricity system*. An engineering assessment carried out for the Renewable Energy Foundation, London. *www.ref.org.uk/images/pdfs/ref.wind.smoothing.08.12.06.pdf*.

SANDIA, (Undated). Personal communication.

Speth, G. (2001). A bridge at the end of the world. New Haven: Yale University Press.

Stern, N. (2006). *Review on the economics of climate change*. H.M.Treasury, U.K.

Transition Initiative. (2009). *www.transitionnetwork.org/*

Toll, S. J. (2006). *The Stern Review of the economics of climate change: A comment*. Hamburg: Economic and Social Research Institute.

Trainer, T. (1985). *Abandon affluence*. London: Zed Books.

Trainer, T. (2006g). Renewable energy: No solution for consumer society. *Democracy and Nature, 3*(3), 1.

Trainer, T. (2006a). The simpler way website. *http://ssis.arts.unsw.edu.au/tsw/*

Trainer, T. (2006b). *War. All Jamie needs to know. http://ssis.arts.unsw.edu.au/tsw/war.html*

Trainer, T. (2006c). *Third world development. http://ssis.arts.unsw.edu.au/tsw/08b-Third-World-Lng.html*

Trainer, T. (2006d). *Our empire. http://ssis.arts.unsw.edu.au/tsw/10-Our-Empire.html*

Trainer, T. (2006e). *The alternative sustainable society. http://ssis.arts.unsw.edu.au/tsw/12b-The-Alt-Sust-Soc-Lng.html*

Trainer, T. (2006f). *The Spanish anarchist workers' collectives; look what we can do! http://ssis.arts.unsw.edu.au/tsw/Spanish.html*

Trainer, T. (2007). Renewable energy cannot sustain a consumer society. *Dordrect: Springer.*

Trainer, T. (2008a). *Renewable energy — cannot sustain an energy-intensive society. http://ssis.arts.unsw.edu.au/tsw/RE.html*

Trainer, T. (2008b). *Solar thermal electricity. http://ssis.arts.unsw.edu.au/tsw/solartherm.html*

Trainer, T. (2009a). *Can the greenhouse problem be solved: The negative case. http://ssis.arts.unsw.edu.au/tsw/CANRE.html*

Trainer, T. (2009b). The Transition Towns movement: A friendly critique. *http://ssis.arts.unsw.edu.au/tsw/TransTowns.html*

Trainer, T. (2010). *The Transition To a Sustainable and Just World.* Sydney: Envirobook.

Trieb, F. (Undated). *Trans-Mediterranean interconnection for concentrating solar power: Final report.* German Aerospace Center (DLR), Institute of Technical Thermodynamics, Section Systems Analysis and Technology Assessment.

von Weizacker, E., & Lovins, A. (1997). *Factor four.* St Leonards: Allen and Unwin. (2008). *WindStats Newsletter. www.windstats.com/.*

World Wildlife Fund. (2009). *The living planet report.* World Wildlife Fund and London Zoological Society. *http://assets.panda.org/downloads/living_planet_report_2008.pdf*

Zittel, W., et al. (2006). *Uranium resources and nuclear energy.* Energy Watch Group, Dec.

Chapter 6

Sustainability: Will There Be the Will and the Means?

Carlo Andrea Bollino and Paolo Polinori*
Department of Economics, Finance and Statistics, University of Perugia (Italy)

Chapter Outline

1. INTRODUCTION

There are increasing concerns about the sustainability of continuing with the type and pace of economic development that the world has enjoyed since the WWII reconstruction period, especially considering the effects of economic growth on the environment. As a few chapters in the book point out, the continuation of business-as-usual economic growth path may not be compatible with long-term sustainability.

In this chapter we argue that a strong effort toward new technologies and investments can allow us to sustain economic growth without endangering the environment. We define this effort toward a *new sustainable paradigm* as consisting of a mix of renewable sources, energy efficiency, and energy savings—issues discussed by others in the present volume. The crucial question, of course, is how much investment is required to achieve a sustainable paradigm and whether it is a plausible sacrifice compared with similar past investments.

Many scholars argue that climate change has a scale and global dimension, which is unprecedented in the history of our planet, and this represents the

* The authors are particularly grateful for the comments and suggestions of Fereidoon Sioshansi and for the excellent research assistance of Gianfranco Di Vaio and Silvia Micheli; all errors remain our own.

greatest challenge and price tag that humanity has ever known. Furthermore, some scholars believe that the challenge is even more daunting, requiring nothing short of decoupling future economic growth from environmental degradation, or even more radical ideas including resorting to no-growth economics, local economics, or other drastic socio-economic *and* lifestyle changes.[1] Other scholars, however, point out that climate change is a typical public good financing problem with the usual tradeoffs: it requires the imposition of immediate and painful private costs in exchange for uncertain future public benefits.[2]

In this chapter, we argue that the difficulty in finding feasible solutions to address climate change is a political, not an economic problem. The lack of political will and willingness to cooperate, however, is not an exclusive characteristic of the climate change, as exemplified by the recent failure of the Copenhagen meeting in December 2009. The lack of cooperation is typical of many vexing global questions related to human rights issues, trade agreements, financial capital taxation, offshore tax havens, property rights, and the safeguarding of biodiversity. Each of these issues requires time before a solution can be found. For example, negotiations to admit China to the World Trade Organization—a rather trivial matter compared to climate change—took over 15 years.

On the other hand, we know that every dollar of additional investment generates at least an additional dollar of GDP, if it is financed by taxation, and even more, if it is financed through private savings, according to the macroeconomic Keynesian multiplier.[3] This simple lesson about the income multiplier effect seems to have been forgotten when politicians discuss the cost of measures needed to implement a proper mix of renewable sources, energy efficiency, and energy savings. This is among the issues examined in this chapter.

The chapter argues that the size, intensity, and scope of policies required to achieve future sustainability are similar to those employed in the past to sustain economic growth, and hence there is no need for radical measures or a major departure from our current socio-economic system. In Section 2, we present a review of economic and legal tools capable of moving toward the aim of sustainable growth with high standards of living, and discuss the monetary costs necessary to achieve it. In Section 3, we make an analysis of historical data, to assess the magnitude of past investment efforts and their effectiveness in promoting growth and development. In Section 4, we evaluate the monetary costs of a transition to a *new sustainable paradigm*, assess its plausibility and welfare implications, and propose effective mechanisms to achieve the desired target. Section 5 concludes.

---

1. In this book's Chapter 7, Moran refers to a "romanticized version of the pre-modern world," while in Chapter 5 Trainer argues for abolishment of our growth-focused, consumer-oriented socio-economic system.

2. Acemoglu et al. (2010) argue that the socially optimal allocation may be reached by using temporary taxes on the use of dirty inputs and subsidies to clean technologies, without halting long-run growth.

3. P. Samuelson, *Economics*, McGraw Hill, 18th edition, 2007.

2. MARKETS, INSTITUTIONS, AND SOCIETY

The new sustainable paradigm involves a comprehensive strategy (Labandeira and Martín-Moreno, 2009) geared to producers, markets, and institutions to shift energy production toward renewable resources, as well as at consumers to adopt lifestyles with low environmental impact (Jackson, 2005).

The issue of consumer habits and lifestyles is crucial and yet controversial. There is considerable debate, for example, on a universal definition of human needs for energy to sustain an adequate living standard. While economic theory considers basic human needs to be finite, reaching a satiation level beyond which diminishing marginal returns applies (e.g., Jackson and Michaelis, 2003), human needs appear to be elastic, expandable, and dependent on social and cultural factors.[4] Hence it is not clear if we can achieve a future world where happiness and wellbeing continue to increase, while using a finite amount of resources and goods.

This leads to other dilemmas. For example, will it be possible to reshape consumption in developed societies to reduce the environmental impact while maintaining a high standard of living while simultaneously increasing the standards of living in developing countries?[5] In our view, these two aims cannot possibly be achieved in a world with no economic growth because we need new investment, and investment needs growth. These issues are further explored below.

2.1. Sustainability, Prices, and Taxes

There is no single satisfactory definition of sustainability, not even the classical definition of the Bruntland Commission.[6] This is partly due to the fact that in the very long term, everything is going to change, adapt, and evolve. What, for example, will be the needs of a future generation in the year 2100? Will these needs require more physical assets and personal possessions or a greater sense of spiritual awareness and living in harmony with the natural environment? In view of the difficulties in defining the concept of sustainability precisely, we propose a practical definition in economic terms.

In a world where there are both traditional resources, such as fossil fuels, and new resources—such as a mix of renewable resources, more efficient energy saving technologies, and so on—we know for sure that in the long run, sustainability means that the share of new resources should approach 100%, as existing resources are finite and therefore doomed to become exhausted.

4. Refer to Chapter 3 by Bartiaux et al.

5. This is the basic idea covered in Chapter 16 by Stulz et al.

6. The Bruntland Commission's definition of sustainability is: "development that meets the needs of the present without compromising the ability of future generations to meet their own need." United Nations General Assembly Report of the World Commission on Environment and Development: Our Common Future (1987). Transmitted to the General Assembly as an Annex to document A/42/427 – Development and International Cooperation: Environment.

In the medium run we can predict the share of renewables as a function of prices because the proportion between traditional and new resources depends on their relative prices and consumer preferences. The reasoning is as follow: As the price of finite resources rises over time, the price of renewables drops in relative terms. This, in turn, makes new resources more attractive, encourages new investments, and leads to new behaviors including the adoption of greater efficiency. Over time, as the adoption of new technologies spreads, the share of new resources increases. If new energy technologies grow faster than total demand, the share of new resources in total consumption accelerates. With economies of scale, the relative price of renewable energy resources falls, leading to faster penetration.

In this context, we define sustainability as increased investment in new resources recognizing that this may lead to a number of sustainable trajectories. Obviously, the higher the level of investment in energy efficiency and renewable resources, the more sustainable the future is likely to be.[7]

To make investment in renewables more attractive, their price must approach those of alternative fossil fuels—the so-called *price parity* concept. To achieve price parity, there are two alternatives:

- Either we wait until fossil fuel prices rise gradually due to growing scarcities. This option entails significant environmental degradation while we wait; or
- We fully internalize the externality costs associated with the use of finite fossil fuels by filling the gap between the private and social costs through taxes. Alternatively, we can capture the *shadow benefits* of renewable resources through subsidies and/or mandatory standards.

Raising fossil fuel prices by including externalities, however, are politically unpopular. History helps to assess to what extent a price increase can be sustained by an economic system before political reaction takes over. Quoting an anecdote from the Italian Renaissance, we know that when the Pope imposed a tax on salt in about 1530, the inhabitants of Florence and Perugia reacted by introducing the famous Tuscan bread, which is unsalted. In Florence, which had partial access to the sea, the sacrifice was bearable. But in land-locked Perugia, the population rioted in the streets, which resulted in the Pope's army taking control.[8]

The lesson may be that a modest price increase may be acceptable if the cause is justified. But how much of an externality price, say a carbon tax, may be bearable to promote new investment in renewable energy technologies and/or

7. This may be viewed as a simplistic definition of sustainability, but we think that throwing certain concentration of greenhouse gases into a computer and claiming that the planet will survive below a certain number of degrees of temperature rise and not any more, is equally simplistic.

8. Examples such as this illustrate the limits of demand reduction in response to a price increase. In cases where the commodity in question is an essential and demand is inelastic, as in the case of salt, there are obvious limits to how much prices can be raised before it becomes unacceptable.

support additional investment in energy efficiency? In principle: "The most appropriate response would be to set up a global infrastructure investment program that gives the appropriate market signals to the private sector and levels the playing field for alternative energy technologies" (Banuri, 2007). In practice, we know that there is the menace of all sorts of distorting administrative procedures, opaque bureaucracies, and stranded costs that continue to hang over the new sustainable development process. And these issues would be amplified on a global scale with highly uneven application of the basic principles.

The key question in this context is whether the world at large is in fact ready to adjust the current prices of fossil-based energy resources to include their full externality costs. Such an adjustment is ultimately needed to spur investment in renewable technologies, as well as more efficient energy using capital stock and in eventually changing our lifestyles and personal habits toward a more sustainable future.

Economic theory can determine the appropriate price for this to happen, and markets are flexible and resourceful if given the necessary time to make the needed adjustments. But if the politicians consider the price adjustments to be too high to be politically acceptable, they will be reluctant to impose them.

2.2. New Resources and Support Schemes

The preceding discussion indicates that reasonable relative cost adjustments are needed to support further investment in renewable resources through policies such as the inclusion of externality prices in tandem with the promotion of energy conservation and reduction of waste.

In a perfect world, with full information and no constraints on government tax policy, the strategy to promote new energy resources consists of setting up a Pigouvian tax,[9] a tax levied on usage of fossil fuels relative to the level of greenhouse gases (GHGs) attributed to their use. This will create the necessary incentives to reduce fossil fuel usage and therefore emissions. The resulting tax revenue can be used to support investments in new resources and energy efficient infrastructure. This is the basic idea behind a carbon tax, simple and elegant in theory but unpopular in practice.

In practice, support mechanisms for new energy resources could be either price-oriented[10] or quantity-oriented.[11] Economic theory has already shown

---

9. A Pigouvian tax is a fee paid by the polluter per unit of pollution, and is set to be exactly equal to the aggregate marginal damage caused by the pollution (Kolstad, 2000).

10. With regulatory price-driven strategies, financial support is given through investment subsidies, soft loans, tax credits, fixed feed-in tariff, or a fixed premium which governments or utilities are legally obliged to pay for renewable energy produced by eligible firms—so-called Green Certificates—or a premium for energy savings actions—so-called White Certificates (Meyer, 2002).

11. With regard to regulatory quantity-driven strategies, governments define the desired level of energy generated from renewable resources though schemes such as renewable portfolio standard (RPS), popular in the U.S.

which is better; it depends on the relative variability, or uncertainty, of the expected costs and benefits. If the uncertainties associated with the implementation costs of new technologies are high, the price mechanism would be preferred. If, on the other hand, there is high uncertainty associated with the benefits to be achieved, then quantity regulation is superior (Nordhaus, 2001).

But are such tax schemes effective in the long run? For empirical evidence, we can examine the effect of energy tax policies of European countries, broadly considered as leaders in the developed world in emission reduction efforts and in having energy efficient economic systems compared to, say, those in the U.S. Historical comparison shows that at the end of the nineteenth century, energy consumption per capita was similar in Europe and North-America: 2.21 toe (tonne of oil equivalent) in the U.K.—more or less in line with the Continental Europe—versus 2.45 toe in the U.S. (Maddison, 2003).

A century later, at the beginning of the twenty-first century, per capita energy consumption stood at 3.89 in the U.K. but 8.15 in the U.S. Even allowing for geopolitical variables including the significant military role of the U.S. as a superpower and the defender of freedom in the world, which entails using more energy than the rest of the Western world, today Americans consume more than double what Europeans consume on a per capita basis, and they do not necessarily enjoy a double standard of living relative to Europe. This is not a moral judgment, but the result of numerous policies including the fact that overall, energy prices in Europe are nearly twice as high due to higher taxes, which encourage energy conservation.

As illustrated in Figure 1, differences in taxation policies among the two continents has resulted in retail gasoline prices in the U.S. being roughly half those in Europe (the gray line in Figure 1).[12] Needless to say, given that oil prices are virtually the same globally, this difference is entirely due to differences in taxation. Over the same period, the relative energy consumption doubled in the U.S. relative to Europe (the black line in Figure 1). There is striking evidence of a negative correlation between the two lines from 1870 to 2009: when relative prices moved down, relative consumption per capita went up. In addition, in periods when relative prices were constant, relative consumption was also stable. At the end of the twentieth century, the relative consumption ratio was higher than 2 and the relative price ratio just below 0.5.

These dramatic differences occurred over a period no longer than 30 years, roughly from 1950 to 1980, as shown in Figure 1, during a period of large-scale structural adjustments induced by price differentials. This time scale, incidentally, is comparable with the future horizon of our debate on achieving sustainability, say the timeframe between the present and 2050.

12. The graph compares the U.S. and the U.K., but similar arguments apply to Continental Europe where petrol and energy taxes are high relative to the U.S.

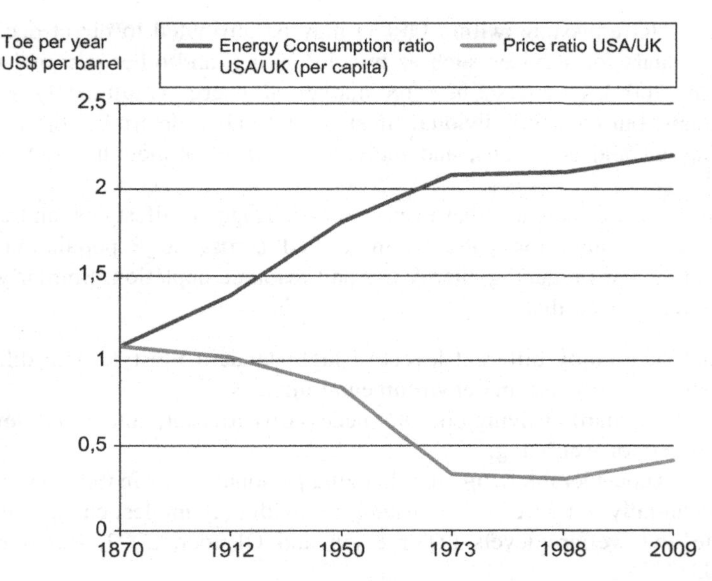

FIGURE 1 US-UK comparison between energy consumption per capita and retail petrol prices (USD per barrel at 1999 prices) during twentieth century. *Sources: IEA (2009) and Maddison (2003). The energy consumption ratio between the US and the UK in 1870, which was 1.10 toe/per capita, is indexed to one; the retail gasoline price ratio in 1870, which was 1.08 USD per barrel, is also indexed to one.*

From our point of view, these figures demonstrate a simple fact: even without a careful analysis of other structural factors, such as population density, income levels and so on, it is possible to induce less energy use by resorting to tax instruments with significant results over a reasonable time period[13] without affecting the standards of living. Stated differently, energy taxes applied consistently over a reasonable period of time can modify the relative prices between fossil fuels—which have considerable externalities—and new renewable energy resources and they can be used as powerful incentives to increase investments to transition to a long-term path of sustainability.

2.3. Lifestyle and Energy Consumption

There is tremendous inertia in our personal habits and lifestyles, and these multiplied by millions of individuals and households explains our massive energy use in aggregate terms, and is among the root causes of the climate change challenge. A recent study, for example, concludes that approximately

13. This contrasts with the book's Chapter 9 by Gray and Zarnikau, which claims that green taxes have modest impact on energy use. The authors take a short-term view and fail to capture the long run structural changes that we have discussed here.

40% of GHG emissions within OECD may be attributed to direct decisions by individuals for services such as travel, heating, and other personal needs (Liverani, 2009). Of course, in a free market, all energy is ultimately used by individuals, but clearly individual lifestyle and habits determine the level of consumption and emissions, and individuals *can* alter their habits to lower both.[14]

The literature indicates that even modest changes in lifestyles, and consequently in consumption patterns, multiplied across large populations can contribute toward reducing energy use and resource depletion. Similarly, it is broadly recognized that:

- People can enjoy different levels of high-standard lifestyle with different levels of energy use and environmental impacts.
- A high standard of living does *not* necessarily translate into a high level of happiness or wellbeing.
- There is considerable scope for changing personal habits in rich countries to substantially reduce energy consumption with only modest changes in lifestyle and welfare levels (Thorgerson and Olander, 2002; Reuss et al., 2003).[15]

Practical examples of lifestyle changes toward sustainability already exist. Jackson and Michaelis (2003), for example, note that, in "... Britain, a variety of movements and networks has developed, in which participants meet in small groups to learn about environmental and social issues, explore lifestyle options, and take collective action. Through mutual support in developing their own culture of consumption, such groups have achieved and sustained significant reductions in household resource use and waste." Referring to a survey conducted in the Netherlands, Biesiot and Noorman (1999) report that, in all income groups, there are households with relatively low energy consumption and related CO_2 emissions. This suggests that there are ample opportunities to reduce energy use across all economic segments and income levels. Adjustments in spending habits in certain key categories—for example, in transport, for vacations, for home insulation, and so on—result in rather drastic changes in primary energy requirements.[16]

In general, the potential for lifestyle change is greater in rich countries where consumption is often disconnected from basic needs and mainly involves

14. For example, see Reusswig, 1994; Lutzenhiser, 1997; and Duchin, 1998.

15. Curran and Sherbinin (2004, p. 118) emphasize that "this has practical importance, since it is often argued that engaging consumers in low-cost activities such as recycling can reduce their incentive to undertake the more costly behavioral changes required for sustainable consumption because they feel that they have already "done enough."

16. As another example, consumption of seasonal fruit, instead of importing fruits out of season, reduces the annual energy impact per household by 15 GJ. Globally, each household can reduce its impact by more than 1500 GJ by simple adjustments without degrading their lifestyle.

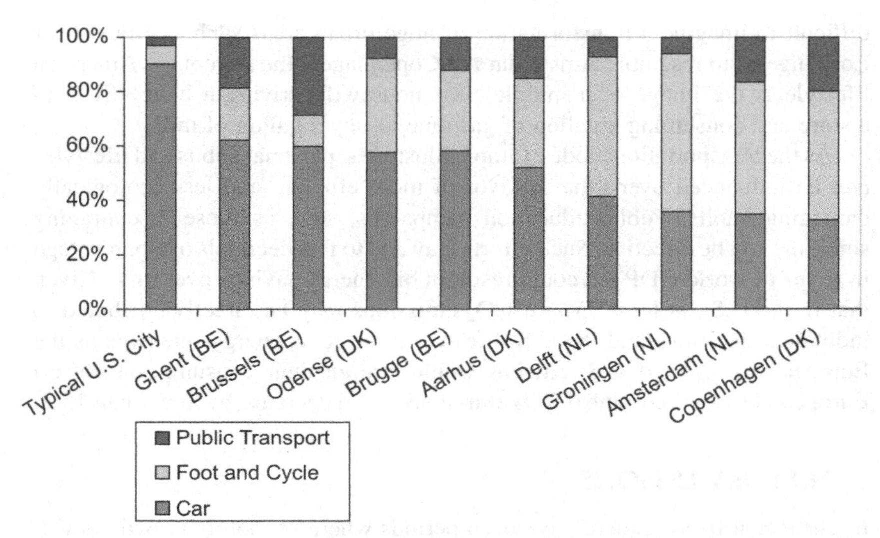

FIGURE 2 Mode split in selected European cities. *Source: Litman (2007)*

positional goods[17] which provide little or no net benefit, because if somebody gains from their consumption, others perceive a symmetric disutility—such as in cases where consumers compete for status (Litman, 2007).

A vivid example of the potential for influencing personal choice with significant energy and environmental impact is the personal car. Consider the composition of different transport modes in selected cities in the developed world (Figure 2). Aside from the unavailability of public transportation options or bicycle lanes, many cultures associate wealth and status with private car ownership,[18] which creates a strong bias toward driving rather than bicycling, walking, or taking public transport, as in the U.S. In bicycle-friendly cities like Amsterdam and Copenhagen, where public transportation systems are reliable and extensive and where social attitudes favor public transport to private cars, citizens can enjoy high levels of mobility and lifestyle with little environmental impact.[19]

The point is that one can *influence* the choice of transport mode through taxation, mass transit, bike lanes, parking fees, and a host of other instruments, including making public transport to be "politically correct." While it would be

---

17. Positional goods play an important role in our current social context. For example, Carlsson et al. (2007, p. 586) conclude that, "People may prefer not only to have a high income and consumption level, but also to have more than others," and "People perceive that they are receiving a disutility if they are surrounded by others who have more than they do."

18. Litman (2007, p. 1) reminds us that in 1986, during a parliamentary debate, the British Prime Minister Margaret Thatcher famously said, "A man who, beyond the age of 26, finds himself on a bus can count himself as a failure."

19. We don't believe that the quality of life in Amsterdam, Copenhagen, or Brussels is lower than in typical U.S. cities.

difficult to imagine a transformation of huge urban areas such as Houston or Los Angeles to resemble Amsterdam or Copenhagen, the stereotype American lifestyle is the image of a middle-class housewife driving a heavy SUV to a store and consuming a gallon of gasoline to buy a gallon of milk.

As the transportation mode example illustrates, personal habits and lifestyles *can* be influenced over time in favor of more efficient and less ecologically damaging habits. Public education campaigns, such as those discouraging smoking, can be effective. Such efforts may add to few decimals of a percentage in terms of world GDP but could result in big energy savings over time. Given that in the U.S., at least 35% of CO_2 emissions may be directly attributed to individual decisions and the U.S. lifestyle is twice as energy-intensive as the European lifestyle, if U.S. citizens would realign their consumption toward European levels, their global CO_2 emissions could decrease by more than 15%.

3. HISTORY LESSONS

In our recent history, there have been periods where economic growth, development, or major social or institutional changes were achieved through the deliberate introduction of policies using the existing socio-economic institutions. Often, these changes required major sacrifices, typically in reduced present personal consumption to finance future economic growth or whatever the goal might have been. Politicians typically led the changes and the public generally supported them because the end justified the means. These efforts usually resulted in lasting changes for the better.

For instance, the Industrial Revolution of the late nineteenth century was certainly a period of decoupling between physical or manual labor and economic growth. The initial sacrifice of society to promote investment in new machinery and equipment resulted in a considerable reduction of hours worked—from 1251 hours worked per capita in the U.K. in 1870 to 657 in 1998—while GDP per man-hour increased from 2.55 to 27.45 in real dollars (Maddison, 2003). Of course, this was accomplished mainly due to large-scale substitution of capital and energy for labor.

The crucial question facing us today is how can we achieve a new industrial revolution, enabling us to decouple future economic growth from increasing energy use and associated GHG emissions? Clearly, this future scenario of a clean and green society must include continued economic growth and a decline of fossil energy use in both absolute and relative terms. The need for acceleration of investment must be flanked by an increase in aggregate savings, which originates from the private sector—households and firms—and the public sector.[20]

20. In macro-economic terms, savings equal investment and it does not matter whether the savings needed to finance investment come from the balance of payments (i.e., from foreign savers), public deficit (i.e., future generations), or the private sector (i.e., current generation).

3.1. The Marshall Plan

The European post-WWII period of reconstruction was one of exceptional development, technological change, and rapid growth. What lay at the root of such economic development? Certainly, one major cause was the Marshall Plan. As described in a World Bank assessment, "The European Recovery Program (the Marshall Plan) after World War II showed how mobilizing resources on a grand scale can build economies and transform enmity into partnership. The architects of the Marshall Plan accepted the challenge of tackling 'hunger, poverty, desperation, and chaos' by rebuilding a continent in the interest of political stability, social development, and a healthy world economy. The Marshall Plan broke a vicious cycle of poverty and regret; it supported economic reconstruction and social order; and it injected money and ideas to rebuild Europe and herald more than 50 years of unprecedented peace, prosperity, and partnership" (World Bank, 2003).

The program, named after the U.S. Secretary of State George Marshall, offered assistance to promote Western Europe's recovery, transferring roughly $13 billion in aid from the United States between 1948 and 1951[21] (Table 1[22]). It played a role in expanding Western European capital stock, financing the reconstruction of infrastructures, and alleviating shortages in resources.

The Marshall Plan did play a key role in "making possible the rapid economic growth of a continent that was devastated by World War II" (Eichengreen, 2007, p. 2), inaugurating the post-war era of economic prosperity and political stability in Western Europe, and it did this mainly in three ways:

- First, it played a role in providing large quantities of aid to the Western countries, to cushion consumption during the years of readjustment and reorganization.
- Second, it helped loosen foreign exchange constraints and improved capacity utilization.
- Third, it encouraged reductions in government spending, relaxation of controls, and the opening up of economies to foreign trade (De Long and Eichengreen, 1991).

However, by most accounts, the greatest success of the Marshall Plan was to accelerate economic growth "by altering the environment in which economic policy was made" and to induce policy shifts which "pushed governments

21. "Between 1948 and 1951 the U.S. transferred $13 billion to the war-torn economies of Europe. (The Administration requested $14.2 billion, Congress authorized $13.4 billion, and $12.5 billion, was ultimately made available. The $14 billion figure frequently cited includes appropriations for economic assistance in Asia, mostly to colonial dependencies of the European participants." (Eichengreen and Uzan, 1992, p. 14).

22. Table 1 does not cover Germany. West Germany in the same period received less aid than Italy, France, or the U.K as % of GDP, but it achieved the greatest growth rate in Europe (Eichengreen and Uzan, 1992, p. 20).

TABLE 1 Marshall Plan Allotments as a Share of GDP*

Country	Marshall Plan ($) 1948-1951*	GDP 1948-1951 ($)	% GDP
Austria	634,000,000	18,905,423,200	3.4%
Belgium	546,000,000	28,684,034,305	1.9%
Denmark	267,000,000	72,027,872,435	0.4%
France	2,576,000,000	64,099,751,615	4.0%
Ireland	146,000,000	2,652,203,773	5.5%
Italy	1,347,000,000	28,730,581,315	4.7%
Netherlands	1,000,000,000	29,603,130,443	3.4%
Norway	241,000,000	53,791,268,949	0.4%
Portugal	50,000,000	2,533,118,906	2.0%
Sweden	118,000,000	101,858,646,527	0.1%
United Kingdom	2,866,000,000	41,202,068,822	7.0%

* *Source: Fauri 2006 and Penn World Table Version 6.3 (Heston et al., 2009)*

toward political and economic orders that used the market to allocate resources and the government to redistribute wealth" (De Long and Eichengreen, 1991, p. 2). These shifts in policy led to the creation of the "mixed economy" that were market- and growth-oriented. Overall, the Marshall Plan can be thought of as a large and highly successful structural adjustment program. It allowed Europe to "… grow rapidly simply by repairing wartime damage, rebuilding its capital stock, and redeploying men drafted into the wartime task of destroying output and productive capacity to the normal peacetime job of creating them" (Eichengreen, 2007). The rapid economic expansion of the early post-war years largely reflected "catch-up growth." Moreover, Europe could sustain its rapid growth by exploiting the backlog of new technologies developed between the two world wars but not yet put to commercial use (ibid).

The success of the Marshall Plan, however, was not limited to its sheer economic size, but rather its political inspirations:

- First, the American generation of the time agreed to sacrifice part of its income to finance capital transfers to another part of the world.
- Second, after the tragedy of the war, Western Europe joined in a common effort to rebuild its economy and society, even though the final results appeared far and uncertain.
- Third, Western European countries laid the foundations for a new organization of mixed public and private institutions, investing in new infrastructure

and technology, improving standards of living including the development of a welfare system, which resulted in self-sufficiency in agricultural production, the eventual creation of the European Community, and industrial policies that created a thriving manufacturing sector while supplying essential services such as clean water, civilian housing, and infrastructure.

3.2. Growth Savings and Investment in Europe

In what sense can we say that the Marshall Plan was an example of a *major stimulus* to development? Immediately after the beginning of the Marshall Plan, in the crucial reconstruction period from the early 1950s to the early 1960s, Europe experienced an acceleration of the private investment ratio to GDP, as shown in Figure 3, where the investment to GDP ratio in the initial year is standardized for major European countries at the level equal to 1. Measuring years along the horizontal axis, we note that after 10 years, investment to GDP ratio was between 1.2 and 1.6 times the initial level. As this increase, on average, was from 15% to 23%–25%, we can state that the acceleration of investment ranged around 8%–10% of GDP.[23]

Starting from the fundamental equation stating that investment must equate the sum of public and private savings plus the balance of the foreign sector, we know that resources available for investment must come either from households and firms, from the public sector through taxation, or the contribution of the balance of payments.[24]

We can estimate that the initial contribution to investment injected into the European economies by the Marshall Plan[25] was around 1% per year in the period 1948–1951. Before the creation of the European Common Market in 1960, the aggregate balance of payments of Europe between 1950 and 1960 was around equilibrium, or at most in surplus around 1%–2% of GDP. In the same period, the public sector was not producing any structural deficit but was in equilibrium. In all, these figures show that foreign and public sector contributions to the investment increase was around 2%–3%.

Thus, given a total investment increase of 8%–10% (and subtracting the 2%–3% contribution calculated above), we conclude that private sector

---

23. In particular, it should be noted that, apart from Greece and other small countries, Italy and the U.K. showed the fastest trend, whereas Norway and Belgium had the slowest. Eichengreen and Ritschl (2009, p. 213) have recently computed private and public savings for Germany in the period 1950–1960. During this period, global savings were on average 16% of GDP and, until 1957, the public component was twice the private one. Only in 1958 the private saving became greater than public saving.

24. We start with the savings-investment identity for an open economy: $S - I = X - M$, where S is savings, I is investment, and X and M are exports and imports of goods and services. This is a consolidated approach (see Chenery and Bruno, 1962; Bacha, 1990; McKinnon, 1964; and Eichengreen and Uzan, 1992).

25. According to some reconstructions of aggregate GDP around 1950, the total Marshall Plan injection was about 3–4% of GDP (1950 values) spread over a period of four years (1948–1951).

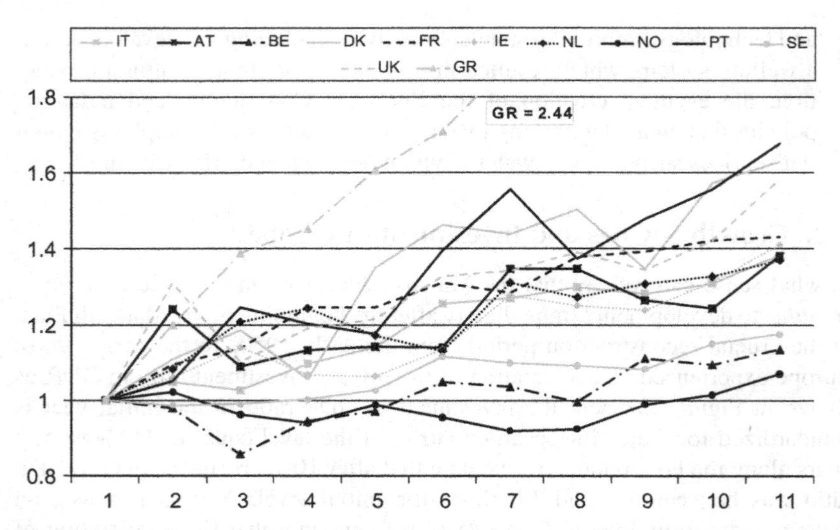

FIGURE 3 Investments to GDP ratio (index number, initial year = 1). *Source: Maddison (2003) and Penn World Table Version 6.3 (Heston et al., 2009)*

contribution to investment was worth roughly 6%–7% of GDP. This implies, given the savings investment identity, that the generation of the 1950s increased the saving rate by 6%–7% of GDP for at least a decade. In other words, European households were able to save that much and lend it to the business sector to build their future development and growth.

Was it worth it? With additional savings of 6%–7% of GDP, which means a sacrifice of current in favor of future consumption from the early 1950s to the early 1960s, the final result was a growth rate of GDP of 5% per year (Table 2); this policy resulted in a doubling of GDP per capita in roughly 17 years—more than double the initial level.

The direct implication of the Marshall Plan was that Europeans were able to rebuild Europe after the devastation of WWII while foregoing more than one-fifth of their current income to future growth. In the process, they managed to build new capital equipment at the rate of one-fourth to almost one-third of GDP per year resulting in new industries, new infrastructure, new consumer goods, and new social institutions that led to today's high living standards and affluent society.

What can we learn from this experience today? We must remember that during that period household lifestyles changed dramatically, entire populations in Europe moved from rural habits, fireplace heating, and water from wells to an urban culture, efficient gas-fired central heating, and drinking-water from taps. More important, Europeans managed to build a new and peaceful union while withstanding considerable threats from the former Soviet Union during the cold war.

There are stark similarities between the efforts and the sacrifices represented by the Marshall Plan and the challenges of sustainability and climate

TABLE 2 Average GDP Growth Rates for Selected European Economies, 1950-1960, in Percent

Countries	1950-1955	1955-1960	1950-1960
Austria	0.06	0.06	0.06
Denmark	0.02	0.05	0.03
Spain	0.07	0.05	0.06
Finland	0.06	0.04	0.05
France	0.04	0.05	0.05
United Kingdom	0.03	0.02	0.03
Greece (1951-1960)	0.04	0.06	0.05
Ireland	0.02	0.00	0.01
Italy	0.06	0.06	0.06
Netherlands	0.04	0.04	0.04
Norway	0.04	0.03	0.03
Portugal	0.05	0.05	0.05
Sweden	0.03	0.03	0.03

Sources: Penn World Table Version 6.3 (Heston et al., 2009).

change today. We are searching for present investment to lower future emissions while building a new climate-friendly society. The question is whether the current generation, globally speaking, is willing to bear the necessary sacrifices required to finance such new investment today, to reap the benefits of a more just, equitable, and sustainable future. Naturally, if we interpret such a sacrifice as "reduction of mileage with a given car fleet," for example, then we have a reduction in energy consumption, which results in a reduction of GDP and therefore a reduction in growth. What would be the future benefit of pursuing such a scheme?

However, this misconception is based on the faulty assumption that the only way to reduce energy consumption is to reduce GDP level.[26] This rejects the possibility that new technological progress can lead to new investment, which enhances energy efficiency. So we can have *more* mileage, with a new car fleet *and* lower emissions. For example, the latest energy-efficient European car models

26. Those who suggest that we must sacrifice 0.5%−1% of global GDP growth to address climate change (e.g., Stern, 2006) are missing the point. Their logic is that decarbonising the energy sector would lead to economic decline.

consume only 3.8 liters of fuel per 100 km, or roughly one gallon for every 60 miles traveled. Experts believe that these numbers can be significantly improved.

3.3. The Need for Sacrifice

The evidence suggests that Western market economies have been able in the past to undertake a sustained, long-lasting burden of resource shift from current to future consumption. The question, therefore, is why are we not able to replicate the miracle growth of the 1950s in Europe by shifting investment toward a new sustainable paradigm as defined earlier in this chapter?

In the case of Marshall Plan, the causal chain was from private savings—funding from American taxpayers to private and public investment in Europe—which resulted in both private and public returns. Private sector profit was primarily driven by extraordinary growth in aggregate demand and productivity and, in addition, there were public returns in terms of political stability, social development, social order, and lasting peace. The combination of private and social returns is what has made European reconstruction an extraordinary event in history.

What is essentially needed today is an initial subsidy to spur investment which can lead to a combination of social and private benefits. But a number of formidable barriers, including the existence of externalities, discourage private investors from undertaking the necessary investments. The surprising paradox is that there are considerable positive benefits accruable to all sectors of the economy in terms of better environmental quality, long-term sustainability, and economic development, not to mention avoiding some of the negative consequences of climate change.

The parallel to the Marshall Plan is to find an initial global subsidy to mobilize forced savings necessary to invest in the new sustainable development paradigm. In this case, public intervention is justified to fully internalize the existing externalities because present market mechanisms alone cannot achieve efficient resource allocation. From an economic perspective, a mechanism is needed to channel additional private investments toward a mix of renewable resources, energy efficiency, and energy savings where the aggregate return is higher than the initial investment.

The problem is that such a plan must be undertaken on a global scale to be effective. Cleaning up one city, state, country, or continent does not address the whole problem. As Nordhaus (2009) stresses, there is a need for coordination to implement such an ambitious scheme worldwide, but cooperation is difficult to achieve and requires time.[27] Many countries, governments, politicians, and stakeholders must be convinced if opportunistic behavior is to be avoided.

27. The construction of the Monetary Union in Europe took more than a decade. The International Monetary System of Bretton Woods collapsed in 1971, and we are still in search of a new, efficient, world payment system.

The order of magnitude of the level of investment required for a coordinated global effort will be discussed in the next section. It suffices to point out that the additional cost imposed by nonparticipants on those who participate increases in a nonlinear fashion. This means that an agreement among a limited group of participants, excluding—or rather excusing—others, would impose an astronomical cost on the participants. This is one reason why the full participation of all countries, or nearly all, is a necessary condition for success.

During the cold war confrontation between the U.S. and the former U.S.S.R., both sides quickly concluded that the only possible outcome of a nuclear war would have been mutually assured destruction and this forced both parties toward self-restraint. Today, the unbearable cost of a unilateral "green solution" in some parts of the globe and the certainty of the "assured destruction" of global natural resources if all the countries in the world continue with business-as-usual, leads us to conclude that a coordinated political solution will eventually be found.[28]

4. THE NEW SUSTAINABLE PARADIGM

This section provides an order of magnitude assessment of the investment required to reach a new sustainable paradigm including a discussion of whether such an investment is feasible in the context of current economic conditions and the reasons behind the failure to reach global consensus in Copenhagen.

4.1. What Level of Investment?

In trying to provide an order of magnitude assessment of the investment level required to reach a new sustainable paradigm, previously defined, we examine several prior studies[29] while assuming a high level of international cooperation and global commitment.[30] We start with three main questions:

- First, what is the desired mix of fuels, that is, the share of renewable resources and its composition in a new sustainable paradigm. This is an important determinant of the required investment costs.
- Second, how much energy saving is envisaged for the new sustainable paradigm, who will be responsible for achieving it, and who will be paying for it,

---

28. In *Fast forward: Ethics & politics in the age of global warming* (2010), authors W. Antholis and S. Talbott reach a somewhat different conclusion, pointing out that the threat of nuclear annihilation is starkly different than climate change, which is invisible and gradual and, hence, can be easily ignored.

29. The European Union, the IEA, the United Nations, G7, the U.S. and China.

30. According to the World Bank, "The many global challenges are deeply linked—to each other and to local concerns. So are their solutions. Managing global spillovers, both environmental and social, and taking advantage of a window of opportunity over the next 20–50 years, will require a big push by global institutions and by national and local institutions." (World Bank, 2003).

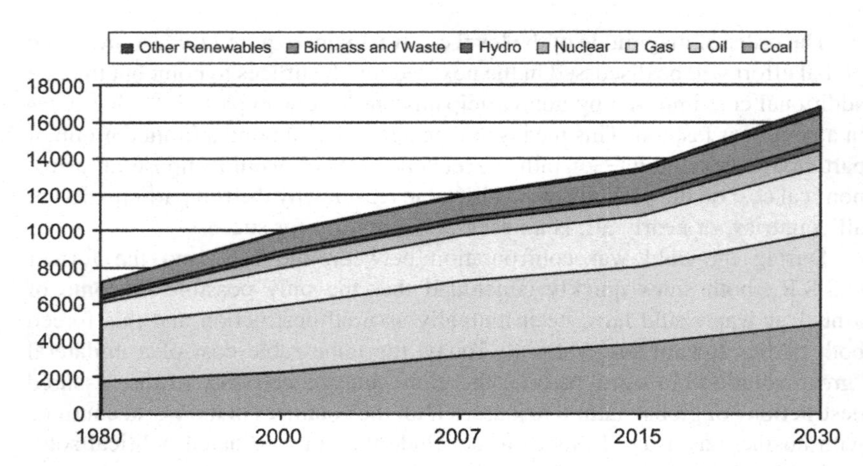

FIGURE 4 World primary energy demand by type of fuel in BAU scenario, in million tons of oil equivalent (Mtoe). *Source: World Energy Outlook, 2009 (IEA)*

including the investment cost for energy-saving efforts and the reduced profits of existing producers?
- Third, what is the required increase in technical efficiency, what is the intensity of technological progress, and how much would it cost?[31]

We build our strategy comparing a business-as-usual (BAU) scenario with a scenario similar to the one described in the International Energy Agency's (IEA's) "450 parts per million (ppm) scenario" (IEA, 2009), which is sufficiently detailed to address all three questions.

The BAU scenario describes a future in which energy markets evolve worldwide without governmental intervention. In the absence of changes to existing energy policies, fossil fuels remain the main source of energy, resulting in growing energy-related CO_2 emissions at similar growth rates until 2050 (IEA, 2009), with serious consequences for the world's climate as documented in other studies (e.g., Stern, 2006). Figure 4 shows the share of primary energy sources in the IEA's BAU scenario.

The IEA target scenario consists of three main components that include new renewable resources, new energy efficient technologies, and behavioral changes and new lifestyles. Broadly speaking, the first component includes increased reliance on hydro, wind, biomass, tidal, geothermal, solar energy, and so on.[32] Energy from renewable resources is capital-intensive and presently more costly than conventional fossil fuel technologies, but there are reasons to expect these

31. For example, refer to P-L Koskimiaki (European Commission Directorate General for Energy and Transport), *Energy savings indicators for policy development in EU*, IEA Energy Efficiency Indicators Workshop, Paris, 21 January 2009.

32. This book's Chapter 13 by Arent et al. describes the prospects for renewable energy technologies.

prices to decline over time for a number of reasons including growing economies of scale and the effects of *learning by doing*. McDonald and Schrattenholzer (2001) show that the estimated learning rates are 1.4% for hydroelectric power plant, 18% for wind power electricity, and 22% for solar PV panels.[33]

The second component in the mix of new resources includes investment in new technology, mainly on the supply side including new fuels such as hydrogen, carbon capture and storage technology, nuclear power, and suchlike.[34]

The third component includes changes in the demand side including behavioral and lifestyle aspects of energy consumption such as switching off lights in empty rooms, adjusting thermostat settings, buying more efficient appliances, cycling to work, investing in home insulation, and so on. Numerous studies have demonstrated that individual behavior can be modified by emphasizing that "climate change is anthropogenic—the product of billions of acts of daily consumption" (Liverani, 2009). The significance of the third component is well-known in OECD countries. Gardner and Stern (2008, p. 3), for example, estimate that U.S. households directly account for 35% of national CO_2 emissions. Other studies have documented the role of new technology. According to Liverani (2009, p.1), "... if adopted, existing efficiency measures[35] for households and motor vehicles can allow energy savings of almost 30 percent, 11 percent of total U.S. consumption."

To estimate the cost of investment required to achieve the target scenario and assess whether it is feasible and sustainable, we start using the same level of resource requirement envisaged in the IEA scenario to stabilize GHG concentrations in the atmosphere at 450 ppm CO_2-eq.[36] The projected global energy investment is estimated to be around $25.6 trillion in 2008 dollars, or roughly 1.4% of global GDP on average between now and 2030[37] (IEA, 2009).

The feasibility of the target scenario is conditional on the willingness of national and international institutions, as well as citizens and businesses, to fundamentally change the way energy is produced and consumed. The relative importance of the three components—renewable, efficiency, and lifestyle—required to reach the target is depicted in Figure 5.

---

33. The authors summarize 26 data sets and consider technological improvement in various countries.

34. A sister volume, *Electricity generation in a carbon constrained world*, edited by F. Sioshansi (2009) covers many low-carbon supply-side options.

35. The role of new technology also arises when we compare the energy saved by curtailment versus that saved by increased efficiency; the latter reduces more carbon emission and saves more energy (Gardner and Stern, 2008, pp. 4–5).

36. Most studies of reductions in CO_2 emissions agree that, to achieve a 450 ppm scenario, world GHG emissions should peak in 2020 at 30.9 Gt and decrease to 15 Gt by 2050.

37. The share is lower in OECD countries (about 0.8% of GDP) and is highest in India, Africa, the Middle East, and Russia (about 3%) (World Energy Outlook, 2009, IEA). Stern (2006) and others have arrived at roughly similar estimates.

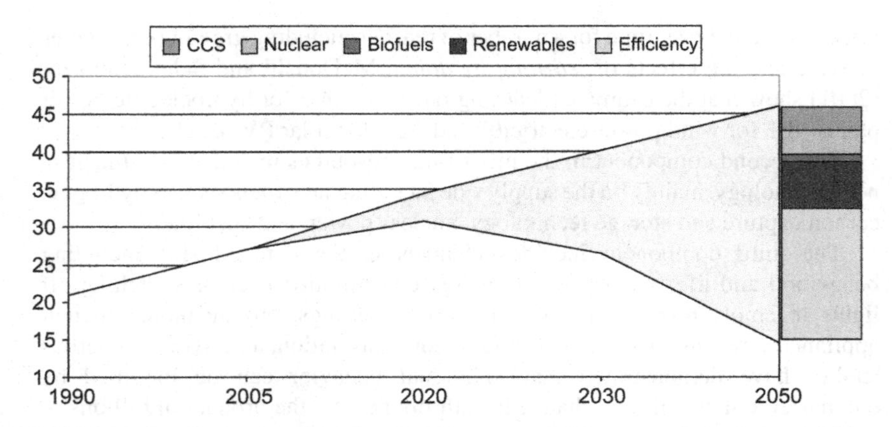

FIGURE 5 Contribution of various abatement measures to 450 ppm scenario in gigatons of CO_2 equivalent (Gt CO_2-eq) *Source: World Energy Outlook, 2009 (IEA)*

While estimating the investment costs is important, IEA and others who have conducted similar studies consistently fail to address the more critical question: How are we going to achieve this level of commitment of financial resources? To answer this question, one must clarify who should pay for what. In the case of IEA cost projections, we believe that the emission abatement cost is underestimated for the following reason. The required household behavioral changes toward energy efficiency and energy savings entail a profound change in consumer lifestyle and preferences, which involves extra costs—both in terms of explicit adjustment costs, and indirect costs such as perceived happiness, psychological costs, and so on.

These *extra* costs, we believe, are not adequately reflected in the IEA's analysis and this aspect cannot be neglected, because people are at the basis of our democratic political system. If consumers—voters—are not convinced, there is no way to achieve political consensus because, in our democratic societies, policy decisions must follow voters' sentiments.[38] To mobilize consensus, education, information, and communication must be applied to convince the masses to change their lifestyle. To reflect these additional costs, we assume that an additional effort equivalent to roughly 5% of the education budget in developed countries will be required. Since the cost of education is approximately 6% of GDP in developing countries, the additional investment may be of the order of 0.3%–0.35% of GDP. Thus, the required level of investment rises from 1.4% to about 1.75% of global GDP, or around $32 trillion on an annual basis through 2030.

38. This book's Chapter 16 by Stulz et al., for example, refers to a referendum in Switzerland where 75% of voters adopted a proposal to reduce their energy and carbon footprint. This citizens' mandate is now reflected in the municipalities' constitution.

TABLE 3 "Target scenario"- Investment Required by Sectors (Sources)

Year	Businesses	Households	Government organizations	Total
2020	41%	40%	19%	100%
2030	49%	37%	14%	100%
2030	$ 15.68 trillion	$ 11.84 trillion	$ 4.48 trillion	$ 32 trillion

Source: Our own calculations based on our scenario and World Energy Outlook, 2009 (IEA)

In practice, investment decisions are made by businesses, households, and governmental organizations. The relative share of each sector may depend on the economic structure of each country, but we can approximately calculate it as follows: 80% must come from private sources (businesses and households) and 20% from public sources (Table 3). Based on this assumption, businesses will bear 49% of total investment in 2030, which amounts to about $15.68 trillion, households 37% or $11.84 trillion, with the remaining 14% or $4.48 trillion coming from the governments.

Table 3 provides a breakdown of the shares of total investment required by the year 2030 by sector. We have constructed Table 4 to be consistent with these assumptions so that its last row has the same figures as the last row of Table 3 for 2030 and its last column is consistent with the investment proportions of Figure 5. In this way, the distribution of the costs to various sectors are represented even though in the final analysis, all costs eventually accrue to individuals through prices and taxes.

In this construct, businesses will collectively bear the lion's share of the financing challenge, roughly burdened by 49% of the total costs. Their efforts

TABLE 4 Investment Required by Targets and Sectors – Year 2030

	Private industrial investment	Household actions	Government intervention	Total share
Energy efficiency	45%		3%	48%
Renewable resources	2%	23%	10%	35%
Energy savings and lifestyle	2%	14%	1%	17%
Total	**49%**	**37%**	**14%**	**100%**

Source: our own calculations based on data of Table 3 and Figure 5 Note: the percentages refer to the total investment required for the target scenario (US $ 32 trillion).

will play a crucial role in developing and investing in energy efficiency technologies, more efficient processes in industry, the development of cleaner transportation options, investment in renewable power plants, purchase of low-carbon commercial vehicles, and myriad other investments. The financing for such business investments will come from many sources, including private cash reserves, bank loans, foreign capital investment, debt issue, and governmental support. Government policies affect corporate investment decisions by means of subsidies favoring lower-emission technologies and penalties for emissions through a carbon tax or cap-and-trade system. These additional costs will ultimately be passed on to consumers in higher prices.

Households are assumed to finance around 37% of the additional investment required. Their effort has two main targets: 23% devoted to new investment in renewable resources and 14% to energy-using consumer goods, such as household appliances and vehicles. In the building sector, energy-efficiency costs are already incorporated into the initial costs of buying the buildings. People's motivation can be activated through incentives, educational and information campaigns, and regulatory regimes implemented by governments.

The government's 14% share will be concentrated on supporting schemes for renewable resources, such as feed-in tariff schemes. Governments play an important function by *indirectly* influencing investment decisions across all sectors of the economy and *directly* through investments in infrastructure including public buildings, the transport sector, nationalized power grids, and other energy-related infrastructures. Moreover, governments typically play an active role in support of research and development and can engage in educational campaigns to sensitize citizens toward achieving environmental and sustainability goals.

The preceding scenario is in line with similar studies conducted by others and is plausible in terms of quantity of resources required and the allocation of investments among different sectors and policies to be implemented. Politicians in democratic societies are aware that there is a need for policy credibility and to tie sustainability to job creation and longer-term economic benefits, as shown by the language used by the European Union (EU): "The opportunities offered by the transition are wide-ranging, given that the eco-industry already accounts, in total, for some 3.4 million jobs in Europe offering particular growth potential.[39]

39. In Europe, renewable energy technologies already account for a turnover of €20 billion and have created 300,000 jobs. A 20% share for renewables is estimated to mean almost a million jobs in this industry by 2020—more if Europe exploits its full potential to be a world leader in this field. In addition, the renewable energy sector is labor-intensive and reliant on many small- and medium-sized enterprises, spreading jobs and development to every corner of Europe. The same is true of energy efficiency in buildings and products. In this context, many view the transition to a low-carbon economy as an opportunity for Europe, rather than a cost (European Union, 2008).

4.2. Political Feasibility

As outlined in Table 4, the required investment to reach a new sustainable paradigm is to be split among businesses, private households, and governments. For the scheme to work, each sector must take necessary steps to finance its share. In theory, and on paper, this appears trivial.

In practice, it is anything but trivial. What is the feasibility of adopting a grand scheme such as the one outlined in the preceding discussion given the current geopolitical and economic realities?[40] Would such an unprecedented level of investment, amounting to 1.75% of global GDP, be acceptable to the public, the business community, and the politicians? Would it be realistic? And most important, how would it be financed?

Four critical questions must be addressed in this context:

- How are we going to achieve this level of funding?
- Would this be realistic?
- Is it unprecedented?
- Is it necessary?

The answer to the first crucial question is simple: governments must act first, imposing the required level of taxation, of the order of 0.65% of GDP.[41] This will spur consistent private behavior with additional resource mobilization equivalent to 1.1% of GDP, to reach the total target of 1.75% of GDP.

Is this realistic? We note that this level of additional taxation is in line with present levels of government intervention in many parts of the world. In the European context, for example, it represents roughly one-third of actual total energy tax revenue—hence it would not be exceptional.

Is this unprecedented? To put it in context, we refer to the Marshall Plan where American taxpayers' money, amounting to roughly 1% of European GDP, resulted in increased private investment of the order of 6%–7% of GDP. This suggests that the required investment is not out of line with historical precedents—but, of course, much bigger in absolute terms since we are dealing with a global issue.

Is it necessary? According to the scientific consensus, it is absolutely necessary to act and to act decisively, which suggests that governments must make the necessary commitments. This is crucial because, as noted in Section 3.3 "The Need for Sacrifice," public benefits associated with a new

40. We also assume long-term structural invariance of balance of payments and public savings worldwide, which means that the basic rules of fair trade and competition throughout the world still apply and that the welfare and healthcare are still financed through public policies.

41. Based on Table 4 figures, we add the 23% needed for renewable resources by household action (additional taxes) to 14% accrued to government intervention to calculate the required total increase in taxation (public savings) equal to 37% of 1.75% of GDP, or 0.65% of GDP.

a sustainable paradigm cannot be internalized by private investors alone. There must be public intervention to spur private investment.

A final question may be how long will it take for a global consensus to emerge, which is briefly addressed in the following section. The short answer is that it may take a while, as exemplified by the U.S. healthcare reform passed in 2010. It took a very long time for this legislation to be enacted. Why should we *not* be optimistic for a similar global resolve on the issue of climate change and sustainability?

4.3. A Post-Mortem of the Copenhagen Conference

At this point, it may be appropriate to ask why bother with all the reasoning and arguments after the inconclusive results of the United Nations' (UN's) sponsored conference held in Copenhagen in December 2009? We believe that our quantitative assessment explains why Copenhagen failed.

We estimate that additional investments of the order of 1.75% of world GDP is required for a prolonged period to achieve a new sustainable paradigm. Participants in Copenhagen agreed to commit to a level of financing roughly equal to 5%–10% of what is considered necessary and for a much shorter period to assist the developing countries to address the climate challenge. This turned out to be off the mark on at least three levels:

- First, the vague financial commitments made by the rich were a fractional down-payment of the actual amount needed. With insufficient public commitment to internalize the difference between private and social benefits, private investment will *not* flow.
- Second, for any commitment to be credible it must apply to both donors as well as the recipients. Treating such an investment as foreign aid is inappropriate since the underlying message was to ask developing countries to reduce the growth rate of their GHG emissions without commensurate commitments by the donor countries that they were going to reduce their own.
- Third, as illustrated in the case of the Marshall Plan earlier in this chapter, short-term injection of resources, while absolutely *necessary*, is not *sufficient* to spur continued growth and sustained investment. By pledging vague commitments to provide financial resources without a strategic plan or a vision to alter the underlying business environment or economic policy, the donor countries lacked credibility.

While this is not intended as a critique of the protracted UN process exemplified by the failure of Copenhagen, it is clear that a new and transparent political commitment, shared by all, is ultimately needed. The sustainability of the world's future cannot be based solely on continual trickles of foreign aid from the rich to the poor countries. In particular, the developed countries have at least three good reasons to support a new political commitment toward a more sustainable future:

- First, making investment to achieve future sustainability targets provides an opportunity to emerge from the current world financial crisis.
- Second, the U.S. has an opportunity to reclaim its technological and moral leadership while maintaining global supremacy in view of the serious challenges from the lower cost emerging economies.
- Third, for the EU, climate challenge offers a unique opportunity to promote not just European culture and products but a more sustainable and healthy lifestyle.

CONCLUSIONS

This chapter concludes on a positive note. Its main message is that we *can* indeed have our cake and eat it too: Sustainability with high standards of living can be achieved provided we are prepared to make an investment of the order of 1.75% of world GDP or, approximately, double the amount for the richest half of the world. This is comparable with the extraordinary effort of the post-WWII reconstruction of Europe.

The required level of investment—while unprecedented in *absolute* terms—turns out to be smaller than the historical examples in *relative* terms. Moreover, we can rely on *existing* technical, legal, regulatory, and—most important—financial and economic systems to make the necessary transition to a sustainable future. The transition will not be painless or cheap and it will take time, but it appears feasible. It does not require a radical change in our socio-economic system, and it does not require a lowering of living standards, nor unacceptable changes in lifestyles.

The chapter's title asks if there will be the means and the will. We believe that the answer to both questions can be given in the positive, with the obvious caveats. Regarding the means, we argue that existing policy instruments are adequate and fall within acceptable ranges, provided that there is the political will.

But what about the will? Our opinion is that politicians can learn from past mistakes—and successes. Contrast the tragic imposition of costly war reparations imposed on Germany at the Treaty of Versailles following the end of WWI to the generosity of Marshall Plan after WWII.

There was no generosity in Versailles: "The Americans had made it clear in the fall of 1918 that they would insist on repayment of the loans they had made to European Allies during the war. They had also decided not to commit Treasury Funds to European Economic restoration. Hence, the British and French came to rely increasingly on German reparation payments to put their countries back on a solid financial footing." (Cipriano Venzon, 1995).

As history shows, the harshness of the repayment obligations imposed on Germany led to bitter resentment and ultimately to WWII. This was not unforeseen. As early as 1919, Keynes predicted that the Versailles Treaty was unjust, not enforceable, and warned about the danger of German economic depression, hyperinflation, and the collapse of European living standards—factors which eventually contributed to WWII (Keynes, 1971, ch. 1).

Contrast this with the generosity—and the vision—of the Marshall Plan after WWII:[42] "This program will cost our country billion of dollars. It will impose a burden on the American taxpayer. It will require sacrifices today in order that we may enjoy security and peace tomorrow. To be quite clear, this unprecedented endeavor of the New World to help the Old is neither sure nor easy."

The question is how did the political will for the Marshall Plan emerge? The answer, in retrospect, is that we learned from the failure of the Treaty of Versailles. In this context, can we learn from the failures of Kyoto and Copenhagen to reach a consensus in the next round of negotiations? Perhaps more important, what are the parallels between the Marshall Plan—to save Western Europe and confront the former Soviet Union—and the threat implied by climate change and lack of long-term sustainability?

We know that history does not grant second chances. If our present politicians do not perceive the threat of climate change and respond accordingly, they will simply not be part of history.

BIBLIOGRAPHY

Acemoglu, D., Aghion P., Bursztyn L., & Hemous D. (2009). The environment and directed technical change. NBER Working Paper No. 15451.

Ardente, F., Beccali, M., Cellura, M., & Lo Brano, V. (2006). Energy performances and life cycle assessment of an Italian wind farm. *Renewable & Sustainable Energy Reviews, 12*(1), 200–217.

Bacha, E. L. (1990). A three gap model of foreign transfers and GDP growth in developing countries. *Journal of Development Economics, 32*(2), 279–296.

Banuri T. (2007). Sustainable Development Agenda for Climate Change, United Nations; Development Policy and Analysis Division, www.un.org/esa/policy/devplan/egm_climatechange/banuri.pdf.

Biesiot, W., & Noorman, K. J. (1999). Energy requirements of household consumption: A case study of The Netherlands. *Ecological Economics, 28*(3), 367–383.

Carlsson, F., Johansson-Stenman, O., & Martinsson, P. (2007). Do You Enjoy Having More than Others? Survey Evidence of Positional Goods. *Economica, 74*(296), 586–598.

Chenery, H. B., & Bruno, M. (1962). Development alternatives in an open economy: The case of Israel. *The Economic Journal, 72*(285), 79–103.

Cipriano Venzon, A. (Ed.), (1995). *United States in the First World War.* New York: Garland Publishing.

42. George Marshall hearings, Senate Foreign Relation Committee, Jan 8, 1948.

Curran, S. R., & Sherbinin, A. (2004). Completing the picture: The challenges of bringing "consumption" into the population–environment equation. *Population & Environment, 26*(2), 107–131.

De Long, B. J., & Eichengreen, B. (1991). The Marshall Plan as a structural adjustment program. *Harvard Institute of Economic Research Discussion Papers Series, N., 1576.* (November).

Duchin, F. (1998). *Structural economics. Measuring change in technology, lifestyles, and the environment.* Island Press.

Eichengreen, B. (2007). *Europe's postwar recovery.* Cambridge: Cambridge University Press.

Eichengreen, B., & Ritschl, A. (2009). Understanding West German economic growth in the 1950s. *Cliometrica, 3,* 191–219, DOI 10.1007/s11698-008-0035-7.

Eichengreen, B., & Uzan, M. (1992). The Marshall Plan: Economic effects and implications for Eastern Europe and the former USSR. *Economic Policy, 14*(4), 4–75.

European Union. (2008). *20 20 by 2020 – Europe's climate change opportunity.* Communication from the Commission to the European Parliament, Brussels, 23.1.2008, COM (2008) 30 final.

Fauri, F. (2006). *L'integrazione economica europea 1947–2006.* Bologna, Il Mulino.

Gardner G.T., & Stern, P.C. (2008). The short list: The most effective actions U.S. households can take to curb climate change. *Environment Magazine.*

Heston, A., Summers, R., & Aten, B. (2009). *Penn world table version 6.3. Center for International Comparisons of Production.* University of Pennsylvania: Income and Prices at the University of Pennsylvania. August.

International Energy Agency (IEA). 2005. *International Energy Outlook.*

International Energy Agency (IEA). 2009. *International Energy Outlook.*

Jackson, T. (2005). Live better by consuming less? Is there a "double dividend" in sustainable consumption? *Journal of Industrial Ecology, 9*(1–2), 19–36.

Jackson, T., Michaelis, L. (2003). *Policies for sustainable consumption.* A report to the Sustainable Development Commission. *www.sd-commission.org.uk/publications.php?id=138*

Keynes, J. M. (1971). *The economic consequences of peace.* New York: Macmillan, St Martin's Press.

Kolstad, C. D. (2000). *Environmental Economics.* Oxford: Oxford University Press.

Labandeira, X., & Martín-Moreno, J. M. (2009). Climate change policies after 2012. *The Energy Journal, 30*(Special Issue 2), 1–5.

Litman, T. (2007). *Mobility as a positional good – implications for transport policy and planning.* Victoria Transport Policy Institute. *www.vtpi.org*

Liverani, A. (2009). Climate change and individual behavior – considerations for policy. *Policy Research Working Paper 5058.* The World Bank.

Lutzenhiser, L., et al. (1997). Social structure, culture, and technology: Modeling the driving forces of household energy consumption. In Stern. (Ed.), *Consumption and the environment: The human causes,* (pp. 77–91). National Research Council.

Maddison, A. (2003). Growth accounts, technological change, and the role of energy in western growth. In Datini, F. (Ed.), *Fondazione Istituto Internazionale di Storia Economica, Economia e Energia,* XIII–XVIII. Florence: Le Monnier.

McDonald, A., & Schrattenholzer, L. (2001). Learning rates for energy technologies. *Energy Policy, 29*(4), 255–261.

McKinnon, R. I. (1964). Foreign exchange constraints in economic development and efficient aid allocation. *The Economic Journal, 76*(301), 170–171.

Meyer, N. I. (2002). European schemes for promoting renewables in liberalized markets. *Energy Policy, 31*(7), 665–676.

Nordhaus, W. (2001). *After Kyoto: Alternative mechanisms to control global warming*. Atlanta: AEA Meeting.

Nordhaus, W. (2009). The impact of treaty nonparticipation on the costs of slowing global warming. *The Energy Journal, 30*(Special Issue 2), 39–51.

Reuss, F., Lotze-Campen, H., & Gerlinger, K. (2003). *Changing global lifestyle and consumption patterns: The case of energy and food*. Paper presented at the PERN Workshop on Population, Consumption and the Environment, 19 October 2003, Montreal, Canada. (Available at http://www.populationenvironmentresearch.org/workshops.jsp).

Reusswig, F. (1994). Lifestyles and ecology. The differentiated ecology of modern societies—with special regard to the energy sector. IKO. Frankfurt/M. *www.populationenvironmentresearch.org/papers/Lotze-Campen_Reusswig_Paper.pdf*

Stern, N. (2006). *Stern review: The economics of climate change*. London: H.M. Treasury.

Thorgerson, J., & Olander, F. (2002). Human values and the emergence of a sustainable consumption pattern: A panel study. *Journal of Economic Psychology, 23*, 605–630.

World Bank. (2003). *World Development Report 2003*. Washington: World Bank.

World Wildlife Fund. (2008). *WWF Living Planet Report 2008*. Gland, Switzerland.

Chapter 7

Is It Possible to Have It Both Ways?

Alan Moran
Institute of Public Affairs, Melbourne, Australia

Chapter Outline

1. INTRODUCTION

Most politicians across the world recognize that measures to reduce carbon dioxide emissions impose costs on their economies. The costs are incurred whether the approach used is a carbon tax, cap-and-trade, or more targeted regulation.

Almost all political leaders have said they accept that should anthropogenic emissions continue to rise, there would be serious, perhaps catastrophic losses to the global economy[1] and to ecologically valued features. Even so, whatever their views of the costs of inactivity, politicians at Copenhagen (December 2009) and Cancun (December 2010) failed to agree on meaningful measures to reduce emissions.

Some leaders may have hoped for an agreement whereby costs would be incurred on their nations' behalf by others. But the most likely explanation for the failure is that political leaders could not accept the costs involved in taking

1. Note however the 2007 Synthesis Report at 5.7 says, "For increases in global average temperature of less than 1 to 3°C above 1980–1999 levels, some impacts are projected to produce market benefits in some places and sectors while, at the same time, imposing costs in other places and sectors. By 2050, global mean losses could be 1 to 5% of GDP for 4°C of warming, but regional losses could be substantially higher." Table SPM 7 puts the costs of stabilization at 535–590 ppm at 1.3% of GDP in 2050.

Energy, Sustainability and the Environment.

defrayment actions or that they consider the benefits from them to be unachievable or inadequate.

Salvaged from the December 2009 Copenhagen Conference and the dozen or so preparatory conferences leading up to it, was a "commitment" to maintain global temperatures below a 2°C increase. The Copenhagen Accord said[2]:

To achieve the ultimate objective of the Convention to stabilize greenhouse gas concentration in the atmosphere at a level that would prevent dangerous anthropogenic interference with the climate system, we shall, recognizing the scientific view that the increase in global temperature should be below 2 degrees Celsius, on the basis of equity and in the context of sustainable development, enhance our long-term cooperative action to combat climate change.

There was no detail of what this might entail in terms of CO_2-e levels globally. Nor were there concrete estimates provided of the measures necessary to effect this by individual countries or country groups. The Accord confined itself to noting that

"deep cuts in global emissions are required according to science."

In line with the Accord, developed countries submitted "quantified economy-wide emissions targets for 2020" by January 31, 2010. Those lodged by major countries are shown in Table 1.

TABLE 1 Quantified Emission Reduction Targets for 2020 from Selected Countries, Percent

Country	Base year	Unconditional commitment	Conditional target
Australia	(2000)	5	15-25
Canada	(2005)	-	17 aligned with US
EU	(1990)	-	20/30
Japan	(1990)	-	25
New Zealand	(1990)	-	10-20
Norway	(1990)	-	40
Russian Federation	(1990)	-	15-25
USA	(2005)	-	17

Source: UNFCCC http://unfccc.int/home/items/5264.php.

2. Decision -/CP.15. The Conference of the Parties takes note of the Copenhagen Accord of 18 December 2009.

The only country registering an unconditional reduction commitment was Australia, and that was just 5%. It was also made in spite of measures the Government considers essential to reach such an abatement target having been rejected by the Australian Parliament. In April 2010, the Australian Government announced a deferral of its emission reduction tax, previously planned to commence in 2010, until at least 2013.

To achieve stabilization of CO_2 and its equivalent would require annual emission reductions for the world as a whole in the region of 60%–80% by 2050—more in per-capita terms. The IPCC reduction scenarios infer global emission levels at 2.2 tonnes per capita[3] to limit temperature rises to 2°C, reductions that would cut North American emissions to 12% of the current year's level and require a halving of China's present (and rapidly growing) emissions. As indicated in Table 1, conditional reductions by developed countries offered for the coming decade are in the 15%–40% range. Developing countries have offered reductions that would, at most, shave their expected business-as-usual emission levels.

For economies seeking to reach living standards comparable to those in OECD nations, radical emission reductions present a challenge that may be impossible to reach. The challenge is hardly less daunting for the developed nations, even those already experiencing lower emission levels, since in many cases these levels have been reached by relying on nuclear and other low-emitting sources, and further replacement by these of other fuel sources is likely to be more difficult.

Of course, if people were to adopt radical changes in their lifestyles, major emission reductions are conceivable. Some contributors to this book, for example Trainer (Chapter 5), are confident that lifestyle changes, changes in behavior, significant investments in energy efficiency, and increasing penetration of noncarbon energy resources will meet the challenge of climate change.

This chapter addresses current annual global emission levels and those required to stabilize atmospheric concentrations. It explores the sources of emissions and the possibilities of abatement across different sectors and uses, and the costs that might be entailed in achieving abatement reductions, culminating in the chapter's main conclusions.

2. CURRENT GLOBAL EMISSION LEVELS

In agreeing to the Copenhagen Accord, political leaders declared that aggregate annual levels of CO_2-e emissions in the years to come must be reduced from current levels that are said to be dangerously high and are rising. The energy component of these is responsible for over three quarters of the total, and globally stood at a level of 4.54 tonnes per capita in 2008.

Australian emissions were higher than those of other large- and medium-sized economies at 21 tonnes per capita, though among smaller economies, Singapore at 35 tonnes, UAR at 43 tonnes, and Qatar at 74 tonnes were higher.

3. *http://co2now.org/index.php?option=com_content&task=view&id=70&Itemid=51.*

U.S. emissions were 19 tonnes and Canada's 17. The EU's emissions stood at 8 tonnes, Russia's 12 tonnes, and Japan's 9.5 tonnes.

Among the most rapidly growing developing countries, emissions were very low in India at 1.3 tonnes per capita, but already exceeded the world average in China at 4.9 tonnes. Africa had just 1.1 tonnes per capita, boosted somewhat by South Africa and Libya (both over 9 tonnes per capita). The many impoverished regions within the African continent had very low emissions, for example Chad at 0.02 tonnes and Uganda at 0.06 tonnes. The pattern in Africa, where the poorest countries generally have the lowest emission levels, is seen elsewhere—the Latin American average was 2.6 tonnes, with particularly low emissions in the poorest countries such as Haiti, 0.023 tonnes, and Paraguay, 0.058 tonnes (see Table 2).

An alternative way of addressing relative emission levels is by examining them in relation to real income levels. For the world as a whole in 2008, there were 0.46 metric tonnes of emissions per thousand dollars (U.S. 2005) of GDP.

TABLE 2 Per Capita emissions of CO_2, Selected Countries and Groups, tonnes per annum

	2004	2005	2006	2007	2008
Canada	19.073	19.301	18.361	17.877	17.268
United States	20.367	20.26	19.803	19.925	19.183
Europe	7.958	7.932	7.974	7.905	7.805
Russia	11.452	11.455	11.823	11.666	12.291
Bahrain	34.104	36.663	39.775	41.03	43.215
Qatar	50.923	66.535	69.364	70.571	74.127
United Arab Emirates	34.167	34.208	36.411	38.461	43.105
Africa	1.156	1.169	1.154	1.163	1.145
Australia	19.407	20.326	20.233	20.321	20.823
China	3.951	4.255	4.461	4.726	4.912
India	1.053	1.09	1.164	1.233	1.311
Japan	9.889	9.766	9.757	9.908	9.539
Singapore	29.031	30.244	31.188	33.864	34.609
Latin America	2.401	2.467	2.52	2.581	2.664
World	4.325	4.406	4.442	4.517	4.54

Source: US Energy Information Administration, http://tonto.eia.doe.gov/cfapps/ipdbproject/iedindex3.cfm?tid=91&pid=47&aid=31&cid=&syid=2004&eyid=2008&unit=MTCDPUSD&products=47

Several European countries—especially those with significant nuclear and hydroelectricity facilities—are at about half of this level and some impoverished nations are even lower. Some countries like Australia and the U.S. with high levels of emissions per capita are much closer to the average in terms of emissions per unit of gross domestic product (GDP) (see Table 3).

TABLE 3 Metric tonnes of CO_2 emitted per thousand dollars (US 2005) of GDP

	2004	2005	2006	2007	2008
Canada	0.54	0.535	0.499	0.478	0.463
United States	0.494	0.482	0.463	0.461	0.443
Latin America	0.311	0.307	0.301	0.294	0.292
Austria	0.292	0.291	0.267	0.251	0.24
Belgium	0.467	0.449	0.411	0.403	0.431
France	0.219	0.216	0.213	0.2	0.206
Germany	0.353	0.342	0.333	0.317	0.312
Italy	0.298	0.297	0.289	0.28	0.279
Poland	0.589	0.56	0.548	0.507	0.485
Portugal	0.301	0.311	0.285	0.278	0.262
Spain	0.338	0.338	0.319	0.313	0.29
Sweden	0.21	0.194	0.183	0.177	0.178
Switzerland	0.171	0.168	0.162	0.151	0.157
United Kingdom	0.286	0.28	0.273	0.256	0.257
Europe	0.332	0.324	0.316	0.304	0.298
Russia	1.027	0.961	0.916	0.832	0.826
Iran	0.671	0.702	0.709	0.678	0.67
Saudi Arabia	0.837	0.823	0.79	0.817	0.842
Africa	0.489	0.479	0.457	0.445	0.425
Chad	0.012	0.013	0.014	0.015	0.015
South Africa	1.638	1.526	1.489	1.466	1.383
Australia	0.579	0.598	0.586	0.573	0.582
China	1.048	1.028	0.971	0.916	0.88

(Continued)

TABLE 3 Metric tonnes of CO_2 emitted per thousand dollars (US 2005) of GDP—cont'd

	2004	2005	2006	2007	2008
India	0.507	0.487	0.482	0.475	0.483
Indonesia	0.463	0.47	0.486	0.509	0.518
Japan	0.332	0.322	0.315	0.312	0.303
New Zealand	0.384	0.383	0.369	0.349	0.355
World	0.502	0.495	0.481	0.471	0.464

Source: US Energy Information Administration, http://tonto.eia.doe.gov/cfapps/ipdbproject/iedindex3.cfm?tid=91&pid=47&aid=31&cid=&syid=2004&eyid=2008&unit=MTCDPUSD&products=47

Based on Table 3, the relationship between income levels and emissions is most evident with regard to the poorest countries. For the rest, nations' economic structures and fuel choice options carry greater weight, and the diversity of outcomes is similar to that illustrated on a per-capita basis in Table 2.

As discussed in the next section, international trade means countries that export energy intensive products incur emissions on behalf of other countries. This tends to reduce the national emission levels of many developed countries while exaggerating those of some developing countries and resource-rich countries like Australia.

3. ANNUAL EMISSIONS LEVELS REQUIRED FOR STABILIZATION

3.1. Overall Requirements

The Stern Report (2006) sought reductions in global emissions of CO_2 by 80% of current levels by 2050. Stern argued that the economic cost will be 1% of world GDP, "which poses little threat to standards of living given that the economic output in the OECD countries is likely to rise by over 200 percent and in developing countries by more than 400 per cent" during this period (P.239).

The Waxman-Markey Bill, which narrowly passed in the House, requires a 20% reduction in U.S. emissions from 2005 levels by 2020 and an 83% reduction by 2050. The Senate version of the same bill envisioned a 17% reduction from 2005 level by 2020, the target mentioned by President Obama at Copenhagen. The longer term level of reduction would bring U.S. emissions to the present world average and is consistent with stabilizing global CO_2 equivalent emissions somewhere between the present 450 and the projected 550 parts per million.

As shown in Table 4, most developed countries' trajectory CO_2-e plans are similar.

TABLE 4 Jurisdictions' proposals for emission reductions at Copenhagen

Comparisons in CO_{-e} levels			
Country	*2020 targets*	*2020 per capita reduction*	*2050 targets*
Australia	5-15 per cent below 2000 levels(4-14 per cent below 1990 levels)	27-34 per cent below 2000 levels(34-41 per cent below 1990 levels)	60 per cent below 2000 levels(60 per cent below 1990 levels)
EU	20-30 per cent below 1990 levels	24-34 per cent below 1990 levels	60-80 per cent below 1990 levels
UK	26-32 per cent below 1990 levels	33-39 per cent below 1990 levels	80 per cent below 1990 levels
US	Return to 1990 levels	25 per cent below 1990 levels	80 per cent below 1990 levels

Source: Carbon Pollution Reduction Scheme: Australia's Low Pollution Future, Australian Department of Climate Change, December 2008

Although the Copenhagen Accord was silent on what might be entailed in its aspirational maximum 2°C increase in temperatures, from additional material made available by the IPCC, it might be inferred that the objective was to progress to an "immediate" 30% reduction in global emissions, which the IPCC suggests is necessary to stabilize CO_2-e levels (see footnote 2).

Targets generally do not take into consideration population increase. By 2050 the world population is expected to be 44% above that of 2005.[4] Given the per-capita levels of 4.5 tonnes of CO_2 prevailing in 2004, if an aggregate 30% reduction was the target, with a 44% population increase this means emissions at 2.2 tonnes per capita as illustrated in Figure 1.

The 2.2 tonnes per capita target is comfortably above the African average of 1.14 tonnes per capita, a level that is accompanied by low living standards; as previously mentioned South Africa, the continent's most advanced country, at 9.8 tonnes has over fourfold the target levels. The target level is also below the current Indian level but, though India has a large and growing middle class enjoying living standards comparable with those of developed countries, this is only 10% of the nation's population. One evocative point about India is that, in 2005, 52% of households did not even have electric lighting in their homes.

An important and under-emphasized feature of country comparisons of emission levels is the degree to which emissions are outsourced. Countries like Australia have major export industries, like aluminum, that are energy intensive

4. *www.npg.org/facts/world_pop_year.htm*

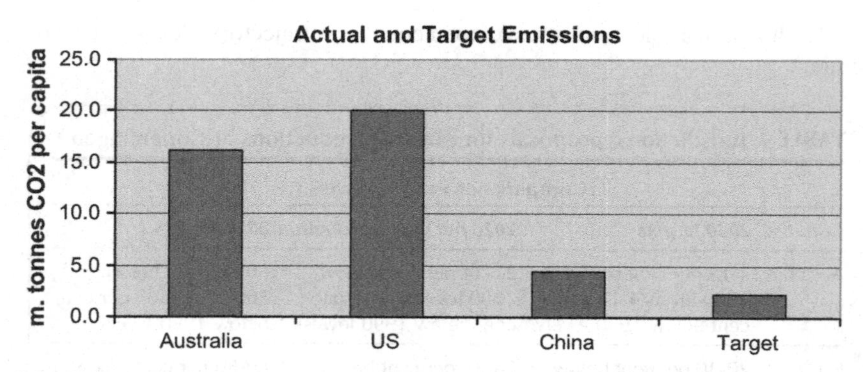

FIGURE 1 Emission goals per capita compared with current levels. *Source: Derived from UNDP, Human Development Report, http://hdr.undp.org/en/statistics/data/hdi2008/.*

and therefore incur the emissions used by other countries. Davies and Caldeira[5] estimate the United States outsources 11% of its total emissions to developing countries, Japan outsources nearly 18%, and European nations from 20%–50%. Switzerland outsources over half of its total emissions.

Confusion about what the targeted emission reductions entail is further illustrated by the statements of political leaders. Thus far the main agreement negotiated at the Copenhagen precursor in L'Aquila, Italy in July 2009 required the developed countries to reduce their emissions in 2050 by 80% and the developing countries by 50%. Present per-capita emission levels of CO_2 are 11.5 and 2.4 tonnes for the developed world and the developing world, respectively. Using simple arithmetic, by 2050 the 80% cut would leave the developed world with 2.9 tonnes of CO_2 per capita and the developing world with less than half of this at 1.2 tonnes per capita. Moreover, this is based on the unlikely event of population growth in the developing countries slowing to the level of that in the developed world.

3.2. Sectors and End-Use Activities Creating Emissions

Energy—in the form of combustion of fossil fuels—is responsible for the great bulk of emissions and these are distributed roughly 30% to both industry and transportation and 20% to commercial and residential activities. Figure 2 illustrates the sources and end uses of emissions globally.

U.S. EPA data demonstrate the extent to which the energy sector (the top portion of Figure 2) dominates the greenhouse gas (GHG) emissions, accounting for 85% of the total.

Looking at end-use emissions, the distribution is different. Land, like the oceans, is both a natural source of emissions and a sink for those emissions. The

5. Steven J. Davis and Ken Caldeira. Consumption-based accounting of CO_2 emissions. *PNAS. www.pnas.org/cgi/doi/10.1073/pnas.0906974107*

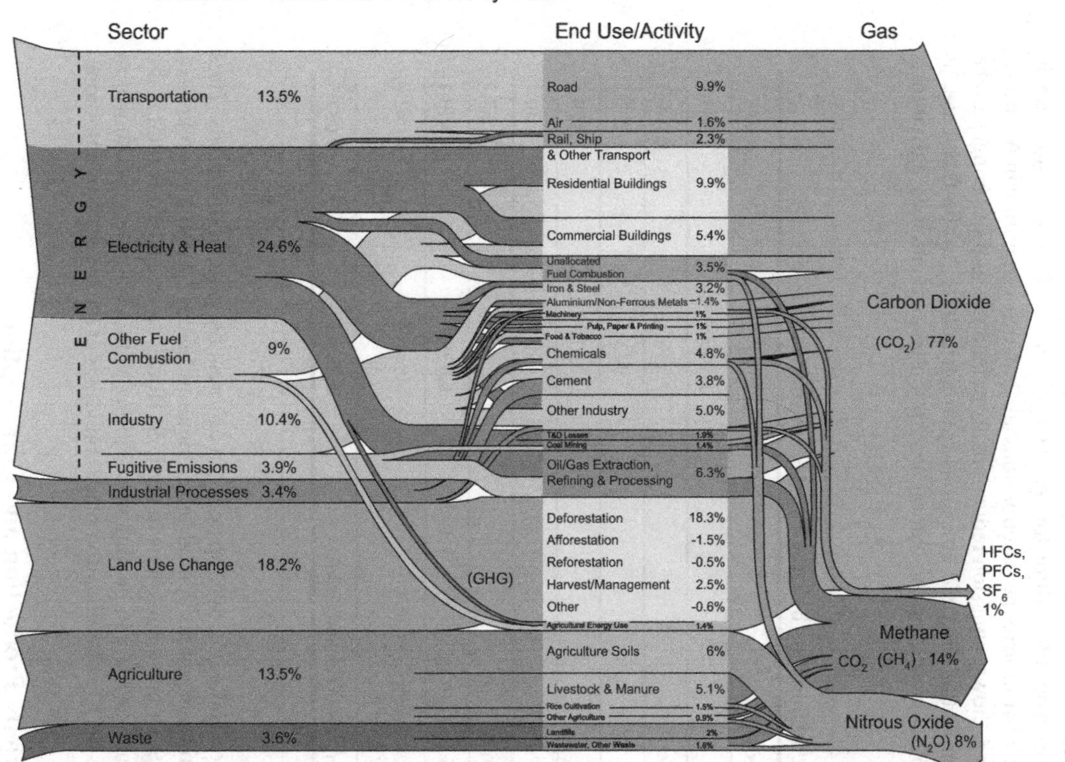

All data is for 2000. All calculations are based on CO_2 equivalents, using 100-year global warming potentials from the IPCC (1995), based on a total global estimate of 41 755 $MtCO_2$ equivalent. Land use change includes both emissions and absorptions. Dotted lines represent flows of less than 0.1% of total GHG emissions.

Source: World Resources Institute, Climate Analysis Indicator Tool (CAIT), Navigating the Numbers: Greenhouse Gas Data and International Climate Policy, December 2005; Intergovernmental Panel on Climate Change, 1996 (data for 2000).

FIGURE 2

extent of this differs country by country depending on the land-use activities, population density, and other variables. The larger, sparsely populated countries tend to be net sinks. Globally, one reputable estimate has land annually absorbing over 9 billion tonnes of CO_2. On the basis of its share of the world land area, even without any positive steps to enhance sinks, Australia is likely to absorb a net 137 million tonnes of CO_2-e a year.[6] This is considerably in excess of the 90 million tonnes of emissions attributed to agriculture in Australia.

As Table 5 shows, emission reductions from land use and its changes in the U.S. also comfortably exceed emissions from agriculture. In the U.S. case, as with many other countries, the natural absorption is augmented by agricultural practices.

TABLE 5 US Emissions by Usage Sector (m tonnes CO_2-e)

Implied Sectors	1990	2007
Industry	2,166.5	2,081.2
Transportation	1,546.7	2,000.1
Commercial	942.2	1,251.2
Residential	950.0	1,229.8
Agriculture	459.2	530.1
U.S. Territories	34.1	57.7
Total Emissions	**6,098.7**	**7,150.1**
Land Use, Land-Use Change, and Forestry (Sinks)	(841.4)	(1,062.6)
Net Emissions (Sources and Sinks)	**5,257.3**	**6,087.5**

There is considerable overlap between different usages classifications. Thus the above estimates put residential and commercial use at 2,480 million tonnes of CO_2-e or 35% of gross emissions in the U.S. However, much of this total is the burning of energy itself in heating and lighting within buildings. As shown in Figure 3, the IPCC's classification places buildings at around 10% of emissions.

3.3. Electricity Production and Consumption

Over the longer term, energy used per unit of GDP has dropped markedly in OECD countries. The U.S. Energy Information Administration data (Figure 4)

6. Pers. Comm. with CSIRO's Dr Michael Raupach.

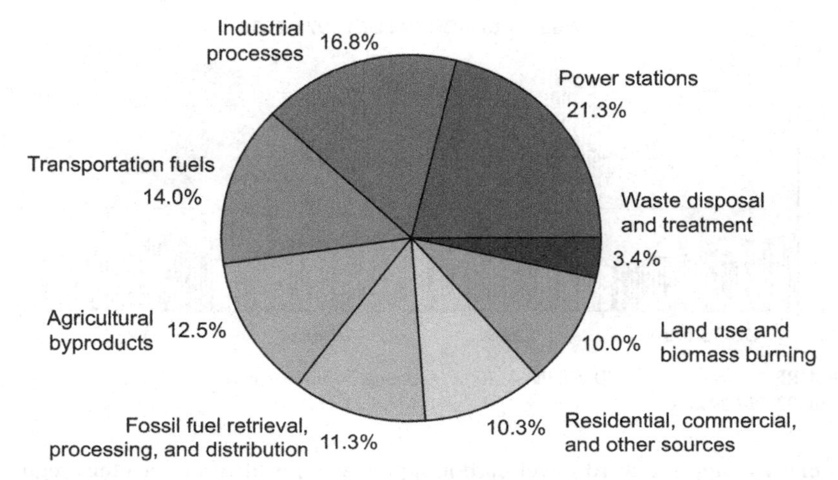

FIGURE 3 Profile of usage. *Source: Kyoto 2001 WG 1 Annual CO_2 Emissions by Sector (Source: Climate Change 2001: Working Group I: The Scientific Basis).*

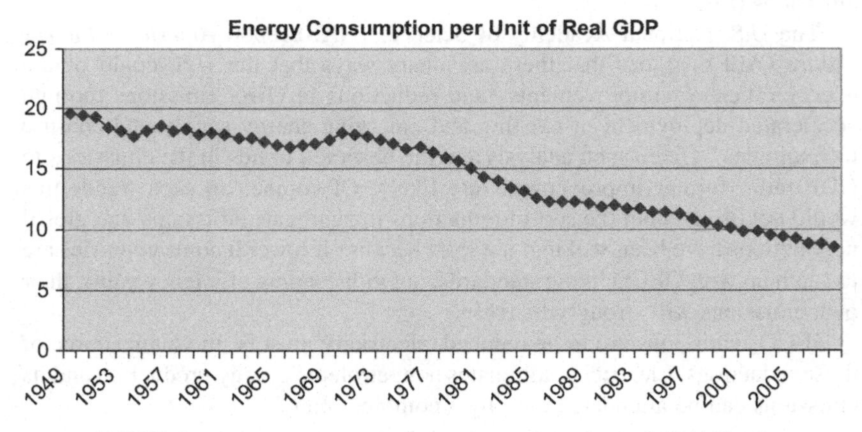

FIGURE 4 *Source: EIA, http://www.eia.doe.gov/emeu/aer/txt/ptb0105.html.*

show world energy in 1949 at 19.47 thousand BTUs per unit of GDP (2000 prices) had fallen to 8.52 by 2008.

Further reductions in energy intensity are likely, though much will depend on energy price trends. Most of the reduction in the EIA energy to GDP ratio came in the years after 1973 when prices increased in real terms. Figure 5 shows similar trends in lower energy usage are evident for developing countries and the world as a whole.

Since 1980 GHG emissions relative to GDP have been on a fairly similar downward trajectory to energy use (see Figures 4 and 5), largely reflecting little change in share between fossil fuels and nonfossil fuels. But a trend in the mix

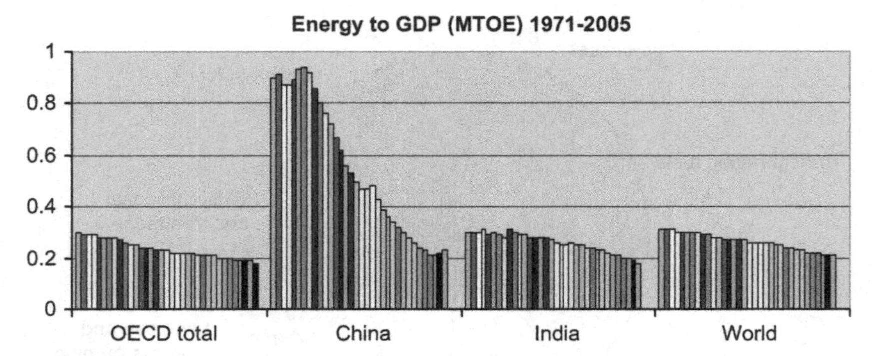

FIGURE 5 *Source: OECD Factbook 2007: Economic, Environmental and Social Statistics - ISBN 92-64-02946-X.*

of energy supply toward lower carbon inputs as a result of tax or other regulatory penalties would bring a lower emissions to GDP ratio by conferring a relative price advantage on fuels with no CO_2 emissions (like nuclear) or low emissions (like gas).

The U.S. National Academy of Sciences (NAS) report *America's Energy Future* (AEF)[7] argues that there are many ways that the U.S. could obtain energy efficiency improvements "and reductions in GHG emissions through accelerated deployment of existing and emerging energy supply and end-use technologies." Given such analysis and the observed trends in the emissions to GDP ratio, further improvements are likely. Of course, on such trends this would not bring about the sort of reductions in aggregate emissions that global negotiations have been seeking, not least because if lower income countries are to catch up with OECD living standards, notwithstanding efficiency gains, their own emissions will strongly increase.

If CO_2 emissions are to be reduced, electricity must be the main source of those reductions. Electricity dominates nonpetroleum energy production and its emissions can be attenuated only by a combination of:

- Reduced usage
- Shifting to low carbon sources

These two topics are further explored below.

4. REDUCED ENERGY USAGE

4.1. Overview

Lower levels of energy usage, if this could be accompanied by high living standards would bring about the reduced emissions that most governments say

7. *America's Energy Future*, summary edition, National Academy of Sciences, 2009.

they are seeking. Heinberg[8] is one writer who concludes that the exotic (largely CO_2-free) renewables are never likely to replace current sources of energy. He argues that lower energy use is compatible with high living standards and points to the U.S. versus European usage rates, saying:

Since Europeans already live quite well using only half as much energy as Americans, it is evident that a U.S. standard of living is an unnecessarily high goal for the world as a whole. Suppose we aim for a global per-capita consumption rate 70 percent lower than that in the United States

Achieving this standard, again assuming a population of 9 billion, would require total energy production of 1800 quads per year, still over three times today's level. Cheap solar panels to provide this much energy would cost $150 trillion, a number over double the current world annual GDP. This scenario is conceivable, but still highly unlikely.

His conclusion that reduced energy usage is the only path to a low carbon economy is based on understandable doubts about the prospects for radical development in technological developments for new energy sources of carbon capture and storage. It is also profoundly and less justifiably pessimistic about prospects for fast breeder nuclear power. Heinberg's view is essentially that resources will set limits to growth and amounts to requiring a return to pastoral societies. These, as he envisages them, involve far less trade activity together with some form of nationalization of energy.

Clearly such solutions would be unattractive outside of a romanticized version of the pre-modern world. Nor is it possible to characterize the European standard as a low energy goal that is widely achievable—South Africa with a standard of living much less than half of Europe's has a comparable level of emissions. Emissions depend very much on the structure of an economy and particularly its industrial profile—a consideration that accounts for China having four times the emission levels of India though the two nations have similar average per capita income levels. China's development is more heavily focused than India's on energy-intensive manufacturing industry; hence China could not adopt India's energy profile without severely damaging its living standards and its government's recognition of this led it to play the leading role in the world's leaders' failure to agree to hard emission reduction standards at Copenhagen in December 2009.

The McKinsey consultancy estimates that there are considerable savings potential from applications of known technologies to promote lower emissions. Based on "market failure" the consultants see a great deal of low hanging fruit ready for plucking if only people are appropriately guided. Figure 6 illustrates the detailed estimates of savings theoretically possible with best practice usage.

---

8. Richard Heinberg, *Searching for a Miracle*, *www.postcarbon.org/new-site-files/Reports/Searching_for_a_Miracle_web10nov09.pdf.*

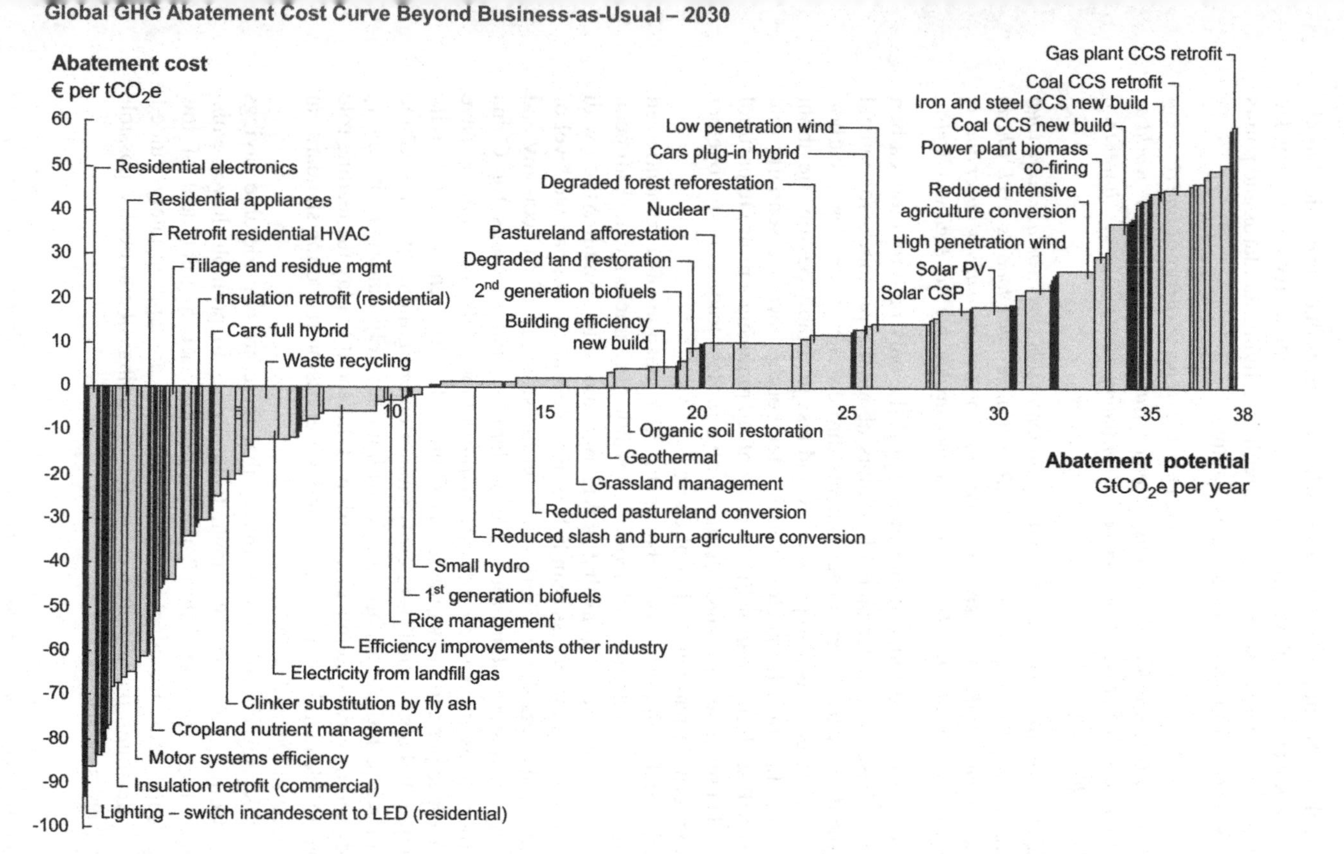

Note: The curve presents an estimates of the maximum potential of all technical GHG abatement measures below €60 per tCO_2e if each lever was pursued aggressively. It is not a forecast of what role different abatement measures and technologies will play.

FIGURE 6 *http://www.climateworks.org/case_studies/PathwaysToLowCarbonEconomy_FullReport.pdf*

Government measures to force consumers to accept these gains amount to a belief that they will not be made simply by relying on consumers' wishes leading to a supply structure that results in optimal value. Especially with regard to the costless measures, regulatory forcing assumes that people could make far better decisions if we were properly informed and, perhaps, if they were more rational. This book's Chapter 8 by Long et al. and Chapter 17 by Rajkovich et al. discuss some of these issues.

Some pitfalls of uncritically accepting these theoretical savings are illustrated by the Australian experience with a subsidized retrofit of ceiling insulation (estimated by McKinsey in Figure 6 to provide a saving of €30 per tonne of CO_2). The Australian scheme was originally estimated to make savings of 50 million tonnes of CO_2 at a total cost of $A2.5 billion. In fact, the over-hasty, now discontinued, roll-out has led to four deaths of contractors and over 100 house fires. And the estimated 50 million tonnes of saving is now considered to be more like 20 million tonnes.[9] This increases the estimated cost of the savings to some $200 per tonne of CO_2.

Under scrutiny, many claims that expert guidance improves individuals' decision-making are found to have little merit. Some acknowledgment of this by governments is seen in them largely opting for a market-based solution to bring about lower emissions. Governments have generally recognised that the most efficient means of emission abatement make use of "economic instruments." These come in two forms.

One sets a ceiling on emissions and allows trading to take place and a price to emerge. This chokes off the uses that less urgently require the inputs from which emissions are a by-product. At the same time it stimulates the search for low-emission alternatives and brings about substitutes within aggregate demand from high-emission uses to those which entail lower emissions.

The alternative "economic instrument" is a carbon tax set at a level the government expects will be optimal in bringing about the required level of abatement for the cost involved. A tax has the same effect as a tradable right system. It also attaches a price to GHG emissions and therefore to energy, energy savings substitutes within energy sources, and products incorporating energy in different concentrations. That price allows the market to bring about the least cost abatement outcomes.

Emission trading schemes are in place or under development in the EU, U.S., Canada, New Zealand, Australia, and South Korea. Only Switzerland has adopted a CO_2 tax and that is an interim measure pending the country's alignment with the EU scheme.

Of course, many countries have taxes on oil-based fuels which have morphed into a highly specific form of carbon tax. In this and other respects, governments are in fact seldom consistent in opting for economic instruments

---

9. *www.theaustralian.com.au/news/features/woolly-claims-on-insulation/story-e6frg6z6-1225834522839*

and in practice involve themselves actively in selective winner-picking policies. These include expensive R&D and demonstration projects, subsidies for certain uses, and have given special support to renewable energy. Such interventions amount to a vote of low confidence in the market mechanisms forcibly advocated in some agendas.

As previously mentioned, many governments justify such intervention on the basis of a deficit of information on the part of consumers. Improving access to information is a central feature of the "new paternalism" literature. New paternalists argue that individuals need to be guided—not ordered—to combat their biases in favor of such things as the status quo, optimism, and lack of willpower or self-control. Rizzo and Whitman[10] describe this as, "claims that careful policy interventions can help people make better decisions in terms of their own welfare, with only mild or nonexistent infringement of personal autonomy and choice."

Among the savings that are said to be available from measures that people would readily accept if well-informed and not prejudiced against by market failure include:

- Smaller more fuel efficient cars
- Greater use of public transport
- Reduced urban sprawl
- Better designed buildings

Let us examine these in order.

4.2. Efficiency of Cars

Vehicle fuel economy is a prominent area where governments are confident that they are better equipped to understand consumers' true needs than those revealed by the purchasing decisions of consumers themselves.

Fuel economy trends in vehicles have mirrored relative price changes for gasoline. From 1935 U.S. fuel efficiency at 14 mpg (5.95 km/l) fell gradually to 11.9 mpg (5.08 km/l) in 1973. Following oil price in that and subsequent years, the efficiency of the total fleet has risen by 42% on 1973 levels to 16.9 mpg (7.18 km/l).[11]

Building on this, many countries have requirements similar to the U.S. Corporate Average Fuel Economy (CAFE) standards, which mandate an average increase in fuel efficiency of cars and trucks by over 40% between 2004 and 2016. This over-riding of consumer choice means the government is mandating a tradeoff between different features of the product. The mandatory

10. Mario J. Rizzo and Douglas Glen Whitman (2009). Little brother is watching you: New paternalism on the slippery slopes, *Arizona Law Review, 51*, p. 685.

11. Michael Sivak and Omer Tsimhoni. (2009). Fuel efficiency of vehicles on US roads: 1923–2006.

nature of the tradeoff means some consumers would not prefer it. The increased priority to fuel economy may come at the expense of initial price, performance, aesthetic features, or even vehicle safety.

It may be that the government is better informed than the manufacturers of the true preferences of consumers, though in competitive markets like automobiles this is unlikely. It is equally unlikely that the government understands the costs and tradeoffs involved. Governments may, however, be basing their regulatory requirements on the notion that without them consumers' divergent preferences will deny the economies of scale necessary to create the priority to fuel savings that consumers at large would prefer. This is a strong call and one likely to shift us along the regulatory path that Rizzo and Whitman (footnote 10), among others, warn against.

Whatever the merits of government regulations on fuel economy, there is considerable evidence that gains beyond those already realized are possible. Thus cars are showing increased economy, as illustrated by Table 6.

TABLE 6 Fuel Efficiency Examples of Selected Models

		Fuel Consumption L/100km			CO_2g/
Vehicle Details	*Fuel Type*	*Comb*	*Urban*	*Extra*	*kmComb*
Toyota Prius 1.8L 4cyl, CVT 1 speed Hatch, 5 seats, 2WD	Elec/Petrol 95RON	3.9	3.9	3.7	89
Volkswagen Golf 103TDI Comfortline 2.0L 4cyl (T), Man 6 speed Hatch, 5 seats, 2WD	Diesel	5.3	6.8	4.4	139
Ford MB Mondeo LX2.3L 4cyl, Auto 6 speed Wagon, 5 seats, 2WD	Petrol 91RON	9.5	14.1	6.8	227

Source: Manufacturers' specifications, 2009.

The Toyota Prius uses only 64% of the fuel of a Volkswagen Golf diesel and only 40% of the fuel of a standard mid-range petrol vehicle. Its GHG saving properties doubtless are responsible for the vehicle achieving a high market share among the "environmentally aware" who see its ownership as a means of demonstrating their credentials. This is assisted by claims that the fuel savings pay for the additional costs of the car within 3–5 years. Other vehicles like the GM Volt are likely to join the Prius as a low-emitting vehicle and the technology itself appears set to further improve on fuel economy.

The IPCC saw considerable scope for further fuel economies with cars by the use of lightweight materials, hybrids, and better aerodynamics.[12]

Based on known technologies, it would be relatively simple to move, though at some price premium, to cars like the Prius that are twice as fuel (and CO_2-e) efficient as the average current stock—and indeed more efficient per passenger mile than most forms of public transport. Doubtless further economies will be made in fuel usage and vehicle design.

4.3. Public Transport

The IPCC placed considerable emphasis on changed energy, urban density, and spatial layouts. Working Group 3 (2007, Table SPM 3) said:

Modal shifts from road to rail and to inland and coastal shipping and from low-occupancy to high-occupancy passenger transportation, as well as land-use, urban planning, and non-motorized transport offer opportunities for GHG mitigation, depending on local conditions and policies.[13]

Many foresee a strong potential for emission reductions if people were to move away from personal car use toward public transport. However, as Table 7 illustrates, the emission reduction potential is not as great as some claim it to be. Highly fuel efficient cars like the Prius are already comparable in emissions per passenger mile with several forms of public transport, though

TABLE 7 US Average CO_2 Equivalent Emissions by Transport Mode/Vehicle

Mode	Passenger Miles	CO2 Equivalent Grams	Grams per Passenger Mile
Automated Guideway	11,647,000	7,892,000,000	678
Commuter Rail	9,470,134,000	1,827,018,000,000	193
Light Rail & Streetcar	1,650,204,000	282,883,000,000	171
Motorbus	19,323,463,000	6,696,660,000,000	347
Metro	14,407,097,000	1,676,605,000,000	116
Trolley Bus	172,982,000	21,480,000,000	124
Average 2006 model car			307
Average light truck			374
2007 Hybrid: Toyota Prius			147

Source: Demographia.

12. *www.ipcc.ch/publications_and_data/ar4/wg3/en/ch5s5-3-1-2.html.*

13. *www.ipcc.ch/publications_and_data/ar4/wg3/en/spmsspm-c.html.*

there is also doubtless scope for further gains in fuel efficiency for some types of public transport.

In addition to these matters, with respect to public transport, there is no easy solution in terms of forcing a modal switch. The U.S. pattern, which is being progressively seen throughout the world, is for a transit market share of only 5% of the average U.S. journey-to-work share (the only segment where it is feasible for transit to serve with any efficiency), and pays for only a quarter of its costs (one half in Europe).[14]

Increasing the share of public transport is likely to bring a more than proportional increase in costs. Population density and concentrated work destinations are crucial to allowing transit to work with any degree of efficiency, and for the most part developed world cities' concentration levels at 2000–4000 people per square kilometer are inadequate to support this. With such concentration levels, the cost, frequency, and origin/destination practicalities give an overwhelming advantage to the flexibility offered by car travel. City authorities may choose to override people's preferences for flexibility by banning cars or heavily taxing their use but this is likely to bring a deterioration of the city's efficiency and attractiveness and may lead to its decline.

The only affluent cities in which public transport retains a financial viability and a high trip share is where population concentrations are high like Hong Kong (25,000 people per square kilometer, and a 90% public transport share) and Singapore (8000 people per square kilometer, with a 63% public transport share).[15] While many commentators applaud these as models for future city developments, people's preferences appear to be for greater personal space including outdoor space and therefore for cities of lower density levels.

Moreover, the nature of production has meant smaller factories and other commercial facilities that are not concentrated in a central business district (CBD) readily served by radial public transport links. No longer do CBDs provide the bulk of urban employment that can be readily served by radial train or bus lines.

Hence, even though telecommunications advances have meant many people need not travel daily to work, the nature of modern cities and workplaces makes the car indispensable. Diverse origin and destination locations mean that cars are the only practical means of serving societies that are affluent enough to be able to afford larger houses on their own land. It is difficult to see preferences for this style of living changing. The corollary is a continuation of centuries-long trends to lower urban densities, epitomized by the first million plus city, ancient Rome, having a density of 57,000 people per square kilometer, compared with 3000 in modern Rome.

The extraordinary difficulty transit systems have in securing high shares of the total transport modal split within modern economies is illustrated by the map of Melbourne (Figure 7). This shows the magnitudes of the task for public

14. *www.ipcc.ch/publications_and_data/ar4/wg3/en/ch5s5-5-3.html.*

15. ibid.

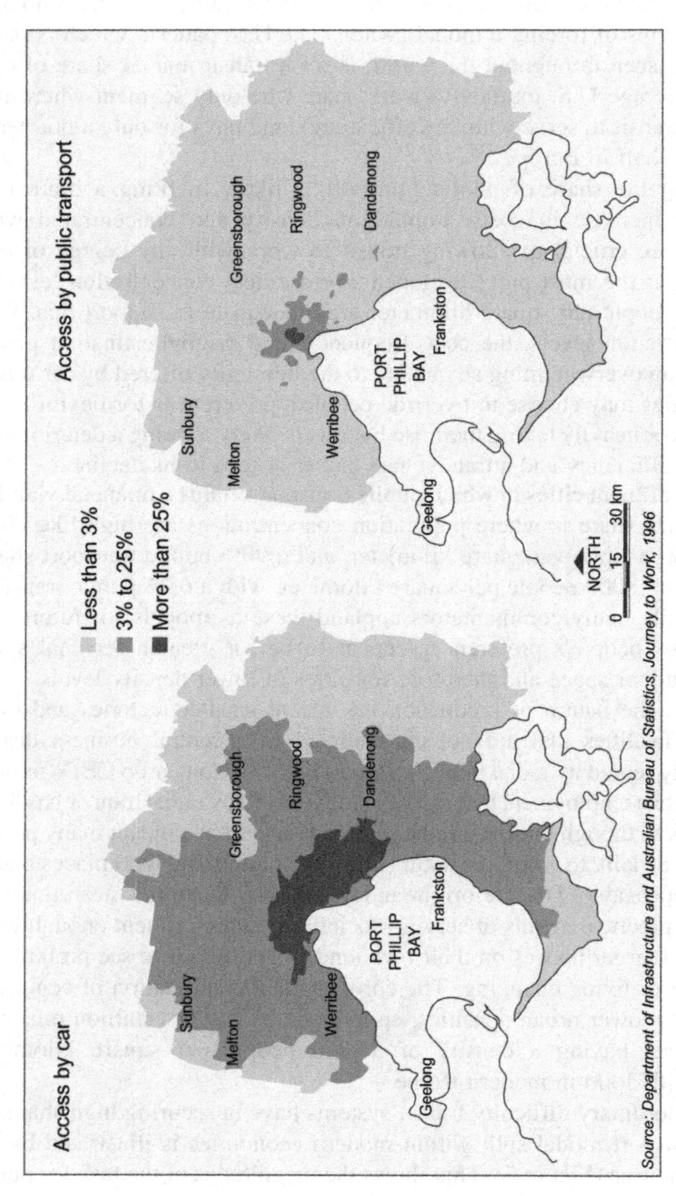

Source: Department of Infrastructure and Australian Bureau of Statistics, Journey to Work, 1996

FIGURE 7 Transport times in Melbourne. *Melbourne, Percentage of jobs accessible within 40 minutes, car and public transport*

transport as a result of consumer preferences for more spacious accommodation and the changed concentrations of work locations. Typically the jobs accessible with a 40-minute trip by public transport are concentrated in CBDs, which comprise a relatively small share (usually around 12%) of total employment. Aside from these areas, access to jobs in acceptable travel times is only possible by using cars.

4.4. Urban Design

Long-standing opposition to "urban sprawl" has recruited the support of GHG concerns. The claim is that if urban areas are more concentrated, we will use less energy for heating and such things as well as needing less energy for transport.

The IPCC argued that:

The potential exists to greatly reduce transport energy use and GHG emissions by shaping the design of cities, restraining motorization, and altering the attributes of vehicles and fuels. Indeed, slowing the growth in vehicle use through land-use planning and through policies that restrain increases in vehicle use would be an important accomplishment. Planning and policy to restrain vehicles and densify land use not only lead to reduced GHG emissions, but also reduced pollution, traffic congestion, oil use, and infrastructure expenditures and are generally consistent with social equity goals as well.[16]

Plausible though this may be, empirically it appears to be false. Careful assessments of GHG emissions per capita have shown that households in the less concentrated outer areas of cities emit less than those in the inner cities.

Thus, the Australian Conservation Foundation has documented emissions by suburb.[17] Typical findings are those for Sydney. Inner Sydney showed 37 tonnes per capita, Burwood in the inner west showed 22 tonnes, and Parramatta in the outer west showed 20 tonnes. Similarly in Melbourne, centrally located Port Phillip had 27 tonnes per capita, inner Darebin had 23 tonnes, and outer Melton had 18 tonnes. The reasons behind such differences are partly due to the inner suburbs having higher income levels but also due to the increased energy spent on clothes drying and heating, elevators, and lighting in apartment blocks.

4.5. Buildings

Buildings are a further area of energy savings identified in reports by McKinsey and others, including Chapter 9 by Gray and Zarnikau and Chapter 17 by Rajkovich et al. in this volume. They account for some 10% of emissions (or perhaps 40% if the use of energy within the buildings is included). The IPCC argued that "By 2030, about 30% of the projected GHG emissions in the building sector can be avoided with net economic benefit."

---

16. *www.ipcc.ch/publications_and_data/ar4/wg3/en/ch5s5-5-1.html.*

17. *www.acfonline.org.au/consumptionatlas/.*

Vattenfall has estimated potential savings from greater efficiency, savings that average 30%–40%[18] as shown in Table 8.

TABLE 8 Estimated Scope for Abatement

Category of energy end use	Potential abatement (% of BAU energy consumption)
Air conditioning	37
Appliances	38
Heating and ventilation	59
Lighting	12
Water heating	28

The consultancy firm CIE, in work undertaken for the Australian Sustainable Built Environment Council, estimated the costs of complying with emission reductions with and without measures that regulate buildings to ensure the use of the "state of the art" energy saving measures.[19] Measured in terms of the tax effect necessary to force the necessary reductions in emissions, it estimated considerable savings would be available. It concluded, "By 2050, the price of GHG emissions is just under $160 per tonne or 14 percent lower when the building sector contributes through energy efficiency."

Doubtless major savings of energy are possible with designs that place a greater priority on this attribute. However many consumers—especially the more affluent—seek out features that make considerable use of energy in requiring areas be permanently heated/air-conditioned and lighted and have panoramic views that militate against energy saving (Figure 8).

Of course, should energy costs show relative increases, in addition to technological improvements that make for cheaper energy services, consumers are likely to demand more substantial savings in selecting building designs.

Central to claims that buildings poorly cater for consumer needs is an assertion that many consumers do not have full access to information and their decisions are taken for them, for example by landlords. As the latter do not pay the energy bills, they are said to be indifferent to this aspect of aggregate costs.

Such claims of principal–agent problems are often the basis for paeans in favor of regulation but are no more plausible in this than in other areas. Thus

18. Vattenfall AB. (2007). *Global mapping of greenhouse gas abatement opportunities up to 2030.* Buildings sector deep-dive.

19. CIE. (2007). *Capitalising on the building sector's potential to lessen the costs of a broad based GHG emissions cut.* Canberra.

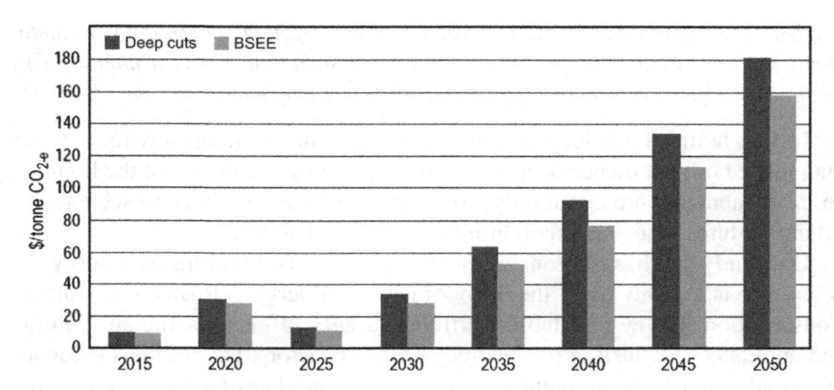

FIGURE 8 Tax effect necessary for designated emission reductions.

people could maintain that "market failure" would frustrate the best value being offered in cars, TV sets, computers, even furniture because the multiplicity of components make decisions too difficult for a consumer to estimate good value for money in terms of fuel economy. Nobody has ever claimed perfect outcomes from markets but their "workable efficiency" seems preferable than outcomes that are imposed even by the most talented regulatory authorities. In addition to competition between suppliers, a wide range of independent expertise is available to assist consumers in making informed choices.

Moreover, notwithstanding the frequent assertion that the derived demand for buildings means that consumers place inadequate priority on the energy costs involved in their use,[20] there is remarkably little evidence to support this. Some surveys have indicated the contrary. Thus according to work conducted by the American Council for an Energy-Efficient Economy and prepared by Arthur D. Little:[21]

Most rental housing incorporates residential grade windows. From strictly an energy-efficiency perspective, landlords have little incentive to invest in upgrading windows in their rental units, since typically the tenant pays the utility bills. However, landlords are motivated by maintaining their properties as modern and aesthetically pleasing, so as to maintain high occupancy rates. They know that windows are one of the most noticeable features in a rental unit. Updated, easy-to-operate windows are a very positive feature, whereas outdated, stuck, or non-functional windows can affect occupancy. In

20. See for example, Carl Blumstein, Betsy Krieg, and Lee Schipper. (1980). Overcoming social and institutional barriers to energy conservation. *Energy, 5*, pp. 355–371; Kenneth Gillingham, Richard Newell, and Karen Palmer. (2009). Energy efficiency economics and policy. *Annual Review of Resource Economics, 1*, pp. 597–619

21. Philip E. Mihlmester, Michael Gibbs, William F. Grimm, and James Stimmel. Millwork 101: Transforming the market for energy-efficient windows. Proceedings of the American Council for an Energy-Efficient Economy, Washington 2000, "*http://www.aceee.org/*"

weatherization programs in New York City, with its very high concentration of multifamily housing, window upgrades are always the single most sought after item by landlords (though they cannot be cost justified on a strictly energy savings basis).

In fact homes built for rent may incorporate more energy-saving features than those built for owner-occupiers. This is because the latter see the home as an investment and are commonly cash-constrained and inclined to see energy-saving features as being expenditure that can be deferred.

One study that has rigorously researched the data on renting and energy use is by Davis,[22] who used the Department of Energy's Residential Energy Consumption Survey to establish different usages of energy-efficient lighting and appliances. Adjusting for income and other factors, Davies finds evidence of a small but significant reduced usage of these products by renters compared to owners.

While this is indicative of an agent–principal problem, as Davis points out, the difference might be because of other factors. Renters for example may be less "green" than homeowners. In addition, the less energy efficient products may be more durable and cheaper, offering lower maintenance costs, which is an important consideration in view of the harsher treatment they are likely to receive from renters.

These caveats aside, the actual savings estimated by the study is that if renters behaved like homeowners for the products analyzed they would consume 0.5% less energy and if the same results applied to total household energy consumption, renters would save a still modest 2% of their energy costs.

Important also in the assessment of the role that buildings can play in bringing lower emissions is their life. Typically only 2%–3% of the building and housing stock is replaced each year.

5. SHIFTING TO LOWER CARBON EMITTING SOURCES

Aside from higher energy efficiency, carbon emission reductions are possible through changing the nature of energy supply to lower carbon emitting sources or collecting and storing carbon emissions.

In the energy share of noncarboniferous fuels, Sweden, Switzerland, and France with nuclear and hydro are at over 40% and most other countries are 10%–20%. This has contributed to the countries where nuclear/hydro is prominent having lower CO_2 emissions per capita—a little over 6 tonnes for France, Switzerland, and Sweden—than others like the U.S. (19 tonnes) and Australia (21 tonnes). Only about 5% of Australian energy is derived from other than fossil fuels. Figure 9 illustrates the nonfossil fuel share of selected economies' energy supply.

---

22. Lucas W. Davis. (2010). Evaluating the slow adoption of energy efficient investments: Are renters less likely to have energy efficient appliances?" *http://ei.haas.berkeley.edu/pdf/working_papers/WP205.pdf*

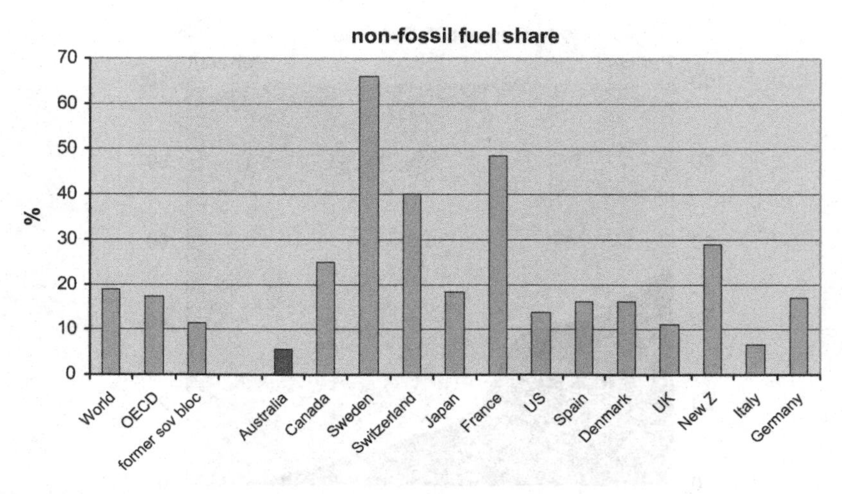

FIGURE 9 *Source: BP http://wwwbp.com/liveassets/bp_internet/globalbp/globalbp_uk_english/reports_and_publications/statistical_energy_review_2008/STAGING/local_assets/2009_downloads/statistical_review_of_world_energy_full_report_2009.xls#'Primary Energy - Consumption'!A1.*

Among supply side measures that can bring lower emissions are shifts to nuclear power and for coal, the dominant source of electricity, and development of carbon capture and storage (CCS). CCS is attracting considerable funding but has yet to demonstrate an ability to operate at acceptable cost.

Vattenfall opened its 30 MW coal plant in 2008, which, with EU support, it hopes to move to commercialization by 2020.[23] Many other projects have also attracted support from the EU and U.S. in particular. (See Sioshansi, 2010.)

There are multitudes of views about the prospects for renewables and other kinds of energy, including this book's Chapter 13 by Arent et al. Although some analysts, including Bjorn Lomborg,[24] maintain that renewables will become competitive, at present they clearly are not. Moreover, Lomborg's own analysis is derived from a simple extrapolation of relative costs and efficiencies of wind power during its infancies at a time when easy gains were readily available.

Many reports by governments and others foresee breakthroughs in low cost, low carbon-emitting energy sources as being imminent, and couch their forecasts accordingly. However, there can be no confidence that novel low-cost dramatic innovations will emerge.

23. Sioshansi F.P. (2010). Generating Electricity in a Carbon Constrained World, Elsevier. *www.vattenfall.com/en/ccs/technology.htm.*

24. Bjørn Lomborg. (2001). *The skeptical environmentalist: Measuring the real state of the world.* Cambridge University Press.

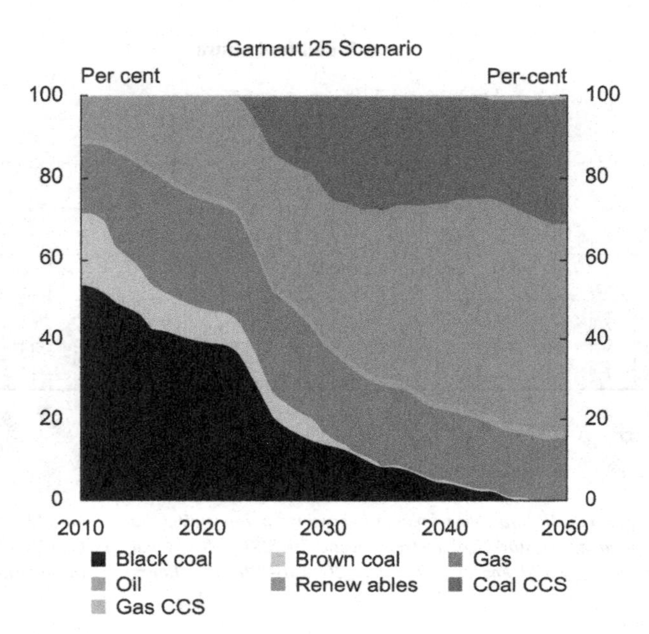

FIGURE 10 Australian Treasury's projected changes in electricity supply shares. *Source:* Australia's low pollution future, *Commonwealth of Australia, 2008, p. 174.*

Even so, heroic assumptions about the rate of technological discovery and adoption are often the basis of government energy composition forecasts. As Figure 10 indicates, the Australian Treasury (which was politically constrained from incorporating a nuclear option) sees renewables—currently dominated by hydro—rising from under 10% to over 50%, almost all of which would be wind. It also sees the unproven technology of carbon capture and storage comprising over 30% of supply. Such projections are, therefore, little more than fantasies, which is not to say that we will not see technologies materialize that at present seem unlikely to be viable or are even currently unknown.

The OECD climate change projections forecast only a miniscule role for renewable energy and are enhanced by envisaging a sizeable increase in nuclear. But the OECD too has CCS playing a major role at some 30%, as shown in Figure 11.

Though the economic modeling driving these numbers is based on empirical observation, uncertainties of their projections going decades into the future are seldom raised. The effects of this deficiency are compounded by government reports greatly inflating the costs of any increase in temperatures while downplaying the economic costs of carbon constraining measures.

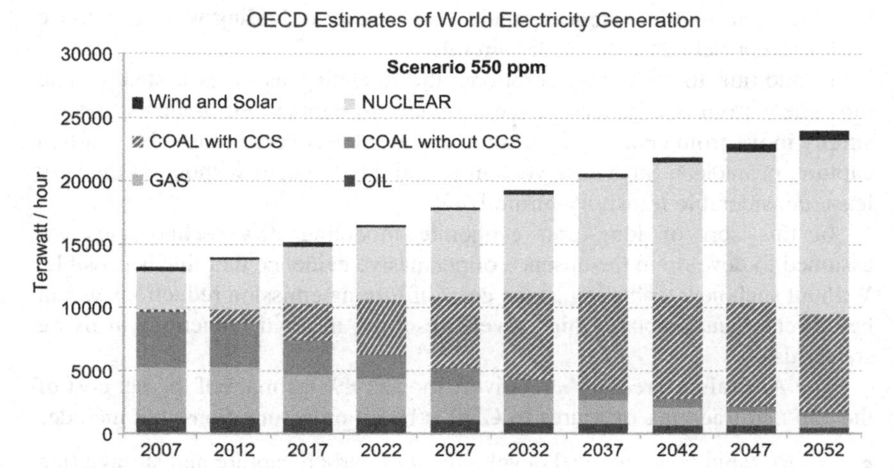

FIGURE 11 *Source: OECD, The economics of climate change mitigation, policies and options for global action beyond 2012.*

6. IS IT WORTH THE PRICE?

Tol[25] points out that for the effects of two and a half degrees warming, ten peer reviewed studies have been published and these have estimated the impacts on global income levels ranging from a positive 1% to −2.5%. But the Stern report (which, as Tol noted, was not peer reviewed) came up with 12.5% loss with this continuing to grow.

Similarly, policy studies by government bodies have tended to understate the effects of the forced change on economies. The Australian Treasury, for example, estimates the net cost to GDP if Australia reduces its emissions by 80%, in line with the goals set out under Waxman-Markey as 8% of GDP per capita by 2050 (see[26]). In the context of a half century of economic activity this is not great since if no mitigation measures were to be taken, per capita GDP would be 66% higher than present; with emissions reduced by 80%, the GDP increase is brought back to 58%.

Present-day energy consumption is extremely reliant on carboniferous fuels. Energy itself is, second to food, the basic building block of all human activities. We have only the flimsiest of experience on which to model the effects of a carbon tax. Unlike the case with oil, which experienced a form of new tax in the OPEC supply restraint in the 1970s, substitutes for the totality of carboniferous fuels do not exist, except for nuclear, and to enable that as

25. Richard S. J. Tol. (2009). *An analysis of mitigation as a response to climate change.* Copenhagen Consensus Centre.

26. *Australia's low pollution future*, Commonwealth of Australia, 2008.

a replacement requires great ingenuity—especially in finding ways to replace oil for motor vehicles, ships, and aircraft.

In addition to such considerations, the modeling assumes a steady state movement from one pattern of the economy to another—it assumes that we simply move from coal to gas to some as-yet-undiscovered renewable, carbon capture, or nuclear. Such a movement is unlikely to occur without, at the very least, considerable transitory turmoil.

In this sort of long-term economic modeling new technologies are assumed to develop in the absence of persuasive evidence that this is possible. Without such new technologies the costs of forcing emission reductions would be driven to unacceptably high levels, resulting in rapid reductions in living standards.

The Australian Treasury's relatively modest 8% estimate of the net cost of the cap-and-trade tax measures to GDP is based on assumptions that include:

- A very rapid technological development of carbon capture and storage (the feasibility about which is questioned by many, including Al Gore)
- Nonhydro renewables comprising half of national supply by 2050, up from less than 1% currently; to achieve this means not only massive subsidized investment but resolving the many issues associated with the intermittent nature of wind and other solar-based power
- A rapid replacement of energy-based businesses with others of similar productivity and an inertia that prevents a rapid relocation of current facilities to lower energy cost locations
- A continued expansion of coal and other energy exports in spite of carbon restraining measures overseas (without which the whole policy is pointless)

It might be argued that energy cannot be that important since it is only 5% of GDP and rather less than this if its distribution costs are excluded. But much the same can be said of food, which in rich countries comprises only some 12% of GDP and most of this is accounted for by distribution and value-added features.

The danger of serious adverse economic outcomes from actions that try to force major structural change suggests a gradual and progressive response to the threat unless such a strategy carries risks of irretrievable damage. The case for a gradual response is reinforced by some of the estimates of alternative policy responses. For Australia, the Treasury's modeling estimates the effects of Australia taking no mitigatory action until 2020 then catching up in the years to 2050 if the need is confirmed and should the necessary huge preponderance of countries also agree to take comparable actions. The net cost to GDP of such a strategy is 0.3%. By 2020 the need for emission reduction policies will be clearer and we will, presumably, have access to all the technological advances that the Treasury claim will be forthcoming.

It is often said that we should "give the planet the benefit of the doubt" by acting to radically reduce GHG emissions, even if the necessity is unclear.

Such a risk-averse approach is often associated with the "precautionary principle."

However, risk has symmetrical features. Focusing only on the possible damage to the environment fails to consider the risks that people—especially in the Third World—will, as a result of forcing lower emissions, fall short of the living standards they seek and which would otherwise seem to be available. We can never be certain about the future and its possibilities, and, if feasible, we should avoid foreclosing opportunities for higher living standards that people appear to want.

Chapter 8

Efficiency First: Designing Markets to Save Energy, *and* the Planet

Noah Long, Pierre Bull, and Nick Zigelbaum
Natural Resources Defense Council (NRDC), San Francisco and New York

Chapter Outline

1. INTRODUCTION

Rapid uptake of energy efficiency is the single largest and most cost effective means of meeting rising global demand for energy services while reducing greenhouse gas (GHG) emissions. For example, a recent McKinsey report estimates that investments in efficiency could realistically cut U.S. energy consumption 23% by 2020, saving consumers $700 billion and creating over 600,000 jobs.[1] Over the longer term, even conservative estimates show immense savings opportunities from efficiency—in the range of 30% over 30 years, which could save Americans $7.5 trillion.[2] Policies that realign market incentives and remove barriers to promote innovation and reward successful technologies and practices will reduce energy use while meeting energy service demands. Efficiency is possible in every industry in a variety of

1. McKinsey & Company. (July 2009). *Unlocking energy efficiency in the U.S. economy. www.mckinsey.com/clientservice/electricpowernaturalgas/US_energy_efficiency/;* see also McKinsey & Company. (2009). Pathways to a low carbon economy: Version two of the global greenhouse gas abatement cost curve, pp. 38–42

2. David Goldstein. (2010). *Invisible Energy: Strategies Rescue the Economy and Save the Planet* (pp. 32–35). Point Richmond: Bay Tree Publishing. Goldstein discusses more aggressive scenarios with higher rates of technological change under which 70–80% savings would be possible (see pp. 38–43).

Energy, Sustainability and the Environment.

forms: improved efficiency of computer servers, industrial process, commercial heating and cooling, video games, televisions and our homes, even location efficiency in land-use planning that can vastly reduce the number of miles traveled by commuters.

Huge progress has already been made in increasing energy efficiency in a number of products and jurisdictions. Some new desk lamps, for example, can now operate with only 10% of the power a common desk lamp would have used—and too many still use—only a few years ago. And in perhaps the most famous example of effective energy efficiency improvement, California efficiency standards spurred innovation that led to vastly more efficient, inexpensive, and feature-laden refrigerators, which displaced massive expansion of fossil and nuclear generation. Whereas entropy and natural resource degradation frame the limits and economic tradeoffs for fossil, nuclear, and even renewable energy generation, efficiency exploration and discovery proves continually dynamic and unbounded: today's efficiency gains serve as guiding pathways toward greater innovation and adoption of new practices with ever deeper energy savings as the result.

The promise of energy efficiency to bring about sustainable economic growth is not a reason for "technological optimism." Technological innovation alone will not solve the climate crisis or lead to a "sustainable" economy.[3] Instead, the focus must shift to the social-institutional arena, framed by the dynamic of government regulation and industry players who ultimately direct the capital that designs, produces, and delivers the goods and services that is valued in our economy. Indeed, the experience of energy efficiency is that well-designed government regulation is key to transforming markets and the behavior of market actors. Well-designed government regulation does not mean *more* regulation; in many cases, it means reduction in regulation and greater dependence on market forces.[4] Rather, well-designed regulations shift the motivations of market actors through changes in incentives and mandates to pursue energy efficiency intensely, without undermining the performance or quality of the product or service provided—more often improving the quality of the products or energy services provided.[5]

While energy efficiency alone will not bring about a sustainable economy, rapid uptake of existing efficiency technologies will facilitate development of further energy saving technologies, which will be key to improving living

---

3. On the impacts of behavior modification and improved building energy efficiency, see this book's Chapter 11 by William Prindle.

4. The best example of less regulation being better may be in increasing location efficiency, where local zoning often forbids mixed uses. In this chapter, much of the regulation discussed is of electric and gas utilities, which are already heavily regulated. This chapter encourages different, not more, regulation of utilities to bring about energy efficiency.

5. Energy efficiency is a prime example of what leading business and media strategist Umair Haque calls "thick value" creation. Umair Haque. (2009). The value every business needs to create now. *http://blogs.hbr.org/haque/2009/07/the_value_every_business_needs.html*

standards and environmental sustainability. Buildings use approximately 40%[6] of the energy consumed in the Unites States and account for 30%[7] of GHG emissions. Appliances make up a large and growing percentage of the energy consumed in buildings, over 40% overall.[8] Meanwhile, electric and gas utilities supply the vast majority of nontransit energy consumed in homes, businesses, and by industry. Utilities have a range of supply options to meet energy demand, most of which have multiple adverse environmental impacts, including contributing to climate change. Efficiency is the least costly, but often least considered, source of "new" energy supply.

This chapter examines the three essential policy components to unlock efficiency in the marketplace, namely,

- Establishing and periodically updating minimum energy performance standards
- Removing market barriers to greater use of efficiency
- Incentivizing high efficiency performance

1.1. Minimum Energy Performance Standards

The far left end of the spectrum in Figure 1 presents the segment of the market that resists progress on efficiency, who we term the "laggards." In the private sector, laggards include noncompliant products and equipment where

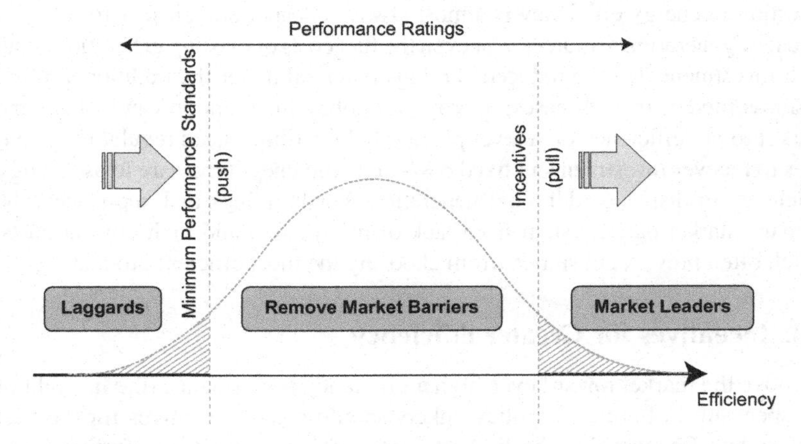

FIGURE 1 A Transformative Market Design for Efficiency *Source: NRDC (2010)*

6. U.S. Department of Energy. (2007). *Annual energy review 2006*.

7. U.S. Department of Energy. (2008). *Building technologies program: Planned program activities for 2008–2012* (pp. 1–5). *www.eere.energy.gov/buildings/publications/pdfs/corporate/myp08 complete.pdf*

8. U.S. DOE. (2009). *Energy information administration buildings databook. http:// buildingsdatabook.eren.doe.gov/TableView.aspx?table=1.1.4*

minimum appliance energy efficiency standards apply, buildings that do not meet minimum energy code requirements, and utilities that fail to utilize cost effective efficiency as an alternative to generation. In the public sector, laggards include regulatory bodies who fail to keep up with legislated schedules for adopting minimum appliance standards—regions without mandatory building code requirements or local jurisdictions lacking appropriate resources to enforce the code, and in the utility realm, the public regulatory authorities that refuse to recognize the value that efficiency can bring to the entire utility system. Well-designed and rigorously enforced minimum performance standards are the best tool for eliminating the private sector laggards in the market. As markets transform, these minimum standards should ratchet upward to keep pace with market-wide efficiency improvements. A suite of advocacy approaches, including education, outreach, top-down political pressure, and legal challenge is often required to motivate public sector laggards.

1.2. Removing Market Barriers to Greater Efficiency

Energy efficiency will only be pushed into the mainstream by removing some of the key market barriers blocking its widespread uptake for the majority of market participants (the middle segment in Figure 1). While efficient buildings and appliances have lower lifetime costs, their upfront cost can be higher, and access to capital to finance efficiency investments can often prove scarce. Similarly, investing in energy efficiency is almost always cheaper and cleaner for utilities than new generation resources—or even the fuel costs of existing generation—but such investments lead to reduced earnings potential under the traditional utility business model. In both cases, appropriate policy intervention can reshape the market to put efficiency on a level playing field. Utilities need regulatory incentives to recover investments in fixed costs and efficiency if they are to use energy efficiency to displace additional generation. For buildings and appliances, the primary market barriers stem from lack of information and high upfront costs, which often prevent consumers from choosing the most efficient product.

1.3. Incentives for Greater Efficiency

To move the market toward ever-higher efficiency performance (the far right of the spectrum in Figure 1), policymakers should create incentives for the best performers. Encouraging the highest levels of efficiency drives further innovation, which leads to higher energy savings. If efficiency is to play a truly meaningful role in preventing a climate disaster and increasing the benefits of energy services in the long run, we must *accelerate* the widespread adoption of today's cutting-edge efficiency technologies and ideas. In practice, this means revolutionizing utility compensation frameworks, directing upstream industry capital flows to value efficiency in buildings and appliances, and continuing to build consumer demand on energy performance.

2. UTILITY REGULATION

Experience shows that while well-regulated utilities can substantially advance the uptake of efficiency, poorly regulated utilities can impede progress on energy efficiency by over-investing in costly supply side resources. Utility regulation does not, by itself, improve the efficiency of any given product, but transitioning utilities from *sellers of energy* to *providers of energy services* can vastly reduce the societal cost of energy services, as well as provide an engine for efficiency investment and innovation in energy end-use technologies.[9]

2.1. The Traditional Utility Business Model

Traditional utility regulation incentivizes utilities to build infrastructure and maximize electricity and gas sales to their customers. While a discussion of the merits of regulated monopolies is outside the scope of this chapter, utilities are mostly large organizations owned either by governments, private investors, or municipalities. For the investor-owned utilities (IOUs) that dominate many markets, including the U.S., the basic bargain struck between regulators and for-profit monopolistic utilities requires utilities to guarantee that they will make adequate investments to keep the lights on for customers within their service territories. In return, regulators allow utilities the opportunity to earn a reasonable return on investments prudently made toward that purpose. Traditional regulation assumed that the only way to "keep the lights on" was to ensure that utilities could generate enough power and maintain enough transmission and distribution infrastructure to serve a forever-growing demand for energy.

Under traditional utility regulation, the cost per unit of energy—that is, "rate"—is generally set at a level to allow utilities to pay back investments in infrastructure and cover fuel costs, plus a "reasonable rate of return" on invested capital.[10] Utilities' fixed costs are typically recovered in volumetric

---

9. The National Action plan for Energy Efficiency recently included the following recommendation as key to pursuit of energy efficiency: "Modify policies to align utility incentives with the delivery of cost-effective energy efficiency and modify ratemaking practices to promote energy efficiency investments." U.S. Department of Energy and U.S. Environmental Protection Agency. (2006). *National action plan for energy efficiency report. www.epa.gov/cleanenergy/documents/napee/napee_report.pdf*; Stow Walker. (2005). Will utility rate reforms open the door wider for distributed generation and demand side management? *CERA Decision Brief*, April 2005. See also: Ralph Cavanagh. (2009). Graphs, words and deeds: Reflections on commissioner Rosenfeld and California's energy efficiency leadership. MIT: *Innovations*, *4*(4 November), 81–89.

10. The fair and reasonable rate depends on the risk and cost of capital; the rates selected vary considerably across jurisdictions. The impact of the rate of return on utility incentives to grow is beyond the scope of this paper. Suffice it to say that high rates of return are more likely to spur investment, but investment itself is not the "problem," since investment can be in clean or dirty generation as well as in technologies that reduce emissions or improve operational or end-use efficiency.

rather than fixed charges. These costs include the capital costs of building new generation, transmission, and distribution, but do not typically include investments in energy efficiency. Once a volumetric charge or rate is set, utilities have a strong incentive to sell as much power as they can—if they sell more than regulators anticipated when they set the rate, each additional unit brings a windfall by over-compensating the utility for nonfuel costs. If they sell less than expected, each unsold unit represents costs eaten either by shareholders of the company, in the case of IOUs, or taxpayers, in the case of publicly held utilities.

The resulting desire to maximize sales is called the "throughput incentive," which results in utilities benefiting from selling more electricity than expected and losing out by selling less electricity than expected.[11] As a result, under traditional utility regulation, the financial health of utilities is inextricably tied to its volume of electric and gas sales.

This model encourages utilities to build infrastructure and sell electricity, it discourages them from investing in or promoting the lowest-cost means of providing energy services to their customers: energy efficiency.

Traditional utility regulation thus aligns utility incentives with increasing sales, and utilities have responded accordingly, often encouraging energy consumption at a rate that even outpaces population growth. Some economists and casual observers believe that energy consumption either drives economic growth or is caused by economic growth, but there is little evidence for either. Energy consumption is expensive and has dire environmental consequences that are often paid for in other sectors of the economy (with health care and environmental degradation being the most obvious).[12] High energy costs also reduce consumer spending and business investment in cleaner and more job-intensive sectors. In fact, California presents a striking example of the opposite: effective efficiency policies have nearly halted energy consumption growth per capita while economic productivity has continued to climb.[13]

Efficiency is often called an "invisible resource" and it is perhaps understandable that in the time when electricity grids were first built, the concept of deploying efficiency as a resource to reduce consumption was not yet

11. See, e.g., Lisa Schwartz. (2009). The role of decoupling where energy efficiency is required by law. *Regulatory Assistance Project Issue Letter, September 2009*; Joseph Eto, Steven Stoft, and Timothy Belden. (1994). *The theory and practice of decoupling*. Energy Division, Lawrence Berkeley Laboratory, University of California Berkeley; Chris Marnay and G. Alan Comnes. (1990). *Ratemaking for conservation: The California ERAM experience*. Applied Sciences Division, Lawrence Berkeley Laboratory.

12. National Research Council. (2009). *Hidden costs of energy: Unpriced consequences of energy production and use*. National Academies Press.

13. Ralph Cavanagh. (2009). Graphs, words and deeds: Reflections on commissioner Rosenfeld and California's Energy Efficiency Leadership. MIT: *Innovations*, *4*(4), 81–89; David Roland-Holst. (2008). *Energy efficiency, innovation and job creation in California*. Center for Energy, Resources, and Economic Sustainability (Ceres), University of California Berkeley.

well-understood. But a century later, when the benefits of energy efficiency are well-documented, utility regulatory structure remains woefully out of date in most jurisdictions. In states where policy frameworks for efficiency are in place, a utility can pay its customers to use more efficient technologies at much lower cost than it can build new resources. Customers benefit in two ways from these investments: first, they are often eligible for financial assistance from utilities for efficiency upgrades, creating upfront benefits in the form of rebates or lower-cost products and longer term energy bill savings; and second, those customers that do not directly participate in the energy efficiency programs also benefit because utilities need to make fewer supply-side investments as demand levels off, lowering the total system costs for all customers.

Another way to think about the importance of energy efficiency is to consider the magnitude of the investment—between $1.5 and $2 *trillion*— utilities will need to make over the next two decades to meet projected customer demand.[14] The U.S. utility electric bill was $364 billion in 2008 (Figure 2). To avoid that bill rising dramatically, it is imperative that utilities direct that investment in clean, low-cost solutions. With the right policy framework in place, it is almost always cheaper to invest in reducing demand through energy efficiency than to increase the supply of energy. As a result, utilities can provide increased energy services with less generation than usually

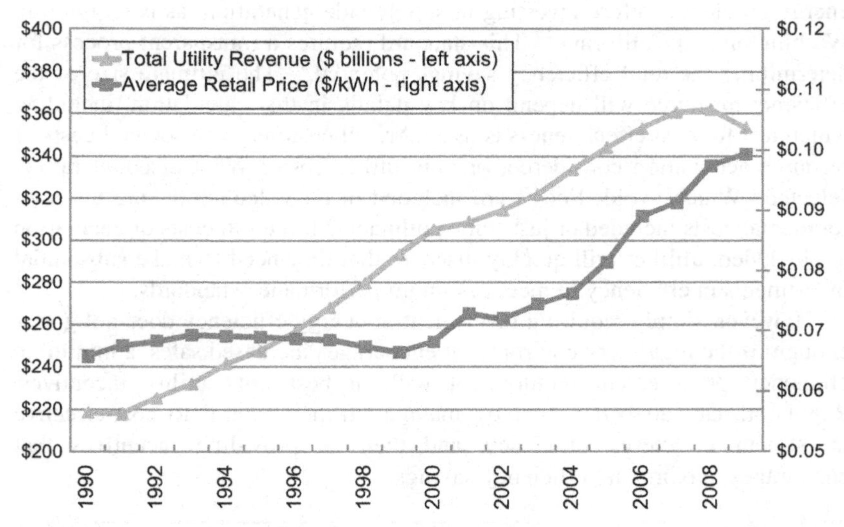

FIGURE 2 Total U.S. Electric Utility Revenue in Real (2005 Reference) Billion $USD and Average Retail Electric Power Rate in Nominal $USD per kilowatt-hour (kWh) from 1990 to 2009 *Source: EIA (2010)*

14. The Brattle Group. (2008). *Transforming America's power industry: The investment challenge 2010–2030* (p. 2). The Edison Foundation.

projected, at lower societal and environmental cost. Customers and utilities will be better off if these resources are leveraged to investment in cost-effective demand side efficiency rather than ever-increasing generation but only with proper regulations that still do not exist in most states.

In rapidly developing countries like India, China, Brazil, and Mexico, where electricity consumption per capita is growing, the scale of required investment to meet increasing demand is even larger. If that investment is turned toward reducing demand through efficiency rather than increased generation, the total energy bill, and total environmental impact of the energy sector, will be reduced substantially.

2.2. Minimum Efficiency Performance Standards

Minimum efficiency performance standards ensure utilities invest appropriately in energy efficiency. Various jurisdictions have attempted different versions of minimum efficiency performance standards, using either investment metrics, for example dollar amounts or percents of revenue, or savings metrics, kW, kWh, and therms. This requirement is often enforced with per-unit penalties for under-performance, providing an established enforcement mechanism and a clear, though incomplete, incentive for utility compliance. Probably the best minimum performance standard is a requirement that utilities acquire all cost-effective energy efficiency before investing in supply-side generation, as is required in Washington and California.[15] This standard requires a transparent process for determining the total efficiency savings potential.[16] The ultimate size of the efficiency resource will depend on key details in this calculation, including which test for cost effectiveness is used: Are all customer and societal costs of resource acquisition considered, or only utility costs? What discount rate is selected? Which avoided costs are included in the calculation—are all environmental costs included or just some pollutants? If the full costs of generation are included, utilities will quickly discover that they need to make substantial investments in efficiency to meet minimum performance standards.

However, simply requiring investment in energy efficiency does not go far enough: if the regulatory environment encourages increased sales, a minimum efficiency procurement requirement will, at best, mix utility incentives. Regulators can do better, first by making utilities neutral to cost-effective investment in energy efficiency, and then by providing incentives that encourages maximizing efficiency savings.

15. Evaluation, measurement, and verification of efficiency savings is also relevant here. Particularly where utilities stand to pay penalties, minimum performance standards should be enforced with a transparent process for program evaluation and savings measurement with a dispute resolution process.

16. Various states require a utility-run Integrated Resource Planning process, while others use independent evaluation of efficiency potential.

2.3. Breaking the Link between Sales and Financial Stability

To make utilities able to achieve energy efficiency requirements, regulators must allow them to recover costs in energy efficiency investment. Whether administered by utilities or third parties, efficiency investments have to be paid for. In the U.S., some jurisdictions allow these expenses to be "capitalized" and added to the rate base, or even to earn a higher rate of return than alternative investments.[17] This policy generally vastly over-compensates utilities for expenses, without regard to the effectiveness of the investment. Other jurisdictions set "system benefit charges" at a certain percent of revenue that must be spent on efficiency and often renewable energy resources.[18] Although this system can guarantee efficiency spending, it does not guarantee that enough funds will be available to pursue all cost-effective energy efficiency. The simplest regulatory fix is to require utilities to achieve all cost-effective energy efficiency and allow them to recover the costs of those investments in their volumetric rates.

In order for a utility to support efficiency without undermining its ability to recover costs, regulators must eliminate the "throughput incentive" that encourages utilities to maximize sales between rate cases. This can be done most effectively through a concept called *decoupling*, which "trues up" expected sales volumes with actual sales volumes, so that utilities recover enough revenue from customers to cover the allowed nonfuel costs but no more. Recent evidence demonstrates that decoupling can be implemented with very modest impacts on customer bills, with rate impacts in the +/− 2% range.[19] Figure 3 shows the states where electric and gas decoupling is either in place or pending.

Decoupling is not the only mechanism, however, with which state utility regulatory bodies have experimented to make utilities neutral to changes in sales volumes. A variety of mechanisms have been used or discussed to eliminate the incentive for utilities to maximize sales, but many of these mechanisms either fail to solve the problem or create other difficulties. Table 1 outlines various mechanisms that are frequently put forward to make utilities amenable to efficiency, with a brief discussion of the advantages and disadvantages of each.

By itself, decoupling will not create incentives for highly effective energy efficiency programs: it will only make utilities neutral to increases or decreases in sales between rate cases. It is an important first step toward changing the business model of utilities, but only a first step. Turning utilities into energy-service companies with incentives for achieving greater efficiency

17. This mechanism is mentioned in the Section 3 of this chapter as a mechanism for incentivizing high levels of investment.

18. Twenty states and the District of the Columbia have some state or local system benefit charge for energy efficiency or renewable energy. Refer to *Database on state incentive for renewables and efficiency*; *www.dsireusa.org/*

19. Pamela Lesh. (2009). *Rate impacts and key design elements of gas and electric utility decoupling: A comprehensive review.*

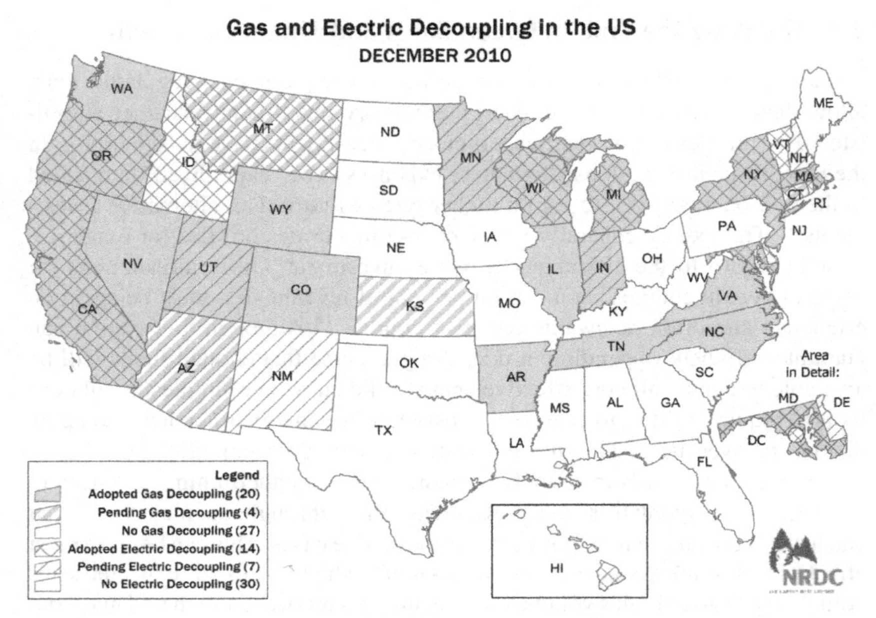

FIGURE 3 U.S. State Adoption of Electric and Gas Decoupling *Source: NRDC (2010)*

in place of new energy generation should be the second step in the transformation. Incentives for efficiency investments without decoupling can lead to perverse impacts: for example, a utility might promote its own energy efficiency programs, but oppose governmental investment in efficiency or improved building codes and appliance standards that would reduce utility sales and revenue.

2.4. Utility Efficiency Incentive Mechanisms

Even with decoupling, however, privately owned utilities have no opportunity to earn a return on investments in efficiency, unlike investments in infrastructure, new power plants, or transmission and delivery networks that accompany additional generation. Thus, even when a utility can recover costs of efficiency investments, and is made neutral to sales impacts of efficiency programs, a utility will likely choose supply side investment over energy efficiency alternatives. This is true even though energy efficiency offers a far less expensive means of ensuring adequate energy to meet customer demand. Regulators should seek more than utility neutrality to efficiency—they should seek to align shareholder interests with customer interests in minimizing system costs and environmental impacts.

Various mechanisms are in use around the country to promote utility activity on energy efficiency and a number of these reform efforts have lead to

TABLE 1 Mechanisms to make Utilities Neutral to Decreases in Sales Volume Between Rate-Cases[1]

Mechanism	Description	Discussion
Decoupling	Regular adjustments in retail rates to allow recovery of fixed (non-fuel) costs based on actual rather than predicted sales.	If implemented properly, this mechanism removes the "throughput incentive to utilities. Utilities recover all allowed investments and fuel, but no more or less. Requires regular true ups, although experience shows impact on customer bills to be minor.
Lost Margin Recovery	Assured recovery of the lost revenues net of avoided variable costs due to energy efficiency programs administered by the utility.	Compensates utilities for lost sales from energy efficiency programs, but does not eliminate incentive to maximize sales to over-recover fixed costs. Requires adequate measurement to assure "lost revenues" actually exist. Also, does not include impacts efficiency actions taken by others, including codes and standards
Straight Fixed-Variable Rate Design	Rate design that recovers all costs that do not vary with sales volume in a fixed charge (volumetric pricing only for fuel/ power costs).	Reduces efficiency and conservation incentives for customers as large and small consumers pay equal fixed charges.
Frequent Rate Cases	Rates are regularly adjusted to reflect current or predicted sales.	Utility incentive to increase sales between rate cases remains. In practice, utilities are always between rate cases.

(Continued)

TABLE 1 Mechanisms to make Utilities Neutral to Decreases in Sales Volume Between Rate-Cases[1]—cont'd

Mechanism	Description	Discussion
Future Test Period	Set rates in a rate case based on "next year" costs rather than on a historical test year basis.	Utility incentive to increase sales between rate cases remains.
Eliminate Purchased Gas Adjustments and Power Cost Adjustments.	Eliminate adjustment of retail rates that pass through changes in power supply or gas supply costs.	If marginal power costs are higher than retail rates, utility will have incentive to reduce sales, but does not permanently eliminate the incentive to increase sales.
Minimum Energy Efficiency Performance Standard	Utilities required to achieve a defined level of energy efficiency	If coupled with penalties for failure, can create a meaningful incentive to promote energy efficiency, up to the specified level. Does not eliminate the throughput incentive.
Reduce Return on Equity[2]	Reduce allowed return on equity to minimize incentive to invest.	Reductions of allowed rates of return will reduce the utilities incentive (or ability) to invest, but does not affect the incentive to maximize sales between rate cases.

[1]*Much of the information in this table was influenced by a collaborative process on utility incentives in the Northwest Energy Coalition. The authors take full responsibility for any errors and for the opinions expressed.*

[2]*Utility regulatory commissions set the allowed return on equity when determining rates for electricity and gas. This mechanism is favored by those who believe utilities cannot be transformed into energy services companies. Whether they select the appropriate rate is always subject to debate (and the uncertainty is emphasized by the range of rates selected by commissions). This mechanism has the negative result of reducing a utilities ability to invest in anything, including technologies favorable to environmental performance and efficiency.*

Source: NRDC (2010)

TABLE 2 State adoption of utility efficiency incentive mechanisms

Incentive Mechanism	State
Shared Savings Incentive	AZ, CA, GA, OK, CO, CT, RI, KY, MA, MI, MN, NH, RI, TX, VT
Rate of Return Incentive	NV, WI
Power Plant Pricing for Energy Efficiency	OH

significant efficiency achievements. Table 2 shows the incentive mechanisms in operation in the Unites States.

Source: The Edison Foundation Institute for Electric Efficiency. (2009). *State energy efficiency regulatory frameworks.*

Unfortunately, some mechanisms miss the mark by encouraging investment rather than efficiency performance. Table 3 describes and briefly lists advantages and disadvantages of various efficiency incentive mechanisms.

Given the potential cost to utility customers of providing an earnings opportunity to utility shareholders for energy efficiency, preferred mechanisms provide an earnings opportunity to utilities based on independently verified customer savings from efficiency programs. Such an earnings opportunity can also be reasonably linked with a risk of penalties for poor performance.

Performance assessment of utility energy efficiency acquisition will require some level of evaluation, measurement, and verification (EM &V) of programs and claimed savings. While a sizeable industry now exists for the purpose of evaluating program administration and implementation and measuring savings results in the United States, the very nature of the field—determining what would have happened if utilities had not intervened—implies a certain level of uncertainty. Jurisdictions vary widely on their protocols for measuring energy savings: some efficiency savings programs are hardly measured at all, while others are subject extensive review. Evaluation of energy programs is very useful for program improvement, but can also lead to controversy when performance evaluation is linked to utility earnings. Successful evaluation programs include transparent and collaborative evaluation plans and meaningful dispute resolution mechanisms to prevent regulatory paralysis.

California's "shared savings" mechanism, approved in 2007, provides the state's IOUs an opportunity to profit from energy efficiency investments by keeping a small percentage of the net benefits from efficiency programs (Figure 4).[20] California's publicly owned utilities were only recently required to

20. CPUC, D. 07-09-043. (2007). *Interim opinion on phase 1 issues: Shareholder risk/reward incentive mechanism for energy efficiency programs.*

TABLE 3 Mechanisms to Encourage Utility Energy Efficiency Resource Procurement[3]

Title of Mechanism	Description	Advantages/ Disadvantages
Rate of Return Incentives	Utility receives a higher return on equity for investment in energy efficiency	Incentivizes investment, rather than actual savings, can lead to very high earnings on efficiency investments
Shared Savings Incentives	Utility receives a share of the net benefits or percentage of program costs based on energy efficiency achievement, or achievement above a certain threshold.	Can require complicated and contentious measurement, but generally aligns shareholder and customer interests for efficiency
Power Plant Pricing for Energy Efficiency	Utility receives a payment for energy efficiency investment that based on the cost of power from a new power plant.	Can lead to much higher payments to utilities than other mechanisms, and ensures adversarial discord over necessarily imperfect measurement of savings
Tax credits to the Utility for Energy Efficiency Investment	Utility receives a reduction in state taxes if it invests in energy efficiency.	Requires state tax subsidies for efficiency spending, which could be covered by utility customers; incentivizes investment not performance
Penalties for Non-achievement of Energy Efficiency Requirements	Utility pays penalty if it falls short of the required level of energy efficiency savings achievement	Solely punitive "incentives" promote utility orientation toward compliance rather than entrepreneurialism; may work better if combined with a meaningful earnings opportunity for high performance

[3]*Much of the information in this table was influenced by a collaborative process on utility incentives in the Northwest Energy Coalition. The authors take full responsibility for any errors and for the opinions expressed.*
Source: NRDC (2010)

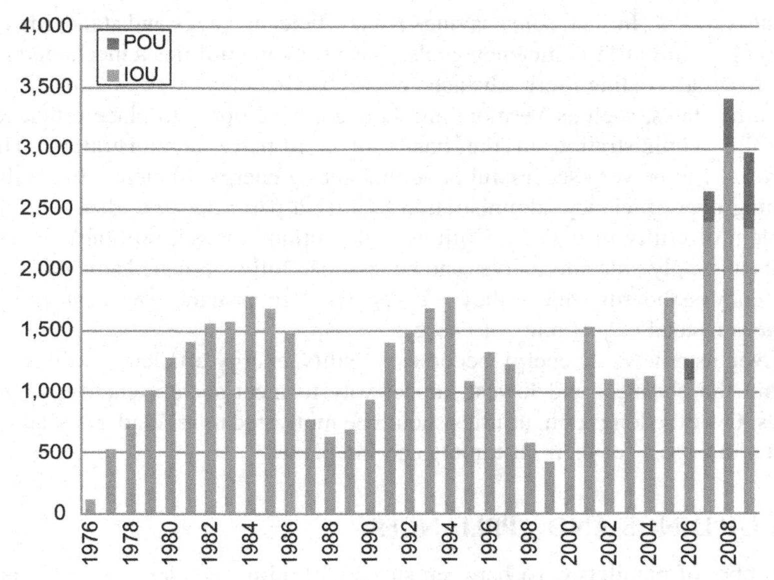

FIGURE 4 Annual California Utilities'—Investor-Owned (IOU) and Publicly-Owned (POU)—Program Energy Savings in gigawatts-hours (GWh) from 1976 to 2008*

**1976-1997 data from the California Energy Commission, provided by Sylvia Bender; 1998-2003 data is from Investor-Owned Utilities' (IOUs) Annual Earnings Assessment Proceeding (AEAP) reports, including savings from the "Summer Initiatives" during California's electricity crisis; 2004-08 IOU annual savings data are calculated by scaling program totals by utility-reported annual savings. PUC,* Energy Efficiency 2006-2008 Interim Verification Report, *Resolution E-4272, Table 8: 2004-2005 Cumulative Savings Estimates, Final VR Ex-Post Evaluation, p.37 (October 15, 2009) (2004-2005 savings). Available at: http://docs.cpuc.ca.gov/PUBLISHED/FINAL_RESOLUTION/108628.htm. PG&E,* Annual Earnings Assessment Proceeding, *Application No. 05-05-001 (May 2, 2005); SCE,* Annual Earnings Assessment Proceeding, *Application No. 05-05-005 (May 2, 2005) ; SDG&E,* Annual Earnings Assessment Proceeding, *(May 2, 2005) ; SoCalGas,* Annual Earnings Assessment Proceeding, *(May 2, 2005) (utility-reported program savings used for scaling purposes only). PUC,* Decision Regarding RRIM Claims for the 2006-2008 Program Cycle, *D.09-12-045, R.09-01-019, Table ES2b: Impacts with Positive and Negative Interactive Effects, 2nd Earnings Claim for PY 2006-2008, p.C3 (December 17, 2009) (2006-2008 program cycle total savings). PG&E, 2008 Annual Report, EEGA, Table 1 (May 1, 2009); SCE, 2008 Annual Report, EEGA, Table 1 (August 3, 2009); SDG&E, 2008 Annual Report, EEGA, Table 1 (August 12, 2009); SoCalGas, 2008 Annual Report, EEGA, Table 1 (September 5, 2009) (utility-reported program savings used for scaling purposes only). Source: NRDC (2009)*

set efficiency savings goals and report savings, but have no mechanisms currently in place to eliminate the throughput incentive, nor is there any statewide policy that requires a positive incentive for delivering effective programs.

In addition to California's success in utility efficiency program administration, the state has some of the most advanced building codes and appliance standards in the United States. These regulatory programs have been advanced and supported by utilities. Managers of California utility support energy saving regulatory programs know that reduced energy sales will not undermine their business model. In fact, some savings from efficiency codes and standards are counted toward utility efficiency goals, thus providing utilities a mechanism to earn from advocating more advanced energy saving codes.

Some states, such as Vermont and Oregon, have opted to place efficiency portfolio administration in the hands of third-party administrators. This approach can be very successful at administering energy efficiency funds, but handing over efficiency administration to a third party does not deal with the problem of utility incentives. Utilities will continue to seek profitable investments in supply-side alternatives and be less than fully motivated advocates for efficiency standards unless they are regulated to provide low cost energy services instead of just units of energy.

Even as renewable energy becomes a reality, energy efficiency is likely to remain the cheapest and lowest impact way to meet society energy service needs. Over the long term, utilities should be motivated to sell and advocate for what works best for their customers and the planet.

3. BUILDINGS AND APPLIANCES

A number of parallels exist between successful utility efficiency policies and policies promoting efficiency for buildings and appliances. However, differences in the market structures and market actors for buildings and appliances require a different regulatory and policy approach than utilities in reaching efficiency objectives.

3.1. Minimum Standards

The process for state or national standard development can be generalized into three steps and can be done either through direct legislation or a regulatory process:

- Understand the market characteristics of the particular product or building type and set an appropriate standard. This step requires looking at specific technical and economic data that includes market size, potential upfront incremental cost to industry and consumers, and energy cost savings.
- Define a suitable test method that tests the product in all relevant usage modes and general consumer behavior patterns. Ideally, this includes energy rating procedures for a given product or building type. This step both allows for apples-to-apples testing and is the first step toward informing the consumer.
- Enforce the standards through certification procedures (building officials and energy raters) and "off-the-shelf" testing for appliances.

3.1.1. Minimum Standards: Buildings

Much like in the utility industry, the building construction industry often ignores energy efficiency if there is no energy efficiency building code, or compliance is not adequately enforced. Builders typically don't consider operating costs during construction because consumers don't look for it and it is often initially more expensive to build more efficiently. Without consideration of operating cost, builders may build the least efficient building possible, even if the purchaser/owner will end up paying a great deal more—when purchase and energy-operating costs are added—over the lifetime of the building. Areas without energy efficiency building codes allow builders to create extremely inefficient buildings. Without a mandate, locally enforced energy efficiency building code, builders often construct the least first-cost buildings they can, without considering operating cost and efficiency.

In the U.S., building energy codes are developed by two bodies: the International Codes Council (ICC) and the American Society of Heating Refrigeration and Air-Conditioning (ASHRAE). The ICC produces the International Energy Conservation Code (IECC) to govern low-rise residential and ASHRAE produces Standard "90.1" to cover commercial and high-rise residential. These are dubbed the *National Model Codes*. Every three years (approximately) the model codes are updated to incorporate current technologies and best practices. The last revision of the IECC (2009) is projected to save 12–15% of the energy demand of a new building over the 2006 version,[21] and the 2004 version of 90.1 saves 14% over the 1999 version.[22] Until recently, the model codes did not make very significant improvements in efficiency. Adoption of these model codes is left up to individual states, where the track record varies considerably.

A similar cycle is at play in the European Union, where energy codes are now required for all member nations. However, these codes are developed and implemented by each member nation as opposed to being centrally created. The disjointed nature of code creation means that some jurisdictions have inadequate codes, leaving room for inefficient new buildings.

In China, there are a set of national building energy efficiency standards, which are currently set at levels of approximately 50% more efficient than a standard 1980 Chinese building. Unlike the U.S., China has selected a benchmark in 1980 and all new standards are compared to that benchmark. The Ministry of Housing and Urban-Rural Development (MOHURD) develops these standards and all cities are mandated to implement the current 50% standard by 2010. Large, more developed cities like Shanghai and Beijing are required to implement a 65% standard and by 2020 all cities should follow the 65% standard, with leading cities utilizing an as-yet

---

21. Energy Efficiency Codes Coalition. (2009). Energy & cost savings analysis of 2009. *IECC Efficiency Improvements*. *www.thirtypercentsolution.org/solution/EECC-Savings_Analysis-Jan-2009.pdf*

22. United States Department of Energy, Office of Efficiency and Renewable Energy. (2008). *Federal Register*, *73*(250, December 30), notices. *www.energycodes.gov/implement/determinations_90.1-2004.stm*

undeveloped 75% standard. These standards apply to both residential and commercial buildings.

India has a similar approach, with less future planning. India has a mandatory building energy efficiency code, the Energy Conservation Building Code (ECBC), which applies to all structures connected to a load greater than 500 kW. That would include high-rise residential and commercial buildings. This code was developed by India's Bureau of Energy Efficiency (BEE). BEE has a broad mandate to reduce wasteful energy consumption, similar to that of the California Energy Commission.

These countries share the understanding that mandatory minimum energy codes save building owners and occupiers money, reduce wasted energy, and catch the laggard builders. Unfortunately, each country struggles with implementation of these codes, which requires more than simply a mandate. Until recently in the U.S., states had little incentive to adopt codes aside from the energy benefit. A similar issue exists in China and India, where regions frequently resist adoption and have vast internal disparities in wealth, making efficient buildings that much more difficult to mandate. Local politics and the drive for economic growth is often a major barrier to even highly cost effective energy building code regulation and implementation, setting up a classic "race to the bottom." The strongest opponents to code adoption and implementation have included real-estate brokers, building developers, and construction industry. This in turn causes local officials who, in the drive to deepen tax-base revenues and show "real signs" of local economic health, go along with developers to expedite new building development and miss huge energy and financial savings opportunities.

States including California, Massachusetts, states in the Pacific Northwest, New York, and Florida have all discovered on their own that mandatory building codes and appliance standards are critical to their energy security. Some Indian states and Shanghai and Beijing have also pushed their own mandatory codes and energy efficiency programs. In the U.S., these programs can cost as little as 2.5 cents/kwhr, out-competing traditional coal generation.[23] Building codes and appliance standards in California have avoided the construction of nearly 24 large power plants over the past 30 years.[24]

In ideal situations, a national agency like the BEE, MOHURD, or at some point in the future, the U.S. Department of Energy (DOE), would be tasked with the development of national model codes to eliminate "wasteful, inefficient, unnecessary, or uneconomic uses of energy"[25]—the mandate of the California Energy Commission. Nationalized codes makes competition among

23. American Council for an Energy Efficiency Economy. (2009). *Saving energy cost-effectively: A national review of the cost of energy through utility-sector energy efficiency programs. www.aceee.org/pubs/u092.htm.*

24. California Energy Commission. (2005). *Integrated energy policy report* (CEC-100-2005-007), p. 70

25. California Public Resources Code, Section 25401.

architecture, construction, and engineering firms across state and country lines easier, with opportunity to scale best practices and lower the time and resources it takes for building industry participants to learn and adapt to ongoing code updates and revisions.

Unfortunately, national mandates on state building codes have not yet materialized. However, the American Recovery and Reinvestment Act of 2009 (ARRA) provides grants to states that agree to update building codes. In addition, the states must achieve 90% compliance with the updated codes in the future.[26] Almost every state committed to updating their code in order to receive State Energy Program funding.[27]

Building codes in the United States, China, Europe, and India are implemented and enforced at the state level, rather than centralized. Each state has the power to minimize energy use in buildings, or waste it. As a result, broad leadership is required to realize the benefits of efficiency through enhanced mandatory state code implementation and enforcement.

If every country developed or simply borrowed and implemented a national, mandatory energy efficiency code, builders would have harmonized requirements and could benefit from shared practices, consumers would have a reliable minimum performance, and the energy savings would be staggering.

According to the Building Codes Assistance Project, if every U.S. state adopted the latest model energy codes, without considering the impacts of future updates and increased stringency, the collective impact in 2030 would amount to 1.2 quads of electricity saved, 80 million metric tons of greenhouse gases avoided, and 9 billion dollars back in the consumers' pockets. Coupling this with regularly updated codes and adequate labeling, the impact would increase dramatically.

3.1.2. Minimum Standards: Appliance

Similarly, without strong minimum standards to push industry to manufacture efficient appliances, consumers will be left with inefficient appliances to carry out essential services both at home and in the workplace.

The technical challenges to improving energy efficiency in large appliances and equipment are in many ways similar to how efficient buildings are designed and built. However, the appliance and equipment market differs from buildings in that, (1) most appliances have lifetimes spanning years not decades, and (2) most appliances are portable; that is consumers have the ability to either physically

26. The 2009 federal stimulus bill, American Recovery and Reinvestment Act, Section 410 required governors to provide assurance that they would seek to update their building codes in order be eligible for State Energy Program Funds. Increased code implementation through inspector training and support (funded through utility or state programs) is an enormously cost-effective energy efficiency investment opportunity.

27. Building Codes Assistance Project. (2009). *http://bcap-energy.org/node/461*. Only one former governor, Sarah Palin of Alaska, refused to provide assurance, but the Department of Energy has not denied funding to any state based on failure to follow through on this commitment.

move the product as their home or work settings might change or secondarily sell the product to another consumer. This aspect of portability presents a set of unique challenges to implementing efficiency for consumer appliances.

Over the latter half of the twentieth century, the U.S. was the global leader in designing and exporting the mass production techniques for a large majority of consumer appliances. Examples of such appliances include the clothes washer, dishwasher, and refrigerator in the 1950s to the mainframe computer of the 1960s to the personal computer in the 1980s. The rapid technological development of consumer appliances occurred at the same time the U.S. and most of the developed world saw unprecedented growth and stability in fossil and nuclear-based energy supplies. Much like in the utility and building sectors, cheap and abundant energy led to complacence among product designers, engineers, and consumers about energy efficiency.

Since many of the leading appliance manufacturers were in the Unites States over this period, the most important appliance efficiency regulatory body was the U.S. Department of Energy (DOE). Up until 1987 with the passage of the National Appliance Efficiency Conservation Act (NAECA), no major U.S. federal legislation existed to mandate minimum appliance efficiency standards. Two earlier federal laws, the Energy Policy and Conservation Act of 1975 that set efficiency "targets" for certain appliances, and the National Energy Policy Conservation Act (NEPCA) of 1980 that authorized DOE to promulgate regulations for 13 products, amounted to ineffective policies to push the appliance market to be more efficient.[28]

After passage of NEPCA of 1980, the Reagan Administration-led DOE chose not to follow mandate to promulgate standards according to the legislated schedule in NEPCA to develop minimum efficiency standards for a number of products. Advocacy groups, led by the NRDC, and several progressive states including California, New York, and Massachusetts, led an effective litigation strategy against DOE through the early 1980s. It was the successful settlement of this NRDC-led legal action against DOE and appliance manufacturers that led to the enactment of NAECA in 1987.

Absent DOE willingness to follow orders to mandate new minimum appliance standards, advocates have found success in setting state-level appliance standards, despite federal preemption for some product types like clothes washers.[29] This efficiency advocacy strategy has proven effective for several product types. In essence, manufacturers, who fear 50 different policies

28. Howard Geller. (1995). *National appliance efficiency standards: Cost-effective federal regulations*. American Council for an Energy Efficiency Economy. *www.aceee.org/pubs/a951.htm*

29. U.S. Court of Appeals for the Ninth Circuit. California Energy Commission (CEC) vs. U.S. DOE and Association for Home Appliance Manufacturers (AHAM) on CEC waiver for preemption to set state-level water efficiency standards for residential clothes washers. Filed October 28, 2009. Before: William C. Canby, Jr. and Kim McLane Wardlaw, Circuit Judges, and David G. Trager, District Judge. Opinion by Judge Canby. No. 07-71576 *www.ca9.uscourts.gov/datastore/opinions/2009/10/28/07-71576.pdf*

regulating their given product, are coerced to join advocates to encourage DOE to set more stringent national standards.

DOE under the Obama Administration has taken more serious steps, including adequate line-item funding to keep up with the revised schedule to set new appliance efficiency regulations.[30] And there is no doubt that Energy Secretary Steven Chu is enthusiastically behind DOE's appliance standards regulatory efforts.[31] Illustrating the strength of minimum federal appliance efficiency standards as a policy tool for reducing U.S. greenhouse gas emissions, Figure 5 shows that by 2030, if DOE keeps to its rulemaking schedule and applies moderately stringent standards, over 150 million metric tons of greenhouse gas emissions will be avoided by 2030.

One area of emerging interest for appliance and equipment efficiency is application into the burgeoning digital infrastructure that is responsible for enabling our growing information-based economy. What differentiates this sector from other consumer appliance products is the sheer scale (global) and the unprecedented change of product innovation and growth; product cycles can be measured in terms of weeks and months rather than years. Keeping up with rapidly changing product cycles is a difficult challenge for governments.

For more traditional appliance products some national governments are working to coordinate to set a single standard, which would serve as *de facto* global standard. Manufacturers who engage on these regulatory proceedings often support them as it provides them with regulatory certainty to globally manufacture and sell certain products.[32] Solid political leadership to authorize new or more stringent standards with assertive regulatory follow-through to implement and enforce the rules is the surest pathway for long-term market transformation to bring up the efficiency floor for appliances.[33]

---

30. U.S. DOE. (2009). *Implementation report: Energy conservation standards activities.* Submitted Pursuant to: Section 141 of the Energy Policy Act of 2005 & Section 305 of the Energy Independence and Security Act of 2007 U.S. Department of Energy, Washington, D.C.: U.S. DOE – EERE. *www1.eere.energy.gov/buildings/appliance_standards/pdfs/seventh_report_congress_aug_09.pdf*

31. In March of 2009, Secretary Chu told *National Geographic*, "Energy efficiency can be improved very quickly.... Appliance standards, ka-BOOM, can be had right away!"

32. International Institute for Energy Conservation (IIEC), 2010

33. The California Public Utilities Commission (CPUC) defines market transformation as: "Long-lasting sustainable changes in the structure or functioning of a market achieved by reducing barriers to the adoption of energy efficiency measures to the point where further publicly-funded intervention is no longer appropriate in that specific market." See: Rosenberg, M., & Hoefgen, L. (2009). *Market effects and market transformation: Their role in energy efficiency program design and evaluation.* CA Institute for Energy and Environment. *http://uc-ciee.org/energyeff/documents/mrkt_effts_wp.pdf*

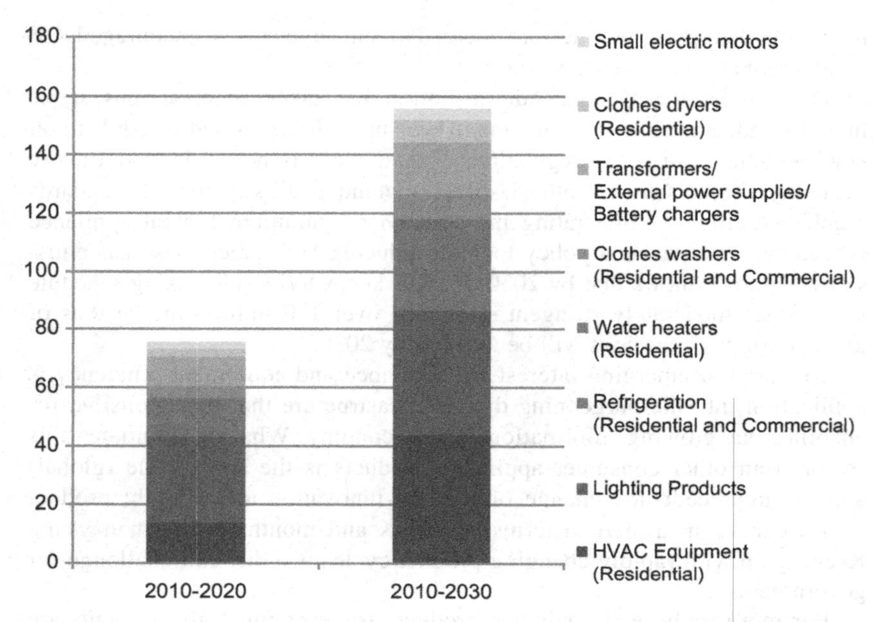

FIGURE 5 DOE Appliance Efficiency Standard Rulemakings Potential Greenhouse Gas Savings in million metric tons CO_2 equivalent (MMTCe) from 2010 to 2030. *Source: Neubauer, M. (2009)*

3.2. Information and Incentives: Overcoming the Market Barriers to Efficiency in Buildings and Appliances

Labels increase the transparency of the buildings and appliance market, creating informed consumers, operators, and owners. Efficiency labels for buildings seek to quantify or compare the efficiency of a given building. Labeling buildings and appliances allows energy efficiency to compete for provision of energy services on a more level playing field.

However, labels alone often do not overcome the impact of "first costs" on consumers, thus incentive programs—to-date most often run by utilities—are necessary to promote greater efficiency levels and promote innovation.

In the long run, labels and incentive programs designed to overcome present-day market barriers and accelerate market adoption of higher performance buildings and appliances must shape market innovation frameworks to give builders and manufacturers continuous reason to deliver better energy services in our buildings and appliances.

3.2.1. Buildings

The U.S. is lagging in this arena by not requiring any energy performance labeling for buildings. Efforts to develop the mechanisms for accurate and low-cost labeling methods are still in development, but not yet ready for widespread market adoption. Labels are essential for consumers to understand what they are consuming and how to buy the healthy choice: an efficient building.

Without an energy efficiency label, there is no way for a potential consumer to know how much energy a new building might demand. Cars, appliances, and some consumer electronics are required to show their relative efficiency. If such information were available for all buildings, consumers could make purchasing decisions based on the total budget impact of various housing options: a cheaper home with higher energy costs may have a higher total cost than a more expensive, but highly efficient one. Transparent information on the energy costs associated with buildings is arguably even more important to potential purchasers than with other products and technologies, because of the slow turnover rate and the relatively high level of energy consumption associated with buildings. According to the Center for Neighborhood Technology, poorly located and constructed buildings can cost as much to operate and commute to and from as the mortgage payments themselves![34] Unfortunately, consumers rarely see these costs when buying a home because of the problem of a lack of transparency for both location and energy efficiency. This section will focus on energy efficiency.

Labels for buildings can be divided into four quadrants, as in Table 6. Normative labeling typically identifies a threshold benchmark. These labels are awarded when a building achieves a performance-based goal and does not differentiate between buildings that meet the goal—even if they meet the goal for different reasons. Incentives or mandates can be tied to normative labeling requirements, which typically use standard testing procedures. These labels are relatively simple and universal and provide a resounding market signal. Examples include ENERGY STAR for homes and commercial buildings[35] and the United States Green Building Council's Leadership in Energy and Environmental Design (LEED)[36] certification.

Informative labels provide a scale by which to judge performance against either peers or a standardized benchmark, such as miles per gallon or the EnergyGuide labels on appliances, discussed below. These labels typically show a scale with poor performance on one end and high efficiency on the other. Unlike the normative labels, the informative versions allow the consumer to compare a broader range of products including those that do not achieve normative ratings and those that go well above and beyond such ratings. However, the informative labels require a more educated consumer to understand the rating, weighing up-front cost against long-term savings. Such informative labels include the Portfolio Manager from ENERGY STAR and the

34. See also: Goldstein, D. (2010).

35. For more information: *www.energystar.gov/index.cfm?c=home.index*

36. For more information: *www.usgbc.org/* and this book's Chapter 9 by Gray and Zarnikau.

Residential Energy Services Network (RESNET). COMNET is a labeling system under development for commercial buildings, similar to RESNET.

Informative labels can be tied to performance-based incentives, discussed later. These labels do not always provide a strong market impact, but they increase transparency to inform consumers of the differences between one building and another. Attaching incentives to an informative labeling program produces a market signal out of what otherwise may be difficult-to-use information. Incentives are a market-pull mechanism that transforms typical construction practices. Although some might argue that adequate transparency and oversight should spark the natural interest of the market in more efficient buildings, incentives are much more likely to transform the market quickly by moving consumers toward the most efficient buildings available.

Linking effective labeling to incentives for higher efficiency magnifies the impact. For example, a tax incentive can motivate more efficient practices for dramatic improvements on conventional practice. Currently a U.S. federal tax credit offers $2000 for a new home that is 50% more efficient than standard, defined as 2004 IECC model code (see Table 4). For commercial buildings, there is a tax deduction of $1.80 per square foot for buildings that exceed the 2001 ASHRAE 90.1 standard by 50%.

The deeper efficiency coupled with a tax-based incentive motivates innovation in construction practices: in 2007, 23,702 homes received the tax credit compared to one-third that number (7100) in 2006. In 2008, only 21,939 homes qualified, but in 2009, the number of homes receiving the credit increased to 37,506, despite the fact that the total number of new homes completed reduced from 485,000 in 2007 to 374,00 in 2008. The result is that by 2008, over 10% of new homes met the qualification, up from less than 1% in 2006.[37]

TABLE 4 Impact of Federal Tax Credit on Efficient Home Construction

Year	Number of Homes Verified for Tax Credit	Total Number of New Homes Sold in the U.S.	% of New Homes Sold Verified for Tax Credit
2009	37,506	374,000	10.0%
2008	21,939	485,000	4.6%
2007	23,102	776,000	3.0%
2006	7,110	1,051,000	0.7%

Source: Communications with RESNET (2010)

37. Communications with RESNET, March 2010

Operational labels refer to the energy or resource use of a building *after construction and during operation*, versus asset-value, which is a calculated measurement through computer modeling. The building is built and occupied for 12 months before it can get an operational rating. This allows for data collection of the real energy demand of that building. These data are then compared with the national average for the same building type and occupancy and a relative, operational rating is produced. Operational ratings are not achieved until after construction and testing under normal conditions. Examples include ENERGY STAR for homes and commercial buildings.

Asset-value labels refer to the "value" of the structure itself, independent of who owns or operates it. The rating is based on a computer simulation model, which compares the new building to a benchmark theoretical building. This provides an apples-to-apples comparison among any labeled home or building and can be deployed before the building is occupied or even built. An example is the Home Energy Rating System (HERS) developed by RESNET.

Operational labels give the occupant an idea of how well they are operating their building while asset-value labels give potential owners an idea of how efficient their building could be operated by any occupant.

Performance-based construction and operation incentives influence near- and long-term practices. A performance-based incentive is tied to an incremental goal above standard practice. Instead of requiring specific components, such as increased insulation or highly efficient equipment, a performance-based incentive considers whole-building energy demand. In other words, a performance-based incentive would require a building to reduce total energy consumption over a year, as compared to a benchmark such as the local energy code requirements. Performance-based incentives allow builders to find creative and cost-effective ways of achieving efficiency. However, prescriptive requirements for efficiency may focus attention on certain well-known energy culprits, they may also tie the hands of builders and stifle innovation if not repeatedly overhauled. In addition, cost-based incentives confuse spending with effective efficiency improvements. Performance-based incentives could also act on operational energy ratings such as ENERGY STAR to determine compliance, thus encouraging the user of the building as well as the design and building team to minimize energy waste.

Each of these four labels are designed for different uses and all are useful (Figure 6). For future policy, the optimal arrangement would be:

- **Operational informative labels** for quantifying real greenhouse gas emissions from a building and/or incentivizing the reductions of those emissions.
- **Operational normative labels** for recognition of outstanding performance. A company could provide an incentive to any of its offices that achieve this

	Informative	Normative
Operational	Energy Star Portfolio Manager	Energy Star for Homes Energy Star for Commercial Buildings
Asset-Value	RESNET COMNET	LEED

FIGURE 6 The Basic Tenets of Energy Labeling* **RESNET: Residential Energy Services Network. COMNET: Commercial Energy Services Network. LEED: Leadership in Energy and Environmental Design. Source: NRDC (2010).*

label. A city or county could do the same. This creates competition and goal-oriented progress.

- **Asset-value informative labels** for informing home and building buyers of the efficiency of a given building. These labels may drive up demand for more efficient construction, which can then be sold at a premium. In addition, these labels could be useful for the mortgage industry to quantify the likely energy savings of a more efficient building and apply that dividend to the buyer's income.
- **Asset-value normative labels** for purchasing new structures. Any entities that own multiple buildings could require new purchases to be of a certain threshold, communicated through this label.

The European Union, China, and Russia each require some form of energy labeling for buildings. The EU requires both asset and operational informative labels on new and existing buildings, to be administered by each member state and to follow general requirements from the Union. China and Russia only require an asset-value informative label, with mixed requirements for application. Although the U.S. is using its labeling programs with various incentive programs to drive the market toward greater efficiency, much remains to be done to standardize labeling across all—new and existing—buildings.

With greater information, incentives can promote innovation and improve market penetration of the most efficient practices. Incentives for improved building efficiency are jurisdictionally spotty and incomplete. Enhanced incentives provide a substantial savings opportunity and could rapidly transform new construction markets. Improving building efficiency should be a particular priority given the long life of most buildings and the resulting "locked in" nature of their energy use. In countries with rapid development, incentives could spark high levels of innovation, leading to vastly reduced energy consumption with equal or improved energy service.

3.2.2. Appliances

As discussed with regard to buildings, appliance energy labels can be generalized into two camps:

- Normative segmentation, which labels higher efficiency products within a given product type
- Energy informative, which reveals how much energy and energy-costs are used over the lifetime of a product

Sufficient awareness and education of product life-cycle energy cost within the scope of consumer preference is required to overcome a primary market barrier—information—that keeps energy efficiency from expanding its reach in consumer appliance markets. Increased consumer awareness on life-cycle energy must avoid "information overload" on product operation and energy use. Too much information can be just as bad as none. Overall, influencing consumer preference toward minimum life-cycle energy costs means capitalizing on the consumer taking a single "glance."

An excellent example of normative market segmentation to build awareness on energy efficiency for consumer appliances and equipment is the trusted and widely known consumer label, ENERGY STAR®, which, for appliances, is administered by the U.S. EPA. Since the 1990s, the ENERGY STAR label program for appliances and equipment has grown from a program that initially targeted computing and office equipment to one that encompasses scores of product types in both residential and commercial markets. Three of every four Americans recognize the label for its simple purpose to distinguish a given brand from another as being more energy efficient, which now covers over 50 distinct product types.[38] Figure 7 shows that the estimated GHG savings from the ENERGY STAR label for appliances is not insignificant, adding up to nearly 900,000 MTCe avoided by 2025.

Like all normative labels, ENERGY STAR for appliances program only provides a single segmentation of the market, which is to say, a product either has the label or not. The biggest risk that this labeling program carries is to fall behind market cycles, thus becoming outdated and watered down by covering excessively high market share of some products. The EPA and DOE must constantly monitor each consumer appliance market and be prepared to ratchet up minimum efficiency thresholds to remain within the 25–50% market share to drive future efficiency innovation.

Coupled with normative labeling such as ENERGY STAR, providing energy information such as the U.S. Federal Trade Commission (FTC) EnergyGuide label is especially useful for consumer appliances whose lifetime energy costs are greater than half of the upfront purchase price. The products that now carry the yellow FTC labels include clothes washers, dishwashers,

38. See *www.EnergyStar.gov*

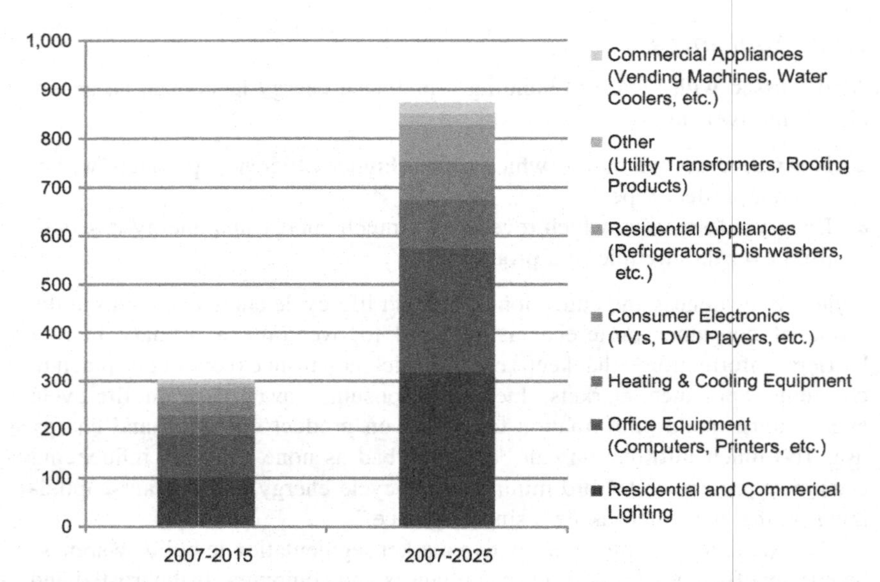

FIGURE 7 Energy Star Appliance Label Program Potential Greenhouse Gas Savings in thousand metric tons of CO_2 equivalent (MTCe) from 2007 to 2025*
**This analysis did not include certain consumer electronic product types and computer servers, which were added to the Energy Star labeling program after 2008. Source: Sanchez, M. (2007)*

refrigerators, freezers, water heaters, window air-conditioners, central air-conditioners, furnaces, boilers, heat pumps, pool heaters, and most recently, televisions. However, despite the information provided by the FTC label, studies have shown that the majority of consumers still fail to recognize or understand the energy use and cost information.[39] Informative labels have to provide enough information, but not too much, or they risk irrelevance. Regardless of whether a label works through market segmentation or directly providing energy use information, as labels are better recognized and understood by consumers, efficiency will play an increasing role in product selection.

As discussed with buildings, informative labeling is also more useful when tied directly to incentive programs.[40] Market incentives for appliances can be an influential consumer-driven policy tool to transform appliance markets

39. Thorne, J., & C. Egan. 2002. *An evaluation of the Federal Trade Commission's EnergyGuide appliance label: Final report and recommendations*. ACEEE, report number A021. *www.aceee.org/pubs/a021execsum.pdf*

40. A less well-known, but still valuable program that works to segment products based on levels of efficiency is the Super-Efficient Home Appliance Initiative administered by the Consortium for Energy Efficiency (CEE). For the Super-Efficient Home Appliance Initiative, CEE maintains a multitiered listing of efficiency levels for various products. The listings are used to design various energy efficiency incentive programs.

toward achieving greater efficiency.[41] Successful programs must first identify where incentives can most effectively be employed in the supply chain.[42] Appliance efficiency incentive programs should align with labeling programs, but the most successful programs go further by fully integrating into the supply chain, and often operate well "upstream" of the final consumer.[43] Monetary-based incentives, particularly to "downstream" consumers and retailers are often less cost-effective than improving information and incentives for efficiency upstream in the appliance market.[44] An example of a successful program is the New York State Energy Research and Development Authority's (NYSERDA) Energy $mart Products Program. The results of the decade-long NYSERDA Energy $mart Program show that it is highly cost-effective, yielding a greater than four-to-one program cost to consumer life-cycle energy savings output (weighted average of all of the products that are incentivized under the program), which is nearly double the program cost-effectiveness ratio for consumer-rebate type programs.[45]

Borrowing many of the successful program elements from the NYSERDA Energy $mart Products Program, a "best in class appliance deployment" (BICAD) program was part of the American Clean Energy and Security (ACES) Waxman-Markey bill that passed out of the House of Representatives in May 2009. The BICAD program leverages two motivations of market

---

41. A key challenge in designing incentive programs to save energy at costs lower than delivering generated energy is ensuring that the savings can be measured and counted as attributable to the program. For example, in many states utility efficiency programs must be able to prove to their regulators that consumers would not have saved energy without the information or incentive provided by the program.

42. Program design involves market-encompassing studies (e.g., market characterization capturing market actors along the supply chain, transaction flows, and information flows) to understand the existing marketplace for a given product.

43. The unfortunate result is that these programs often appear "invisible" and are therefore not well-known or understood among the wider audience of key policymakers and utility bill payers who have the ultimate say in directing the funding to procure necessary resources to carry out such programs.

44. See W. Golove and J. Eto (1996).

45. The NYSERDA Energy $mart™ Products program rewards retailers who sell a higher proportion of ENERGY STAR-labeled products than nonqualifying high-efficiency products. Retailers have used any variety of creative and innovative aims to get consumers to purchase the more efficient products. Common retailer sales tactics that have emerged include eye-catching energy and environmental promotional pieces, coupons, and price markdowns for the efficient products, consumer education via signs showing product energy savings, and favorable shelf positioning for efficient appliances. The program does not discriminate based on store or company size—everything from local "ma and pop" hardware stores to Home Depot, Lowe's, and Best Buy are able to take part and benefit from the program. See: Engel, V. (2007). *New York Energy $mart Products Program market characterization, market assessment and causality evaluation.* Final report. NYSERDA. *www.nyserda.org/Energy_Information/ContractorReports/Summit%20Blue/2007%20Reports/Final%20MCAC%20NYE$%20Products%20Program.pdf*

participants: the drive for retailers to increase sales and manufacturers to gain visibility.

The program is structured around three types of incentives targeted at each major market actor in the general residential and commercial appliance marketplace.

An incentive to retailers for increasing sales of best-in-class product models.

- DOE would determine best-in-class models, defined as up to the best performing tenth of the competing products in a class and review the standards annually to keep up with changing market conditions.
- Retailers would obtain a bonus for each sale of a best-in-class model.
- The size of the bonus would be based on the given product's lifetime energy cost savings of the best-in-class product model compared to the average product in the product class.

An incentive to retailers for the retirement and recycling of an existing inefficient but functioning product.

- Bounties would be based on the difference in energy costs between the retired product and the deemed energy costs of an average new product in the product class, discounted for the retired product's estimated remaining life.
- The legislation also includes a provision to establish standards for environmentally responsible methods of recycling.

An incentive to manufacturers for introduction of super-efficient best-in-class products.

- The Department of Energy would determine the highest efficiency product that could plausibly be mass-produced and provide a bonus to manufacturers for introducing products meeting that standard.
- This "golden carrot" approach has been successful in the past and will spur the creation of more efficient products while making those products more affordable.

Incentives for appliance efficiency can be more efficiently applied in the marketplace to achieve market transformation—where the market share of high and super-high efficiency products moves the curve forward. Programs such as BICAD at the national level would be a promising start.

CONCLUSIONS

If utilities are to meet the future energy needs of homes, industry, and businesses while decreasing emissions, the traditional utility business model must change. The financial health of utilities should be tied to how well they meet their customers' energy service needs at least cost, not how much energy they sell. An economy-wide carbon cap would provide an incentive for all sources to cut emissions, but capping emissions will not be enough, on its own, to overcome the barriers facing energy efficiency. Utility-specific regulatory changes will also be necessary to put energy efficiency on a level playing field and transition utilities toward becoming energy efficient service providers. In particular, policymakers should:

- Require utilities to pursue all cost-effective energy efficiency—all energy efficiency measures that are cheaper than supply-side sources, accounting for the full societal costs of generation
- Allow efficiency investment cost recovery and make utilities indifferent to changes in energy sales through "decoupling"
- Provide earnings opportunities for energy efficiency performance

With these reforms, regulators can reshape the market structure to enable utilities to deliver reliable energy services at minimum cost to society.

In the buildings and appliance sectors, there are three policies at the national and state level that are critical to improving efficiency performance.

- First, robust and frequently updated mandatory building codes and appliance standards, set state- or nationwide at the technology-neutral "floor" for performance.
- Second, expanded and improved labeling and rating programs based on lifetime energy costs. In buildings, all four types of labels should be deployed, as discussed above, but the greatest void is currently in the commercial asset-value field. Labeling is essential to provide information to consumers and assist compliance and enforcement mechanisms for both mandatory codes and incentives.
- Third, incentives through both tax policy and direct subsidy. Appliance incentives should target upstream retailers, distributors, and manufacturers in the industry supply chain to expand the share of energy efficient appliances and equipment for consumers.

These three strategies can be employed in any country and have been used to varying degrees in numerous jurisdictions. With the leadership to implement such strategies at the local and state level, these reforms will eventually transform the building and appliance markets, promote innovation, and reduce the global environmental and social costs of meeting our growing demand for energy services.

Taken together, these reforms unlock market demand for energy efficiency by appropriately aligning incentives toward minimizing the full societal impact of energy services. Implemented aggressively, these policies can drive down the use of energy even as demand for energy services continues to rise.

BIBLIOGRAPHY

American Council for an Energy Efficiency Economy. (2009). *Saving energy cost-effectively: A national review of the cost of energy through utility-sector energy efficiency programs. www.aceee.org/pubs/u092.htm.*

The Brattle Group. (2008). *Transforming America's power industry: The investment challenge 2010–2030* (p. 2). The Edison Foundation.

Building Codes Assistance Project. (2009). *DOE approves 50 state and territory energy plans under Federal Recovery Act. http://bcap-energy.org/node/461*

Building Codes Assistance Project. (2010). *Energy code calculator. http://bcap-ocean.org/resource/energy-code-calculator*

California Energy Commission. (2005). *2005 integrated energy policy report.*

California Public Utilities Commission, D. 07-09-043. (2007). *Interim opinion on Phase 1 issues: Shareholder risk/reward incentive mechanism for energy efficiency programs.*

California Public Utilities Commission. (2009). *Decision approving 2010 to 2012 energy efficiency portfolios and budgets* (p. 167).

California Public Utilities Commission, Energy Division. (2010). *Phase 1 report: Residential new construction (single family home) market effects study* (page xi).

Calwell, C. (2010). *Is efficient sufficient?* European Council for an Energy Efficient Economy. *www.eceee.org/sufficiency*

Cavanagh, R. (2009). Graphs, words and deeds: Reflections on commissioner Rosenfeld and California's energy efficiency leadership. MIT. *Innovations, 4*(4, November), 81–89.

Consortium for Energy Efficiency. *www.cee1.org/*

Database on State Incentive for Renewables and *Efficiency. www.dsireusa.org/*

Edison Foundation Institute for Electric Efficiency. (2009). *State energy efficiency regulatory frameworks.*

Energy Efficiency Codes Coalition. (2009) *Energy & cost savings analysis of 2009 IECC efficiency improvements. www.thirtypercentsolution.org/solution/EECC-Savings_Analysis-Jan-2009.pdf*

Energy Star. *www.energystar.gov/index.cfm?c=home.index*

Engel, V. (2007). *New York Energy $mart Products Program market characterization, market assessment and causality evaluation.* Final report. NYSERDA. *www.nyserda.org/Energy_Information/ContractorReports/Summit%20Blue/2007%20Reports/Final%20MCAC%20NYE$%20Products%20Program.pdf*

Eto, J., Stoft, S., & Belden, T. (1994). The theory and practice of decoupling. Energy Division. *Lawrence Berkeley Laboratory.* University of California Berkeley.

Geller, H. (1995). *National appliance efficiency standards: Cost-effective federal regulations.* American Council for an Energy Efficiency Economy. www.aceee.org/pubs/a951.htm

Goldstein, D. (2010). *Invisible energy: Strategies to rescue the economy and save the planet.* Bay Tree Publishing.

Golove, W., & Eto, J. (1996). *Market barriers to energy efficiency: A critical reappraisal of the rationale for public policies to promote energy efficiency.* No. LBL-38059, UC-1322. LBNL. *http://eetd.lbl.gov/ea/EMS/reports/38059.pdf*

Lesh, P. (2009). *Rate impacts and key design elements of gas and electric utility decoupling: A comprehensive review.*

Marnay, C., & Comnes, G.A. (1990). *Ratemaking for conservation: The California ERAM experience.* Applied Sciences Division, Lawrence Berkeley Laboratory.

McKinsey & Company. (2009). *Pathways to a low carbon economy: Version two of the global greenhouse gas abatement cost curve.*

McKinsey & Company. (2009). *Unlocking energy efficiency in the U.S. economy. www.mckinsey.com/clientservice/electricpowernaturalgas/US_energy_efficiency/*

Neubauer, M. et al. (2009). Ka-BOOM! *The power of appliance standards. www.standardsasap.org/state/2009%20federal%20analysis/ka-BOOM%20overview.html*

Roland-Holst, D. (2008). Energy efficiency, innovation and job creation in California. *Center for Energy, Resources, and Economic Sustainability (Ceres)*. University of California Berkeley.

Rosenberg, M., & Hoefgen, L. (2009). *Market effects and market transformation: Their role in energy efficiency program design and evaluation*. CA Institute for Energy and Environment. *http://uc-ciee.org/energyeff/documents/mrkt_effts_wp.pdf*

Sanchez, M. et al. (2007). 2008 status report: Savings estimates for the ENERGY STAR voluntary labeling program. LBNL. *http://enduse.lbl.gov/info/LBNL-56380%282008%29.pdf*

Schwartz, L. (2009). *The role of decoupling where energy efficiency is required by law*. Regulatory Assistance Project Issue Letter.

State of California, Public Resources Code, Section 25401.

Tax Incentives Assistance Project. *Builders and manufacturers incentives. http://energytaxincentives.org/builders/new_homes.php*

Thorne, J., & Egan, C. (2002). *An evaluation of the Federal Trade Commission's EnergyGuide appliance label: Final report and recommendations*. ACEEE, report number A021. *www.aceee.org/pubs/a021execsum.pdf*

U.S. Court of Appeals for the Ninth Circuit. California Energy Commission (CEC) vs. U.S. DOE and Association for Home Appliance Manufacturers (AHAM) on CEC waiver for preemption to set state-level water efficiency standards for residential clothes washers. Filed October 28, 2009. Before: William C. Canby, Jr. and Kim McLane Wardlaw, Circuit Judges, and David G. Trager, * District Judge. Opinion by Judge Canby. No. 07−71576. *www.ca9.uscourts.gov/datastore/opinions/2009/10/28/07-71576.pdf*

U.S. DOE and U.S. EPA (2006). *National action plan for energy efficiency report. www.epa.gov/cleanenergy/documents/napee/napee_report.pdf*

U.S. Department of Energy. (2007). *Annual energy review 2006*.

U.S. DOE, Office of Energy Efficiency and Renewable Energy. (2008). *Federal Register, 73*(250), Notices (December 30). *www.energycodes.gov/implement/determinations_90.1-2004.stm*

U.S. DOE. (2008). Planned program activities for 2008−2012. *Building Technologies Program* 1−5. *www.eere.energy.gov/buildings/publications/pdfs/corporate/myp08complete.pdf*

U.S. DOE. (2009). *Energy information administration buildings databook. http://buildingsdatabook.eren.doe.gov/TableView.aspx?table=1.1.4*

U.S. Department of Energy, Energy Efficiency and Renewable Energy. (2009). *Implementation report: Energy conservation standards activities. www1.eere.energy.gov/buildings/appliance_standards/pdfs/seventh_report_congress_aug_09.pdf*

Walker, S. (2005). *Will utility rate reforms open the door wider for distributed generation and demand side management?* CERA Decision Brief.

Part II

Technological Fixes

Chapter 9

Getting to Zero: Green Building and Net Zero Energy Homes

Meredith Gray,* and Jay Zarnikau**
*City of Austin Office of Homeland Security and Emergency Management, **The University of Texas at Austin and Frontier Associates LLC

Chapter Outline

1. INTRODUCTION

Today's average home in the developed world consumes more energy than homes did 50 years ago, despite modern innovations and technologies. The relatively cheap cost of average home construction materials has led home-buyers to cool and heat rarely used floor space and adopt home designs that are energy-inefficient for the local area. The general increase in dwelling size has been another significant trend over the past half century, contrasting with a decrease in the average family size in many developed nations. For example, this book's Chapter 14 by Bauermann and Weber describes how higher income levels and bigger homes are correlated in Germany. Similarly, the U.S. average house size increased 41% between 1978 and 2008.[1] To move toward a more sustainable future, energy consumption must be reduced—and a great place to start is at home.

1. U.S. Census Bureau. *Median and average square feet of floor area in new one-family homes sold by location. www.census.gov/const/C25Ann/soldmedavgsf.pdf.*

This chapter examines residential energy consumption, emerging home construction and appliance technologies, and the types of dwellings that would be necessary to achieve significantly lower levels of energy consumption in homes throughout the developed world. Current energy use is profiled to provide a benchmark. We review today's state-of-the-art technologies. To explore how dwellings and lifestyle could be altered to meet energy reduction goals, we survey some of the more innovative green building programs, zero net energy (ZNE) home subdivisions, and "deep green" community designs. These initiatives reveal what could be achieved today through the application of advanced building and community design principles that stress energy conservation, water conservation, minimization of waste, recycling, indoor air quality, and related measures to minimize environmental impacts and provide a blueprint for a large-scale response to climate change. Good building design needs to be at the forefront of any major energy consumption overhaul, but the behavior of building occupants cannot be overlooked. As this book's Chapter 11 by Prindle and Finlinson highlights, occupant behavior and operational aspects of energy use may be much more effective than technological fixes.

Moreover, designing and constructing new homes to achieve aggressive energy efficiency goals is much easier and less expensive than retrofitting the existing building stock. While the widespread dissemination of today's most efficient technologies and practices would fall short of achieving negligible energy consumption, the remaining gap may be overcome by relying on small-scale on-site distributed renewable energy resources. Yet, the costs and physical limitations of renewable energy resources suggest that sole reliance on renewable energy resources to fill the gap could prove impractical, at least in the near-term. Consequently, changes in behaviors and lifestyle may also be required.

The following section profiles home energy use in today's developed economies. The design of effective strategies to reduce energy use requires attention to the considerable differences in residential energy use among different nations and climates. Section 3 describes some leading initiatives designed to promote more energy efficient homes. These green building programs, government policies, and zero energy home projects demonstrate that thoughtful design and construction can improve the performance of new homes. Section 4 considers the measures that would be required to reduce energy consumption in existing homes by reviewing the results of recent energy efficiency potential studies. Section 5 discusses how policies, programs, and enabling technologies might contribute to the achievement of the potential efficiency gains by overcoming some key barriers. The analysis presented in Section 6 suggests that a goal of negligible net home energy use may be hypothetically possible but is unlikely to be practical in the near-term based merely on physical measures, such as improvements in building shell characteristics, more-efficient appliances, and the widespread deployment of distributed renewable energy systems. Rather, the analysis shows that behavioral and lifestyle changes will be necessary. The chapter's conclusions are provided at the end.

2. RESIDENTIAL ENERGY USE

To provide a benchmark, we review household energy use in the world's developed economies. After examining *direct* or *on-site* energy use, we comment on the *indirect* or *embodied* energy use associated with the production or delivery of consumer products and services prior to their use.

2.1. Direct Energy Use

Direct per capita energy consumption varies greatly among the world's developed nations. Among OECD countries for which data are available, Canada has the highest level of per capita energy consumption, followed by the U.S. and Finland. If adjustments by the International Energy Agency (IEA) to take into consideration differences in climate are taken into account, the U.S. would have the highest level of per capita residential energy consumption, as indicated in Figure 1.[2] Levels of per capita energy consumption in the U.S. are at least twice the levels of per capita consumption in Spain and New Zealand on either a normalized or unadjusted basis. The explanation for this discrepancy is related both to lower energy costs and to behavioral energy use patterns.

Among 19 large nations for which detailed data are available,[3] space heating constitutes the largest energy end use, accounting for roughly 53% of total household energy use in terms of heating units in 2005.[4] For example, in Germany the share of household energy used for space heating is as high as 80%, as described in Chapter 14 by Bauermann and Weber. Additionally, there is virtually no air-conditioning in the residential sector in Germany—typical of colder European climates—whereas the reverse is true in the U.S.

As stated in this book's Chapter 1 by Sioshansi, consuming energy provides no utility by itself. The demand for energy resources is derived from the demand for various services provided by appliances and equipment, which rely on energy to operate. The difference in energy use among various nations reflects their relative level of prosperity, climate, lifestyles, energy prices, real estate prices, home construction practices, the age of the housing stock, government policies, utility energy efficiency program activity, appliance stocks and efficiencies, and a variety of other factors including habits, cultural, lifestyle, and awareness of environmental consequences of energy use and production.[5]

---

2. IEA. (2007). *Energy use in the new millennium: Trends in OECD countries* (p. 77).

3. Australia, Austria, Canada, Denmark, Finland, France, Germany, Greece, Italy, Japan, Korea, the Netherlands, New Zealand, Norway, Spain, Sweden, Switzerland, United Kingdom, and the United States.

4. IEA. (2008). *Worldwide trends in energy use and efficiency* (p. 46).

5. In this book's Chapter 3, Bartiaux et al. describe some of the cultural and behavioral determinants of energy use within and among different societies.

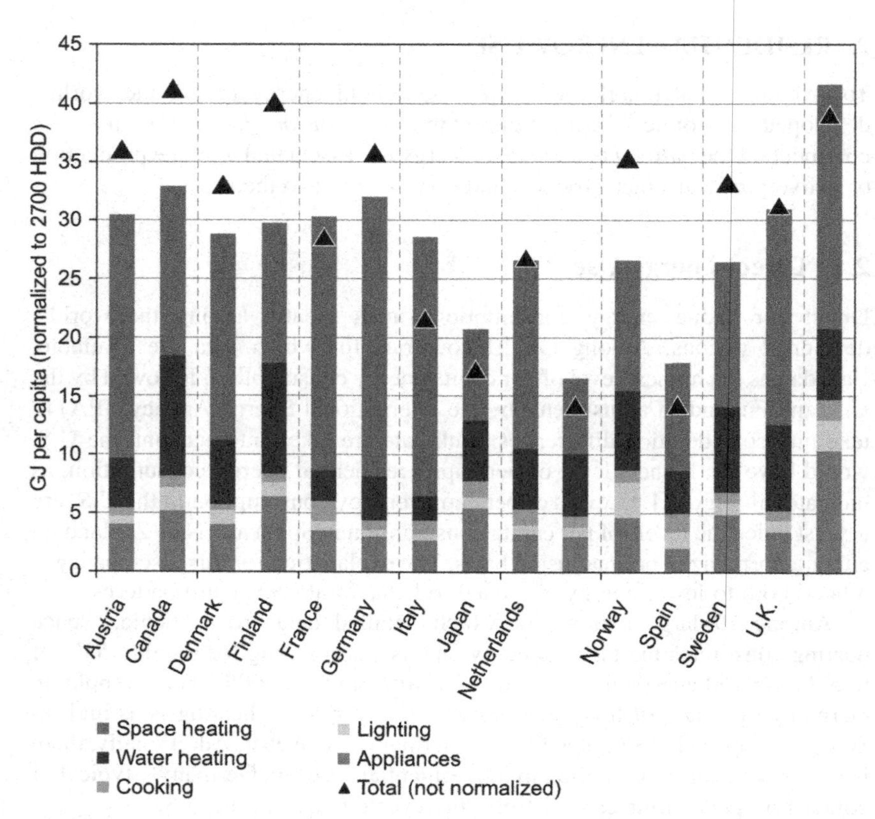

FIGURE 1 End-Use Composition of Per Capita Energy Use Among Developed Nations in 2004. *Source: Adopted from ©OECD/IEA (2007), Energy Use in the New Millennium: Trends in OECD Countries, p. 77. Data are from OECD/IEA Indicators Database. Triangles indicate the actual level of energy consumption prior to any normalization. The colored bars indicate consumption levels and end-use breakdowns once energy use is normalized to the same climate conditions.*

Much of the gap in household energy consumption between the U.S. and Canada and other developed nations may be traced to differences in the average size of homes. The average sized dwelling in the U.S. is nearly 50% larger than the average home in all other developed nations, in part due to lower land prices in the U.S. and Canada.[6] The average U.S. home today is 50% larger than one built 25 years ago.[7] While housing square footage did increase dramatically throughout the 1990s and 2000s, this may have reached a saturation point as energy prices began to rise in the late 2000s, as noted in Figure 2. Meanwhile, the average household size in the U.S. declined from 3.1 people in 1970 to 2.6

6. IEA (2007), p. 72.

7. Avid Home Studios. (2009). *The house of tomorrow from the recession today. www.avidhomestudios.com/blog/2009/01/05/the-house-of-tomorrow-from-the-recession-today.*

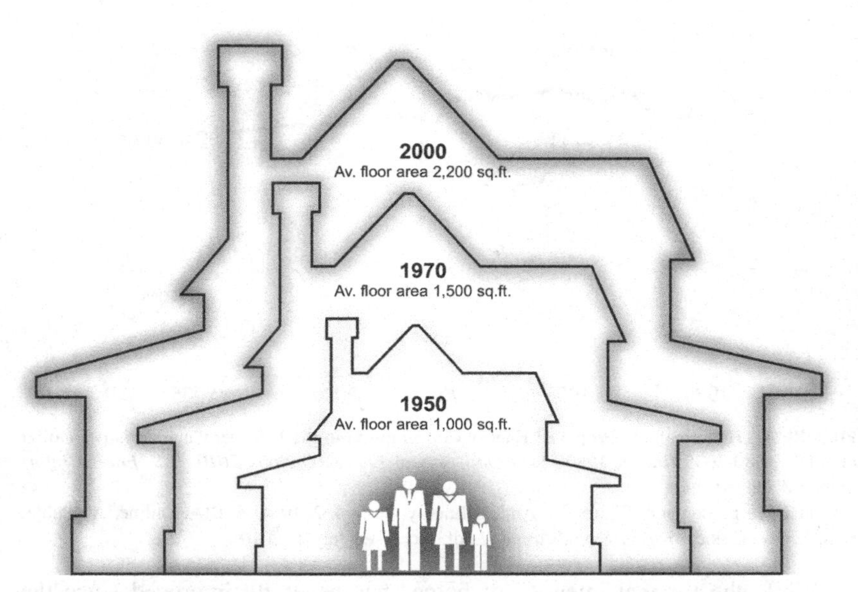

FIGURE 2 Growth in the Average Home Size in the U.S. Over Time. *From: U.S. Census Bureau, National Association of Home Builders.*

people in 2007[8]—a trend that is happening in a number of other industrialized countries as well.

This dramatic increase in housing size combined with factors such as population growth and increasing appliance usage led to an increase in U.S. energy consumption. However, noting the 2009 drop in energy use (to 310 million BTU per person), the U.S. Energy Information Administration (EIA) predicts that energy use per capita will begin declining in 2013, further decline by 0.3% per year on average, resulting in a 2035 projection of 293 million BTU in 2035 (Figure 3).[9] This decline in future consumption is also based on a fall in energy intensity (BTU of energy use per dollar of real GDP) as a result of structural changes and efficiency upgrades.

The composition of household energy use in advanced economies is vastly different from the end-use breakdowns found in the developing world. For example, cooking accounts for over 60% of total household energy consumption in India[10] and Mexico, much higher than in OECD.[11]

8. U.S. Census Bureau. (2009). *America's families and living arrangements: 2007. www.census.gov/prod/2009pubs/p20-561.pdf.*

9. U.S. EIA. (2010). *Annual energy outlook 2010. www.eia.doe.gov/oiaf/aeo/pdf/trend_2.pdf.*

10. World Business Council. (Undated).

11. Adriana Zaarias-Farah and Elaine Geyer-Allely. (2003). Household consumption patterns in OECD countries: Trends and figures. *Journal of Cleaner Production, 11*, 819–827.

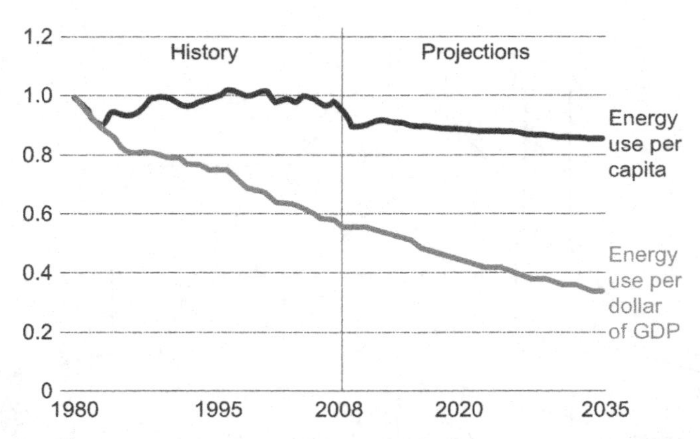

FIGURE 3 Historical and Projected Energy Consumption in the U.S. per Capita and per Dollar of GDP, 1980-2035 (index, 1980=1). *From: Annual Energy Outlook 2010. U.S. Energy Information Administration.**

*Annual energy demand. Figure 39. Annual Energy Outlook 2010. U.S. EIA. Online. Available: http://www.eia.doe.gov/oiaf/aeo/pdf/trend_2.pdf. Accessed July 1, 2010.

While the thermal integrity of homes has generally improved since the energy crises of the 1970s, this improvement has been offset by other factors resulting in increased per capita household energy use in most developed nations in recent decades.[12] This growth in average per capita household energy consumption can be traced to a variety of causes. Rising per capita income in developed countries has resulted in demand for larger homes and more energy-intensive services and appliances.[13] Average dwelling size per capita increased by 17% in 15 of the world's most prosperous nations[14] from 1990 to 2004. Meanwhile, occupants per dwelling declined from 2.8 in 1990 to 2.6 in 2004, with the declines most pronounced in Japan and Spain.[15] As a result, living space per person has increased. Moreover, higher levels of comfort have become the norm over time. For example, indoor temperatures in

12. In this chapter, we shall adopt the usual convention of using heating values (e.g., BTUs and Joules) to measure energy consumption so that we may employ many of the readily-available government statistics. It should be noted, however, that use of measures of energy consumption that rely on heating value metrics to monitor achievements in energy efficiency suffer from many problems and can result in many misleading findings. See, for example: Berndt, E. (1985). Aggregate energy, efficiency, and productivity measurement. *Annual Review of Energy, 3*, 225–73; Jay Zarnikau. (1999). When different types of energy resources are aggregated for use in cconometric studies, does the aggregation approach matter? *Energy Economics, 21*(5), 485–492. Jay Zarnikau. (1999). Will tomorrow's energy efficiency indices prove useful in economic studies? *The Energy Journal, 20*(3), 139–146.

13. IEA (2008), p. 47.

14. Austria, Canada, Denmark, Finland, France, Germany, Italy, Japan, the Netherlands, New Zealand, Norway, Spain, Sweden, United Kingdom, and the United States.

15. IEA (2007), p. 72.

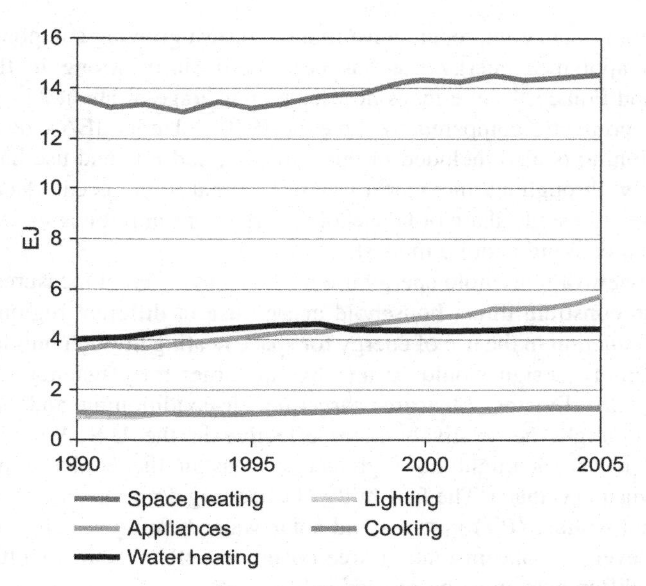

FIGURE 4 Recent Trends in Total Household Energy Use in Developed Nations, in exajoules. *Source: Adopted from ©OECD/IEA (2008),* Worldwide Trends in Energy Use and Efficiency. *Data are from OECD/IEA Indicators Database.*

the U.K. have increased by 3°C over the past decade.[16] There has been increasing reliance on energy to save time and labor through automation—mechanical clothes washing and drying, for example—and to improve hygiene—for example, cleaner clothes and more-frequent bathing.[17]

Electricity use has been rising due to its versatility or *form value* as an energy resource.[18] In contrast, per capita end-use consumption of natural gas, the other predominant household energy source in developed countries, has been declining since the 1980s.

Appliances other than water heaters, the fastest growing category of end-uses, now account for 21% of the total in developed nations as indicated in Figure 4.[19] Refrigerators, freezers, washing machines, dishwashers, and televisions account for half of the electricity use in this category.[20] Dramatic efficiency gains have been achieved in refrigerators, freezers, washing machines, and dishwashers over the past couple decades. But miscellaneous appliances, including computers, flat screen televisions, mobile phones, and

16. World Business Council (undated), p. 26.

17. OECD. (2002). *Towards sustainable household consumption.* Chapter 3.

18. Phillip Schmidt, Tom Sparrow, John Vanston and Jay Zarnikau. (1994). *Neo-electrification in the Information Age.* Edison Electric Institute.

19. IEA (2008), p. 46.

20. IEA (2008), p. 48.

home entertainment equipment, constitute the fastest growing component. The growth in appliance energy use has been particularly strong in the U.S., Finland, and France.[21] Americans now have an average of about 25 consumer electronic products, compared to three in 1980.[22] Under IEA's definitions, air-conditioning is also included in this category, and this end use has grown significantly throughout the world's hotter climates in recent decades. In developed nations, the share of household energy consumption associated with lighting and cooking is quite modest.

This review of household energy use reveals that different measures will be required to constrain direct household energy use in different regions of the world. A reduction in the use of energy for space heating through building shell and community design would be an obvious target for efficiency efforts in Europe's cold climates. Measures targeting air-conditioning and appliance energy use might be relatively more effective in the U.S. Water heating efficiency measures might have greater success in the cooler climates of Scandinavia and Canada. The feasibility of offsetting domestic energy use with on-site photovoltaic (PV) systems and solar water heating systems will vary based on levels of solar insolation, tree cover, and other factors. Cultural and economic differences must be recognized, as well.

2.2. Indirect Energy Use

The indirect effects of consumer decisions must also be considered if our goal is to reduce overall energy consumption. Indirect energy consumption refers to the energy embodied in consumer goods and services such as clothing, food, and building materials.[23] Thus consumption decisions regarding the purchase of a wide range of consumer products, as well as technology choices, government policies, and other factors, have energy impacts which tend to be reflected in government statistics as direct energy consumption in other sectors, such as manufacturing, transportation, and agriculture.

Modeling approaches designed to quantify indirect energy consumption in households or throughout the economy have been developed by a number of researchers.[24] Their findings suggest that in many countries indirect energy

---

21. IEA (2007), p. 67.

22. Jad Mouwad and Kate Galbraith. (2009). Plugged-in age feeds a hunger for electricity. *New York Times*, 20 September 2009.

23. This book's Chapter 4 by Meyer et al. also discuss indirect energy use.

24. For example: Christoph Weber and Adriaan Perrels. (2000). Modeling lifestyle effects on energy demand and related emissions. *Energy Policy*, *28*, 549–566; Robert Costanza. (1980). Embodied energy and economic value. *Science*, *12*(210, 4475, December), 1219–1224; Carey King, Jay Zarnikau, and Phil Henshaw. (2010). Defining a standard measure for whole system EROI, combining economic top-down and LCA bottom-up accounting. *Proceedings of Energy Sustainability 2010 Conference*, Phoenix.

consumption is at least as great as the direct or on-site household energy use that is reported in government statistics. Indirect energy use or the energy embodied in housing operations, transportation operations, food, and apparel may be twice the level of direct household energy use in the U.S.[25] Among 11 member states of the European Union (EU) for which data are available, indirect energy accounts for between 36% and 66% of each state's total household energy consumption.[26] However, it should be noted that the quantification of indirect effects is fraught with difficulties, including boundary problems,[27] issues pertaining to attribution, and the inherent limitations of input-output models.

In the remainder of this chapter, we focus on *direct* energy consumption, but we are nonetheless mindful of the importance of the indirect component. Our analysis of opportunities to reduce energy consumption begins with an examination of efficiency opportunities in the design and construction of new homes.

3. NEW HOMES WITH ENERGY EFFICIENCY AS A GOAL

How can the baseline levels of energy consumption presented in the previous section be realistically lowered? A first, and relatively cost-effective, step involves improved design and construction of new buildings. The initiatives described in this section confirm that significant efficiency gains can realistically and cost-effectively be achieved through today's technologies when supplemented with behavioral changes.

Consider that half of all buildings constructed today will still be in use in the middle of this century.[28] Additionally, buildings—including commercial buildings—use 40% of the total energy in both the U.S. and the EU.[29] As the world

---

25. Shui Bin and Hadi Dowlatabadi. (2005). Consumer lifestyle approach to U.S. energy use and the related CO_2 emissions. *Energy Policy, 33*, 197–208.

26. A. H. M. E. Reinders, K. Vringer, and K. Blok. (2003). The direct and indirect energy requirement of households in the European Union. *Energy Policy, 31*, 139–153; Kees Vringer and Kornealis Blok. (1995). The direct and indirect energy requirements of households in the Netherlands. *Energy Policy, 23*(10), 893–910.

27. The boundary problem refers to the question of "how far" to trace energy impacts. Is tracing the energy consumption of a toaster to a coal mine enough? Or, must one consider the energy used in mining the coal, or in manufacturing the equipment that mines and transports the coal? This analysis could be carried on endlessly. This problem is commonly encountered when a "life cycle analysis" approach is adopted.

28. Kaplan, Siena. (2008). *Building an energy efficient America: Zero energy and high efficiency buildings*. Environment American Research & Policy Center. *www.environmentamerica.org/uploads/kx/Im/kxImeu7IZctnMBnd2AZJQg/AME-efficiency.v6.pdf*.

29. U.S. Department of Energy. (2007). *Annual Energy Review 2006*. *www.eia.doe.gov/emeu/aer/contents.html*.

population desiring modern technology in their homes increases, efficient building design becomes a critical consideration.

The types of initiatives discussed in this section generally fall into three categories:

- Programs focused on energy efficiency through traditional building shell and heating, ventilation, and air-conditioning (HVAC) equipment upgrades, resulting in a modest improvement in energy efficiency beyond code requirements
- Holistic green building programs
- Zero energy efficient buildings

3.1. Traditional Building Shell Plus Basic Energy Efficiency Upgrades

In 1992, the U.S. Environmental Protection Agency (EPA) introduced the ENERGY STAR program as a voluntary market-based partnership to reduce greenhouse gas (GHG) emissions through increased energy efficiency. This program was expanded in 1995 to introduce labels for residential heating and cooling systems and new homes. Homes that earn the ENERGY STAR qualification meet strict guidelines for energy efficiency set by the EPA. To receive the ENERGY STAR designation, homes must be at least 15% more energy efficient than homes built to the 2004 International Residential Code (IRC),[30] although energy-saving features such as effective insulation, high performance windows, more efficient heating and cooling equipment, and ENERGY STAR qualified lighting and appliances can often lead to 20%–30% savings. Twelve percent of new housing built in the U.S. qualified as ENERGY STAR.[31] However, a voluntary market-based incentive may not work as effectively as a mandatory energy performance requirement would.

3.2. Green Buildings

Green building programs strive to promote overall resource conservation, through attention to water conservation opportunities, minimization of construction waste, recycling, improvements in indoor air quality, and efficient building operation in addition to energy conservation. The current leader in setting stringent standards for sustainable construction is the U.S. Green Building Council (USGBC), a nonprofit established in 1993 to promote

---

30. U.S. Environmental Protection Agency. *ENERGY STAR new homes. www.energystar.gov/index.cfm?c=new_homes.hm_index.*

31. U.S. Environmental Protection Agency. (2006). *2006 annual report: ENERGY STAR and other climate protection partnerships. www.epa.gov/cpd/annualreports/annualreports.htm.*

LEED Energy Credits Emphasize Demand-Side Solutions

Energy and Atmosphere		35 Possible Points
☑ Prerequisite 1	Fundamental Commissioning of Building Energy Systems	Required
☑ Prerequisite 2	Minimum Energy Performance	Required
☑ Prerequisite 3	Fundamental Refrigerant Management	Required
☐ Credit 1	Optimize Energy Performance	1–19
☐ Credit 2	On-site Renewable Energy	1–7
☐ Credit 3	Enhanced Commissioning	2
☐ Credit 4	Enhanced Refrigerant Management	2
☐ Credit 5	Measurement and Verification	3
☐ Credit 6	Green Power	2

FIGURE 5 LEED Energy Credits Emphasize Demand-Side Solutions. *From: LEED 2009 for New Construction and Major Renovations. U.S. Green Building Council.*

sustainability in building design, construction, and operation. The USGBC is known for its Leadership in Energy and Environmental Design (LEED) third-party rating system, which awards points to residential and commercial construction development to achieve a certification level of certified, silver, gold, or platinum. There are now LEED rating systems available for new construction; existing buildings; commercial interiors; core and shell redesign projects; schools, retail, and health care; and the most recent addition of LEED for neighborhood development.

Does green building integrate ZNE principles and practices? Under the revised LEED 2009 (Version 3) certification system, there are seven areas in which buildings and construction projects can receive credits, including one section titled "Energy and Atmosphere," which mandates fundamental refrigerant management and fundamental building commissioning evaluations (Figure 5).[32]

Additional points are awarded for on-site renewables, optimization of energy performance, and purchase of green power credits. However, green building certifications such as LEED do not require a building to have *net zero energy* use, only to reduce energy use to a level below the minimum required by law.[33] The risk is that designers may be tempted to achieve only the minimum energy savings necessary or implement strategies inappropriate for a project or site to garner LEED credits. LEED-certified buildings do generally use resources more efficiently than conventional buildings that are simply constructed to comply with the minimum required by local or state building code. Yet lifetime building performance of "green" buildings may not actually result in overall energy savings compared to a standard building.

32. U.S. Green Building Council. *LEED for new construction. www.usgbc.org/DisplayPage.aspx?CMSPageID=220.*

33. U.S. Green Building Council. *LEED for new construction. www.usgbc.org/DisplayPage.aspx?CMSPageID=220.*

Green building practices consider the entire building envelope including semiheated spaces, lighting, HVAC systems, and electrical systems. The LEED rating system does require meeting the requirements of ASHRAE 90.1-2004, with extra points given for exceeding these requirements. It can cost significantly more up front to build "green," but often home developers choose to pursue LEED certification to market a project as more sustainable and therefore be able to charge a price premium for those homes.

Although green buildings strive to achieve energy efficiency and other sustainability measures, ZNE buildings take the crucial further step of eliminating the need to be connected to an energy grid as well as the need to consume a nonrenewable resource. Today's green buildings standards may provide a useful bridge to a future where ZNE buildings become the norm, and technologies developed for green buildings such as programmable thermostats and design elements may be easily adapted to ZNE buildings as well.

3.3. Green Building Home Case Studies

Green buildings, while not ZNEs, are designed to use energy more efficiently and may be able to be modified to use less energy still as technology improves. The green building program in Austin is highlighted here to exemplify how citywide programs can influence construction practices. Many cities and communities throughout the U.S. and internationally[34] now have successful green building campaigns or policies in place, though we also examine case studies in areas that do not benefit from formal initiatives.

3.3.1. Austin, Texas

Austin, one of the fastest-growing cities in the U.S., is home to perhaps the oldest and largest green building program in the U.S. The city launched ambitious energy conservation programs in the 1980s in hopes of "building a conservation power plant"—that is, a demand-side resource which could displace the city-owned utility's next baseload capacity addition. In 1990, the city began to transition its new home construction energy conservation program into a holistic green building rating system, which promotes water conservation, minimization of construction waste, recycling, improvements in indoor air quality, and efficient building operation in addition to energy conservation. This market transformation program provides technical support to homeowners, architects, designers, and builders in the design and construction of sustainable homes and buildings. Rating tools with up to 130

34. This book's Chapter 16 by Stulz et al., for example, describes similar initiatives in Switzerland.

green building specifications have been developed for single-family and multifamily homes, as well as commercial buildings in this hot and humid climate.[35]

Since 1991 Austin Energy's Green Building Program has rated more than 7000 single family homes, 60 commercial buildings comprising more than 3 million square feet of space, and 57 multifamily projects containing 8,81 units.[36] In 2008 alone, 1022 single-family homes and 1611 multifamily homes participated in Austin's program, resulting in an estimated 19 MW of demand reduction and 47,200 MWh of energy savings.[37] A home receiving the program's highest rating is expected to consume less than half of the consumption of an average home in Austin.[38]

As noted in this book's Chapter 18 by Jennifer Clymer, Austin's green building program has become part of a larger Climate Protection Plan with the goal of making Austin the leading city in the U.S. in the fight against global warming. The broader objectives are to make all new homes built in Austin "zero energy capable" by 2015, to make all nonresidential buildings 75% more energy efficient by 2015, to require energy audits and efficiency improvements on existing homes and buildings, and to develop carbon neutral ratings for homes and buildings.[39]

The City of Austin also leads the U.S. on zero energy homes.[40] In 2007, Austin's Zero Energy Capable Homes Task Force, created by the City Council, set forth an action plan. The task force was particularly successful because of its public-private partnership in code development, and members included a variety of stakeholders. The City of Austin's Energy Code amendments include improvements to the building shell to ensure that homes are built tightly, testing air-conditioning ducts to ensure they do not leak more than 10%, testing the air flow of air-conditioners to ensure cool air goes where intended to go, requiring a Radiant Barrier System to stop radiant heat before it penetrates the home, and requiring that 25% of home lighting be high-efficiency lighting.[41]

---

35. Clinton Climate Initiative. (Undated). *Austin's green building program facilitates the construction of sustainable buildings*.

36. Clinton Climate Initiative. (Undated). *Austin's green building program facilitates the construction of sustainable buildings*.

37. Richard Morgan. (Undated). *Green building in Austin, TX*, presentation.

38. Richard MacMath. (2007). *Austin energy green building program*. SA06 Austin Energy's Zero Energy Home Design Competition, presentation (May 5).

39. Richard Morgan. (Undated). *Green building in Austin, TX*, presentation.

40. City of Austin. (2010). *Austin leads the nation on zero energy homes*. City of Austin press release. *http://www.austinenergy.com/About%20Us/Newsroom/Press%20Releases/2010/zeroEnergy.htm*.

41. Ibid.

3.3.2. Other Notable Green Building Initiatives

Within the U.S., a number of other communities have launched notable green building initiatives, including a variety of local programs in California;[42] Atlanta, Georgia;[43] Scottsdale, Arizona;[44] Portland, Oregon;[45] and Boulder, Colorado.[46] In Europe, many green buildings use far less energy than green buildings in the U.S. with comparable certifications.[47] Key standards and assessment programs in Europe include Passivhaus in Germany and Minergie in Switzerland.[48] These European programs rely on an integrated design process with the architectural design, which results in ultra-low energy use buildings which rely on passive heating and cooling.[49]

3.4. Zero Net Energy Efficiency Buildings

ZNE efficient buildings are those that either do not draw any power from the energy grid—net zero site energy use—or contribute enough back to the grid through on-site power generation over the course of a year that they end up with a net zero energy use balance—net zero source energy use.[50] ZNE buildings provide numerous benefits, including a reduced total net monthly cost of living for occupying building space, insulation from future energy price increases, and higher home or building resale value.

In general, it is easier to save energy than produce energy. Consequently ZNE buildings tend to focus primarily on energy efficiency, and then use renewable energy sources available on-site as much as possible, as further described in this book's Chapter 17 by Rajkovich et al.

A major difficulty in quantifying energy savings and net zero potential is whether or not energy is achieved *on-site* through conservation, *off-site* through credits, or at the source through generation. A mix of all three strategies seems ideal for cost-saving and building with future innovation in energy efficient

---

42. *Build it green, directory of GreenPoint rated ordinances in California. www.builditgreen.org/find-greenpoint-rated-ordinance.*

43. Southface Institute. *www.earthcrafthouse.com/About/newhomes.htm.*

44. See: *www.scottsdaleaz.gov/greenbuilding.*

45. See: *www.sustainableportland.org/bps/index.cfm?c=41481.*

46. See: *www.bouldercolorado.gov/index.php?option=com_content&;task=view&id=208&Itemid=489.*

47. Yudelson, Jerry. (2009). *Green Building Trends: Europe*. Island Press.

48. Passive House Institute. *www.passivhaustagung.de/Passive_House_E/passivehouse_definition.html.*

49. Minergie. *www.minergie.com/home_en.html.*

50. Torcellini et al. (2006). *Zero energy buildings: A critical look at the definition*. National Energy Renewable Laboratory (NREL). *www.nrel.gov/docs/fy06osti/39833.pdf*

technologies in mind. In zero energy building projects, on-site energy sources involve the use of solar panels, while energy conservation measures often include passive solar building design techniques. Geothermal heating and cooling systems using the earth as a heat sink can also provide energy efficiency as well as hot water by-products, reducing the need for heating additional water for daily use. Additionally, economies of scale exist for entire neighborhoods or larger communities to meet ZNE targets in aggregate, rather than merely as individual units within the community.

Wind resources are usually limited because of structural and noise challenges and distance from an optimal wind corridor location, though some may be placed within a project's boundary, such as in an adjacent parking lot, and still be considered as part of the on-site energy generation. Energy conversion devices such as daylighting or combined heat and power devices cannot be considered on-site production, but are instead efficiency measures.[51]

Renewable sources imported to the site—such as wood pellets, ethanol, or biodiesel—are not technically on-site renewable but may be useful in reducing embedded energy present in a structure.[52] Older housing stock and older commercial buildings will not lend themselves to ZNE consumption and zero carbon emissions annually, which is why most plans for net zero construction aim to improve code for *new* buildings. A last option for supply-side renewable energy sources would be for the homeowner or building manager to purchase "green credits" for renewable sources such as wind power or utility PV systems that are available to the electrical grid—off-site of the zero energy building.

The City of Austin has a plan to require all construction to be zero energy compliant by 2015. This goal may not be met by 2015, but having a standard such as this for designers and construction managers to strive for is a major component in reducing energy costs, and at the very least helping homeowners and other building occupants to consider energy conservation an important part of energy efficient design.

President Barack Obama signed an executive order in October 2009 that mandated that federal buildings and agencies must meet certain sustainability targets, including "implementation of the 2030 net-zero-energy building requirement."[53] These goals were inspired by Architecture 2030's "2030 Challenge" goals, which are set by the nonprofit and endorsed by the

---

51. Kaplan, Siena. (2008). *Building an energy efficient America: Zero energy and high efficiency buildings*. Environment American Research & Policy Center. *www.environmentamerica.org/uploads/kx/Im/kxImeu7IZctnMBnd2AZJQg/AME-efficiency.v6.pdf*.

52. Ibid.

53. The White House. (2009). *Federal leadership in environmental, energy, and economic performance*. Executive Order (October 5). *www.whitehouse.gov/assets/documents/2009fedleader_eo_rel.pdf*.

American Institute of Architects. The 2030 Challenge goals will attempt a dramatic reduction in GHG emissions by changing all aspects of the building process from planning and design to construction.[54] The state of California has proposed that all new homes be ZNE by 2020 and commercial buildings by 2030, as further described in this book's Chapter 17 by Rajkovich et al. The California Energy Commission has recommended that Title 24, the energy efficiency section of California's building codes, be revised to include the ZNE goal.[55] This goal could be met through on-site clean distributed generation, energy efficient design, and "no net purchases from the energy or gas grid."

Research into ZNE building is currently being supported by the U.S. Department of Energy (DOE) Building America Program, which forms research partnerships with residential building industry members, with a goal of reducing the energy use of housing by 40–100%.[56] The World Business Council for Sustainable Development (WBCSD) supports zero energy building activities and aims for buildings to "reduce demand by design, be highly efficient and generate at least as much energy as they consume." [57]

Similarly, the Canadian Net-Zero Energy Home (NZEH) Coalition formed in 2004 to promote a ZNE home initiative which has now been folded into the Canada Mortgage and Housing Corporation's Equilibrium Sustainable Housing Demonstration Initiative.[58] Its vision is for all *new* home construction to meet a ZNE home standard by 2030; however, net-zero is not mandatory in Canada. In the United Kingdom, the Code for Sustainable Homes produced by Communities and Local Government mandates that homes must be built to "Level 6" standards by 2016. The zero-carbon Level 6 is the highest level awarded and involves the use of solar panels, biofuel boilers, or wind turbines that replace entirely the energy taken from the national grid.[59]

---

54. Architecture 2030. *The 2030 challenge*. *www.architecture2030.org/2030_challenge/index.html*. The 2030 Challenge has now been formally adopted by the U.S. Conference of Mayors; the states of New Mexico, Washington, Illinois, and Minnesota; the National Governors Association; the National Association of Counties; and the cities of Santa Fe, New Mexico, Richmond, Virginia, and Santa Barbara, California.

55. California Energy Commission. (2007). *2007 integrated energy policy report*. *www.energy.ca.gov/2007_energypolicy/index.html*.

56. U.S. Department of Energy. *Building technologies program*. *www1.eere.energy.gov/buildings/building_america*.

57. World Business Council for Sustainable Development. *www.wbcsd.org/templates/TemplateWBCSD5/layout.asp?MenuID=1*.

58. Net Zero Energy Home Coalition. *www.netzeroenergyhome.ca/*.

59. Communities and Local Government. *Code for sustainable homes*. *www.planningportal.gov.uk/uploads/code_for_sust_homes.pdf*.

3.5. Zero Energy Home Case Studies

In this section, we highlight a number of initiatives, with a focus on efforts in North America in light of our greater familiarity with activities in that region. The Masdar City Initiative is covered in this book's Chapter 19, by Sovacool. This project in Abu Dhabi is interesting for its proposed use of only zero-energy homes powered entirely by solar power and other forms of renewable energy.[60] Table 1 provides a summary of case studies comparing best practices, unique solutions, and kWh saved (where known).

3.5.1. Austin, Texas

Austin has become a leader as well for affordability in design of zero-energy homes with its Solutions Oriented Living (SOL) development, designed by architecture firm KRDB. SOL is a 5.5 acre, mixed-income development of 40 homes in East Austin that aims for its homes to generate as much energy as they draw from the grid.[61] Net zero capability will be accomplished through passive design including passive ventilation and daylighting, thermally efficient windows, structurally insulated panels for a framing system, modular construction, geothermal heating and cooling systems to achieve a seasonal energy efficiency rating of 27, and PV panels in arrays of three to six kilowatts, depending on house size.[62] SOL was able to design this development with a mixture of affordable and market-rate homes. SOL completed three homes in October 2009, with two other homes under construction.[63]

3.5.2. Houston, Texas

The concept of green communities is in use at Discovery at Spring Trails, which claims to be "Houston's first solar-powered hybrid community."[64] The developer, Land Tejas, has partnered with General Electric and Masco in developing construction standards and a suite of residential technologies and appliances into the GE Masco "Ecomagination"™ builder program. This project uses renewable energy savings from solar power and whole-home energy efficiencies[65] to guarantee that energy bills will be lower than comparably sized ENERGY STAR homes.[66] All homes use solar power: "green" and

---

60. Masdar City. *www.masdarcity.ae/en/index.aspx.*

61. Solutions Oriented Living. *www.solaustin.com/.*

62. Ibid.

63. SOL Austin blog. *www.solaustin.com/blog/?m=200910.*

64. Discovery at Spring Trails. *http://discoveryatspringtrails.com.*

65. Allison Wollam. (2008). GE to light up Land Tejas development. *Houston Business Journal.* (18 January 2008). *www.bizjournals.com/houston/stories/2008/01/14/daily43.html.*

66. *Discovery at Spring Trails—do the math. http://discoveryatspringtrails.com/technology-2.*

TABLE 1 Case Study Highlights

City	Development Name	Best Practice	kWh Saved (if known) and/or Photovoltaic (PV) Generation Capacity
Austin, Texas	SOL (40 homes designed by KRDB)	• passive ventilation and daylighting • thermally efficient windows • structurally insulated panels for framing system • modular construction	• 3-6 kW solar arrays on each house[1]
Houston, Texas	Discovery at Spring Trails (by Land Tejas and General Electric)	• wall-mounted home touchscreen to control energy efficiency technology	unknown[2]
Boulder, Colorado	Solar Village Prospect (16 lofts built by All American Homes)	• passive solar design • icynene foam	• 2.5 kW solar arrays • savings of 67,400 kWh per year (according to design team)[3]
Chicago, Illinois	EcoPower Project (100 homes planned, 7 constructed; designed by Environmental Resource Trust, Inc.)	• use of residential energy credits (RECs) to make low-income housing more affordable	• met goal of 75%-ZEH (33% of that met with solar power)[4]
Sacramento, California	Premier Gardens (95 homes by Premier Homes)	• tankless water heater • 25 year limited warranty on solar panels • spectrally selective glass	• 2 kW AC solar electric home power[5]

TABLE 1 Case Study Highlights—cont'd

City	Development Name	Best Practice	kWh Saved (if known) and/or Photovoltaic (PV) Generation Capacity
Berkeley, California	SolarMap, Solar Cities designation	• SmartSolar program with free energy consulting • Solar America City (by DOE) designation provides financial assistance	• 142 kW of PV systems for residential developments[6]
Freiburg, Germany	Solar Settlement (designed by Rolf Disch)	• "PlusEnergy" concept – buildings produce a positive energy balance	• homes use 10-15 kWh • homes produce 3-12 kWh

[1] *Solutions Oriented Living. Homepage. Online. Available: http://www.solaustin.com/. Accessed April 21, 2009.*
[2] *Discovery at Spring Trails. Home. Online. http://discoveryatspringtrails.com.*
[3] *Solar Village Prospect. Online. Available: http://www.solarvillagehomes.com/multi_family/prospect/index.php.*
[4] *Environmental Resources Trust, Inc. Final Report of the Zero Energy Homes for Chicago EcoPower Project. Dec. 2005. Online. http://www.carbonfund.org/Documents/ERT%20Final%20Report.pdf.*
[5] *Case Study: Premier Homes – Premier Gardens. Online. Available: http://www.bira.ws/projects/files/BA_Solar_CS_Premier.pdf.*
[6] *Solar America Cities - Berkeley. Online. Available: http://www.solaramericacities.energy.gov/cities/berkeley/.*

"deep green" homes will have a minimum of 1 kW and 3 kW PV generation capacity, respectively. Deep green homes will also be equipped with battery storage capable of discharging for several hours at 2 kW, and electric vehicle charging stations—residents of these homes will have access to some form of electric vehicle. The first homes being built in the Discovery at Spring Trails community will be equipped with a wall-mounted, touchscreen GE Energy Monitoring Dashboard to monitor residential energy and water use. Future homes may be equipped with future generations of the GE dashboard or other devices, which will likely include energy management functionalities such as the ability to shut off or reduce the energy usage of home appliances in response to a price signal.[67]

Discovery at Spring Trails will also offer some interesting energy features at the community level. A PV farm will have 500 kW of generation capacity. It will provide power to the community's water treatment plant, and will also be connected to a 250 kW battery and an electric vehicle charging station. At this location, a supervisory control and data acquisition (SCADA) system will coordinate solar generation, battery charging/discharging, and charging of electric vehicles.

3.5.3. Boulder, Colorado

Solar Village Homes are considered "near zero" homes. While not entirely self-sustaining, this project has features that may be replicable for ZNEs elsewhere. The Solar Village Zero is the first of a line of homes ranging in size from 1100 to 2457 square feet.[68] Energy efficient features include passive solar design, icynene foam insulation to minimize conductive and convective air movement, fiberglass windows, as well as solar hot water and 2.5 kW of solar panels on the roof. These homes are built using modular construction by All American Homes, which recycles virtually all its construction materials at the factory, thereby reducing embedded energy needed to produce new materials. Furthermore, Solar Village Prospect is a "village" concept with 16 condos and lofts as well as retail and dining, and was named the 2006 Exemplary Solar Building of the Year.[69] Optimum solar orientation is also achieved so that panels face true south. Based on modeling using DOE-2 software, the design team estimates that these condominiums will save 67,400 kilowatt-hours (kWh) per year of electricity.[70]

---

67. Chapter 10 by Ehrhardt-Martinez et al. covers similar beyond-the-meter initiatives.

68. Solar Village Homes. *www.solarvillagehomes.com/about/index.php.*

69. Solar Village Prospect. *www.solarvillagehomes.com/multi_family/prospect/index.php.*

70. 2006 Colorado Renewable Energy in Buildings Awards. Colorado Renewable Energy Society. *www.cres-energy.org/reba_2006_svp.html.*

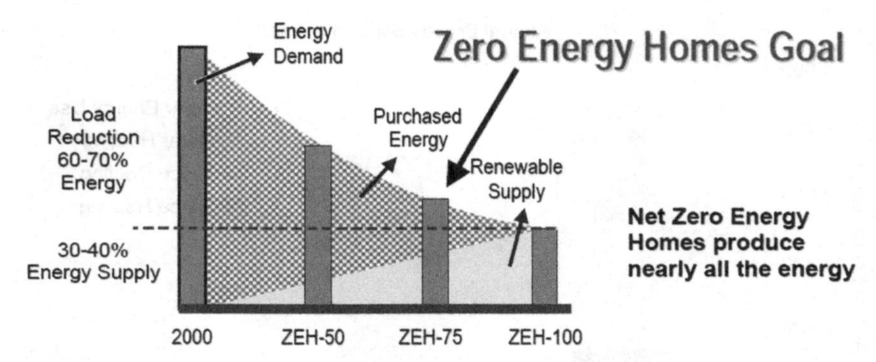

FIGURE 6 Chicago EcoPower Zero Energy Homes Goal. *From: Environmental Resources Trust, Inc. Final Report of the Zero Energy Homes for Chicago EcoPower Project. Dec. 2005.*

3.5.4. Chicago, Illinois

The Zero Energy Homes for Chicago's "EcoPower" Project were designed by Environmental Resource Trust, Inc. (ERT), to utilize renewable energy certificates (RECs) as a mechanism for making low-income housing more affordable by employing them as residential solar generation stations. The original proposal was for 100 solar homes to be built in Chicago's west side, but only seven homes were recently constructed.[71] The overall goal of the Zero Energy Solar Home was to exceed the USDOE minimum standard for Zero Energy Homes (minimum is ZEH-50%). ERT had a goal set of ZEH-75%, and achieved 67% of that goal with energy efficiency and 33% of it with solar power. As shown in Figure 6, the idea is to increase reliance on renewable supply as a home approaches true Zero Energy compared to standard design.

3.5.5 Sacramento, California

Premier Gardens is a 95-home zero-energy community certified by Sacramento Municipal Utility District (SMUD) as a Solar Advantage Home and an Advantage Home, exceeding California Title 24's energy cooling requirements by 50%.[72] Additionally, these homes have a 2 kW AC solar electric home power system, a 6.5 gallon-per-minute tankless water heater, a mechanically designed heating and air-conditioning system, spectrally selective glass windows, and tightly sealed air ducts. ConSol, a U.S. Department of Energy Building America Team Partner, estimates that homeowners in Premier Gardens would pay $600 less annually on their energy bills than homeowners in U.S. standard

---

71. Environmental Resources Trust, Inc. (2005). *Final report of the zero energy homes for Chicago EcoPower project. www.carbonfund.org/Documents/ERT%20Final%20Report.pdf.*

72. U.S. Department of Energy. *Premier Gardens: Moving toward zero energy homes. www.consol.ws/files/Premier_Gardens.pdf.*

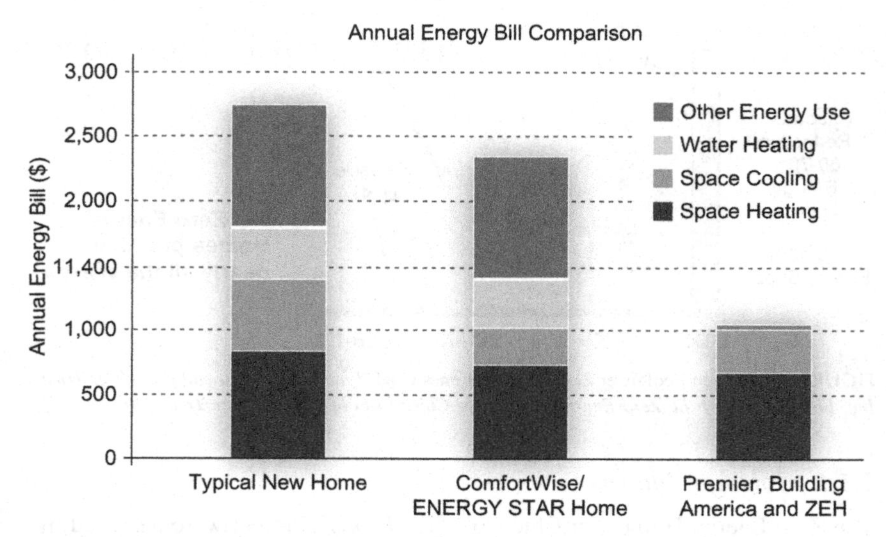

FIGURE 7 Annual Energy Bill Comparison Between Typical New Home, Energy Star Home, and Premier Gardens/Zero Energy Home (ZEH). *From: Building American Best Practices Series. U.S. Dept. of Energy Research Toward Zero Energy Homes.**
*Case Study: Premier Homes – Premier Gardens. Online. Available: http://www.bira.ws/projects/files/BA_Solar_CS_Premier.pdf.

construction homes and $400 less annually than those in U.S. ENERGY STAR homes (Figure 7). As shown, almost zero dollars of the ZNE annual bill are spent on space cooling.

3.5.6. Berkeley, California

The U.S. EPA has an online five-step toolkit for communities seeking to reduce their environmental footprint.[73] This program consists of a Community Assessment, Trends Analysis, Vision Statement, Sustainable Action Plans, and Implementation. The most useful section for communities is likely to be the Sustainable Action Plans section, which provides case studies from Berkeley's city planning recommendations for improving citywide sustainability. Berkeley is of note particularly because of its Solar Map, an interactive tool for viewing the locations of existing solar installations in Berkeley and calculating the benefits of solar power. The SmartSolar program also provides free solar energy consulting services to residents and businesses.[74]

---

73. U.S. Environmental Protection Agency. *Green communities. www.epa.gov/greenkit/index.htm.*

74. City of Berkeley. *SmartSolar program. Energy and sustainable development. www.ci.berkeley.ca.us/ContentDisplay.aspx?id=37808.*

The Berkeley Climate Action Plan was adopted in 2009, and the Building Energy Use chapter plans to improve energy efficiency in all buildings but also recommends achieving ZNE performance buildings for all new construction by 2020.[75] Berkeley is one of 25 U.S. cities to be designated by the DOE as a Solar America City, meaning that the DOE will provide financial and technical assistance to accelerate the adoption of solar technologies.[76] As the first phase of this partnership, Berkeley will develop and implement a pilot turnkey program to install 142 kW of PV systems and 10 solar hot water and solar air-heating systems in the residential building sector.[77]

3.5.7. Freiburg, Germany

The Solar Settlement in Vauban outside of Freiburg, Germany, is an interesting concept. This housing community is the first in the world where all buildings produce a positive energy balance, known as "PlusEnergy." These Plusenergiehaus® buildings were built between 2000 and 2005 and use only 10–15 kWh/m^2, a tenth of the energy requirement of a conventional house, while possessing solar panels capable of producing 3–12 kW.[78] This solar energy surplus is sold back into the citywide energy grid. Freiburg's ability to produce buildings capable of positive energy balance while still maintaining a high "developed world" standard of living should be of note to other regions seeking to reduce energy dependence. Additionally, the City of Freiburg, Germany, instituted a 15 kWh per sq meter ordinance, as mentioned in the previous case study section. An energy use "diet" of this type, which drastically reduces the amount of energy the average residential user consumes, would be achievable in other developed countries as well. Energy efficiency measures are needed as well as behavior modifications, such as using passive cooling and setting the thermostat higher in summer instead of turning on air-conditioning, or simple habitual changes such as turning lights off when not in use.

3.5.8. Quantifying Benefits of Zero Net Energy

ZNE buildings attempt to be more energy efficient than standard buildings or even regular "green" buildings. However, zero energy efficiency projects can only be said to be zero emissions post-construction, because obtaining

---

75. City of Berkeley. *Climate action plan. http://www.ci.berkeley.ca.us/uploadedFiles/Planning_and_Development/Level_3_-_Energy_and_Sustainable_Development/Berkeley%20Climate%20Action%20Plan.pdf.*

76. Solar America Cities. *www.solaramericacities.energy.gov/.*

77. Solar America Cities—Berkeley. *www.solaramericacities.energy.gov/cities/berkeley/.*

78. Energy Consumption. PlusEnergieHaus. Rolf Disch, architect. *http://www.plusenergiehaus.de/index.php?p=home&pid=11&L=1&host=1#a502.*

and transporting construction materials uses nonrenewable energy. Additionally, these projects assume that any technologies used, such as PV panels or off-site wind energy, will be available over the life of the building. This might be particularly difficult to determine when generation facilities are separated from the actual building, as they may be superseded by future development.

The benefits of zero energy buildings are clearly the long-term return on investment and reduced energy usage overall. One major remaining challenge is making them affordable for the average homebuyer. Zero energy buildings do cost more to build, and payback period varies largely depending on materials used and system upgrades required. One Canadian study from 2007 found a payback period of about 31 years for the upgrades to the entire home, all systems included, assuming no government financial incentives or tax breaks.[79] Specific technologies, such as spectrally selective glass, may have a shorter payback period of two to three years, however.

A City of Austin task force found average annual energy savings of 2515 kWh of electricity and 4 therms of gas after implementing the following requirements for new homes:

- Requirement for building thermal envelope testing
- Requirement for installation of a radiant barrier system
- Requirement for testing of duct system leakage
- Requirement for submittal of HVAC sizing calculations
- Requirement for testing of air balancing of HVAC systems
- Requirement for system static pressure testing
- Revision of the restriction on electric resistance water heating
- Requirement for 25% of lighting to be high efficacy[80]

These requirements are estimated to increase the cost of building an Austin-area home by $1179 with a payback period of 5.2 years. The gap between the Canadian payback period and the Austin payback period is striking—yet the City of Austin study includes some government financial incentives that subsidize the true cost. Regardless, any ZNE homes need to be built to last longer than the average building to justify additional costs.

The preceding discussion suggests that there are a variety of ways of getting to zero or near-zero in direct residential energy consumption. The appropriate approach depends on climate, local construction practices, the involvement of government entities, and the creativity of home builders, architects, and engineers. If ZNE homes were to be adopted through a U.S.

79. Humphrey Tse and Alan S. Fung. (2007). *Feasibility of low-rise net zero energy homes for Toronto*. 2nd Canadian Solar Buildings Conference. *http://www.solarbuildings.ca/c/sbn/file_db/Feasibility of low rise net zero energy houses for toronto.pdf*.

80. Austin Zero Energy Homes Task Force. (2007). *Clean energy for Texas*. *www.cleanenergyfortexas.org/austin.html*.

national mandate for all new development, the housing stock in 2050 would look much different than it does today. The sooner the U.S. transitions to ZNE home building, the better for future energy use. In the year 2050, the U.S. is projected to have about 420 million residents.[81] If a national mandate was passed in 2010 that required all new homes to be constructed to a ZNE standard, by 2050 roughly half of the nation's projected 150 million homes would not be net energy consumers, and might actually even become energy producers, adding energy back to the grid. Assuming that future homes use about the same amount of energy as today's homes—roughly 10,000 kWh annually—this national mandate would have the effect of saving 750 billion kWh annually or nearly a 60% kWh savings over current U.S. home energy consumption of 1280 billion kWh.

4. WHAT WOULD IT TAKE TO TRANSFORM THE NEW RESIDENTIAL BUILDING STOCK?

Studies consistently suggest that total energy consumption in the residential sector in a developed nation could be cut between one-quarter and one-half through the application of readily-available modern energy efficient technologies to the existing building stock and by promoting better building construction practices. This section reviews various studies which have examined the potential for efficiency improvements.

The 2009 U.S. Academy of Sciences report *America's Energy Future* states that buildings use 73% of electricity and 40% of all U.S. energy consumed and points out that the building sector has the greatest potential for energy efficiency.[82] This report also found that although consumer demand is expected to increase, using currently available or emerging efficiency technologies in buildings could still lower energy use by 25%–30% by the year 2030 compared to predictions reflected in the EIA reference case (Figure 8). This is a dramatic result, namely, that future energy use can be *less* than it is today if the building sector is transformed over the next decade so that new construction is vastly more energy efficient than today's homes.

Similarly, analyses by the IEA conclude that energy use in buildings in the developed economies could be cost-effectively reduced by about one-half over time relative to a baseline projection using measures that are feasible today.[83]

81. U.S. Census Bureau. (2002). *Demographic trends in the 20th century.* Census 2000 Special Reports, CENSR-4, Table 5 (November). U.S. Census Bureau. (2004). *U.S. interim projections by age, sex, race, and Hispanic origin* (November 18).

82. National Academy of Sciences and the National Academy of Engineering. (2009). *America's energy future: Technology and transformation. http://sites.nationalacademies.org/Energy/Energy_043338.*

83. International Energy Agency. (2008). *Energy efficiency requirements in building codes, energy efficiency policies for new buildings.* IEA Information Paper (p. 79).

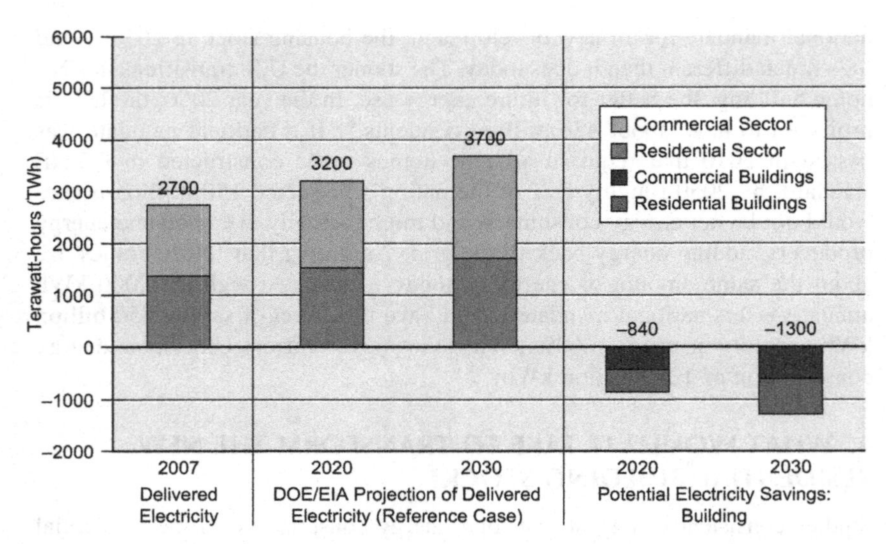

FIGURE 8 Building Sector Delivered Electricity, Future Delivered Electricity Projection, and Potential Electricity Savings, in TWhrs. *From: America's Energy Future: Technology and Transformation. 2009. National Academy of Sciences and the National Academy of Engineering.*

The IEA's roadmap to achieving carbon dioxide (CO_2) reduction targets by 2030 relies on buildings-related energy efficiency to provide a one-third contribution toward meeting the CO_2 targets, with 38% of the efficiency savings coming from potential space heating improvements, 20% from water heating, 10% from cooking, and over 20% from improvements in appliances and miscellaneous end-uses.[84]

A widely-cited study by the consultancy firm McKinsey & Company contends that residential energy consumption in the U.S. can be reduced by 28% from its projected value by 2020, relative to a business-as-usual scenario.[85] A meta-analysis by the American Council for the Energy Efficient Economy (ACEEE) of the results of energy efficiency potential studies for areas of the U.S. suggests that it is technically possible to reduce electricity consumption by 33% and to reduce natural gas usage by 40% in the U.S.[86] Assuming reasonable payback periods, the economic potentials for electricity and gas were estimated to be 20% and 22%, respectively. Over 100 studies of the potential for energy efficiency savings have been completed for various

84. International Energy Agency. (2008). *Towards a sustainable energy future* (p. 22).

85. McKinsey & Company. (2009). *Unlocking energy efficiency in the U.S. economy* (July, p. 29).

86. Steven Nadel, Anna Shipley, and R. Neal Elliott. (2004). The technical, cconomic, and achievable potential for energy efficiency in the US—a meta analysis of recent studies. In *2004 ACEEE Summer Study on Energy Efficiency in Buildings.*

TABLE 2 Types of Residential Measures (Existing Technologies) Commonly Evaluated

Category or End Use	Market/Measures
Thermal Efficiency and Comfort	• Air infiltration reduction • Attic ventilation and vapor barriers • Ceiling insulation to R-38 for houses and R-26 for mobile homes • Duct sealing and insulation • ENERGY STAR® windows • Floor insulation for mobile homes and pier and beam foundations • Solar screens and film on existing windows • Reflective roofs • Wall insulation • Radiant barrier
High Efficiency Cooling	• High efficiency air conditioning • High efficiency heat pump • Ground source heat pump
Electric Water Heating, ENERGY STAR® Appliances, and Efficient Lighting	• Clothes washer • Dishwasher • Refrigerator • Room air conditioner • High efficiency electric water heater • Tankless water heaters • Water heater insulation and pipe wrap • Low-flow showerheads and faucet aerators • Compact fluorescent light bulbs
Green Building	• Residential new construction and remodeling promoting sustainability and energy efficiency

Source: Prepared by the authors for this chapter from various sources.

utility service areas or states and provinces in North America by consulting firms and advocacy groups.[87] Regardless of the region studied or its climate, the resulting estimates of savings potential are remarkably similar. Table 2 lists the types of efficiency measures typically considered in these studies for a Southern U.S. climate.

Impact evaluations of the green building programs and ZNE communities reviewed earlier in this chapter—while limited in scale—provide some support

87. One of the earlier studies was Alan Meier, Janice Wright, and Arthur Resenfeld. (1983). *Supplying energy through greater efficiency.* University of California Press.

for these claims. For example, Austin Energy has found that a new green building home which receives a 5-Star rating—the highest rating awarded in their program—will consume about one-half as much energy as a typical existing home in Austin, Texas. Requiring such ambitious standards would result in savings for the average consumer over their home's lifespan.

Trade organizations may lead the way in developing better construction standards where federal and statewide measures do not exist. The American Society for Heating, Refrigeration, and Air-Conditioning Engineers (ASHRAE) Standard 90.1 – Energy Standard for Buildings Except Low-Rise Residential Buildings is the benchmark for building energy performance. The Illuminating Engineering Society of North America (IESNA) is a joint sponsor of this standard. However, this standard is not legally enforceable, and does not cover energy efficiency in single family home construction. In 2008, the U.S. Department of Energy determined that requiring 90.1-2004 would generate site energy savings of about 12%, and recognized 90.1-2004 as the new national energy standard effective two years later in 2010.[88] However, not all states have a commercial energy building policy in place, much less a residential energy building policy.

Indeed, impressive gains have been achieved in many end-uses and technologies in recent years, and adoption of the more-efficient commercially-available technologies can result in energy savings. Lighting provides some obvious examples.[89] Compact fluorescent bulbs use one-quarter of the energy of incandescent bulbs for the same lumen output. LED bulbs provide further savings. Refrigerators provide another example. In 1979, the average new refrigerator in the U.S. consumed over 1300 kWh per year.[90] Federal efficiency standards enacted in 1990, 1993, and 2001 required energy use in medium-sized refrigerators to be reduced to 900 kWh, 700 kWh, and 500 kWh, respectively. Today's federal standard of 500 kWh is the same level of energy consumption used by refrigerators in the 1950s, prior to design changes designed to improve their storage capacity, reduce condensation, add ice-making features, and include automatic defrost capability.[91]

Recognition of interactive effects among efficiency measures and opportunities to better match sources and uses of heat within a home may illuminate new savings opportunities. Obviously, improvements in the thermal integrity of building shells through greater insulation, air leakage reduction, reflective

---

88. Department of Energy. (2008). *Federal Register: FR Doc E8-30975, 73*(250). *www.thefederalregister.com/d.p/2008-12-30-E8-30975.*

89. See Lamp History in the Nutshell, at: *www.lampreview.net/.*

90. Average calculated from D&R Database of Refrigerator Energy Use as described in U.S. Department of Energy (2009). Refrigerator Market Profile. Available at: *http://www.energystar.gov/ia/partners/manuf_res/downloads/Refrigerator_Market_Profile_2009.pdf.*

91. See: EcoMall, *The new wave of energy efficient refrigerators*, at *www.ecomall.com/greenshopping/icebox2.htm.*

roofs, thicker walls, passive solar design, and improved windows affects space heating and air-conditioning needs. Building energy use simulation models are typically used to quantify these effects. But further efficiencies may be gained when equipment works as a system. A geothermal heat pump may be used to heat water in addition to providing space heating and air-conditioning. Heat exhausted from a clothes dryer may be used for space heating. Deciduous trees may offer shade during summer afternoons, thereby reducing air-conditioning loads, while permitting warming heat during winters.

If the application of existing commercially-available technologies can cost-effectively yield reductions of between one-quarter and one-half of contemporary household energy usage in a modern society, why do such opportunities remain untapped? Part of the answer is that many of the traditional barriers to energy efficiency must be overcome. Technologies must advance further. Home design and construction practices must improve. Government policies can play a prominent role. Improved energy pricing and consumer information can contribute. For many consumers, lifestyles will be affected. To bring net energy consumption down to negligible levels, it is likely that distributed renewable energy resources will be needed to fill the remaining gap. The following section discusses the potential contribution that each of these factors may play.

5. WHAT ELSE NEEDS TO HAPPEN?

Our review of energy efficiency potential studies suggests that energy usage in the household can be reduced by one-quarter to one-half using existing commercially-available technologies and practices. The combination of efficient neighborhood design, thermally-efficient building construction, and the use of high efficiency appliances and equipment can reduce household energy use considerably. Energy use can be reduced significantly through proper building design and construction. Provided we permit PV systems and other types of on-site renewable energy generation to fill the gap, the goal of negligible *net* energy consumption is achievable in a technical sense. But can the few isolated case studies described in this chapter become the predominant technologies and practices of tomorrow? In this section, we identify some key impediments, and discuss how advances in technology, government policies, utility programs, better information, and changes in consumer behavior could assist in overcoming the impediments. There is hope for the residential building sector as key players recognize the cost savings from more energy efficient homes.

We face some enormous challenges in establishing these practices and technologies as the norm. Key barriers include:

- *Initial cost.* The most efficient technologies generally require premium initial costs to consumers. The on-site renewable energy generation needed to reduce net consumption down to negligible levels is presently more expensive than the purchase of power from utility grids. The extra cost could be financed with longer term mortgages or mortgages that give a preferential rate to more efficient homes.

- *Longevity of the building stock.* Today's energy-inefficient building stock will continue to provide shelter for many years into the future. Retrofitting the existing building stock involves significant additional costs and challenges, and therefore the focus of ZNE efforts should be on new homes. Older homes will be eventually phased out or the cost of retrofits will be seen as necessary.
- *Credible information.* Presently, consumers have scarce, and often misleading, information about the energy-related consequences of their purchase and operation decisions with respect to homes and appliances. For more information about labeling buildings for energy efficiency, see this book's Chapter 8 by Long et al.

Additionally, a long list of other barriers have been reported, including a divergence of financial interests between landlords and tenants, homebuilders and homeowners, and other "principal—agent" problems;[92] a lack of product or service availability; and a lack of financing to overcome first-cost hurdles. Homeowners face tradeoffs between cost, convenience, traditional design features, and green features. Further, the presence of building codes and homeowners' associations tend to restrict innovation. Inefficiency must be banned somehow, whether it's by building code or federal government mandate, for retrofits to make sense and new homes to be built more efficient from the design stage onwards.

These impediments may be difficult to overcome, though numerous areas for progress exist. Federal and other incentives can reduce the payback period for homeowners to implement energy efficient technologies. Given the recent focus on reducing energy consumption, there are some important new technologies being developed in research laboratories or just reaching commercialization today. The amount of hot water required by washing machines, for example, may be nearly eliminated through the use of nylon beads[93] or ozone.[94] LED household lighting continues to improve. Energy efficiency improvements in new dwellings are being achieved by returning to some older technologies such as earth sheltered underground homes, straw bale homes, and adobe homes as well as through the application of new technologies including spectrally-selective low-emissivity windows, radiant barriers, window shades that respond to the need for more or less exterior light inside the dwelling, and modern heat pumps.

---

92. American Council for an Energy-Efficient Economy. (2007). *Quantifying the effects of market failures in the end-use of energy.* Final draft report to the International Energy Agency.

93. Washing without water. (2009). *The Economist*, Sept. 5—11, p. 12—14.

94. See *http://web-japan.org/trends/science/sci060313.html.*

Government policies establishing building codes, minimum appliance efficiency levels, and tax policies to encourage energy conservation have had an impact on residential energy use in recent decades.[95] Given the number and diversity of policy experiments, we now have a decent appreciation of what works—for example, carefully-designed appliance efficiency standards, building codes, labeling programs, and some educational messages—and what doesn't work—mandating specific technologies and certain market-distorting subsidies, for example. A noteworthy example of a progressive government policy at the federal level is the U.K. Sustainable Communities Act of 2007.[96]

Green taxes and higher energy prices can get us part of the way there. This strategy has been taken most seriously in Denmark, where a variety of products and services that adversely affect the environment are heavily taxed. In Denmark, green taxes account for 5% of overall GDP, 50% of the price of electricity, and 44% of the price of natural gas.[97] These taxes have been credited with spurring the development of renewable energy in Denmark.[98] As noted in this book's Chapter 4 by Meyer et al., simply reducing the subsidies provided to conventional energy providers would move us in the right direction.[99]

But the demand for energy is price-inelastic, suggesting that green taxes on energy can have only a modest impact.[100] Efficiency needs to be mandated in every way that is cost effective and benefits the entire building sector and future residents. The limited response of energy consumption to changes in energy prices is in part due to its limited overall cost to consumers. The share of disposable income spent on energy in the developed world tends to be between 2% and 3% and has generally declined since 1990.[101] Thus energy costs may not be high enough to attract the attention of many consumers in developed economies.

While changes in the level of prices may have some impact, changes in the structure of energy prices may have a more pronounced effect. While real-time pricing and critical peak pricing of electricity tends to elicit a reduction in

---

95. For a review of country-specific policies, see *Progress towards energy sustainability in OECD countries* at: *www.helio-international.org/Helio/anglais/reports/oecd6.html*.

96. See: *www.opsi.gov.uk/acts/acts2007/ukpga_20070023_en_1*.

97. IEA (2007), p. 76. See also, *OECD Observer*, at: *www.oecdobserver.org/news/fullstory.php/aid/497/Green_taxes.html*.

98. Mikael Andersen. (1994). Green tax reform in Denmark. *Environmental Politics, 3*(1), 139–145.

99. IEA estimates that the 37 largest developing countries spent $557 billion in 2008 on energy subsidies.

100. World Business Council. (Undated). *Energy efficiency in building: Transforming the market* (p. 7).

101. IEA (2007), p. 74.

consumption during high-priced periods, some overall conservation effect may be present.

While appliance efficiency standards tend to limit consumer choices and may raise initial appliance prices, they have proven to be an effective means of achieving efficiency gains as witnessed by the success of refrigerator and air-conditioner efficiency standards in North America. Proposed television efficiency standards in California are expected to result in $8.1 billion in energy cost savings.[102] Efforts to mandate minimum fuel efficiency levels for automobiles can deliver much greater savings.

New home construction standards and building design are making a large contribution toward energy savings. Largely as a result of building codes, energy use in new homes in the Netherlands is about 40% of the usage level of a home built in the 1960s.[103]

Greening the large existing stock of buildings is considerably more difficult than designing and building for efficiency. Some communities have introduced requirements that when homes change hands, they must have an audit and certain minimum efficiency standards must be met. This has been implemented in Austin, Texas, and Berkeley, California, to improve the energy efficiency of the existing stock of homes. A national system for rating all residential dwellings has been implemented in Australia to provide homebuyers with information pertaining to the energy performance of a home, and similar labeling requirements have been imposed in the EU.[104]

Utility-sponsored energy efficiency programs have also proven very successful in promoting energy efficiency. Through a combination of aggressive energy programs, building codes, and complimentary efforts, California has succeeded in reducing per capita household energy use in recent years without sacrificing economic growth.

Better information regarding the energy-related consequences of consumer choices can provide a contribution. Today, it is difficult to discern which building materials are more environmentally benign than others. "Greenwashing" with unfounded or confusing claims regarding environmental benefits is common. The direct costs of various goods and services may not accurately include environmental costs, as noted in other chapters in this volume.

Approaches designed to provide energy consumers with greater information regarding their in-home energy usage via energy monitors and other technologies show great promise. Past studies on real-time direct feedback devices

---

102. California Energy Commission. (Undated). *Just the facts: The truth about proposed TV standards. www.energy.ca.gov/appliances/TV_Standards_Facts.pdf.*

103. OECD. (2002). *Towards sustainable household consumption* (p. 46). *http://books.google.com/books?id=oh9Pw5fcd70C&printsec=frontcover&source=gbs_navlinks_s.*

104. Charles Ries, Joseph Jenkins, and Oliver Wise. (2009). *Improving the energy performance of buildings: Learning from the European Union and Australia.* RAND Corporation.

have indicated potential savings from 5%–15%.[105] Labeling the energy consumption associated with appliance use can make a contribution.

While the promise of smart grid technology is probably overblown, home automation and controls through smart grid technologies may eventually make a significant contribution. The operation of appliances and equipment may be optimized using in-home controls, hopefully leading to a more efficient use of energy-intensive equipment. Time-differentiated pricing could be implemented on a residential energy consumers, leading to efficiency gains. Consumers could control and program the operation of appliances from remote locations, as discussed in this book's Chapter 10 by Ehrhardt-Martinez et al.

Actual achievements in resource conservation and cost savings rely heavily on behavioral factors. Simple behavioral changes can also help, such as switching appliances to low-power modes when not in use, reducing the use of space conditioning when a dwelling is not occupied via a programmable thermostat or manual means, and scheduling clothes drying during cooler periods of the day. Coordination among electric utilities, water utilities, and solid waste districts to promote overall resource conservation programs is complicated and coordination failures may result in suboptimal results.

Any success in reducing household energy cost may introduce a new challenge. Consumers are likely to operate energy-intensive equipment more, maintain more-comfortable thermostat settings, and neglect additional conservation opportunities if their energy costs decline. Thus, bounce-back or elasticity impacts must be considered.[106]

There has been considerable debate regarding whether a higher standard of living necessarily requires greater consumption of energy or electricity use, in particular.[107]

---

105. Darby, Sarah. (2006). *The effectiveness of feedback on energy consumption*. Environmental Change Institute, University of Oxford; Parker, Danny, et. al. (2008). *Pilot evaluation of energy savings from residential energy demand feedback devices*. Florida Solar Energy Center; Mountain, Dean C. (2006). *The impact of real-time feedback on residential electricity consumption: The hydro one pilot*. Mountain Economic Consulting and Associates, Inc.; and Mountain, Dean C. (2007). *Real-time feedback and residential electricity consumption: British Columbia and Newfoundland and Labrador pilots*. Mountain Economic Consulting and Associates, Inc.

106. Linda Baker. (2007). On the rebound. *Scientific American* (August).

107. See the following, for example: John Kraft and Arthur Kraft. (1978). On the relationship between energy and GNP. *Journal of Energy and Development*, Spring, 540–552. Ali Akarca and Thomas Long. (1980). On the relationship between energy and GNP: A Re-Examination. *Journal of Energy and Development*, Spring, 326–331. Donald Murray and Gehunag Nan. (1992). The energy consumption and employment relationship: A clarification. *Journal of Energy and Development, 16*(1), 121–130. Milton Searle. (1986). Trends in the electricity and economic growth relationship. *Public Utilities Fortnightly*, 30 October. Danilo Santini. (1994). Verification of energy's role as a determinant of U.S. economic activity. In John Moroney (Ed.), *Advances in the Economics of Energy and Resources*, Vol. 8, pp. 159–194. Eden Yu and Jai-Young Choi. (1985). The causal relationship between energy and GNP: An international comparison. *Journal of Energy and Development, 10*(2), 249–272.

Some studies have claimed to have uncovered a relationship. However, there has recently been evidence of a "decoupling" between energy use and economic prosperity.[108] Many of the studies which claimed to demonstrate that economic development could be "caused" by increases in energy consumption have been found to have suffered from statistical problems and deficiencies in the manner in which energy consumption was measured.[109] The experience of California in recent years suggests that aggressive energy efficiency programs can succeed in reducing per capita household energy use without sacrificing economic growth.[110] One potential model to consider is the Swiss 2000 Watt Society, further described in this book's Chapter 16 by Stulz et al.

Getting those formerly unconcerned with energy usage to be receptive to "extreme" measures such as net zero building may require a rebranding campaign or mandatory restrictions. Individuals must decide that saving energy is a priority for them. Although more research is certainly needed in this area, long-term or permanent attitudinal shift will require a multidimensional approach to changing attitudes and impacting behavior.[111] The Metropolitan Group, a Washington, D.C.-based social change agency, endorses a five-phase approach to building public will: 1) framing the problem; 2) building awareness; 3) becoming knowledgeable/transmitting information; 4) creating a personal conviction; and 5) evaluating while reinforcing.[112] Because the energy efficiency behavioral change desired is on a monumental scale, it may be necessary to win public will and approval through such a measured approach.

In designing effective implementation strategies, we must avoid focusing on the conservation of energy while neglecting other inputs. Energy, as an input to living activities, is not separable from other inputs. Decisions that focus solely on minimizing energy inputs may be economically inefficient. That is, they may result in excessive costs for capital equipment or greater labor requirements. Or minimizing energy could lead to a lower level of comfort or productivity.

---

108. OECD (2002), p. 45.

109. Jay Zarnikau. (1996). A re-examination of the causal relationship between energy consumption and GDP. *Journal of Energy and Development.*

110. Skip Laitner. (2009). *The positive economics of climate change policies, what the historical evidence can tell us.* ACEEE (p. 20).

111. Metropolitan Group. (2009). *Building public will: Five-phase communication approach to sustainable change. www.metgroup.com/assets/612_bpwarticle0209.pdf.*

112. Ibid.

6. THE BOTTOM LINE: IS IT PRACTICAL AND CAN WE AFFORD IT?

Assuming the many practical barriers to achieving energy efficiency were successfully overcome, what might it cost to achieve zero net household energy purchases on a widespread basis? Some analysis is offered here.

The residential sector of the U.S. is analyzed, given the authors' familiarity with energy efficiency opportunities in that country and the availability of energy efficiency potential estimates. As a starting point, we adopt the conclusions of the McKinsey report, which finds that residential energy use could be reduced by 28% by 2020 through cost-effective measures,[113] and we make the assumption that *all* of those efficiency measures are adopted for the sake of this hypothetical example. These measures are cost-effective, so there is no *net* cost to implementing them on a net present value basis. In fact, there is a net savings of $395 billion to American households. A remaining 8240 trillion BTUs would need to be met through some combination of efficiency measures which were not likely to prove cost-effective relative to a benchmark avoided cost of $13.80 per MMBTU[114] or through on-site distributed renewable energy resources (Figure 9).

PV systems would provide a feasible—in many situations—though costly means of bridging much of the remaining gap. If the average home in the U.S. had a 15.6 kW[115] PV system producing 1450 kWh per year per kW of installed capacity, the energy needs of America's residential sector that would remain following the adoption of cost-effective energy efficiency measures would be met. Here it is assumed that solar energy would be used to displace natural gas and heating oil, as well as electricity purchases. The average household would export power when the output of the solar system exceeded on-site needs and would purchase power from the grid at other times, but net energy purchases would be negligible. The cost of PVs is likely to establish the cost of reaching energy efficiency goals, since this could be the technology that fills the gap.

It is assumed that advances and economies of scale in manufacturing would reduce the installed cost of PV systems including the cost of invertors from the current cost of about $6[116] to an average $4 per Watt over the 2010 to

113. McKinsey & Company, *Unlocking Energy Efficiency in the U.S. Economy*, July 2009, p. 29.

114. Presumably, this benchmark avoided cost would change if the demand for fossil fuels and purchased electricity dropped by the magnitude envisioned in this scenario. Nonetheless, we shall retain this value to maintain consistency with the McKinsey report.

115. The BTU value was adjusted downward by 1% to account for the need to go from natural gas water heating (where applicable) to electric (powered with PV systems). Electric water heaters tend to use fewer BTUs.

116. This is a median value, based on the cost of nearly all projects completed in Texas in 2009 that received a rebate from an investor-owned utility. Installed costs (including the inverter) ranged from $4 per Watt to $13 per Watt.

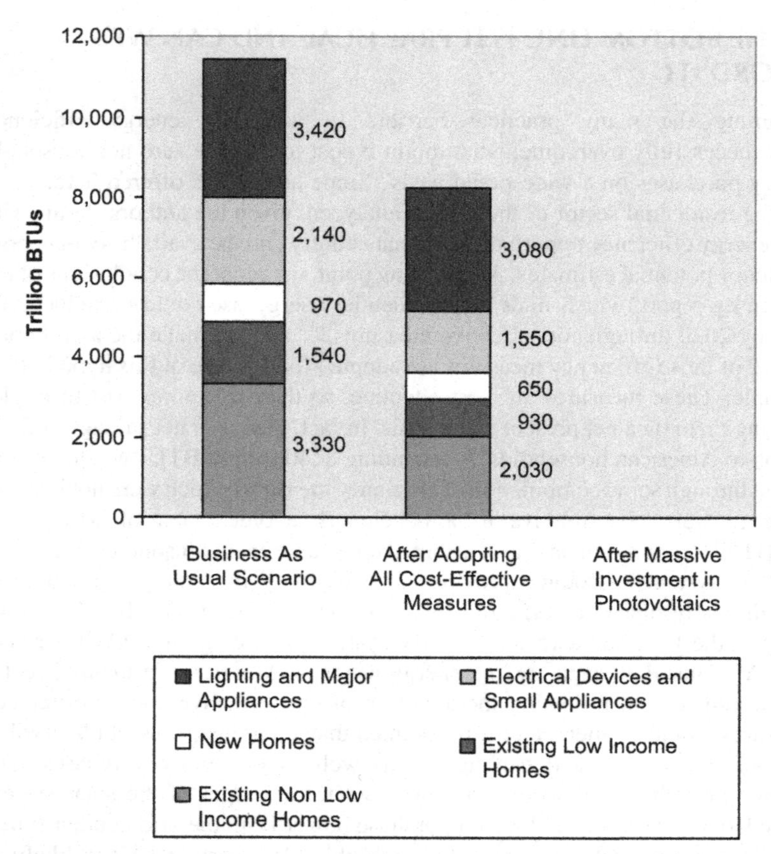

FIGURE 9 Hypothetically Moving the Entire U.S. Residential Sector to Net Zero. *Source of data on second bar: McKinsey & Company, 2009.*

2020 period.[117] Thus, the average home would invest $59,956 in a PV system. On a national basis, this initial investment would total $6595 billion.[118] Assuming a 25-year life for PV systems, a 6% discount rate, and the avoided cost values assumed in the McKinsey report, the net present value of the cost—that is, the initial cost of the equipment exceeds the economic value of

117. Much more optimistic projections of the cost of photovoltaic panels are available, including projections that solar energy shall reach "grid-parity" with the cost of electricity purchased from utilities within three to five years in some U.S. states. See Christopher Mims. (2009). The no-money-down solar plan. *Scientific American*, December, pp. 50–51.

118. To place this figure in perspective, this cost is almost ten times the size of the budget of the Troubled Asset Relief Program (TARP) established by the U.S. government to improve the financial health of financial institutions following the crises of 2008-2009. It is also nearly ten times the direct cost spent by the U.S. on the war in Iraq from 2001 to 2009.

the energy generated, since no tax credits are assumed—of the solar system would be $45,015 per home. If the net savings from the cost-effective energy efficiency measures were also considered, the *net incremental cost* to reach ZNE would decline to $41,424 per home, which represents a 15% premium given that the average cost of a U.S. home in 2009, including the land, was $270,900. This premium, however, would be recovered through savings in energy costs in as few as 5.2 years, as discussed earlier.

This scenario hints at some practical constraints. Will residential home builders be willing to pay more initially for energy efficiency improvements and design? Even with government subsidies, the prospect of putting costly PV equipment on the roof of one's house or building geothermal energy technology into the cost of construction can seem financially daunting when the return on investment might not be seen for 5 or 10 years. There are physical limitations regarding the quantities of PV panels that could be placed on rooftops, in yards, or community areas. Implementation will be site-specific because areas with considerable cloud cover or high tree cover may not be good sites for solar energy collection.

In the end, behavioral changes will also be needed. The physical changes are likely to be too expensive and impractical in the short-term, the period in which the housing stock is largely "fixed," particularly in the existing building stock.

Housing preferences have evolved in recent years, from a preference for "McMansion"-style suburban homes with extra bedrooms and bathrooms to more scaled-down versions without large garages and recreation rooms. As energy prices increase, consumers are more likely to prefer homes they can heat and cool cheaply. However, the existing building stock may pose a problem as it may be cheaper to buy an older larger home with more square footage than desired than build a new energy-efficient one. Subsidies and standards will be needed to push toward new ZNE homes and modifications for existing homes.

CONCLUSIONS

What would it take to reduce direct residential energy consumption to a negligible, if not zero, net level? It would begin with a bold vision—bold because our present energy use habits need to change. Achieving a ZNE goal would require truly worldwide cooperative input: initiatives at the community and regional level, as well as national or international energy efficiency policies.

Placing society on a path toward negligible energy use presents a set of informational, institutional, and behavioral challenges. Present patterns of energy resource use depend on climate, local building practices, lifestyles, and a host of other factors. Research and development, consumer education, training of building trade professionals and craftsmen, greater automation and optimization of energy use through new technologies, and better price signals can all play a role. On-site renewable energy will likely become more affordable over time, making this an option to fill the gap.

We have outlined some serious limitations to ZNE homes entering the market, but a review of leading green building programs, zero home energy use initiatives, and deep green communities suggest that it is possible to reduce net direct energy purchases to negligible levels in new homes and communities. Our ability to dramatically improve existing dwellings is less certain, and may rely more on behavioral changes of those residents to actually approach "near zero" net energy use. Lifestyles will need to change so that homeowners downscale their energy use habits and redefine "living comfortably." There is evidence of a mindset shift occurring in the developed world as we become aware of our ballooning share of the world's consumption of natural resources. Homebuilders are realizing that future generations may want housing that does not consume finite resources for heating, cooling, and lighting.

Many homes are energy inefficient today, the result of decades-old building practices being followed. Because residential homes are generally built to last, the worst-case scenario is that housing will be built today that is both energy inefficient and built to last—meaning that future owners will be stuck paying the bill for years to come. Reducing energy use without sacrificing lifestyle or comfort has always been the goal of the energy efficiency industry, but it is difficult to get to negligible energy consumption without making some sacrifices.

Our overall outlook on the ZNE building front is optimistic, because there are signs that we are heading down the right path. Technology is catching up to demand, demand from consumers is beginning to drive the market, and market changes are influencing design so that once-outlandish ideas are becoming more feasible. Switching to a new paradigm will not be simple, but the sooner that the developed world gets on that path toward a zero-energy future, the better.

BIBLIOGRAPHY

Akarca, A., & Long, T. (1980). On the relationship between energy and GNP: A re-examination. *The Journal of Energy and Development, (Spring)*, 326–331.

American Council for an Energy-Efficient Economy. (2007). *Quantifying the effects of market failures in the end-use of energy.* Final draft report to the International Energy Agency.

Austin Energy Green Building. (2008). Guide to the Single-Family Home Rating. *http://www.austinenergy.com/energy%20efficiency/Programs/Green%20Building/Participation/singleFamilyHomeRatingGuide.pdf.*

Anderson, D. (1998). *Progress toward energy sustainability in OECD countries. www.helio-international.org/Helio/anglais/reports/oecd6.html.*

Andersen, M. (1994). Green tax reform in Denmark. *Environmental Politics, 3*(1), 139–145.

Architecture 2030. *The 2030 Challenge. www.architecture2030.org/2030_challenge/index.html.*

Avid Home Studios. (2009). *The house of tomorrow from the recession today. www.avidhomestudios.com/blog/2009/01/05/the-house-of-tomorrow-from-the-recession-today.*

Baker, L. (2007). On the rebound. *Scientific American*, (August).

Berndt, E. (1985). Aggregate energy, efficiency, and productivity measurement. *Annual Review of Energy, 3*, 225–273.

Bin, S., & Dowlatabadi, H. Consumer lifestyle approach to U.S. energy use and the related CO_2 emissions. *Energy Policy*, 33, 197-208.

California Energy Commission. (2007). Integrated energy policy report. *www.energy.ca.gov/2007_energypolicy/index.html.*

California Energy Commission. (2005). Just the facts: The truth about proposed TV standards. *www.energy.ca.gov/appliances/TV_Standards_Facts.pdf.*

Case Study: Premier homes—premier gardens. *www.bira.ws/projects/files/BA_Solar_CS_Premier.pdf* 2007.

City of Austin. (2010). City Council Adopts New Energy Code Amendments; Action Moves Austin Closer to "Zero-Energy" Homes. City of Austin press release. *http://www.austinenergy.com/About%20Us/Newsroom/Press%20Releases/2010/zeroEnergy.htm.*

City of Berkeley. (2009). Climate Action Plan. *http://www.ci.berkeley.ca.us/uploadedFiles/Planning_and_Development/Level_3_-_Energy_and_Sustainable_Development/Berkeley%20Climate%20Action%20Plan.pdf.*

Community Energy Services Corporation. (2010). SmartSolar. *http://www.ebenergy.org/smartsolar.*

Clinton Climate Initiative. (Undated). *Austin's green building program facilitates the construction of sustainable buildings.*

Communities and Local Government. *Code for sustainable homes. www.planningportal.gov.uk/uploads/code_for_sust_homes.pdf.*

Costanza, R. (1980). Embodied energy and economic value. *Science, 12, 210*(4475), 1219–1224.

Darby, S. (2006). *The effectiveness of feedback on energy consumption. Environmental Change Institute*. University of Oxford.

Disch, R. (2010). *Energy Consumption. PlusEnergieHaus. http://www.plusenergiehaus.de/index.php?p=home&pid=11&L=1&host=1#a502.*

Discovery at Spring Trails. http://discoveryatspringtrails.com.

Discovery at Spring Trails – do the math. http://discoveryatspringtrails.com/technology-2.

Eberhard, J., et al. (2002). *Steps toward a 2000 watt society.* Novatlantis.

EcoMall *The new wave of energy efficient refrigerators. www.ecomall.com/greenshopping/icebox2.htm.*

Environmental Resources Trust, Inc. (2005). *Final report of the zero energy homes for Chicago EcoPower project. www.carbonfund.org/Documents/ERT%20Final%20Report.pdf.*

International Energy Agency. (2007). *Energy use in the new millennium: Trends in OECD countries.* International Energy Agency.

International Energy Agency. (2008). *Energy efficiency requirements in buildingcodes, energy efficiency policies for new buildings*. IEA Information Paper. International Information Agency.

International Energy Agency. (2008). *Towards a sustainable energy future*. International Information Agency.

International Energy Agency. (2008). *Worldwide trends in energy use and efficiency.* Location: International Information Agency.

Kaplan, S. (2008). *Building an energy efficient America: Zero energy and high efficiency buildings.* Environment American Research & Policy Center. *www.environmentamerica.org/uploads/kx/Im/kxImeu7IZctnMBnd2AZJQg/AME-efficiency.v6.pdf.*

King, C., Zarnikau, J., & Henshaw, P. (2010). Defining a standard measure for whole system EROI, combining economic top-down and LCA bottom-up accounting. *Proceedings of Energy Sustainability 2010 Conference.* Phoenix.

Kraft, J., & Kraft, A. (1978). On the relationship between energy and GNP. *Journal of Energy and Development (Spring)*, 540–552.

Laitner, S. (2009). The positive economics of climate change policies, what the historical evidence can tell us. *ACEEE*, 20.

Lamp History in the Nutshell. *www.lampreview.net/*.

MacMath, R. (2007). *Austin energy green building program*. SA06 Austin Energy's Zero Energy Home Design Competition, presentation, (May 5).

Masdar City. *www.masdarcity.ae/en/index.aspx*.

McKinsey & Company. (2009). *Unlocking energy efficiency in the U.S. economy*.

Meier, A., Wright, J., & Resenfeld, A. (1983). *Supplying energy through greater efficiency*. University of California Press.

Metropolitan Group. (2009). *Building public will: Five-phase communication approach to sustainable change*. *www.metgroup.com/assets/612_bpwarticle0209.pdf*.

Mims, C. (2009). The no-money-down solar plan. *Scientific American, December*, 50–51.

Morgan, R. (Undated). *Green building in Austin*, TX, presentation.

Mountain, D.C. (2006). *The impact of real-time feedback on residential electricity consumption: The hydro one pilot*. Mountain Economic Consulting and Associates, Inc.

Mountain, D.C. (2007). *Real-time feedback and residential electricity consumption: British Columbia and Newfoundland and Labrador pilots*. Mountain Economic Consulting and Associates, Inc.

Mouwad, J., & Galbraith, K. (2009). Plugged-in age feeds a hunger for electricity. *New York Times*, (20 September).

Murray, D., & Nan, G. (1992). The energy consumption and employment relationship: A clarification. *Journal of Energy and Development, 16*(1), 121–130.

Nadel, S., Shipley, A., & Elliott, R.N. (2004). The technical, economic, and achievable potential for energy efficiency in the US—a meta analysis of recent studies. In *ACEEE Summer Study on Energy Efficiency in Buildings*, August.

Net Zero Energy Home Coalition. *www.netzeroenergyhome.ca/*.

OECD. (2002). *Towards sustainable household consumption*. Location: OECD.

Parker, D., et al. (2008). *Pilot evaluation of energy savings from residential energy demand feedback devices*. Florida Solar Energy Center.

Reinders, A. H. M. E., Vringer, K., & Blok, K. (2003). The direct and indirect energy requirement of households in the European Union. *Energy Policy, 31*, 139–153.

Ries, C., Jenkins, J., & Wise, O. (2009). *Improving the energy performance of buildings: Learning from the European Union and Australia*. RAND Corporation.

Santini, D. (1994). Verification of energy's role as a determinant of U.S. economic activity. In J. Moroney (Ed.), *Advances in the economics of energy and resources, 8*, 159–194.

Schmidt, P., Sparrow, T., Vanston, J., & Zarnikau, J. (1994). *Neo-electrification in the information age*. Edison Electric Institute.

Searle, M. (1986). Trends in the electricity and economic growth relationship. *Public Utilities Fortnightly*, October.

Solar Village Homes. *www.solarvillagehomes.com/about/index.php*.

Solutions Oriented Living. *www.solaustin.com/*.

The Alley Flat Initiative. *http://thealleyflatinitiative.org*.

Torcellini, P., et al. (2009). *Zero energy buildings: A critical look at the definition*. National Energy Renewable Laboratory (NREL). *www.nrel.gov/docs/fy06osti/39833.pdf*.

U.S. Department of Energy. (2007). *Annual energy review 2006*. *www.eia.doe.gov/emeu/aer/contents.html*.

U.S. Department of Energy. (2008). *FR Doc E8-30975. Federal Register, 73(250). www.thefederalregister.com/d.p/2008-12-30-E8-30975.*

U.S. Department of Energy. (2010). *Building technologies program. www1.eere.energy.gov/buildings/building_america.*

U.S. Department of Energy. (2007). Premier gardens: Moving toward zero energy homes. *www.consol.ws/files/Premier_Gardens.pdf.*

U.S. Environmental Protection Agency. (1997). *Energy conservation. EPA 905-F-97-011. www.epa.gov/reg5rcra/wptdiv/p2pages/energy.pdf.*

U.S. Environmental Protection Agency. (2006). *Annual report: ENERGY STAR and other climate protection partnerships.*

U.S. Environmental Protection Agency. (2010). *ENERGY STAR new homes. www.energystar.gov/index.cfm?c=new_homes.hm_index.*

U.S. Environmental Protection Agency. (2009). *Green communities. www.epa.gov/greenkit/index.htm.*

U.S. Green Building Council. (2010). *LEED for new construction. http://www.usgbc.org/leed/nc/.*

U.S. White House. (2009). *Federal leadership in environmental, energy, and economic performance.* Executive Order (October 5). *www.whitehouse.gov/assets/documents/2009fedleader_eo_rel.pdf.*

Vringer, K., & Blok, K. (1995). The direct and indirect energy requirements of households in the Netherlands. *Energy Policy, 23*(10), 893–910.

Washing without water. (2009). *The Economist*, (5–11 Sept.), 12–14.

Weber, C., & Perrels, A. (2000). Modeling lifestyle effects on energy demand and related emissions. *Energy Policy, 28*, 549–566.

Wollam, A. (2008). GE to light up Land Tejas development. *Houston Business Journal. www.bizjournals.com/houston/stories/2008/01/14/daily43.html.*

World Business Council. (Undated). *Energy efficiency in building: Transforming the market.*

World Business Council for Sustainable Development. (2010). *http://www.wbcsd.org.*

Yu, E., & Choi, J.-Y. (1985). The causal relationship between energy and GNP: An international comparison. *Journal of Energy and Development, 10*(2), 249–272.

Yudelson, J. (2009). *Green building trends: Europe.* Location: Island Press.

Zaarias-Farah, A., & Geyer-Allely, E. (2003). Household consumption patterns in OECD countries: Trends and figures. *Journal of Cleaner Production, 11*, 819–827.

Zarnikau, J. (1992). A re-examination of the causal relationship between energy consumption and GDP. *Journal of Energy and Development, 21*(2), 229–240.

Zarnikau, J. (1999). When different types of energy resources are aggregated for use in econometric studies, does the aggregation approach matter? *Energy Economics, 21*(5), 485–492.

Zarnikau, J. (1999). Defining 'total energy use' in economic studies: does the aggregation approach matter? *The Energy Journal, 20*(3), 139–146.

Chapter 10

Beyond the Meter: Enabling Better Home Energy Management[1]

Karen Ehrhardt-Martinez,* John A. "Skip" Laitner,† and Kat A. Donnelly**
*University of Colorado, †American Council for an Energy-Efficient Economy,
**emPower Devices and Associates

Chapter Outline

1. INTRODUCTION: RESIDENTIAL ENERGY CONSUMPTION AND ENERGY INVISIBILITY

Residential buildings currently account for more than one-fifth of total energy demand in the United States, approximately 22%, and residential energy use is on the rise. In the past decade alone, residential energy use grew by 23%, and projections through the year 2030 indicate another 8% increase (EIA, 2010). Much of the projected growth is expected to result from a combination of factors, including a growing number of people and households and an increased demand for energy services. Given the ongoing concerns over climate change and energy security, and the need for significant improvements in our current level of energy efficiency throughout the economy, the

1. Much of the information presented in this chapter is drawn from a much larger study: Ehrhardt-Martinez et al. (2010). *Advanced metering initiatives and residential feedback programs: A meta-review for economy-wide electricity-saving opportunities.*

Energy, Sustainability and the Environment.

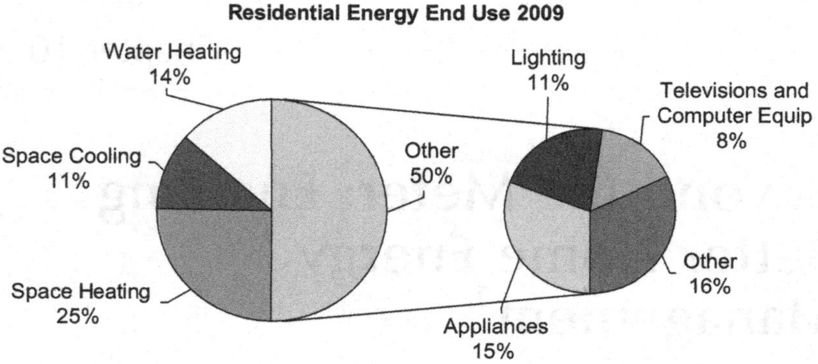

FIGURE 1 Pattern of U.S. Residential Energy Use. *Source: EIA (2009).*

continued growth in residential sector energy demand is particularly worrisome.

In the U.S. residential sector, energy resources are used for a variety of purposes. As shown in Figure 1, most energy resources are devoted to space heating (25%), followed by water heating (14%), and air-conditioning (11%). The remaining 50% is used by appliances, lighting, electronics, and other miscellaneous devices (EIA, 2009).[2] Among home appliances, refrigerators are one of the largest users of electricity although energy consumption for televisions and set-top boxes is projected to increase at the greatest rate—1.9% annually—followed by personal computers and related equipment at 1.7% (EIA, 2009). Of particular note, these and other home electronics consume power not only when in use, but also when they are in standby and off mode. Their power supplies alone can draw significant loads even when disconnected from the appliance, the so-called *phantom load.*

Over the course of the last 40 years there have been dramatic improvements in the energy efficiency of many residential buildings and household technologies including refrigerators, furnaces, and air-conditioners. However, many of these efficiency gains have been offset by preferences for larger houses, increased air-conditioning use, and penetration of a greater variety of new appliances and electronics. For example, the past 10 to 15 years have witnessed a dramatic growth in acquisition and use of a range of home electronics—ranging from iPods, cell phones, and computer games to set-top boxes and larger flat-screen televisions. These gadgets are expected to contribute an increasing percentage of home energy use.

Given these trends, it is clear that efforts to reduce household energy consumption will be severely constrained unless citizens are actively engaged

2. These proportions vary by region such that heating comprises a larger proportion of energy use in cooler climates while air-conditioning comprises a larger proportion in warmer climates.

in more proactive home energy management activities. But such endeavors are not as easy as one might imagine. The use of electricity and natural gas in modern households is largely invisible to residential energy consumers. Consumers see the benefits of a brightly lit home and experience the convenience of modern appliances, but they don't see the energy required to achieve these ends. The very invisibility of energy and its costs makes smart management—whether conservation measures or cost-effective energy efficiency improvements—difficult.

Compared to the use of wood or coal, natural gas and electricity flow effortlessly, seamlessly, and silently into our homes as they fuel our furnaces, power our air-conditioners, and meet a wide variety of other energy service demands; doing so without any notable trace of their presence. For most people, the only measure of their energy use is the bill that they receive at the end of the month, up to 45 days after consumption.[3] Unfortunately, the typical utility bill—even for the best energy detective and the most energy-conscious consumer—is an inadequate tool for managing energy use. Monthly bills may report the number of kilowatt-hours (kWh) of electricity or therms or cubic meters of natural gas consumed, and the resulting costs that are incurred, but they don't indicate which end-uses are responsible for the most energy. Neither do they tell consumers just how energy-hungry, nor how energy-(in)efficient their existing appliances might actually be. Nor do they indicate how changes in consumer choices and behavior could enhance or offset energy demands associated with changing weather patterns, new appliances, and the acquisition of new electronic equipment. In short, most households in the United States and throughout the world are among the energy blind; unable to see the energy that they consume and the impact that it has on both the economy and the environment.

The dysfunctionality of our current energy system has been recognized for many years. More than a quarter century ago, Willett Kempton and Laura Montgomery (1982) illustrated the paradox of consumption that occurs without meaningful information in the following way:

[Imagine a grocery] store without prices on individual items, which presented only one total bill at the cash register. In such a store, the shopper would have to estimate item price by weight or packaging, by experimenting with different purchasing patterns, or by using consumer bulletins based on average purchases.

The invisibility of modern energy resources also impedes the establishment of social norms concerning what might be considered "appropriate" levels of energy consumption.[4] Not only are most energy consumers blind to their own

3. In some countries, such as the U.K., the bill is quarterly and is often estimated, so the disconnect is even worse.

4. This book's Chapter 3 by Bartiaux et al. also discusses social and cultural dimension of energy use.

level of energy consumption, but they are equally unaware of the amount of energy consumed by their neighbors and other people living in the same city or region of the country. Without an appropriate frame of reference, individuals and households have a hard time determining whether their particular patterns of energy consumption are excessive or moderate and whether some type of intervention is warranted. Yet, access to this kind of information, particularly through a dynamic and near real-time feedback process, can be a big motivator in modifying energy consumption practices in ways that save money, decrease energy waste, and reduce environmental impacts.

This chapter explores the potential role of smart meters and feedback devices and programs to identify how households, individuals, and communities can empower themselves, reshape current energy consumption patterns, and potentially reconfigure existing energy production systems. In the next section, we begin with a discussion of the growing proliferation of smart meters and the need to pair these with in-home displays or other feedback mechanisms that allow energy consumers to access their own energy use information. We then describe a variety of different in-home feedback technologies that are currently available to consumers to help them better manage their energy resources. In Section 3, we discuss the historical evidence linking different types of feedback to behavioral change and energy savings. The section includes estimates of the range of savings that have been achieved through the application of each of the five types of feedback. Section 4 looks inside the household to explore the actions that people take in response to feedback—the actions that are responsible for generating household energy savings—and how those actions vary across households. The penultimate section of this chapter discusses the degree to which feedback-induced energy savings in households might contribute to reductions in national energy consumption trends, by estimating the magnitude of potential feedback-induced energy savings in 2030. The chapter concludes with a discussion of potential future benefits of an integrated smart grid, including opportunities for enhancing energy efficiency, renewable energy, and distributed generation; providing the means for plug-in vehicles; and expanding on the mechanisms to induce better home energy management.

2. SMART METERS AND IN-HOME DISPLAYS

While the invisibility of energy consumption (and production) has clearly impeded the ability of households to manage their energy consumption, recent innovations offer the possibility of energy savings in the near term. Many of these innovations rely on a range of increasingly ubiquitous information and communications technologies (ICT). In particular, the combination of faster Internet connections, new web-based interfaces, advanced utility meters, in-home energy displays, smart phones, smart appliances, and the phenomenal data storage and manipulation capacity of today's computers offer the opportunity to easily and conveniently manage household energy consumption in a whole variety of ways.[5] The application and integration of ICT throughout systems of energy production, distribution, and consumption could result in tremendous energy savings and reveal important patterns and variations in residential energy use. This information could also provide consumers with a context for evaluating their own energy consumption patterns relative to their actual energy service needs. Perhaps even more intriguing, the active management of household energy consumption could open up larger efficiency gains throughout all levels of the economy.

A growing number of recent programs and projects are implementing and studying the impact of ICT and advanced meter infrastructure (AMI) on reducing energy waste. Their findings suggest that when ICT is used to provide energy consumers with detailed and timely feedback, they are often successful in achieving significant energy savings. One recent study by Sarah Darby (2006), for example, estimates that direct forms of feedback can generate household electricity savings in the range of 5%–15%. And a more recent meta-review of more than 50 related studies by Ehrhardt-Martinez et al. (2010) generally confirmed this range of home electricity savings suggesting that average feedback-related savings during the period 1995 to the present has ranged between 4% and 12%. The combination of advanced utility energy meters, in-home displays, and well-designed programs are among the most innovative approaches that have recently expanded the opportunity for greater levels of residential energy savings.

The Federal Energy Regulatory Commission (FERC, 2008) defines advanced metering as a "system that records customer consumption (and possibly other parameters) hourly or more frequently and provides for daily or more frequent transmittal of measurements over a communication network to a central collection point." Unlike old-fashioned utility meters with their distinctive rotating disks, advanced meters are digital devices that provide the opportunity for two-way communications between the utility and consumers. Some advanced utility meters can be enabled to also communicate directly with household appliances and devices. When combined with in-home displays and well-designed programs, advanced meters hold the potential of providing energy consumers with convenient, real-time energy consumption data and energy cost information. In other words, these new devices and applications can be effective tools for making energy visible and for providing consumers with the resources that they need to effectively manage their household energy consumption.

Recently, governments and utilities throughout the European Union, North America, Australia, and Japan have begun to invest in the development of smarter electricity grids. One of the first tasks has been to install advanced utility meters in homes and businesses. As of December 2008, roughly 4.7% of

5. See Ehrhardt-Martinez et al. 2010 for a more detailed discussion.

the nearly 143 million electric meters in the United States had been replaced by advanced meters, bringing the total number of advanced meters to 6.7 million (FERC, 2008). The pace of change has been rapid and is likely to accelerate. Eighty-seven percent of all advanced meters in place as of the end of 2008 were installed during the prior two years. However, the distribution of advanced meters across the United States has been uneven. As shown in Table 1, penetration in Pennsylvania, Idaho, and Arkansas has been the highest at 23.9%, 13.8%, and 11.3%, respectively. Also noteworthy is that fewer than 20 states

TABLE 1 Penetration of Advanced Meters – States with at Least 3% Penetration

State	AMI Meters	Total Meters	Penetration
Pennsylvania	1,443,285	6,036,064	23.9%
Idaho	105,933	769,963	13.8%
Arkansas	168,466	1,488,124	11.3%
North Dakota	33,336	375,473	8.9%
South Dakota	41,191	475,477	8.7%
Oklahoma	161,795	1,875,325	8.6%
Texas	868,204	10,870,895	8.0%
Florida	765,406	9,591,363	8.0%
Georgia	342,772	4,537,717	7.6%
Missouri	204,498	3,098,055	6.6%
Vermont	20,755	375,202	5.5%
Alabama	139,972	2,774,764	5.0%
Kentucky	105,460	2,161,142	4.9%
South Carolina	114,619	2,373,047	4.8%
Kansas	61,423	1,426,832	4.3%
Wisconsin	117,577	3,039,830	3.9%
Wyoming	12,268	318,282	3.9%
Arizona	96,727	2,810,224	3.4%
North Carolina	143,093	4,771,479	3.0%
Total	4,946,780	59,169,258	144.1%

Source: FERC (2008).

have penetration rates of 3% or higher. Perhaps surprisingly, the list of top penetration rates does not include California—despite that state having a very large number of meters (14.6 million), and despite the fact that it is usually ranked as among the more energy-efficient states. In general, the largest number of meters has been installed in some of the most populated states including Texas (10.9 million meters), Florida (9.6 million), New York (7.8 million), and Pennsylvania (6.0 million).

According to FERC, the planned deployment of nearly 52 million advanced meters is scheduled to take place during the next 5 to 7 years. That represents nearly an eightfold increase from current installations. When combined with the 6.7 million advanced meters already deployed, the total penetration of advanced meters in the near term will approach 40%. Notably, however, only 11% of advanced meters are currently used in conjunction with price-based demand response programs, and less than 1% in conjunction with home area networks or with other technologies that are likely to result in energy savings. In other words, up to now, utility investments in advanced metering installations are primarily driven by utility concerns associated with reducing the costs associated with meter readings, improving service reliability through automated outage detection, and creating the means for the future implementation of real-time energy pricing structures.

While advanced metering technologies have achieved a relatively low level of market penetration to date; utilities have committed to a large-scale deployment in the near future, primarily as a result of the potential peak electricity and cost savings that may be directly captured by the utilities. Importantly, however, utility-based decisions about the ways in which these technologies are implemented are likely to shape future energy savings as well as determine which technologies and players will benefit the most.

Of critical importance in this new emerging market is the distinction between the utility returns on their investments, primarily through peak load shifting and reductions in operating costs, and the consumer energy bill savings, and how the distribution of benefits is likely to impact the overall scale of total energy savings. To maximize both energy savings and consumer benefits, new advanced utility meter technologies must be accompanied by in-home feedback devices or well-designed home energy reports. In-home feedback can take several different forms, including:

- Third-party presentation of energy use data where a third party is responsible for presenting the data to household energy consumers through paper reports or via the Internet.
- Third-party home networking systems where third-party providers install in-home energy displays that provide occupants with real-time energy consumption data.
- Do-it-yourself feedback devices where homeowners purchase and install in-home energy displays that provide real-time energy consumption data.

An important benefit of third-party presentation of indirect (after consumption) energy use data is that it is often relatively inexpensive and generally doesn't require any *additional hardware deployment.* Currently, there are a handful of service providers that leverage existing utility and other data to provide content-rich feedback to residential consumers. One such company, OPOWER (formerly known as Positive Energy), incorporates numerous behavioral principles to provide rich and consumer-targeted feedback. OPOWER applies advanced analytical tools to leverage available data, such as utility back-end data along with outside data sources, including parcel, tax assessor, demographics, and so on, to create customized household profiles (Kavazovic, 2009). Their expertise lies in "cleaning up" the utility and other existing data, and then transforming them into useful information for the consumer (ibid.). By providing information and feedback specific to a given consumer, and suggesting how energy consumption in given individual household compares to its neighbors or households with similar incomes or family size, consumers are more likely to respond to tips that might cost-effectively reduce their energy use.

A similar approach is being taken by Efficiency 2.0, a newer entrant into the energy efficiency software market (Fehrenbacher, 2009) that is currently pursuing partnerships and pilots with utilities to provide energy management feedback to consumers. Efficiency 2.0's software platform, Community Connect^SM, has been specifically designed with behavior change best practices in mind. It provides personalized energy information and advice to consumers via the web and includes a focus on goal-setting and follow-through. The Efficiency 2.0 platform also integrates several forms of social pressure. Users can enter as little or as much information as they would like about their own energy practices, lifestyles, and behavior, depending on their desired level of involvement. Efficiency 2.0 focuses on interpreting the data to provide customized energy recommendations to drive behavior changes (Frank, 2009).

Google.org is another organization interested in leveraging existing sources of energy consumption data with the goal of energy savings. The Google.org PowerMeter interface project creators believe that personal energy use data belong to the consumer and that they should be available in a standard, nonproprietary manner. Like the other examples, Google.org plans to harvest data from utilities, but also coordinates with an in-home energy display, and plans to eventually coordinate with other home automation technologies to make energy information more accessible and useful to end-users. In its early stages, Google.org is using an energy display device to measure, monitor, and report electricity consumption for approximately 200 of its employees, with the product recently available to the general public. Although it doesn't directly disaggregate the overall household electricity use into specific end uses, Google.org is training the software to recognize energy demand peaks and patterns and correlate them with specific appliance usage like the dryer or refrigerator (Olsen, 2009). As more frequent and appliance-specific data become available, in effect translating the data stream into a format that identifies specific end uses, the PowerMeter promises to provide more detailed information to the consumer. The PowerMeter software platform was recently released to the public.

Home automation networks are another means of providing residential energy consumers with home energy consumption information in addition to control of appliances and other devices. Home automation networks can be thought of as "the last mile" of the smart grid. They provide a variety of home energy management options through a wide selection of mostly interoperable products and services that, when integrated, act as a single system. It ranges from piecemeal parts of a network to a full-fledged interoperable network of water, gas, and electricity devices that can communicate with the utility. A complete home automation network can result in a system that optimizes household performance based on supply conditions and time-of-use market prices, as well as consumer price, comfort, and environmental preferences, and includes the following components:

- *In-home smart devices and appliances*: Networking and/or communicating chips embedded in and attached to appliances and devices that allow for wireless and/or wired automation.
- *Advanced network systems and software*: Wireless mesh networks and/or disambiguation algorithms that provide measurement and feedback of appliance-specific data.
- *Potential for two-way communication with the utility*: Interface tools that analyze and display data from smart meters and utilities to in-home energy displays, smart thermostats, web, TV, mobile phone, and so on.

The simplest home automation network begins with a smart thermostat that controls heating and air-conditioning equipment and that communicates with a central computer and/or the utility's metering system with the capability to add incremental components to grow the home energy network. With a sophisticated home automation network, the smart thermostat, in-home energy display, and an Internet, mobile phone, or TV interface can manage uses, such as space heating and cooling, water heating, pool pumps, lighting, individual electrical devices, and so on, and provide opportunities to set up comfort or savings targets, or targets somewhere in-between. This enables the systems to appeal to the distinct and varied concerns that motivate household energy savings whether they are concerns regarding energy costs, the environment, and/or social-conscious motivations. For instance, several companies describe a layered approach to developing a "rulebook" made up of algorithms based on the consumer's preferred comfort levels, for example, target temperature in the weekly schedule. The customer can also set up acceptable hourly prices, or budgets, and the system automatically adjusts heating and cooling conditions. In most cases, the customer has the choice to override the system at any time or to simply "set and forget" about it and let the home area network optimize household energy use.

Most home automation systems involve the use of both action-based tactical messages, such as: set back the thermostat four degrees, as well as objective-based messages that indicate the need for immediate individual energy conservation, because "X" is happening in the electric grid. The individual then chooses how they want that event to affect their lives. For instance, they can ignore the event and pay higher peak rates where applicable, or they might choose to cycle the freezer or pool off for a couple of hours. The customer, in a sense, chooses if they want to be engaged in the information, and if so, they will continue to gain awareness and can eventually participate in a much more proactive way. In addition, most home automation companies are at the beginning stages of providing community and social networking platforms designed to facilitate community comparisons and challenges as part of their service offerings.

In addition to automation technologies, some home automation companies also apply important behavior principles in designing their feedback systems. Such behavioral principles include the use of social norms, goal setting, competitions, social networks, comparisons, special pricing structures, and actionable tips and recommendations. One such company, Control4, illustrated how enabling-technologies can help eliminate household energy waste by providing the "energy efficiency cruise control for the home" (Nagel, 2009).

3. SOME HISTORICAL EVIDENCE

Utility advanced meters and in-home feedback devices provide a modern means by which residential energy consumers can become more knowledgeable about their energy consumption practices and become active managers of their energy resource use. Historically, energy use feedback has taken a variety of different forms but all forms are based on longstanding research in the fields of psychology, sociology, communications, and marketing. Most current and past feedback initiatives are based on the notion that individual and household behavior can be shaped by providing people with information and motivation. Information elements address current and past levels of energy consumption as well as tips and recommendations to help people to consume less. Motivational elements provide the meaning and the context that provide the reason for people to change their current energy consumption practices. Sometimes this is as simple as attaching positive consequences to energy-wise consumption behaviors thereby making those behaviors more attractive to consumers, or attaching negative consequences and making unsound behaviors much less desirable (Abrahamse et al., 2005). In the case of household energy consumption, negative consequences often take the form of higher energy bills associated with higher levels of consumption. New electricity rate structures provide even greater negative economic consequences for households that fail to manage their energy consumption during peak periods or periods of high demand.

Other characteristics of feedback initiatives also appear to correlate with program effectiveness as measured by higher participation rates and energy savings. Of particular interest are:

- The frequency of the feedback
- Whether the feedback is direct (real-time) or indirect (after consumption)
- Whether or not the feedback provides a contextual framework by which individuals can evaluate their performance

3.1. Frequency and Types of Feedback

Feedback can range from continuous to infrequent and from direct to indirect. As illustrated in Figure 2, a recent report by the Electric Power Research Institute (EPRI, 2009) distinguishes between four types of indirect and two types of direct feedback. Provided after consumption, indirect feedback includes standard billing, enhanced billing, estimated feedback, and daily/weekly feedback, while direct feedback includes real-time feedback as well as real-time plus feedback. Not surprisingly, standard billing tends to be the least costly to implement but also provides the least amount of information to consumers. At the other end of the scale are real-time plus systems that work with home automation networks, providing frequent energy use data that are disaggregated by each specific end-use.

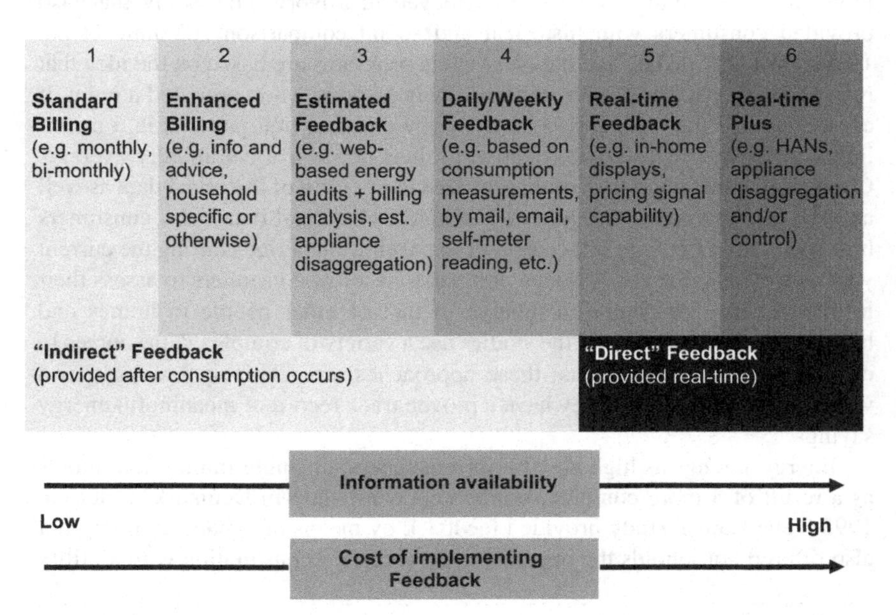

FIGURE 2 Types of Feedback and Effects. *Source: EPRI (2009).*

Past studies suggest that direct forms of feedback and feedback that is provided with greater frequency tend to be more effective (Darby, 2006; Fischer, 2007; Abrahamse et al., 2005). For example, studies from as far back as the late 1970s have shown that in-home energy displays have successfully reduced energy consumption by as much as 12% compared to a control group without the in-home device (McClelland and Cook, 1979–80; van Houwelingen and van Raaij, 1989). More recent studies indicate that even greater savings are possible (EPRI, 2009) but that feedback-related energy savings from direct feedback generally fall within the range of 5%–15% (Darby, 2006).

How important is the type of feedback in determining residential energy savings? The effectiveness of five different kinds of feedback is discussed in more detail below. These include: (1) enhanced billing, (2) estimated feedback, (3) daily or weekly feedback, (4) aggregate real-time feedback, and (5) disaggregated real-time feedback.

3.1.1. Enhanced Billing

A review of nine enhanced billing studies published between 1976 and 2009 showed energy savings ranging from 2.5%–8% (Table 2). For example, the Sacramento Municipal Utility District (SMUD) commissioned OPOWER to develop a program using social norms and monthly home energy reports to reshape the behavior of its residential electricity customers. This effort resulted in a savings of 2.5% over an 18-month period (Ehrhardt-Martinez, 2009). Even more dramatic savings (8%) were achieved in a Norwegian study that also provided consumers with historical and social comparisons (Wilhite et al., 1999). Both the SMUD and the Norwegian programs are based on the idea that residential energy consumers will positively respond when provided a point of comparison so they can assess their energy consumption patterns in a meaningful context—for example, relative to their peers or a community average. Comparative information is often provided in the form of historical data as well as social comparisons with other households. Historical data show consumers how their current energy bill compares to past billing periods during the current year as well as prior years. Social comparisons allow consumers to assess their level of energy consumption relative to that of other people in homes and household like theirs. While the studies use a variety of complex data sources to calculate social comparisons, these approaches are relatively low cost, and when designed correctly, they have a proven track record of meaningful energy savings.

Energy savings as high as 10% were achieved in single-family households as a result of a more complex, multicomponent study in Denmark[6] (Nielsen, 1993). The Danish study provided feedback by means of enhanced billing but also offered households the opportunity to receive a consultation with a utility

6. The same study revealed savings of just 1% for apartment dwellers.

TABLE 2 Enhanced Billing Feedback Studies

Study	Energy Savings	Location
1 Ehrhardt-Martinez 2009 and Summit Blue	2.5%	United States (CA)
2 IBM 2007	6% overall savings (conservation by group = 6% for TOU, 4.7% for CPP, 7.4% for critical peak rebate)	Canada
3 Kasulis et al 1981	not specified	United States (Oklahoma)
4 Nielsen 1993	10% for single family HH, 1% for flats (8% and 7% for groups 2 and 3)	Europe (Denmark)
5 Seaver and Patterson 1976	not reported	United States (PA)
6 Staats et al 2004	5% immediately following test period, 8% 2 years later (with no subsequent intervention)	Europe (Netherlands)
7 Wilhite and Ling 1995	7.6% after 1st year, 10% after 2nd year	Europe (Netherlands)
8 Wilhite et al 1999	n.a.	Europe (Netherlands)
9 Wilhite et al 1999	8% after 2 years	Europe (Netherlands)

Source: Ehrhardt-Martinez et al. (2010).

representative to assess potential means of achieving energy savings and provide financing opportunities. Another study found that the use of customized recommendations outlining ways that households could save energy also played an important role in eliciting feedback-related savings (Seaver and Patterson, 1976). Finally, at least one study suggests that the savings achieved by enhanced billing are likely to persist over time, when combined with commitment strategies, group interventions, and social interaction (Staats et al., 2004). The study found energy savings of 5% after seven months of feedback and savings of 8% after two years despite the lack of any subsequent intervention.

3.1.2. Estimated Feedback

Web-based tools have become an increasingly popular means of providing household energy consumers with estimated feedback. Estimated feedback relies on data provided by the individual or household as opposed to the utility

TABLE 3 Estimated Feedback Studies

Study	Energy Savings	Location
1 Abrahamse et al 2007	5.1% (control group used 0.7% more energy)	Europe
2 Benders et al 2006	8.5% in direct energy consumption; change in indirect energy consumption not sig.	Europe (Netherlands)
3 Elliott et al 2006	Not significant	United States (CA)

Source: Ehrhardt-Martinez et al. (2010).

or a third-party provider. Highlighted in Table 3, early research on the effects of this type of feedback suggests moderate levels of associated savings, on the order of 5%–8.5% (Abrahamse et al., 2007; Benders et al., 2006). A third study that explored the effect of this type of feedback on peak energy savings found that while the use of estimated feedback did result in overall energy savings, it did not result in peak period savings above and beyond the preexisting peak rate structure (Elliott et al., 2006). Notably, the savings that were achieved were not limited to peak events, but instead tended to be distributed somewhat evenly across time.

3.1.3. Daily/Weekly Feedback

The last form of indirect feedback (daily/weekly) has historically been relatively low-tech in its implementation (summarized in Table 4). Most studies have relied on the use of feedback cards, door hangers, and other hand written methods to inform participants of their actual energy consumption patterns and savings. As such, this approach has been relatively labor-intensive and difficult to scale up as historically implemented. Nevertheless, energy savings have been notable, ranging from 4%–19%. Given these substantial savings, it is important to consider the ways in which web-based technologies could potentially facilitate implementation of this type of feedback program and allow for larger scale studies.

Savings of 10% or more are commonly reported by programs using this type of feedback. Among those studies with higher levels of energy savings, most combined multiple approaches. For example, two studies that combined pricing rebate schemes with daily/weekly feedback achieved energy savings of 18% and 11%–12%, respectively (Hayes and Cone, 1977; Battalio et al., 1979).

Another study achieved 12% savings by combining the delivery of daily/weekly feedback with the use of comparative and historical norms (Brandon and Lewis, 1999). And a relatively large scale study[7] in California achieved

7. One thousand households participated in the California study.

TABLE 4 Daily/Weekly Feedback Studies

	Study	Energy Savings	Location
1	Battalio et al. 1979	11 and 12% for groups receiving feedback and rebates; no sig. savings for feedback only group	United States (Texas)
2	Bittle et al 1979	4%	United States (IL)
3	Bittle et al 1979-80	9.6% for high electricity consumers	United States (IL)
4	Brandon and Lewis 1999	4.3% reduction compared with pretest; 12% lower compared to control group	Europe (UK)
5	Hayes and Cone 1977	33% from payments, 18% from daily feedback, 19% from information	United States (WV)
6	Kantola et al 1984	3-14% depending on the treatment (11.3-13.8% for dissonance group; 3-3.7% for feedback group; 4-11.6% for tips)	Other (Australia)
7	Nolan et al 2008	10% during the month that door hangers were distributed (significant); 7% in the following month (not significant)	United States (CA)
8	Robinson 2007	0% saving associated with feedback - measured after TOU pricing was already in effect; i.e. no net impact of feedback	Canada
9	Schultz et al 2007	descriptive norm group = decline of 5.7% for HHs consuming above avg, increase of 7.9% for HHs consuming below avg; when injunctive norm was added, low consumers remained low	United States
10	Seligman et al 1978	10.5%	United States (NJ)
11	Seligman et al 1978 and Becker 1978	13.0%	United States (NJ)

(Continued)

TABLE 4 Daily/Weekly Feedback Studies—cont'd

	Study	Energy Savings	Location
12	Staats and van-Leeuwen 2000	6%	Europe (Netherlands)
13	Winett et al. 1982 [Summer]	Summer Savings: 15% for electricity, 34% for electricity for cooling	United States (VA)
14	Winett et al. 1982 [Winter]	Winter Savings: 15% for electricity, 25% for heating	United States (VA)
15	Haakana et al 1997	6% savings for district heating (3%-9% range) while control group increased 1%-2%, 17%-21% decrease in electricity consumption	Europe (Finland)

Source: Ehrhardt-Martinez et al. (2010).

savings of roughly 10% by combining daily/weekly feedback with descriptive norms (Nolan et al., 2008). Importantly, this same study compared the energy savings associated with a variety of potential motivating factors and found that energy saving behaviors were most heavily influenced by social norms, for example, communicated through reports showing how much energy other people were consuming, as opposed to concerns for environmental protection, benefits to society, or saving money. According to Nolan and colleagues, normative social influence produced the greatest change in behavior compared to information highlighting other reasons to conserve, even though respondents rated the normative information as least motivating. These results illustrate the potential power of normative messages for reshaping behavior despite the fact that their influence is often under-detected by individuals themselves.

Goal setting is another means of motivating consumers to save energy that, when combined with daily/weekly feedback, can result in enhanced energy savings. Interestingly, the size of the goal seems to play an important role in determining subsequent energy savings. In one study (Seligman et al., 1978), households were divided into two groups. The first group was given a relatively easy savings goal of 2%, while the second group was given a much more difficult savings goal of 20%. Notably, the group with the difficult savings goal was the only group that achieved significant energy savings, 13% on average. However, the effects of goal setting don't seem to be important in all contexts. For example, a later study of goal setting and its application in conjunction with a variety of other potential motivating factors (Winett et al., 1982) found that goal setting was not successful in generating energy savings. Instead, this study found that the effects of behavioral modeling were much more important. For instance, as part of the study, participants were shown videotapes demonstrating different conservation strategies. When used in conjunction with feedback, behavior modeling was effective in generating energy savings of 15% on average.

3.1.4. Aggregate, Real-Time Feedback

As shown in Table 5, energy savings associated with aggregate (entire household), real-time feedback vary widely, but typically fall somewhere between 0.5% and 18%, depending on the characteristics of the feedback device and its use in combination with innovative program designs. Regardless, household-level savings above 10% are somewhat unusual.[8] Nevertheless, an early study using an outdoor device designed to notify participants when to turn off their air-conditioning achieved energy savings of 15.7% (Seligman et al., 1978) . The device monitored outdoor air temperatures and notified participants when the outdoor air temperature fell below a predetermined temperature threshold, prompting customers to turn off their air-conditioning. Another study was successful in achieving 12% savings through a combined approach that used an in-home device called "The Indicator," which targeted energy savings goals, and general energy use information (van Houwelingen and van Raaij, 1989). Two more recent studies in North America were also successful in achieving energy savings in excess of 10% (Mountain, 2008; Pruitt, 2005). In the first, 58 Canadian households were given a Blue Line Power Cost Monitor. At the conclusion of the 18-month study, households that were given the monitor were consuming 18% less electricity. The study also found that people who began the process with favorable attitudes toward energy conservation and efficiency were likely to conserve more. Another study worked with 2600 households in Arizona to test the effects of the Salt River Project (SRP) M-power Monitor. This study combined in-home feedback with a pay-as-you-go program. Preliminary results suggest average savings of 13%.

Notably, the lowest levels of energy savings associated with aggregate, real time feedback devices seem to be associated with programs focused on reducing peak demand rather than generating overall energy savings. For example, in a study of 480 California households participating in a time-of-use pricing structure, households with continuous display electricity use monitors actually *increased* their overall electricity use by 5.5% during the 12-month study period (Sexton et al., 1987).[9] Another more recent example is a study designed to test the impact of the Ambient Energy Orb on 1500 Maryland households (Case et al., 2008). The study used both critical

---

8. Less than 15% of studies of aggregate, real-time feedback revealed savings of 10% of greater.

9. These same households were successful in shifting electricity use from peak to off-peak periods, which was the goal of the pilot.

TABLE 5 Aggregate Real-Time Feedback Studies

	Study	Energy Savings	Location
1	Seligman et al. 1978	15.7%	United States (NJ)
2	Carroll et al. 2009a	5.5% (ranged from 0 to 48%)	United States (NV)
3	Carroll et al. 2009b and Parker et al. 2008	7.4% (Savings range: +9.5% to −27.9%)	United States (FL)
4	Case et al. 2008 (power point presentation) and Faruqui 2009 (BG&E pilot)	0.50%	United States (Maryland)
5	Hutton et al. 1986	4.1% compared to control group in Quebec, 5% compared to control group in B.C. (for natural gas only), no savings found in CA	Canada (Montreal and Vancouver) and United States (Dallas, TX and Vacaville, CA)
6	Hydro One Networks Inc 2008	3.3% with TOU rates (7.6% with in-home devices and TOU), 6.7% for HHs with display but not TOU	Canada
7	MacLellan 2008 [Nstar]	3%	United States
8	Martinez & Geltz 2005	Not specified	United States (CA)
9	McClelland and Cook 1979 (see EPRI reference)	12%	United States (NC)
10	Mountain 2006	6.5%	Canada
11	Mountain 2008	18.1%	Canada (New Foundland)
12	Mountain 2008	2.7%	Canada (BC)

13	Parker et al 2006	As high as 56% using the feedback device as a diagnostic tool	United States
14	Parker et al 2008	7% (weather adjusted savings ranged from an increase of 9.5% to a savings of 27.9%) 11 homes showed savings; 6 showed increases	United States (FL)
15	Peterson et al 2007	32%	United States (OH)
16	Pruitt 2005	12.8% (13.8% in summer, 11.1% in winter)	United States (AZ)
17	Scott 2008 and Sipe and Castor 2009	Savings not significant	United States
18	Sexton et al 1987	Electricity demand increased 5.5% overall and 12% in off-peak periods; demand declined 1.2% during peak periods	United States (CA)
19	Sulyma et al 2008	8.6%	Canada
20	van Houwellingen and van Raaij 1989	HH w/display:12.3%, HH w/monthly feedback: 7.7%, HH that self monitored: 5.1%, HH w/info only: 4.3%	Europe (Netherlands)

Source: Ehrhardt-Martinez et al. (2010).

TABLE 6 Disaggregated Real Time Plus Feedback Studies

Study	Energy Savings	Location
1 Dobson and Griffin 1992	12.9%	Canada
2 Horst 2006	not measured	United States (MI)
3 Karbo and Larson 2005		Europe (Denmark)
4 Ueno et al 2006a	9%	Other (Japan)
5 Ueno et al 2006b	12% (Total electric consumption decreased by 18%, total gas consumption decreased by 9%)	Other (Japan)
6 Wood and Newborough 2003	15% (31 HHs saved more than 10%, 6 HHs saved more than 20%)	Europe (UK)

Source: Ehrhardt-Martinez et al. (2010).

peak pricing and peak time rebates. In addition, one of the study groups was also using an air-conditioner (AC) switch where the utility was able to turn off a participant's AC unit during peak times. Peak savings ranged from 17%–33% across the study groups. Total energy savings, however, were only 0.5%. These study results were fully consistent with the more narrow focus of peak time reductions.

3.1.5. Disaggregated, Real-Time Plus Feedback

Perhaps the most innovative and exciting of the various feedback devices are those that provide disaggregated, or end-use specific, real-time feedback. These high-tech gadgets offer the promise of providing households with the most detailed and timely energy consumption data, however they also tend to be among the most costly approaches to feedback and only a handful of studies have been done to assess their effectiveness thus far.[10] Nevertheless, preliminary research suggests a range of potential savings of 9%–18% (Table 6). In Canada, a computer-based feedback device called the Residential Electricity Cost Speedometer was found to generate savings of nearly 13% (Dobson and Griffin, 1992) while two Japanese studies of the Online Energy Consumption Information System (Ueno et al., 2006a and 2006b) found savings of 9% and 18%, respectively.[11] A similar study evaluated the effect of the Energy Consumption Indicator in 20 U.K. households and found that it was successful in reducing energy consumption by 15% on average.

10. All but one of the studies on real-time disaggregated feedback have been performed outside of the United States (two in Japan, two in Europe [U.K. and Denmark], and one in Canada).

11. The first study involved only nine households.

The overall findings of Ehrhardt-Martinez et al. (2010) suggest that feedback programs performed between 1995 and 2010 have generated average household energy savings of 4%–12%, depending on the type of feedback and related program characteristics. Among the various types of indirect feedback, enhanced billing strategies resulted in average savings of 2.5%–8%, while estimated feedback and daily/weekly feedback strategies resulted in average savings of 5%–8.5% and 4%–19%, respectively. Among the direct forms of feedback, real-time aggregate feedback resulted in energy savings of 0.5% to as much as 18%, while real-time plus strategies resulted in savings of 9%–18%. Despite the information provided by past studies, the future savings potential from feedback remains somewhat unclear. Longer-term studies with larger sample sizes are needed to adequately evaluate the long-term savings potential from feedback. In addition, energy savings are likely to be enhanced by better program designs that build on the knowledge gained and lesson learned from past initiatives.

4. THE IMPACT ON HABITS, LIFESTYLES, AND CHOICES

While it is clear that utility-driven advanced metering initiatives and other programs that provide residential consumers with feedback regarding their energy consumption can result in significant reductions in energy use, few studies have studied the actions that residential consumers are engaging in so as to bring about the energy savings associated with feedback initiatives. In a 2004 study of the impact of a pilot residential time-of-use pricing program in Sacramento, California, researchers explored this question in some detail (Wood et al., 2004). Although the survey results are not based on a representative sample, the study's findings provide some preliminary insights as to the ways in which people choose to change their habits, lifestyles, and choices in ways that result in energy savings.

Participation in the program was voluntary and most participants chose to participate either because they wanted to save money (88%) or because they wanted the ability to control their energy usage (54%). In addition, roughly one-third indicated that their participation was motivated by a concern for the environment. In terms of actual energy savings, the study's findings showed that 86% of participants used less energy during high or critical periods and that 67% of participants used less energy overall. Energy use during critical price periods declined by 16%, while overall energy use declined by 4%. But how did people achieve these savings?

As shown in Table 7, households engaged in a variety of activities to save energy. Nearly all participants (95%) reported engaging in new habits to minimize energy use during critical price periods. The principal strategy involved shifting usage to nonpeak periods. In particular participants were less likely to use air-conditioners, dishwashers, and clothes washers during peak periods. They also reported taking fewer showers or baths during these periods and cooking indoors less often.

TABLE 7 Categories of Change and Behaviors

Type of Change	Behavior	Percent
New Habits	Shifted usage	95%
New Habits	Checked thermostat display for critical periods	83%
Energy Stocktaking	Repaired air ducts	8%
Energy Stocktaking	Changed default temperatures on thermostat	42%
Low-cost Investments	Installed CFLs	59%
Higher-cost Investments	Replaced single with dual-pane windows	11%
Higher-cost Investments	Replaced inefficient refrigerator	9%
Higher-cost Investments	Replaced inefficient air conditioning	5%
Higher-cost Investments	Installed ceiling or wall insulation	5%

Source: Wood et al. (2004).

Respondents also reported engaging in energy stocktaking behaviors including repairing air ducts (8%) and changing the default temperatures on their thermostats (42%). Among the respondents who saved the most energy overall were those that invested in energy efficient products. More than half of all participants (59%) invested in compact fluorescent light bulbs. A smaller proportion of households invested in more costly energy efficiency upgrades including new windows (11%), a new refrigerator (9%), a new air-conditioner (5%), or added insulation (5%).[12]

These findings contrast with an earlier and larger study of conservation behavior by residential consumers during and after the 2000–2001 California energy crisis (Lutzenhiser et al., 2003). The 2003 study used data obtained from 1666 in-depth telephone interviews with randomly selected residential households in five major California utility service territories. Some interesting findings from the 2003 study indicate that "more than 75% of households participating in the survey reported taking one or more conservation actions," and that reductions in energy demand were largely due to changes in behavior (65%–70%) as opposed to investments in hardware solutions or on-site generation projects (25%–30%). Table 8 shows reported conservation behaviors. Note that the top three behaviors involved changes in habits and routines as opposed to consumer investments or energy stocktaking behaviors.

12. Higher-cost investments were relatively rare despite the fact that the sample population was found to have higher incomes compared to the general population in the same geographic area. More specifically, 50% of pilot participants had annual incomes over $100,000 per year compared to 12% of people in the general population.

TABLE 8 Behaviors as a Function of Technology Categories

Type of Behavior	Description	Percent
Lights Behaviors	Behaviors related to turning off lights or using fewer lights	65.5%
Other Heat/Cool Behaviors	Behaviors related to heating and cooling other than not using the AC at all (e.g. using AC less, using ceiling fans, changing thermostat, etc.)	48.5%
Small Equipment Behaviors	Behaviors related to household appliances (using them less, turning them off and unplugging them)	32.2%
Light Bulbs	Hardware related purchase/use of CFLs or other energy saving bulbs	22.2%
Peak Behaviors	Behaviors related to using energy during off-peak hours	20.0%
H_20 Behaviors	Behaviors related to using less water or using less hot water (e.g. shorter showers, wash in cold/warm water, turn water heater down, etc.)	12.2%
Appliances	Hardware-related purchased/use of new non-fixed appliances (e.g. refrigerator, washer/dryer, window AC, fans, etc.)	10.4%
Turning off AC	Behavior related to not using the AC at all	9.6%
Shell Improvement	Hardware related to one-time improvements to the house (e.g. windows, insulation, a new piece of fixed equipment such as water heater, AC, furnace, etc.)	7.9%
Large Equipment Behaviors	Behaviors related to pools, spas, irrigation motors (e.g., turn off, use less often)	6.0%

Source: Lutzenhiser et al. (2003).

Another important difference between the two studies involved the question of motivation. In the 2003 study (summarized in Figure 3), survey respondents reported that their conservation efforts were motivated by a wide variety of factors. While minimizing energy costs was among the principal motivators, respondents also reported being motivated by their desire to avoid blackouts (82%), use energy resources as wisely as possible (77%), do their part to help Californians (73%), and protect the environment (69%). According to the report, "qualifying for a utility rebate was the least common

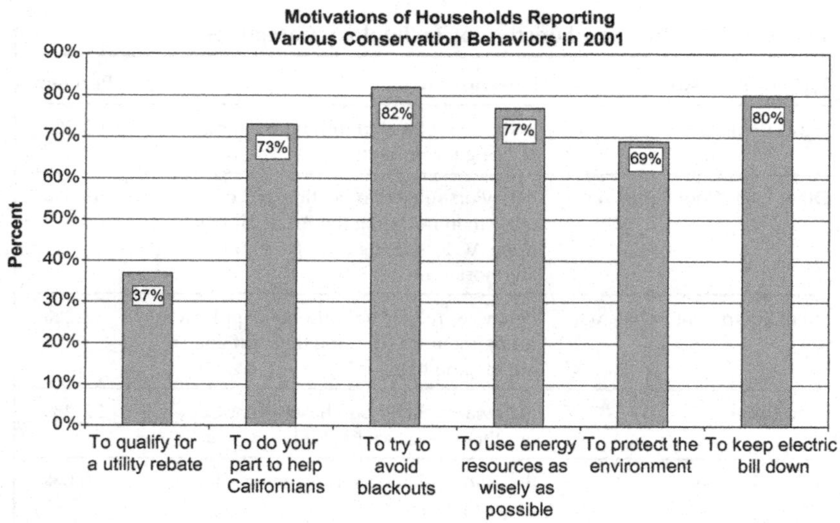

FIGURE 3 Motivations and Household Behaviors. *Source: Lutzenhiser et al. (2003).*

motivation, and available utility rebates were not relevant to most of the actions consumers took."

The findings from these two studies suggest that large, behavior-related energy savings opportunities are available in the residential sector and that feedback is likely to be an effective mechanism for unlocking potential energy savings. Among the types of energy efficiency and conservation behaviors, investments in new equipment and appliances appeared more likely within more affluent populations and are generally undertaken in conjunction with a change of residence or a remodel or part of a stylistic, as opposed to functional, upgrade (Lutzenhiser et al., 2003). For the larger population, households appear to be more likely to reduce energy consumption through changes in habits and routines or through energy stock-taking behaviors. Importantly, these energy-conservation behaviors are likely to be motivated by a variety of factors including self-interest (energy bill savings) as well as civic concerns and altruistic motives. These findings suggest that narrowly defined energy-efficiency programs aimed at the installation of new, more energy-efficient technologies alone (the practice of traditional utility programs) are likely to realize only a small fraction of potential behavior-related residential energy savings. Similarly, programs that limit their appeal to self-interest alone are unlikely to leverage the broad range of factors that motivate people to action.

5. THE MAGNITUDE OF POTENTIAL SAVINGS

How large are potential feedback-related energy savings? Currently the prevailing assumption is that advanced meters are likely to be installed in nearly all customer premises by 2030, and much sooner in some states or for some utilities. However, barring the implementation of specific policies or standards, their near-term use is likely to be limited to achieving utility interests in managing the grid, reducing peak demand loads more efficiently, and for reducing utility costs associated with monitoring, tracking, and billing customers. In other words, despite the potential energy savings made possible by the installation of advanced meters, it is entirely possible that these devices won't be used to encourage the more efficient use of electricity throughout the entire year and across all customer end uses.

Despite these concerns, several researchers recently worked together to develop several estimates of potential feedback-induced energy savings at the national level (Ehrhardt-Martinez et al., 2010). Their estimates are based on a Monte Carlo exercise that explored three alternative electricity consumption scenarios and allowed for a random increase in annual customer participation in year-round efficiency feedback programs. Notably the study was based on the presumption that by 2030, somewhere between 60% and 95% of all customers—whether residential, commercial, or industrial users of electricity—might be participating in such programs. At the same time each scenario assumes a different range of conservation and/or efficiency savings: first, a low-range savings of 2.5%–8% of annual consumption per customer; next, a medium-range savings of 5%–12%; and finally, an aggressive or high-range savings of 9%–18% of annual consumption. Given these assumptions, the Monte Carlo exercise ran a total of 10,000 simulations for each of the three scenarios to explore the potential impact over 2010–2030 time horizon.

The first critical insight from the Monte Carlo simulations is that advanced metering together with active customer participation in well-designed utility feedback programs could provide consumers and businesses with a significant net savings on their electricity bills. Depending on the breadth and effectiveness of program design, individual consumer savings might range from roughly 4%–12% electricity savings annually, while the economy-wide savings might range from 4%–10% annually. Over the 20-year time horizon, the net present value of technology and program costs might range from roughly $50–$60 billion dollars while saving the economy a total of $70–$200 billion.

Using a total resource cost test—in effect, examining total economy-wide costs and total economy-wide energy bill savings—the benefit cost ratio appears to range from about 1.2 to as much as 3.7. The lower ratio reflects a less-effective program implementation and negligible economies of scale that would otherwise lower technology costs. The higher ratio, on the other hand, captures a much higher level of market penetration, a reduction in both technology and program costs over time, and a greater savings per unit of technology and program cost. Although not explored in their analysis, the emerging results from this innovative study suggest that well-designed feedback programs are likely to reduce greenhouse gas emissions from the production and use of electricity by as much as 10% at a substantial net benefit to the U.S. economy. As explained below, well-designed feedback programs, enabled by smart grid and smart technologies throughout the economy, could form the backbone of a much more energy-efficient economy that might cause energy consumption to drop by 50% or better by 2050 (Laitner, 2009)—but to achieve these savings we need to increase the rate of investment in such programs and technologies. The good news is that productive investments in these programs and technologies might also strengthen overall economic activity (Laitner, 2009).

CONCLUSIONS

As the smart grid connects more and more energy users into a true energy network, the potential benefits for households, utilities, and society are also likely to expand in terms of increased energy efficiency, expanded use of renewable energy, better energy management practices, and cleaner transportation options. The installation of new utility advanced meters and the application of well-designed feedback initiatives can potentially catalyze these benefits. In short, the smart grid offers the opportunity to transform traditional energy structures in ways that benefit households, utilities, society, the economy, and the environment—if it's done well. We conclude this chapter with a brief discussion of three specific ways in which the smart grid might transform our traditional energy system into a more robust, energy-efficient economy.

Energy Efficiency, Renewable Energy, and Distributed Generation

Compared to the current electricity system in which energy flows in one direction (from the utility to the consumer) and an information exchange is limited to monthly consumption and billing data, the Smart Grid provides the means for multidirectional energy flows as well as the development of more complex and more energy-efficient information systems. As a greater number of homes, offices, and other buildings become connected to the Smart Grid, everyday buildings that become more energy-efficient could also be transformed into energy generators. Whether through the use of photovoltaic systems, mini wind turbines, or other mechanisms, houses and commercial buildings will have the opportunity to produce as well as consume electricity (see, for example, Griffith et al., 2007, which explored such possibilities in our nation's commercial buildings).

Such a transformation would have enormous implications for energy efficiency, carbon emissions, and the role of households and utilities in the energy system. For instance, large efficiency gains would result from a microgrid system in which small scale electricity production and consumption occur in close proximity, significantly reducing transmission-related energy losses, and giving households ownership over their electricity generation and use. Of course this type of system would also greatly expand renewable sources of electricity production and reduce the level of carbon emissions associated with electricity generation. Moreover the implementation of microgrid generation systems offers households the possibility of becoming energy suppliers by selling surplus energy back to the grid. According to a recent article in *Fast Company* (Kamenetz, 2009), "The microgrid is all about consumer control—aligning monetary incentives, with the help of information technology, to make renewables and efficiency pay off for the average homeowner, commercial developer, or even a town." Ultimately, microgrid systems provide the opportunity for an electricity system that is redundant, resilient, and secure.

Plug-in Vehicles and Energy Security

The redundancy, resilience, and flexibility of the Smart Grid may also provide the means of enabling the development and use of Vehicle to Grid (V2G) technologies in which plug-in electric vehicles will both draw energy from, and contribute energy to, the electricity grid. Forthcoming electric and hybrid electric vehicles can be plugged in at night (when electricity is abundant and cheap) to draw electricity from the grid and recharge their batteries. During the day (when electricity is scarce and expensive), the same vehicles can be plugged in to provide extra power to the grid during times of high demand. As such the V2G technologies help benefit utilities by helping to overcome electricity supply constraints associated with times of peak demand that occur during the day when industrial plants, commercial enterprises, and air-conditioning result in maximum demand for electricity (Kempton and Tomic, 2005). In fact, new electric passenger vehicles may store enough energy to power several homes for hours, and utilities may be interested in paying vehicle owners to plug-in their vehicles to supply needed energy. When combined with microgrid technologies, the potential benefits are likely to be enormous, including the possibility of transportation options that can greatly reduce emissions.

Personal Carbon Budgets and Climate Change

In a system of Smart Grids, innovative feedback interventions, microgrid generation, and V2G technologies, today's energy consumers will become tomorrow's energy managers and the only constraints on personal energy consumption will be the costs of energy and the limits to carbon emissions. In order to limit overall carbon emissions and encourage households and others to constrain the consumption of more carbon-intensive energy sources, one innovative policy proposes the implementation of personal carbon budgets and trading schemes (Parag and Strickland, 2009). Such a policy would provide individuals with "an annual carbon emissions 'budget' for their personal use" and would implement a "downstream carbon cap and trade policy instrument." People would be responsible for the carbon emissions associated with household energy use, private transport, and aviation and would be able to buy additional emissions or sell their surplus credits in the personal carbon market.

This type of reconfiguration of our energy system could provide an energy revolution in all end-use sectors and for all fuels that so many of today's leaders are calling for. Such a reconfiguration could also provide for a much more decentralized and democratic energy system in which the line between producers and consumers is increasingly blurred, in which households become capable energy managers, and in which the environmental impacts of energy production and consumption are minimized. How big is the potential? With smart infrastructure as a foundation, coupled with well-designed, behavior-savvy feedback programs that inform and integrate consumer involvement into the energy market, new ICT-based systems might open up much greater levels of energy efficiency improvements than are typically reflected in policy assessments. The evidence suggests that such people-centered policies and programs could open the way toward a 50%–60% energy savings by 2050 (Laitner, 2009) with renewable energy technologies powering the balance of our energy needs (Jacobson and Delucchi, 2009). Hence, the opportunities for more positive social, economic, and environmental outcomes exist in the short term, but the critical choices that will drive this set of beneficial outcomes have yet to be made. And such choice could begin by enabling better home energy management using new Smart Grid and feedback technologies.

BIBLIOGRAPHY

Abrahamse, W., Steg, L., Vlek, C., & Rothengatter, T. (2005). A review of intervention studies aimed at household energy conservation. *Journal of Environmental Psychology, 25*, 273–291.

Abrahamse, W., Steg, L., Vlek, C., & Rothengatter, T. (2007). The effect of tailored information, goal setting, and tailored feedback on household energy use, energy-related behaviors, and behavioral antecedents. *Journal of Environmental Psychology, 27*, 265–276.

Battalio, R. C., Kagel, J. H., Winkler, R. C., & Winett, R. A. (1979). Residential electricity demand: An experimental study. *Review of Economics and Statistics, 61*(2), 180–189.

Becker, L. J. (1978). Joint effect of feedback and goal setting on performance: A field study of residential energy conservation. *Journal of Applied Psychology, 63*(4), 428–433.

Benders, R. M. J., Kok, R., Moll, H. C., Wiersma, G., & Noorman, K. J. (2006). New approaches for household energy conservation—in search of personal household energy budgets and energy reduction options. *Energy Policy, 34*(18), 3612–3622.

Bittle, R. G., Valesano, R., & Thaler, G. (1979). The effects of daily cost feedback on residential electricity consumption. *Behavior Modification, 3*(2), 187–202.

Bittle, R. G., Valesano, R., & Thaler, G. (1979-80). The effects of daily feedback on residential electricity usage as a function of usage level and type of feedback information. *Journal of Environmental Systems, 9*(3), 275–287.

Brandon, G., & Lewis, A. (1999). Reducing household energy consumption: A qualitative and quantitative field study. *Journal of Environmental Psychology, 19*(1), 75–85.

Case, M., Butts, M., & Harbaugh, W. (2008). *AMI/Smart Energy Pricing (SEP) update presentation* (October 15). Baltimore Gas and Electric.

Darby, S. (2006). *The effectiveness of feedback on energy consumption: A review for DEFRA of the literature on metering, billing, and direct displays*. Environmental Change Institute. University of Oxford. *www.eci.ox.ac.uk/research/energy/downloads/smart-metering-report.pdf.*

Dobson, J.K., & Griffin, J.D.A. (1992). Conservation effect of immediate electricity cost feedback on residential consumption behavior. In *Proceedings of the ACEEE 1992 Summer Study on Energy Efficiency in Buildings, 10*, 33–35. American Council for an Energy-Efficient Economy.

Ehrhardt-Martinez, K. (2009). *Evaluation of positive energy's home energy report program implemented in conjunction with the Sacramento Municipal Utility District*. American Council for an Energy-Efficient Economy.

Ehrhardt-Martinez, K., Donnelly, K. A., & Laitner, J. A. (2010). *Advanced metering initiatives and residential feedback programs: A meta-review for economy-wide electricity-saving opportunities*. American Council for an Energy-Efficient Economy.

Electric Power Research Institute (EPRI). (2009). *Residential electricity use feedback: A research synthesis and economic framework*. Report #1016844. B. Neenan and J. Robinson, principal investigators.

Elliott, J., Martinez, M., Mitchell-Jackson, J., & Williamson, C. (2006). The California bill analysis pilot: Using web-based bill analysis as a tool to reduce on-peak demand. In *Proceedings of the ACEEE 2006 Summer Study on Energy Efficiency in Buildings, 2*: 94–105. American Council for an Energy-Efficient Economy.

Energy Information Administration. (2010). "Annual Energy Outlook 2010." *http://www.eia.doe.gov/oiaf/aeo/index.html*. Washington, D.C.: U.S Department of Energy, Energy Information Administration.

Federal Energy Regulatory Commission (FERC). (2008). *2008 assessment of demand response and advanced metering*. FERC.

Fehrenbacher, K. (2009). Energy-use software to rival Google's Efficiency 2.0 keeps a low profile as it develops elaborate algorithms-its "secret sauce"-to curb energy use and rival the big guys. *Business Week. www.businessweek.com/technology/content/jul2009/tc20090728_008663.htm.*

Fischer, C. (2007). Influencing electricity consumption via consumer feedback: A review of experience. In *Proceedings of the ECEEE 2007 Summer Study on Energy Efficiency, 9*, 1873–1884. European Council for an Energy-Efficient Economy.

Frank, Andy. 2009. June 3, 2009 In-depth Interview with Andy Frank, Executive Vice President, Business Development, Efficiency 2.0.

Galvin Electricity Initiative. (2007). *The path to perfect power: New technologies advance consumer control.*

Griffith, B., Long, N., Torcellini, P., Judkoff, R., Crawley, D., & Ryan, J. (2007). *Assessment of the technical potential for achieving net zero-energy buildings in the commercial sector.* National Renewable Energy Laboratory.

Haakana, M., Sillanpää, L., & Talsi, M. (1997). The effect of feedback and focused advice on household energy consumption. In *Proceedings of the ECEEE 1997 Summer Study on Energy Efficiency, 4*, 1–11. European Council for an Energy-Efficient Economy.

Hayes, S. C., & Cone, J. D. (1977). Reducing residential electrical energy use: Payments, information, and feedback. *Journal of Applied Behavior Analysis, 10*(3), 425–435.

Horst, G.R. (2006). *Woodridge energy study & monitoring pilot*. Whirlpool Corporation. *http://uc-ciee.org/dretd/documents/Woodridge%20Final%20Report.pdf.*

Jacobson, M. Z., & Delucchi, M. (2009). A path to sustainable energy by 2030. *Scientific American, (November)*, 29–65.

Kamenetz, A. (2009). Why the microgrid could be the answer to our energy crisis. *Fast Company.* (July).

Kavazovic, O. (2009). In-depth interview with Ogi Kavazovic, Positive Energy, director of marketing, (28 April).

Kempton, W., & Montgomery, L. (1982). Folk quantification of energy. *Energy, 7*(10), 817–827.

Kempton, W., & Tomic, J. (2005). Vehicle-to-grid power implementation: From stabilizing the grid to supporting large-scale renewable energy. *Journal of Power Sources, 144*(1), 280–294.

Laitner, J. A. (2009). *Climate change policy as an economic redevelopment opportunity: The role of productive investments in mitigating greenhouse gas emissions.* ACEEE Report E098. American Council for an Energy-Efficient Economy.

Laitner, J. A., & Ehrhardt-Martinez, K. (2008). *Information and communication technologies: The power of productivity; how ICT sectors are transforming the economy while driving gains in energy productivity.* American Council for an Energy-Efficient Economy.

Lutzenhiser, L., Kunkle, R., Woods, J., & Lutzenhiser, S. (2003). *Conservation behavior by residential consumers during and after the 2000–2001 California energy crisis.* California Energy Commission.

McClelland, S. W., & Cook, J. (1979). Energy conservation effects of continuous in home feedback in all-electric homes. *Environmental Systems, 9*(2), 69–173.

McClelland, L., & Cook, S. W. (1979–80). Energy conservation effects of continuous in-home feedback in all-electric homes. *Energy Systems, 9*(2), 169–173.

Mountain, D. (2008). *Real-time feedback and residential electricity consumption: British Columbia and Newfoundland and Labrador pilots.* Mountain Economic Consulting and Associates Inc.

Nagel, P. (2009). In-depth interview with Paul Nagel, vice president of strategic development, and DeAnn Zebelean, public relations director, Control4. (23 April).

Nielsen, L. (1993). How to get the birds in the bush into your hand: Results from a Danish research project on electricity savings. *Energy Policy, 21*(11), 1133–1144.

Nolan, J. M., Schultz, P. W., Cialdini, R. B., Goldstein, N. J., & Griskevicius, V. (2008). Normative Social Influence is Underdetected. *Personality and Social Psychology Bulletin, 34*(7), 913–923.

Olsen, Kirsten. 2009. April 17, 2009 In-depth Interview with Kirsten Olsen, Google.org, Product Manager.

Parag, Y., & Strickland, D. (2009). *Personal carbon budgeting: What people need to know, learn, and have in order to manage and live within a carbon budget, and the policies that could support them. Working paper.* U.K: Energy Research Centre.

Parker, D.S., Hoak, D., Meier, A., & Brown, R. (2006). How much energy are we using? Potential of residential energy demand feedback devices. In *Proceedings of the ACEEE 2006 Summer Study on Energy Efficiency in Buildings, 1*, 211–222. American Council for an Energy Efficient Economy.

Petersen, J., Shunturov, V., Janda, K., Platt, G., & Weinberger, K. (2007). Dormitory residents reduce electricity consumption when exposed to real-time visual feedback and incentives. *International Journal of Sustainability in Higher Education, 8*(1), 16.

Pruitt, B. (2005). *Salt River project m-power.* Presentation at the Second Annual Workshop for Energy Efficiency Program Design and Implementation in the Southwest. Southwest Energy Efficiency Project.

Robinson, J. (2007). The effect of electricity-use feedback on residential consumption: A case study of customers with smart meters in Milton, Ontario. *Unpublished master's thesis.* ON, Canada: University of Waterloo.

Schultz, P. W., Nolan, J. M., Cialdini, R. B., Goldstein, N. J., & Griskevicius, V. (2007). The constructive, destructive, and reconstructive power of social norms. *Psychological Science, 18* (5), 429–434.

Seaver, W. B., & Patterson, A. H. (1976). Decreasing fuel oil consumption through feedback and social commendation. *Journal of Applied Behavior Analysis, 9*(2), 147–152.

Seligman, C., Darley, J. M., & Becker, L. J. (1978). Behavioral approaches to residential energy conservation. *Energy and Buildings, 1*(3), 325–337.

Sexton, R. J., Brown-Johnson, N., & Konakayama, A. (1987). Consumer response to continuous-display electricity-use monitors in a time-of-use pricing experiment. *Journal of Consumer Research, 14*(1), 55–62.

Staats, H., Harland, P., & Wilke, H. A. M. (2004). Effecting durable change: A team approach to improve environmental behavior in the household. *Environment and Behavior, 36*(3), 341–367.

Staats, H., van-Leeuwen, E., & Wit, A. (2000). A longitudinal study of informational interventions to save energy in an office building. *Journal of Applied Behavior Analysis, 33*(1), 101–104.

Ueno, T., Sano, F., Saeki, O., & Tsuji, K. (2006). Effectiveness of an energy consumption information system on energy savings in residential houses based on monitored data. *Applied Energy, 83*(8), 166–183.

Ueno, T., Tsuji, K., & Nakano, Y. (2006a). Effectiveness of displaying energy consumption data in residential buildings: To know is to change. In *Proceedings of the ACEEE 2006 Summer Study on Energy Efficiency in Buildings, 7*, 264–277. American Council for an Energy Efficient Economy.

van Houwelingen, J. T., & van Raaij, W. F. (1989). The effect of goal setting and daily electronic feedback on in-home energy use. *Journal of Consumer Research, 16*, 98–105.

Wilhite, H., Hoivik, A., & Olsen, J. (1999). Advances in the use of consumption feedback information in energy billing: The experiences of a Norwegian energy utility. In *Proceedings of the ECEEE 1999* Summer Study, 2, 02. European Council for Energy Efficient Economy.

Winett, R. A., Hatcher, J. W., Fort, T. R., Leckliter, I. N., Love, S. Q., Riley, A. W., et al. (1982). The effects of videotape modeling and daily feedback on residential electricity conservation, home temperature and humidity, perceived comfort, and clothing worn: winter and summer. *Journal of Applied Behavior Analysis, 15*(3), 381–402.

Wood, V., Erickson, J., Lutzenhiser, L., Lutzenhiser, S., Ozog, M., & Sylvia, B. (2004). What goes on behind the meter: real customer response to residential time-of-use pricing. In *the proceedings of the 2004 ACEEE Summer Study on Energy Efficiency in Buildings.* Washington, DC: American Council for an Energy-Efficient Economy.

Chapter 11

How Organizations Can Drive Behavior-Based Energy Efficiency

William Prindle* and Scott Finlinson†
**ICF International, †NORESCO*

Chapter Outline

1. INTRODUCTION

The debate on behavior and culture change is heavily tilted toward individuals and households as the primary focal points for study and action. Yet organizations use more than twice as much energy than households: In the U.S., commercial and industrial energy users accounted for about 51 quads of primary energy while households accounted for about 21 quads in 2007 (EIA AEO, 2010). Moreover, leading organizations are demonstrating significant, measured changes in energy use, much of it through behavior and culture change; the record is less clear in the general consumer or household arena. Organizations do this by developing strategies that overcome the barriers to efficiency investments and behaviors that many organizations contain.

Energy, Sustainability and the Environment.

A traditional economics view might suggest that employees are less motivated to save energy at work than at home, because the monetary rewards are less direct at work. However, leading organizations have made energy performance a priority for everyone, connecting individuals' success in saving energy with the organization's success. They are also moving beyond extrinsic monetary rewards, and are finding ways to tap their employees' intrinsic motivations to reduce pollution, improve productivity, and innovate. They are finding that many employees, when invited to make a difference in this way, respond with creative and energetic solutions that deliver substantial savings in energy and other resources.

This chapter focuses on organizational behavior, distinct from household or individual behavior, as the most promising means for realizing significant near term reductions in energy use and greenhouse gas (GHG) emissions. It surveys field experience from behavior-change driven efficiency programs, reviews the limited research literature in the field, and describes in more depth the experience of specific organizations and program models.

The chapter consists of the following sections:

- An overview of U.S. nonresidential energy usage, with the primary focus on variables that affect measured energy performance. It shows in various ways that traditional technical factors, such as age or efficiency technologies, are not correlated with measured energy performance.
- The record on behavior-based energy savings programs, with the primary focus on building retro-commissioning (RCx) and related studies, which tend to show that substantial energy savings are available from building operations, and that the quality of building management and operations is more strongly correlated with energy performance than is the presence of energy efficiency technologies—a surprising finding with important implications.
- Research and analytical underpinnings of energy use and energy performance, with the main focus on studies and experience that attempts to explain how occupant and organizational behavior can best affect energy use and support improved energy performance.
- Examples of leading corporate energy efficiency strategies that make extensive use of behavior-based efficiency measures, including selected case studies from IBM, Toyota, and PepsiCo.
- An approach to behavior-based energy savings in university settings using a performance contracting business model, with measured results.
- A look forward at how behavior-based energy efficiency strategies in organizations can lead to dramatically-lower energy use *and* carbon emissions over the longer term.
- A conclusions section.

2. OVERVIEW OF U.S. NONRESIDENTIAL ENERGY USAGE

We use the term "nonresidential" in this chapter because we are shifting the focus from the building level or the end-use level, which have traditionally driven U.S. data collection and analysis efforts, to the organizational level. In the U.S. today, organizations are becoming larger and more diverse, so attempting to understand and control energy use based on building type or end-use is less important than understanding how organizations can measure and manage performance across a wide range of building types and end uses. A company like IBM, for example, developed in the twentieth century as a manufacturer of adding machines, then mainframe computers. In the twenty-first century, its manufacturing operations account for only 20% of its energy use; 47% is consumed in buildings, and 33% in data centers.

Shifting the focus to the organization from the building type or end-use technology has the effect of shifting the analysis and the action emphasis from specific technology solutions to broader management methods and operational solutions. This is appropriate because U.S. national survey data show that the ways buildings are *operated* and *managed* has a stronger effect on energy performance than age, technology, or other factors that engineering-based analyses typically consider—a surprising and highly significant finding.

A review of the Commercial Buildings Energy Consumption Survey (CBECS) data shows a number of counter-intuitive findings. While building type accounts for the greatest variability in Energy Use Index (EUI) values[1] (Table 1), other factors one might expect to explain differences in energy consumption do not bear out such expectations.

Table 1 shows not only that energy use per square foot varies by as much as sixfold between building types, but also that what CBECS calls "commercial" space is actually more than one-third public or nonprofit-occupied space. Education, health care, houses of worship, public assembly, and government operations account for at least one-third of buildings that are generically called commercial.

CBECS data refute the notion that buildings are getting better from an energy efficiency perspective as we construct newer buildings incorporating more efficient designs. Figure 1 shows that buildings built since 1990 use *more* energy per square foot than those built before 1920. One might counter with the argument that pre-1920 buildings may lack modern energy systems such as central air-conditioning—yet buildings built in the 1990s use as much energy as those built in the 1960s. The year 1970 is considered a tipping point in environmental awareness and energy efficiency design: Yet, if we compare the average EUI for all buildings built before 1970 with those built since, the newer buildings use 7% *more* energy than the older ones.

The argument could be made that older buildings can be made more efficient through renovation—indeed, one would expect that lighting, mechanical equipment, and envelope components would become more efficient as older

1. The Energy Use Index (EIU) is measured at total BTU of energy consumption at the building site, per square foot of conditioned floor space. EUI is a widely used metric for building energy performance.

TABLE 1 U.S. Commercial Building Stock by Building Type

Principal Building Activity	Number of Buildings (thousand)	Floorspace (million square feet)	Sum of Major Fuels (Trillion BTU)	Energy Use Index (BTU/SF)
Education	386	9,874	820	83,046
Food Sales	226	1,255	251	200,000
Food Service	297	1,654	427	258,162
Health Care	129	3,163	594	187,796
Inpatient	8	1,905	475	249,344
Outpatient	121	1,258	119	94,595
Lodging	142	5,096	510	100,078
Retail (Other Than Mall)	443	4,317	319	73,894
Office	824	12,208	1,134	92,890
Public Assembly	277	3,939	370	93,932
Public Order and Safety	71	1,090	126	115,596
Religious Worship	370	3,754	163	43,420
Service	622	4,050	312	77,037
Warehouse and Storage	597	10,078	456	45,247
Other	79	1,738	286	164,557
Vacant	182	2,567	54	21,036
Percent Floor Area in Non-Commercial Uses		33.7%		
Percent Energy Use in Non-Commercial Uses			35.6%	
Ratio of Highest to Lowest EUI				6

Source: EIA (2003).

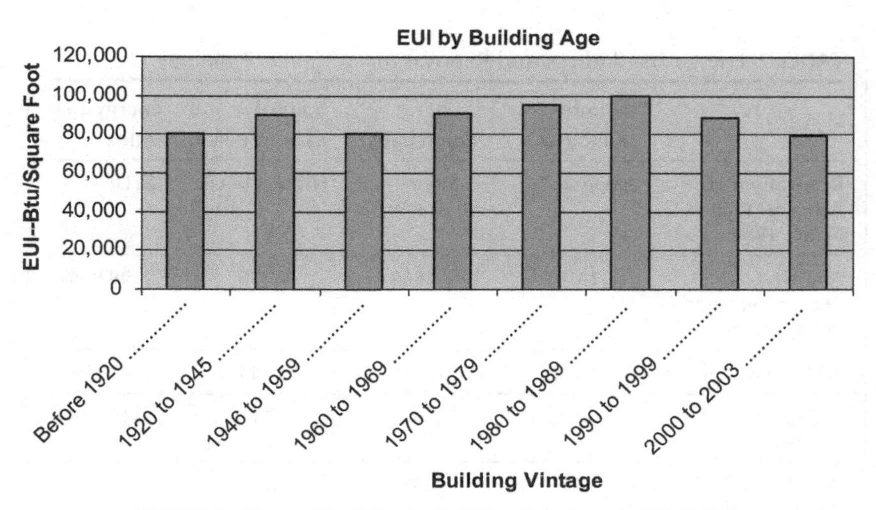

FIGURE 1 Energy Use Index by Building Age. *Source: EIA (2003).*

components and systems are replaced. Yet CBECS data again show an opposite pattern. As shown in Table 2, buildings built before 1980 that have since been renovated use 25% *more* energy than similar-vintage buildings with no renovations. Buildings with specific energy system renovations, such as HVAC or lighting, use 35% and 33% *more* energy, respectively, than similar buildings with no renovations. Moreover, the average EUI for all buildings built since 1980 is 15% *higher* than that for older buildings. These trends are confounding—one would expect older buildings with upgraded lighting or HVAC to be more efficient than those without upgrades. Yet the data show a dramatically opposite picture. The clear inference is that energy technology upgrades overall do *not* produce improvement in energy performance, and in fact appear to *increase* energy use indices substantially.

CBECS' rich data set allows a still deeper examination of the effects of efficiency technologies in building energy use. Table 3 compares the EUIs of buildings containing various mechanical, envelope, and lighting efficiency features with the EUIs of buildings in which equipment usage is reduced when the building is not in use. Comparing the average EUIs of buildings with efficiency features against those of buildings with usage reduction shows that buildings with efficient technology features use 26%–31% *more* energy than those applying usage-reduction methods. While the data categories in this table may overlap, and thus the EUI differences may not be as pronounced as the table indicates, Table 3 nonetheless adds to the evidence that the presence of energy efficiency technology *does not by itself* predict building energy performance.

The CBECS data do not provide explanations for these counter-intuitive results. One could hypothesize that newer buildings are "over-designed," with

TABLE 2 Energy Use Indices and Renovations of Older Buildings

	Number of Buildings	Floorspace	Sum of Major Fuels	Energy Use Index
Renovations in Buildings Built before 1980	(thousand)	(million square feet)	(Trillion BTU)	(BTU/SF)
Any Type of Renovation Since 1980	1,018	17,844	1,766	98,969
Addition or Annex	256	6,551	733	111,891
Reduction In Floorspace	22	1,012	117	115,613
Wall or Roof Replacement	370	8,070	777	96,283
HVAC Equipment Upgrade	442	10,768	1,156	107,355
Lighting Upgrade	455	10,275	1,085	105,596
Window Replacement	310	6,354	613	96,475
Insulation Upgrade	227	4,015	381	94,894
Other Renovation	19	523	50	95,602
No Renovations Since 1980	1,710	18,714	1,482	79,192
Building Newer than 1980	1,917	28,225	2,573	91,160

Source: EIA (2003).

more energy services provided per square foot. For example, older buildings may lack central air-conditioning, and so spaces may be cooled through small, packaged units that serve only part of the conditioned space. This could result in less energy used for air-conditioning for the building as a whole. One could also posit that newer buildings are used more intensively, with more people and equipment per square foot, and more hours of use in a given period. However, the available data do not provide clear explanation along these lines. This does suggest, however, as indicated in Table 3, that operating practices can influence EUI-measured performance more strongly than the presence of a given set of efficiency technologies.

TABLE 3 **Energy Use Indices of Buildings with Efficiency Technologies vs. Buildings with Usage Management Strategies**

HVAC Conservation Features	Number of Buildings (thousand)	Floorspace (million square feet)	Sum of Major Fuels (Trillion BTU)	Energy Use Index (BTU/SF)
Variable Air-Volume System	466	19,597	2,380	121,447
Economizer Cycle	508	21,108	2,589	122,655
HVAC Maintenance	2,581	51,163	5,170	101,050
Energy Mgmt System (EMCS)	252	15,630	1,782	114,012
Building Envelope and Lighting	Average	26,875	2,980	110,895
Multipaned Windows	2,201	38,910	3,929	100,977
Tinted Window Glass	1,323	29,887	3,098	103,657
Reflective Window Glass	308	8,544	927	108,497
External Overhangs or Awnings	1,233	17,242	1,737	100,742
Skylights or Atriums	331	12,546	1,307	104,177
Daylighting Sensors	74	2,868	377	131,450
Specular Reflectors	928	26,118	2,829	108,316
Electronic Ballasts	2,577	46,882	4,746	101,233
EMCS For Lighting	60	4,781	538	112,529
Equipment Usage Reduced When Building Not in Use	Average	23,011	2,456	106,749
Heating	2,878	42,722	3,740	87,543
Cooling	2,761	43,205	3,844	88,971
Lighting	3,685	46,987	3,818	81,257
Office Equipment	1,504	19,397	1,465	75,527
	Average	**38,078**	**3,217**	**84,478**

Source: EIA (2003).

These empirical results do not suggest that energy efficiency technologies are not necessary or not effective in improving building energy performance. They suggest, rather, that efficiency technologies may be necessary to *enable* energy performance, but they are not sufficient to *deliver measured* energy performance over time. In other words, it may be the ways in which building occupants and managers operate the facilities that determine measured energy performance, more than the nominal efficiency characteristics of specific technologies or components. The authors draw the inference that both are needed—sound design, efficient technology, and effective operation and management of the building. We also infer that the last of these—the effectiveness of building operation and management—is more salient in predicting measured energy performance. That is, an efficient building can be poorly managed and use too much energy, while an average-efficiency building, if managed well, can achieve above-average energy performance.

This review of macro-level nonresidential building energy-use data sets the stage for the remainder of the chapter, which focuses more specifically on analyses and case studies that illustrate the effects of superior energy management practices in producing significant energy performance improvements.

3. THE RECORD ON OCCUPANT/OPERATOR BEHAVIOR AND ENERGY USE IN ORGANIZATIONS

In more than three decades since the energy crises of the 1970s first made energy efficiency an organizational priority, the role of occupant and operator behavior has come to light in various forms. The federal Institutional Conservation Program, which gave grants to schools and hospitals for energy efficiency retrofits, completed its first evaluation in 1983. That analysis, based on 200 site visits, found that institutions that were rated best in the quality of their energy managers experienced average energy savings of 20%, while grantees with less-stellar energy managers averaged 9% savings. Similarly, institutions that were observed to be regularly following efficiency operations and maintenance procedures that were recommended in their energy audits saved an average of 16%, compared with 3% for grantees that were not supporting the recommended operation and maintenance (O&M) procedures (Synectics, 1983).[2]

More recently, behavior-based school energy efficiency programs have evolved into more concerted efforts. The Alliance to Save Energy's Green Schools program, for example, working with multiple California school districts, achieved savings of as much as 30% simply by managing energy use with sustained focus (Alliance to Save Energy, 2009). By engaging students, teachers, administrators, and custodians in a coordinated effort, Green Schools can reshape energy use patterns dramatically. At one school, in addition to

2. The Synectics Group. (1983). *An evaluation of the Institutional Conservation Program: Results of on-site analyses, final report*. Prepared for the U.S. Department of Energy.

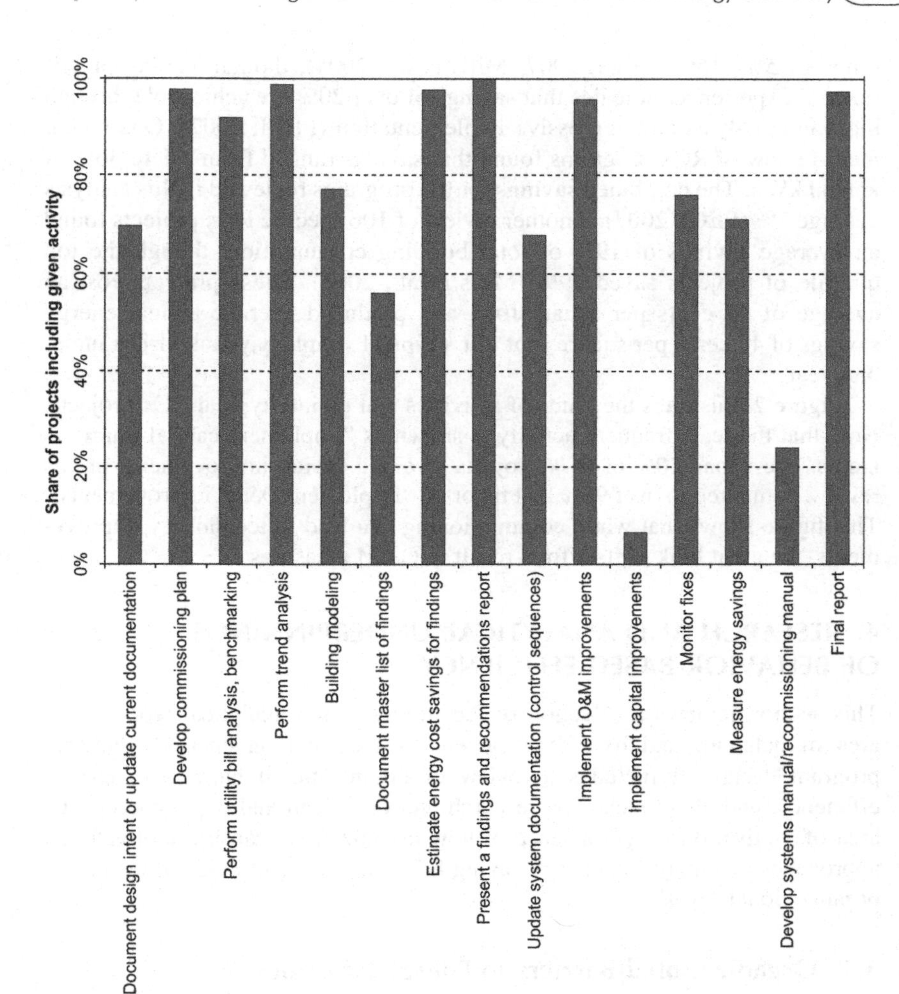

FIGURE 2 Scope of Retro-Commissioning Projects. *Source: Mills et al. (2004).*

unplugging and shutting down electronic devices at the end of the day, the custodians placed green reminder tags on any devices left on to remind teachers and students to shut them off later, turned air-conditioning systems off at the end of the day, and installed timers on outdoor lights to avoid unnecessary use.

The rapidly-growing field of RCx emphasizes tuning up building systems to run more efficiently, training operators and occupants to follow better O&M practices, and developing monitoring and reporting systems to keep building energy performance on track. Capital investments are typically small portions of RCx efforts—the majority of the analysis and implemented measures consist of adjustments to existing equipment, and changes in the way equipment is operated and maintained. Studies have found savings in the range of 5%–15% (PECI, 2007; Mills et al., 2004), though leading practitioners' experience indicates that savings of over 20% are achievable through thorough analysis and aggressive implementation (PECI, 2007). One recent meta-review of RCx programs found that savings ranged from 3% to 36% of annual kWh. The combined savings for the programs reviewed in this analysis average 7% (PECI, 2007). Another review of 106 specific RCx projects found an average savings of 19% of total building consumption, though the top quartile of projects saved 29% (Mills et al., 2004). These projects cost an average of 41 cents per square foot, and produced average annual energy savings of 42 cents per square foot, for a typical simple payback of just under one year.

Figure 2 illustrates the kinds of activities that go into typical RCx projects. Note that the least-frequent activity is shown as "Implement capital improvements." Less than 10% of RCx projects involved capital improvements in this review, compared to over 90% that reported "Implement O&M improvements." This figure shows that while commissioning can lead to technology improvements, the great bulk of its efforts result in O&M measures.

4. RESEARCH AND ANALYTICAL UNDERPINNINGS OF BEHAVIOR-BASED EFFICIENCY

This section reviews a selection of the literature and analytical work in the area of behavior and its effects on energy use, and on energy efficiency program design. It includes a review of organizational barriers to energy efficiency, and also touches on research and organizational experience in the area of motivation, to get a sense of how organizations can use motivational approaches to supporting energy-saving behavior, at the individual, team, and organizational levels.

4.1. Organizational Barriers to Energy Efficiency

The energy research field has produced extensive analyses of market barriers to energy efficiency investment, and to changes in building occupant behavior (IEA, 2007). Most of this work has focused either on broad market barriers between different market participants, for example, between home builders and home buyers, or on cognitive/behavioral factors affecting individuals, such as risk perception. Less work has been done on the barriers that exist within organizations, but barriers have been observed inside organizations that are analogous to those found to exist in generalized market situations. To examine barriers within organizations, one must step away from the classical economic view that firms, like individuals, are assumed to act in monolithic and completely rational fashion.

One of the most plausible intra-organizational barriers is a form of the principal–agent problem acknowledged by classical economics. This barrier involves an agent that makes decisions on the energy efficiency of a piece of equipment at purchase, or on the operational efficiency of a facility or energy-using system. The agent "should" be acting in the rational interest of the principal, defined as the person or entity that has the most direct interest in saving energy. In generalized markets, an agent could be a home builder, or a rental property owner, and the principal could be a home buyer or an apartment renter who pays the energy bill. Within an organization, classical theory might assume that all parties would act in the firm's rational interest, which would be minimizing life-cycle costs of energy systems.

However, an examination of the characteristics of organizations in practice reveals glaring principal—agent problems. For example, facility operators may want efficient equipment and operations, but procurement departments may use lowest-bid buying policies that do not select efficient models. Or facility managers may be called up to operate their buildings efficiently, but may see no direct benefit from so doing if energy bills are paid by another department.

Organizations resolve these barriers through policies and information systems that align the interests of different actors within the organization. While organizational change across an entire economy is slow and uneven, leading organizations as illustrated in this chapter are breaking through these barriers. Procurement policies, for example, may be changed to specify high-efficiency criteria or otherwise take operating as well as capital costs into account; individuals' performance goals and reward systems are aligned with their facilities' energy performance. Information systems are established that collect, track, and report energy performance, and create feedback and accountability loops that reinforce actions to improve performance. The Pew Center report on best practices in corporate energy efficiency describes these systems in detail (Pew, 2010).

4.2. Literature Review of Behavioral Factors

The research literature on behavioral factors in building energy use, compared to the extensive body of research, analysis, and program experience centered on technological factors, is quite limited. Nonetheless, some recent trends indicate a rapidly rising focus on behavior as distinct from technology as a factor in energy use in buildings. The California Institute for Energy Efficiency (CIEE) has launched an ambitious series of research papers under the heading of "Energy Behavior and Decision-Making." CIEE's recent work has been driven in part by the state of California's commitment to dramatic reductions in carbon emissions under its AB 32 legislative mandate. While California has supported some of the world's most aggressive building energy efficiency policies, planners recognize that to realize AB 32's goals, behavior changes as well as technology improvements will be needed.

CIEE has recently issued a number of white papers on the topics related to behavioral factors in program and policy design. Two of these (Lutzenheiser,

2009 and Sullivan, 2009) focus on the residential and nonresidential issues. Both papers begin by defining and critiquing the "physical-technical-economic-model" (PTEM) paradigm that underlies traditional energy efficiency program design. In the PTEM paradigm, consumer behavior and choice are seen as instrumental, rational, and—most importantly—secondary to the *physical* devices that are seen as the actual users of energy. In this paradigm, energy efficiency programs seek to substitute efficient devices as *technical* solutions, providing equivalent "energy services" like lighting or comfort at lower levels of energy use that result in reduced *economic* costs. Energy efficiency solutions in this paradigm are based on the "correct" *technical/economic* choice, in which energy users make *physical* changes in their energy devices that reduce energy consumption.

This research explores alternative paradigms that could better explain the ways in which people use energy, drawing on behavioral economics, cultural anthropology, and sociology disciplines. It shows how the extreme variability in real-world energy use can be associated with consumer demographics, cultural backgrounds, and local social influences. They posit that most energy-using behavior is governed by unconscious habitual actions and routines. For example, someone may stand in front of an open refrigerator door, thinking only of where to find a particular food item, not of the energy use impacts of holding the door open.

The CIEE research also points out that energy use is rarely *individual* in nature, but is more often performed in groups living together. For example, thermostat settings are often a function of compromise among household members, or lighting consumption is a function of how many people are using lights in various rooms. In addition, human action is *culturally* determined—that is, behaviors, devices, and buildings have wider meanings to persons and groups, beyond their energy use attributes. These alternative paradigms suggest that energy efficiency policies and programs should focus as much on *energy services* as on *physical devices,* on the hypothesis that the energy "service" a user wants is not a fixed phenomenon, but is socially defined and maintained.[3]

In this socially-determined framework of energy service demands, the level of energy services can be reshaped to reduce the demand for energy. For example, in the U.S., building simulation models typically assume that every cubic foot of space is uniformly conditioned to well-established temperature levels and airflow volumes. But in other countries, such as Japan, central heating and cooling has not been a traditional building design or lifestyle practice. Comfort conditioning is often provided only in certain spaces, and there may be no common assumption of what constitutes a "comfortable" air temperature. Today's sensing and control technology can provide this kind of

3. This book's Chapter 3 by Bartiaux et al. discusses the social, cultural, and anthropological underpinnings of energy use.

adaptive comfort, delivering lighting, airflow, and temperature to specific spaces only as needed. Many people leave lights on in unoccupied areas, under the incorrect, culturally transmitted assumption that turning lights off and on increases energy use. Correcting that cultural misapprehension, and possibly automating it with sensors, can dramatically reduce lighting consumption.

4.3. Analytical Work on Simulation of Behavioral Effects

The research on socially-determined energy consumption is so new that it is mostly heuristic, and has spent more time deconstructing the PTEM paradigm than in building new ones to the point of being ready for incorporation in field programs. However, some work has been done on behavioral impacts in the realm of building simulation modeling. Dean et al. (2006), for example, published a paper on the creation of an Occupant Energy Index (OEI), using a 0–100 scale based on a statistical average set of operating conditions. By varying assumptions about thermostat settings, and usage patterns of lights and appliances, the authors simulated a series of cases for typical prototype homes in cold and warm climates. They found that, as one might expect, thermostat settings could account for variations of 40%–50% in total annual energy use. They used standard setpoints of 68°F in winter and 78°F for the reference case, 74-winter and 72-summer for a "constant comfort" case, and 62-winter and 84-summer for a "maximum conservation" case. The constant-comfort case increased energy used by 23%–30%, while the maximum conservation case reduces it by 17%–20%. The authors conclude it could be straightforward to use these simple behavioral factors to simulate energy savings impacts of behavior changes, and to create a rating scale for building operation that parallels the more traditional rating scale based on physical characteristics.

4.4. Advances in Motivational Research and Practice

A new development in motivation theory is emerging, with particular applicability in the organizational world. Summarized and popularized in Pink (2009), the generic term used for it is *intrinsic motivation*. Psychology researchers have long operated in a paradigm that entertained only two general types of motivation: biological and extrinsic. Biological motivation drives us to find enough to eat, stay warm, reproduce, and so on. To motivate us to do more than the minimum, theorists have long held that extrinsic rewards, most often monetary, are necessary. Yet a growing body of research indicates that for complex tasks and those requiring creativity and perseverance, intrinsic motivation can be more effective in achieving and sustaining success.

This intrinsic motivation factor is beginning to appear in large organizations' energy efficiency and sustainability programs. In the Pew Center on

Global Climate Change's recent report on best practices in corporate energy management (Pew, 2010), respondents to the project survey reported that one of their biggest surprises in mounting their programs was the enthusiasm and creativity with which employees responded to the program. A sample of verbatim responses includes the following:

"Our biggest surprise was the broad employee interest in energy and environmental action."

Cummins Inc.

"The backbone of our strategy is employee engagement. Without that even the best capital projects can fail."

Citi

"There is a wealth of creative ideas at every level of the organization. All we need to do is provide opportunities for those ideas to surface and grow."

PepsiCo

"It is important that teams on the ground feel empowered to implement initiatives, and are given the support and resources to take initiative on their own."

News Corporation

"An increasing number of employees are starting to care passionately about the Group's energy and GHG performance."

Rio Tinto

These examples illustrate the phenomenon of *resonance*, an emerging concept in the organizational development field that refers to states of increased morale, energy, and productivity in organizational settings. People working in resonant organizations are more motivated, more connected to the mission, more interested in focusing their talents on collective goals. As an example, PepsiCo began holding Sustainability Summits, bringing together hundreds of employees and suppliers from around the globe to share and apply concrete ideas. Drawing on landmark analytical work such as the worlds' first full-product-cycle carbon label for its Walkers Crisps brand, PepsiCo people were encouraged to think "outside the box." In their largest Asian division, for example, they began by looking at boiler efficiency, but soon began to examine boiler fuels as well. This led them to shift from imported fossil fuels to local rice-hull sources for fuel. The boiler upgrades not only increased efficiency, they reduced fuel purchase costs, and reduced reliability risks by using locally-grown fuels not subject to global price volatility or supply vulnerabilities.

To date, analytical methods have not been developed for quantifying the energy and carbon impacts of innovative, behavior-based solutions beyond the incremental PTEM solutions that form the basis of most energy efficiency potential studies. These studies typically look at discrete, incremental technology substitutions device by device and end-use by end-use. But they don't capture synergies beyond the incremental.[4] More to the point, they don't capture the potential for energy performance improvement that could flow from organizational efforts to more actively manage consumption, and to continuously look for improvements across multiple performance indicators.

4.5. Redefining "Energy Services"

Conventional energy efficiency analyses also use standard definitions for *energy services*. For example, they define "comfort" as maintaining air temperatures and humidity levels within specified ranges, on a continuous basis, for every cubic foot of space in a given building. "Lighting," likewise, is typically defined by engineering standards based on uniform illuminance levels for an entire space. However, comparable levels of comfort can be achieved in a wider range of nominal air temperature/humidity combinations, based on the thermal performance of the building shell and on the culturally-determined expectations. Lighting can also be designed on a task basis such that general illumination levels can be much lower. Occupancy-based comfort and lighting systems can provide variable levels of comfort/lighting to different spaces/tasks at only the levels and times they are needed. As the CBECS data indicate, altering the level of energy services provided during unoccupied hours can dramatically reduce energy use.

This new set of views on motivation and the definition of energy services could drive a new wave of energy use reductions, by going beyond traditional incremental efficiency analysis to examine underlying assumptions about energy service demands. The combined effects of using motivational strategies to manage energy more actively, redefine energy service demands, *and* apply advanced technologies to serve those redefined energy service demands more efficiently could drive radically lower levels of energy consumption.

5. LEADING ENERGY EFFICIENCY STRATEGIES IN THE CORPORATE WORLD

Leading corporations have begun to act on reducing their energy and carbon footprints. Corporate sustainability programs have spread rapidly in the past decade, and as they conduct environmental footprinting analyses, companies are radically rethinking the importance of their energy use. A survey of 48 leading companies, including heavy manufacturers, auto manufacturers, computer and chip makers, financial institutions, and apparel companies, found that while corporate energy use accounted for less than 5% of revenues on average, it accounted for the great majority of companies' carbon footprint

4. For example, the traditional PTEM approach would tend not to examine radical building shell improvements that dramatically downsize or eliminate HVAC system components, which in turn reduces total capital costs and energy use, beyond any results that would flow from a parametric, incremental analysis.

(Pew, 2010). This realization brought home in a concrete way the emerging perception among corporate leaders that the era of cheap energy and free GHG emissions is over, and that they face an era of carbon-constrained energy markets. It shifted energy in the eyes of chief executives from a minor cost issue to a major environmental and sustainability issue. As reported in the survey results, reducing carbon emissions was the top reason for companies' launching their efficiency strategies. Moreover, the survey reported that CEOs were the most important champions for the success of companies' efficiency efforts.

The recent Pew Center report on corporate energy management has documented the actions leading companies have taken to drive down their energy use as part of energy efficiency and GHG mitigation strategies. Through surveying dozens of companies and conducting in-depth case studies on six, this project distilled what it terms the Seven Habits of Highly-Efficient Companies. These are:

1. **Efficiency is a core strategy**—For companies with leading strategies, energy efficiency is an integral part of corporate strategic planning and risk assessment. It is not treated like just another cost management issue, or as a sustainability "hoop" to jump through. It has become part of an ethos and a corporate culture in which energy efficiency is essential to a thriving enterprise in the twenty-first century.
2. **Leadership and organizational support is real and sustained**—When energy efficiency is really a core part of the organization, its leaders can talk about it without notes. Beyond what CEOs and other leaders say in speeches, leadership commitment to efficiency shows up in many forms, in multiple media, and such communications are frequent and prominent, inside the company and out.
3. **The company has SMART energy efficiency goals**—SMART is a well-known acronym for Specific, Measurable, Accountable, Realistic, and Time-bound, and has been used in mission and goal-setting exercises for years. What makes the SMART concept specific and unique to energy efficiency strategies is that goals are:
   - Organization-wide
   - Translated into operating/business unit goals
   - Specific enough to be measured
   - Have specific target dates
   - Linked to action plans for achieving them in all business units
   - Updated and strengthened over time
4. **The strategy relies on a robust tracking and performance measurement system**—The adage that "you can only manage what you measure" applies to energy efficiency strategies as much as any critical cost factor or performance indicator. Creating an effective energy tracking and performance system requires collecting a lot of data from a lot of different and sometimes

disparate sources, and often bridging the disconnects between organization units. In an exemplary system:

- Data are collected regularly from all business units
- Data are normalized and baselined
- Data collection and reporting is as "granular" as possible
- The system tracks performance against goals in a regular reporting cycle
- The system includes feedback mechanisms that support corrective action
- Performance data are effectively visible to senior management
- Energy performance data are broadly shared, internally and *externally*
- The system is linked to a commitment to continuous improvement

5. **The organization puts substantial and sustained resources into efficiency**—Any effective effort in an organization requires resources—people, capital, systems, and so on. The leading companies in this field, however, have not necessarily relied primarily on capital investments to drive energy efficiency results, but rather have obtained substantial savings through operational practice changes and moderate-cost technologies. The three key ways in which leading companies invest resources in their efficiency strategies are:
   - The energy manager/team has adequate operating resources
   - Business leaders find capital to fund projects
   - Companies invest in human capital
6. **The energy efficiency strategy shows demonstrated results**—The leading companies in the Pew study not only conceived and launched their strategies, they followed through with operational changes and capital projects, and captured the results through their energy performance measurement systems. The key features of leading programs in this respect include:
   - The company met or beat its energy performance goal
   - Successful energy innovators are rewarded and recognized
   - Resources are sustained over a multiyear period
7. **The company communicates energy efficiency results**—The best companies in this field make energy efficiency a living part of their story; they place efficiency successes prominently in both internal and external communications. Energy efficiency becomes a part of the story the company tells about itself, part of its identity and its culture. The best programs include both an internal communications plan that raises awareness and engages employees, and external efforts that document commitments and successes.

Figure 3 illustrates the Seven Habits, predominantly behavioral in nature; they describe how the organization behaves, organizes and acts on data, communicates internally and externally, and creates goals and rewards. Perhaps surprisingly, in only a few places do capital investment and technology

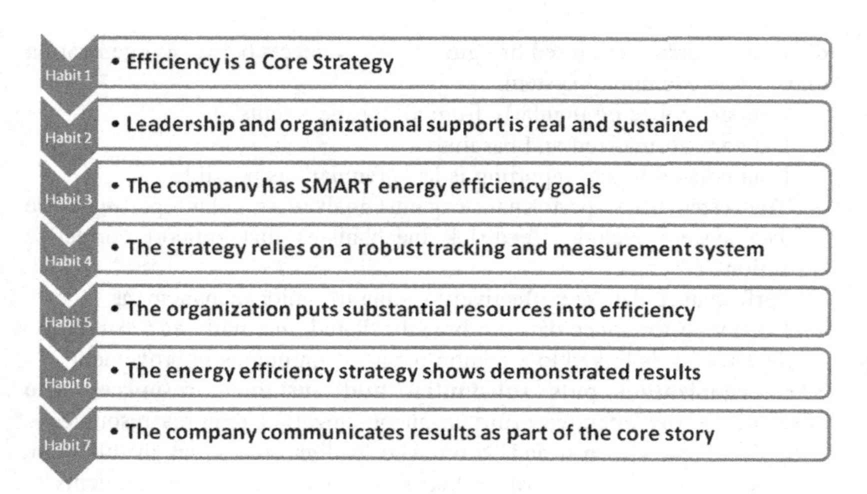

FIGURE 3 The Seven Habits of Highly-Efficient Companies. *Source: Pew (2010).*

improvements become central to the successful implementation of energy efficiency strategies.

The Pew study involved 48 survey respondents, and six case study participants. The subsections that follow highlight some aspects of the six case studies included in the Pew Center report, emphasizing the behavior-oriented approaches these companies are taking, on the operational level as well as the strategic level.

The examples below focus on relatively short-term results, and so may seem to show less-than-dramatic results. To illustrate what can happen when these practices are sustained over more than a decade, however, the United Technologies case study documents savings accumulated over 14 years. In 1996, the company set a goal of reducing energy used per dollar of revenue 25% by 2006. By 2001, it had already met this goal, and so reset it at 40%. By the target date of 2006, actual savings were 53% per dollar of revenue—well beyond the revised 40% target—and had reduced absolute energy use by 19%. Based on this experience, the company set a new goal of 12% carbon emission reduction target by 2012—but this time in *absolute* terms, not normalized by revenue. As of 2010, the company had already reached this goal—with the *cumulative effect of reducing its carbon emissions by some 30%* while more than doubling its revenue. Without going into the details of its practices, which are similar to those described in the three examples below, United Technologies' experience illustrates the deep impacts that can be attained through concerted effort over an extended period.

5.1. Corporate Case Study: IBM

IBM's popular image is typically associated with technology—computers and information systems. Yet less than 20% of the company's 2008 revenues came from hardware sales; most of its customer offerings consisting of various integrated solutions involving services, software, and hardware. And while IBM still makes semiconductors and other electronics components and products, it has only three major production facilities out of 1100 facilities worldwide. Most of its energy efficiency strategy thus focuses less on manufacturing technology investments than on more basic operation improvements in its far-flung real estate operations. In the new corporate efficiency program launched in 2006, behavior change is a major focus. Examples include:

- **Vacuum pump scheduling**—In its clean-room semiconductor manufacturing operations, the traditional practice has been to run equipment continuously, under the assumption that quality control and productivity required it. In this spirit, the company kept vacuum pumps for cleaning running 24/7. Looking closer at this operation, the plant energy staff asked the production staff when the equipment was actually used; this resulted in an 86% reduction in runtime and energy use for these pumps.
- **O&M checklists**—Because of the large number of facilities IBM manages, and because most of them have similar lighting, HVAC, and other energy systems, it drives energy performance at a large fraction of facilities mainly by developing simple checklists, that typical facility staff can understand, and asking facility managers to verify that these procedures have been followed as part of the IBM monthly energy reporting system.
- **Continuous commissioning**—For its largest 24 facilities, IBM maintains a "granular" monitoring system that provides operating data and control capabilities for lots of specific equipment and systems. The system can also produce graphic summaries and comparative analyses that help the energy team detect anomalies down to individual pieces of equipment. For example, it enabled them to detect a boiler with malfunctioning controls that came on every day at 6 A.M.—easily fixed with little or no cost. This system has enabled many such no-cost energy performance improvements.

IBM set a goal for its program of achieving energy savings equal to 3.5% of previous year energy use, company-wide annually from 2006 to 2010. In 2008, using the kinds of improvements described above and others, the company achieved savings exceeding 6%. This amounted to more than $30 million in energy bill savings, and the company spent less than $9 million on efficiency investment, providing a simple payback of 3–4 months. Only a behavior-driven program could achieve that kind of cost-effectiveness.

5.2. Corporate Case Study: Toyota

Toyota is known for its fuel-efficient vehicles like the Prius hybrid, but it also strives to be the most energy-efficient automaker in the world, to support its sustainability goals and also to remain competitive in world automobile markets. Being a major manufacturer, one would think that its goal would be realized mainly through technology investment. But while Toyota does upgrade equipment, like the air compressors that drive many of the tools in its production shops, it focuses equally if not more intently on reducing wasted energy in plant operations. For example:

- **Compressed air shutdown**—Working closely with its production shop teams, the energy management organization provides data, down to the one-minute level, on compressed air operations. As a result, crews now shut off their tools at every break and at the end of every shift.
- **Treasure hunts**—Instead of energy audits, which have a negative connotation that implies wrongdoing, Toyota calls its plant assessments "treasure hunts." Much of the focus of these events are on operating efficiency, not technology efficiency. Treasure hunt teams composed of peers from other plants typically arrive on a Sunday, and check for equipment that is left on. They then observe Monday's first and second shift operations, and present a report to plant management and shop captains on Tuesday.
- **Race for the greenest**—Toyota has evolved ways to encourage energy-saving behavior. In its Kentucky plant, every month there is a meeting of shop captains and plant staff, around a large magnetic board shaped like a racetrack. Each shop is represented on the board by a miniature vehicle. As the energy team leader reads off the month's performance results, production teams' cars advance or fall back on the track. Conducted with humor and accompanied by other rewards, this "race for the greenest" has become part of the company culture, driving attention and action that changes behavior.

One might ask what rewards are provided for such performance improvements. This is where the Toyota Kaizen culture proves to be a subtle but powerful asset. Kaizen is a culture in which each individual is responsible for taking responsibility for continuously improving their job, whatever it may be. This continuous-improvement ethic, combined with data and technical assistance, can lead to dramatic savings over time. Toyota also applies the Kaizen ethic in a collective and practical way, by creating a companywide database of energy and other technical innovations. These database entries, called "Kaizens," are available to all company personnel. So beyond establishing individual responsibility, Toyota encourages idea-sharing and collaboration.

Toyota's energy management program set a goal in 2002 to reduce the energy used to produce a vehicle by 30% by 2011. It met that goal in 2007, driving consumption down from over 9 million BTUs per vehicle to 6.3 million. Much of that success came from the behavior-driven improvements illustrated above.

While these behavior-based methods of continuous improvement are the core of Toyota's day-to-day efficiency strategy, they are part of a larger

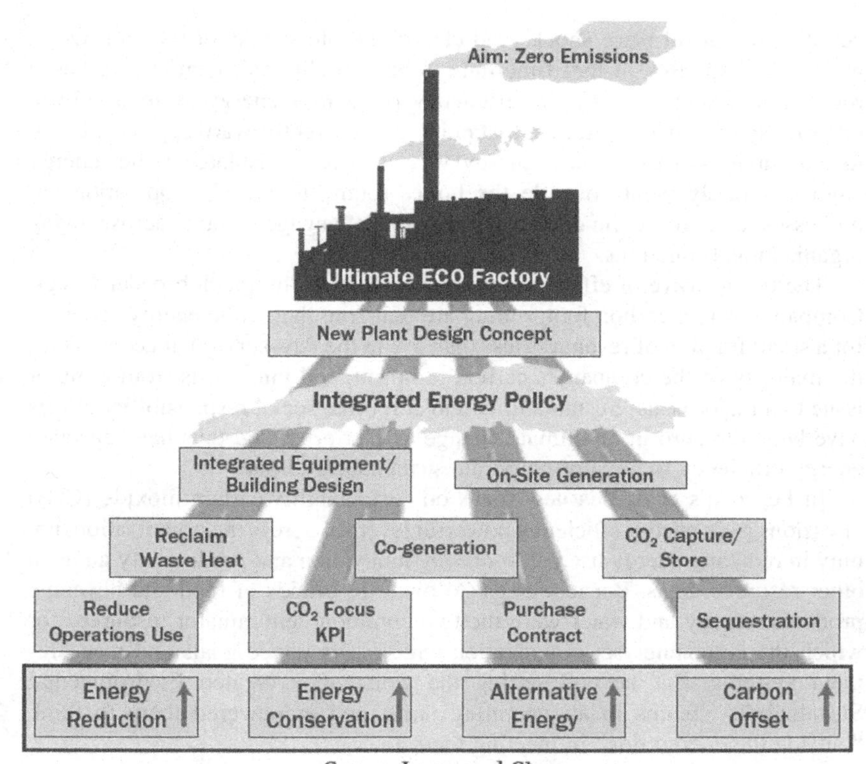

FIGURE 4 Toyota's "Ultimate Eco-Factory. *Source: Pew (2010).*

framework of sustainability goals leading to a zero-emissions manufacturing environment. Figure 4 illustrates this longer-term goal. Note that two of the four foundational "blocks" in this figure are "Energy Reduction" and "Energy Conservation," and are based largely on improving operational performance. By driving down energy requirements to the minimum, Toyota sees these elements as critical to making a zero-emissions plant feasible; if energy needs are low enough, some combination of solar, green power, biofuels, and other low-emission supply technologies may be able to serve the plant's energy needs affordably.

5.3. Corporate Case Study: PepsiCo

PepsiCo exemplifies the paradigm shift described above as "getting out of the energy box." In the first wave of corporate energy management beginning in the 1970s, energy was a cost factor that had leaped into prominence by the combination of oil price shocks and electricity rate shocks of that era. Organizations' responses in that timeframe took a traditional, engineering-based route: assess and improve the efficiency of facility energy systems. Given historically low energy prices to that point, this straightforward approach found lots of savings. But because energy was seen as an isolated issue, energy engineers rarely went "outside the boiler room" to redesign operations or processes in a more fundamental way, or to engage people across many organizational functions.

The recent wave of efficiency strategies is driven by much broader forces. Companies doing carbon footprinting are realizing that while energy accounts for a small fraction of revenues (less than 5% in the Pew survey), it accounts for the majority of the company's carbon footprint, and thus shifts from a minor issue to a major issue. Sustainability and corporate social responsibility efforts have begun to zero in on climate change as a priority, and thus have elevated energy efficiency to a major corporate sustainability issue.

In PepsiCo's case, this new focus on sustainability/carbon dioxide (CO_2) emissions gave energy efficiency powerful leverage across the organization, not only in reducing energy use but in driving innovation and productivity gains in other resource areas. Because PepsiCo owns 18 brands of food and beverage products, energy and water were the two common-denominator resources for which the companies set organization-wide performance goals and measurement systems. But beyond setting the goals, they created Environmental Sustainability Teams in all operating units, and empowered them to think "outside the energy box" in meeting their goals.

This approach produced results that not only met the basic goals, but generated other innovation and productivity gains. Two examples:

- **Tortilla chips**—Beginning in Frito-Lay's tortilla chip operation, the first step was installing damper draft controls on the ovens. This reduced heat loss, but also improved heat distribution and hence the quality of the chips. The team then added heat recovery technology, further reducing energy use, and increasing total throughput. So energy efficiency not only reduced costs, but also improved product quality and total production. These innovations stemmed from the energy and water savings goals, which forced production staff to look at these resources across their operations.
- **Total sustainability innovation**—At an Arizona facility, the first sustainability action was to treat wastewater and apply it to adjoining croplands to grow alfalfa. The next wave was to upgrade treatment to the point that water can be reused for plant supplies, freeing up cropland for renewable energy production. This multiwave strategy reduced the plant's dependence on scarce water supplies, and its dependence on volatile energy supplies, reducing fundamental business risks as well as energy and water costs.

PepsiCo's energy strategy also exemplifies another powerful new trend in large organizations: supply chain management. Like many diverse product companies, PepsiCo uses contract manufacturers for much of its product volume. In recent years it has asked major suppliers to report some of the same energy and water data it requires of its internal operating units, using its Measure Up online tracking system. It has also asked (and provided technical assistance to) suppliers to engage them in energy efficiency programs like ENERGY STAR Buildings and Plants programs. It has established a scorecard system based on the percentage of suppliers engaged in such activities.

Supply chain strategies are potentially powerful in terms of total potential impact. When Wal-Mart first calculated its carbon footprint, it found that 90% lay in its supply chain. When PepsiCo conducted the Walkers' Crisps life cycle carbon labeling study, it found that 70% of the carbon footprint lay outside the manufacturing process. Nike found that only 8% of its footprint came from internal operations: 63% came from contract manufacturers and 26% from shipping. As major organizations (and they need not be just corporations—institutions like universities, hospitals, and governments can do the same) focus on driving energy efficiency and other sustainability practices up their supply chains, they can greatly magnify their impacts.

6. BEHAVIOR CHANGE IN A PERFORMANCE CONTRACTING FRAMEWORK

Energy savings performance contracting is a turnkey service that provides customers with a comprehensive set of energy efficiency and conservation measures, with guarantees that the savings produced by the service will fund the project. The energy service companies that provide performance contracting have just recently begun to embrace behavior change as an energy conservation measure, even though behavior-based measures can be very cost-effective. This can be attributed in part to the difficulty in measuring and verifying any energy savings derived through behavior change. Within an organization, individuals' energy consumption is most always measured at the building level, sometimes only at the organization level, very different from the residential sector. Adding to this level-of-analysis issue is the fact that the organization's buildings are often being retrofitted as the behavior change is occurring within, augmenting measurement challenges. There are few established and accepted metrics for assessing individual energy consumption, regardless of residential or organizational environments. Compounding the confusion, little research has been conducted on energy consumption behaviors in an organizational setting. In sum, organizational energy consumption behavior change programs often suffer because of their perceived "soft" or nonexistent measurement and verification component. Hence, a behavior change-based efficiency measure can be more difficult to "sell" within a performance contract mechanism.

6.1. Do We Need to Measure Behavioral Savings?

Measuring behavioral energy savings in a residential setting does not generalize to measuring behavioral energy savings within an organization. This is because there are significant differences between organizational and residential energy consumption. For example, residential energy consumption is most always examined at the individual or family level. However, organizational energy conservation interventions aimed at reducing energy consumption may impact at the individual, group, department, division, or organization level. Hence, measuring organizational energy consuming behaviors—and their associated savings with any behavioral change—entails multiple levels of analysis.

Another difference between residential and organizational settings involves the technologies and the participatory decision-making—or lack thereof—among the individuals involved. Some technologies share similarities with respect to installation and function in both settings, such as installing energy efficient lighting, low-flow faucets, and water-saving toilets. But while the technologies may be similar, the information that guides their use often is not. Homeowners, because they typically pay for efficiency measures themselves, can be expected to have some basic understanding of how the device functions and its expected benefits, and can be expected to use the device effectively. By contrast, organizational members may have little knowledge or instruction on the use or benefits associated with newly installed devices. This lack of information may lead to differing acceptance levels of efficient devices among organizational users, resulting in nonuse, misuse, or other undesirable behaviors associated with any newly installed energy efficient technologies.

Perhaps more importantly, people in organizational settings often have no feedback on their usage behavior related to energy conserving behaviors or energy-efficient technologies, whereas residential users typically see some feedback through their energy bills. In this way, individuals in organizational settings are rarely directly accountable for energy costs. This lack of accountability and feedback concerning organizational energy consumption can enable or even encourage energy-wasting behaviors. Thus, a focus on "the human dimension" of organizational energy consumption, distinct from a technology-only perspective, is necessary to achieve zero-net energy performance.

6.2. Defining Behavior Strategies that Work in Universities and Colleges

As described above, Americans use more energy as organizations than as households. This is especially true for universities, which incur energy expenditures through at least two key groups: their employees, and the students who reside on campus. This two-pronged impact strengthens the case for organization-level energy conservation programs in university settings, especially for resident students, who consume energy up to 168 hours a week during school terms, while employees may only be on campus 40 hours a week. Because students rarely pay energy costs directly, promoting energy conservation behaviors in a university setting is fundamentally different than promoting energy conservation behaviors in family residences.[5] Accordingly, we examine two different strategies to address a residential student energy conservation program:

Target just three substantial energy-wasting behaviors, instead of many.

Change intrinsic motivations instead of using external rewards to achieve behavior change. As described above, the emerging research on intrinsic motivation shows great promise to drive energy-conserving behaviors.

Unlike students, families living in traditional single-family owned units typically bear direct financial responsibility for their energy consumption. But in organizational settings, until recently there has been little direct financial responsibility for energy use, although the leading corporate case studies described above indicate a hopeful trend toward changing this. Not surprisingly, many students report a sense of entitlement to use as much energy as they wish. This stems from a common perception that tuition, room, and board covers all costs of running a university, including their utility consumption. Fortunately, a majority of both university students and staff report that they are willing to engage in environmentally-beneficial behaviors, but are often confused or uninformed as to which specific conservation behaviors are actually beneficial, and how to pursue them on a practical daily basis.

Many existing programs use a "Top 10" approach, where ten, or more, energy saving tips are put forth to effect behavior change. The issue is that such tips have varying levels or magnitudes of energy savings associated with them. Because there is little understanding of how much energy a specific behavior actually saves, the tips approach typically produces very small energy savings (e.g., unplug a cell phone charger to reduce phantom power) compared to other tips that can produce substantial energy savings (e.g., taking a 5-minute shower instead of the 14-minute college average). This suggests that behaviors having a significant impact should be identified and targeted to maximize the effectiveness of behavior interventions, and to produce significant energy savings and carbon footprint reductions. By the same token, behaviors that have little impact can be ignored to reduce program costs and avoid information overload.

The second strategy rejects competitions as a motivator to effect behavior change. To date, most colleges and universities have used a competition approach in their energy conservation initiatives, providing rewards to the groups that "win." However, rewards given to individuals for behavior

---

5. The same phenomenon applies in many other types of buildings where individual occupants are virtually disconnected from bearing the direct energy costs through the institutional setting.

measured at the group level can be problematic. For example, rewards can be seen as trivial and thus not effective as motivators, as is often the case when distributed among large groups of individuals. Will students really change their behaviors for a slice of pizza every so often?[6] Conversely, if the rewards are more substantial, the costs can approach the amount of savings generated, eroding the bottom line financial impact. Either way, giving awards creates "losers." Group-level behavior is made up of individuals, whose individual motivations and actions may vary greatly within the group, and therefore there will be "losers" who worked very hard to change their behaviors, and there will be "winners" or free-riders who did nothing at all to change. Providing extrinsic rewards on a group basis may thus not be an effective way to affect individual behavior, which in turn limits this approach's effectiveness for achieving maximum reductions at the group level.

Emerging research on extrinsic versus intrinsic motivation indicates that external rewards may undermine the more powerful potential effects of individuals' intrinsic values or motivations. For example, a behavioral change program based on competition between residence halls may devalue or even ignore key motivators, such as wanting to be a "good environmental citizen" or conserving energy because "it is the right thing to do." Moreover, external rewards often cannot be sustained over time, either due to prohibitive costs or other reasons. Research also shows (Pink, 2009) that extrinsic rewards, to be effective over time, must be periodically increased—not a sustainable model for an organization with limited resources. Therefore, with dwindling resources and generally pro-environmental leanings of traditional college-aged residential students, an energy conservation program based on intrinsic motivations is the preferred and more effective strategy.

6.3. Designing and Assessing a More Effective Organizational Energy Conservation Program

Using focus groups and survey data, we identified the range of voluntary energy consuming behaviors that occur in a residential student building. We then chose three specific behaviors with the greatest magnitude of energy consumption and the greatest variance and hence opportunity for change: length of showers, turning off HVAC when leaving the dorm room, and powering down computers. Next, we generated motivational factors driving the three specific behaviors, both intrinsic and extrinsic that were considered barriers to conserving energy. The results of the subsequent focus groups and survey responses show a lack of information regarding the actual energy consumption of the chosen behaviors, and in many instances, misinformation, or myths, also surrounded these behaviors. Additionally, survey respondents reported that

6. "Pizza parties" for the winners are often provided as a prize for on-campus competitions.

they engaged in these three behaviors less than "half the time," indicating significant potential for change. Lastly, these three behaviors all have a significant impact on the organization's energy expenditures. Utilizing all identified factors, an organizational energy conservation intervention was developed and pilot tested within a subset of the organization to assess effectiveness.

6.4. Measuring the Impact of Behavior Change in Residential Student Buildings

We established two compelling reasons to calculate a savings metric attributable to behavior change alone within a performance contracting framework:

To enhance flexibility in designing a performance contract by allowing longer payback energy conservation measures within the overall project.

To estimate the carbon footprint and GHG reductions associated with behavior change.

Hence, using the pilot study results, savings can be estimated for other similar populations considering a performance contract or investigating strategies to reduce their carbon footprint. For this pilot study, two similar sets of buildings were assessed with respect to their building-level electrical and water consumption during a six-week period compared to the same six-week period the previous year. To assess the effectiveness of the conservation program targeting internal factors driving three energy consuming behaviors, students whose Resident Advisors (RAs) attended the training session were compared to students whose RAs attended a "placebo" training session comprised of content not related to energy conservation. The complex of buildings receiving the energy conservation program showed a 9% reduction in electricity consumption, and a 9% reduction in water consumption, compared to the control buildings not receiving the energy conservation program.

This pilot study was then repeated at different buildings, with different RAs, different students, and a different trainer and training session occurring in a different season, with similar results. Based on results from these two pilot studies, this behavioral change program was projected to save an average of $14.26 in energy costs per academic year per student compared to the control group. Based on the success of the pilot studies, the program was implemented campus-wide at the beginning of the next academic year, while the residence halls were being retrofitted through a performance contract. At the end of the academic year, the client determined that their residence halls' utility cost avoidance had "over-performed" the projected retrofit savings by approximately the amount of the behavioral savings. From this, it is reasonable to infer that the students contributed an additional 9% energy savings over and above the savings realized from the building retrofits. When considering that students control only a small fraction of their building's total energy consumption, a 9% reduction at the building meter from changing just three energy consuming behaviors is substantial. When comparing the cost of a behavior change program to the cost of many technology retrofits, this "people-based" energy conservation measure can produce some of the shortest paybacks within a performance contract or other program portfolio.

The intrinsic motivation strategy focusing on substantial energy consuming behaviors shows promise, especially compared to an external reward/competition approach with its costly prizes, problematic creation of "losers," and lack of sustainability. This strategy could be effective with most any organizational population. Identifying specific behaviors associated with substantial organizational energy costs, identifying specific factors that drive those behaviors, and then targeting only the most influential factors appears to be a robust program model. The two pilot studies also offer a way to estimate behavioral savings in similar organizations that do not have baseline data or building-level meters. From these treatment-versus-control evaluations, it may be reasonable to estimate behavioral savings based on the pre-versus post-program survey data incorporated into the pilot studies. Asking similar questions in a similar survey methodology, one can extrapolate expected savings where meters don't exist. In one specific situation where this strategy was implemented, an entire master-metered campus "over-performed" the building retrofit savings projections by amounts consistent with the behavior-savings estimates.

CONCLUSIONS

Leading organizations, like those described in the Pew study and universities using behavior-change models to augment their energy-savings programs, are beginning to shift the focus of energy efficiency and carbon emission reduction strategies from solely-technological efforts to embrace behavior-based strategies. This shift is happening on more than one level. In the narrowest sense, it involves straightforward operational changes like equipment scheduling; but on a broader level, it involves changing how the entire organization thinks and behaves toward energy and GHG emissions. As organizations assess their carbon footprints, the role of energy use shifts fundamentally, from a minor cost issue—typically less than 5% of revenues—to a major footprint factor—typically more than 80% of internal footprint. The survey results in the Pew study bear this out; as organizations realize that energy use is the largest cause of their carbon footprint, it rises dramatically in prominence as a performance issue. Other surveys confirm this trend: *The Economist*'s Intelligence Unit conducted a 2009 survey (*The Economist*, 2009) of 538 senior executives from a diverse set of companies around the globe; it found that 54% had established GHG reduction strategies, and 62% had established energy efficiency programs, even though few of these companies were directly subject to GHG regulation schemes.

This realization can shift the organization's awareness and action focus, moving energy "out of the boiler room" into an issue that the entire organization must address. Beyond narrow changes in specific operations aimed at meeting specific performance targets, leading organizations are engaging more people, and more aspects of their business, in their sustainability efforts. PepsiCo's experience described above exemplifies the kinds of broader thinking and action that can dramatically change the way energy and carbon are treated in organizations, leading to major shifts in energy use and carbon footprints.

Utilities and other public-interest-driven efficiency programs are also beginning to shift their focus from the PTEM model described above to include more behavior-based models that use performance measurement for the whole building more than incentives for specific products or projects. The retrocommissioning work described above has become more common in more program portfolios. Utilities like PG&E are beginning to focus on not just whole buildings, but whole organizations for fleets of buildings, through their More Than a Million program, using ENERGY STAR benchmarking and other techniques. The EPA recently launched the Building Performance with ENERGY STAR program model to support program managers wanting to move in this direction. Using the ENERGY STAR benchmarking and performance measurement approach, it offers a "pay for performance" rather than a "pay for widgets" approach. Rather than provide defined incentives for specific technologies, building owners can use a variety of methods to improve performance, but their incentives are tied to their measured performance, not to the specific measures they install.

The "Seven Habits" features of exemplary organizational energy efficiency strategies are soon to be codified in a new ISO standard. Referred to as ISO 50001, it is expected to be completed by 2011 or 2012. It defines the practices, embodying many of the "Seven Habits" that an organization must adopt to be deemed in compliance. It is expected that specific system standards, based more on engineering design and energy performance, for systems such as steam, compressed air, or motor systems, will become embedded within ISO 50001. The U.S. Department of Energy (DOE) is serving as the secretariat during the development phase. In parallel with this effort, DOE is supporting the Superior Energy Performance program, which incorporates both the management practices in the "Seven Habits" and specific technology and performance metrics. Superior Energy Performance is expected to become an independent nonprofit organization, and to evolve a formal recognition/certification process.

The question remains open as to what role behavior-based organizational energy efficiency and related emission-reduction strategies will play in realizing a zero-carbon future for buildings and industrial facilities. Looking at many of the initial efforts, total impacts are modest, driving energy savings in the range of a few percentage points annually. However, as United Technologies' program shows, over 13 years they have reduced absolute energy use by 30% while more than doubling in revenue. Peering into a zero-carbon future requires some imagination; but there are some present-day examples that hint at such a future:

- Toyota's Ultimate-Eco-Factory vision, which rests on continuous improvement in energy efficiency as it works toward incorporating renewable and other features of a carbon-free future. Driving down energy use to the minimum required levels is a core strategy for enabling renewable energy to serve a facility's energy needs.
- Staples' strategy of investing in cost-effective energy efficiency, and using the reduced energy bills to finance lease payments for solar PV on store roofs. Pursuing in cycles over time, Staples foresees driving down energy use to the point that PV will be able to meet all its electricity needs in some stores.
- PepsiCo's approach of starting with energy efficiency in the product manufacturing process, extending the analysis to water savings, reducing energy and water needs to the point that recycled water and locally-produced renewables can serve all the plant's needs.

The key to these examples is a twofold behavioral strategy: (1) focus on practical, short-term, cost-effective efficiency measures, pushing this way of thinking and behaving across the entire organization; (2) as they become practical, integrate other resources into building and plant operations, looking at the entire resource picture. This requires constant attention to detail in the present while thinking holistically for the longer term. What leading organizations can teach us is that the first element is the most important, and never ends.

Any quantitative conclusions to be drawn from the relatively limited record of behavior-based energy efficiency programs must be tempered by the relative infancy of this field. But we feel confident in the following qualitative conclusions:

- Measured energy performance in buildings appears to correlate better with operation and management practices than with the presence of specific efficiency technologies.
- Programs based solely on behavior modification have achieved results comparable to the better technology-based programs.
- Emerging research on behavior and energy use is likely to produce helpful new guidance on driving behavior-based programs to still greater levels of success.
- Leading organizational energy efficiency strategies today use behavior change not as an afterthought but as a core element of their programs, and their results are driven as much by behavior change as by technology change.
- Behavior-based programs can be delivered by for-profit as well as nonprofit organizations.

As in most in-depths of energy efficiency as it plays out in the real world, we do not find a single "silver bullet" solution in behavioral approaches to saving energy and reducing carbon emissions. Rather, we observe many "silver BBs," which if fired in a well-aimed and sustained way, with learning effects incorporated on a continuous-improvement basis, can become the core of long-term efforts that ultimately lead to zero-energy building performance. This process takes time, but what leading organizations show is that the driving force is the continuing attention to detail on the behavioral side, combined with technology improvements at intervals, that leads to dramatic improvements in the long run.

BIBLIOGRAPHY

Alliance to Save Energy. (2009). *Students leading the way: Energy-saving success stories from Southern California.*

Dean, B., et al. (2006). *Using an occupant energy index for achieving zero-energy homes.* ICF International.

Economist Intelligence Unit. (2009). Countdown to Copenhagen: Government, business, and the battle against climate change. Survey published by *The Economist magazine*.

EIA. (2003). *Commercial buildings energy consumption survey.* U.S. Energy Information Administration. *www.eia.doe.gov/emeu/cbecs/cbecs2003/detailed_tables_2003/detailed_tables_2003.html#consumexpen03*.

EIA AEO (2010). *Annual Energy Outlook 2010.* U.S. Energy Information Administration. *www.eia.doe.gov/oiaf/aeo/index.html*.

IEA. (2007). *Mind the Gap—Quantifying Principal-Agent Problems in Energy Efficiency. http://www.iea.org/w/bookshop/add.aspx?id=324, http://www.iea.org/w/bookshop/b.aspx*.

Lutzenheiser, L., et al. (2009). *Behavioral assumptions underlying California residential sector energy efficiency programs.* California Institute for Energy and Environment.

Mills, E., et al. (2004). *The cost-effectiveness of commercial-buildings commissioning: A meta-analysis of energy and non-energy impacts in existing buildings and new construction in the United States.* Lawrence Berkeley National Laboratory.

PECI. (2007). *A retrocommissioning guide for building owners.* Portland, OR: Portland Energy Conservation, Inc. Funded by U.S. EPA.

Pew. (2010). Prindle, W.R., *From shop floor to top floor: Best business practices in energy efficiency.* The Pew Center for Global Climate Change.

Pink, D. (2009). *Drive: The surprising truth about what motivates us.* Riverhead Books.

Sullivan, M. (2009). *Behavioral assumptions underlying energy efficiency programs for businesses.* California Institute for Energy and Environment.

Synectics. (1983). *An evaluation of the institutional conservation program: Results of on-site analyses.* The Synectics Group, Inc. Prepared for the U.S. Department of Energy.

Chapter 12

Reinventing Industrial Energy Use in a Resource-Constrained World

Marilyn A. Brown*, Rodrigo Cortes-Lobos† and Matthew Cox**
Georgia Institute of Technology

Chapter Outline

1. INTRODUCTION

In an increasingly competitive and resource-constrained world, improving the energy efficiency of industry is essential for maintaining the viability of manufacturing, especially in a world economy where production is shifting to low-cost, less regulated, developing countries. With the rapid growth of manufacturing and energy-intensive production in expanding economies such as China, India, and Brazil, there is an opportunity for new facilities to deploy the latest energy-saving and carbon-reducing technologies and practices. In the U.S. and many other industrialized economies, there is a substantial existing infrastructure of older, less efficient manufacturing facilities that need to be upgraded. The variable energy intensity of manufacturing processes across countries reflects these differences and suggests the potential for further improvement (IEA, 2009b).

This chapter describes the progress made to date and the magnitude of the remaining opportunities, stemming both from broader use of current best practices and from a range of possible advances enabled by emerging technologies and innovations. It begins by focusing on the potential for improving energy efficiency in several major energy-consuming industries. After describing the principal barriers to deployment of energy-efficient technologies particularly in the U.S., it explores policy innovations that have successfully transformed industrial practices in five countries: the Netherlands, Denmark, India, Japan, and China. The goal is to identify lessons that can shift industry toward greater efficiency across the globe, thereby becoming part of the climate solution.

2. RECENT TRENDS IN ENERGY PRODUCTIVITY

Industry is the largest energy-consuming sector in most countries of the world, accounting for 37% of primary energy use worldwide (IPCC, 2007, p. 453). Large enterprises dominate most energy-intensive industries across the globe, especially in industrialized countries. In contrast, small- and medium-sized enterprises (SMEs) play greater roles in emerging economies. In India, for example, SMEs have significant shares in the metals, chemicals, food, and pulp and paper industries, and they account for 50% of China's asset value and 75% of its exports. These SMEs face special challenges when attempting to upgrade their energy efficiency due to limited technical and financial resources.

U.S. industrial energy use represents approximately one-third of total U.S. energy consumption and about 8% of global energy use. A majority of this is consumed by five energy-intensive industries: chemicals, oil refining, iron and steel, pulp and paper, and cement (Figure 1). Less energy-intensive industries include the manufacture and assembly of automobiles, appliances, electronics, textiles, food, beverages, and other products. Since energy is a smaller portion of their overall costs, historically these industries tend to pay less attention to finding ways to cut energy use. However, current evidence shows this may be changing with an increased focus on reducing carbon footprints.

The production of energy-intensive goods is likely to continue to increase worldwide, as populations and standards of living grow. However, an expanding proportion of this production is likely to be located in developing countries. For example, while the U.S. remains the world's largest producer of bulk chemicals and refined petroleum products, China has become the world's largest producer of steel, aluminum, and cement (IPCC, 2007, p. 451). Global competition for export markets, foreign investments, and raw materials is intensifying. The International Energy Agency (IEA) projects global industrial energy demand will more than double by 2030 (IEA, 2009c). Moreover, the IEA projects a convergence between developed and developing countries in terms of energy intensity by 2050 (IEA, 2003).

The significant shift to offshore manufacturing to meet the demands of U.S. markets means that the U.S. is actually responsible for approximately 5 quads of additional industrial energy use: products imported into the United States in

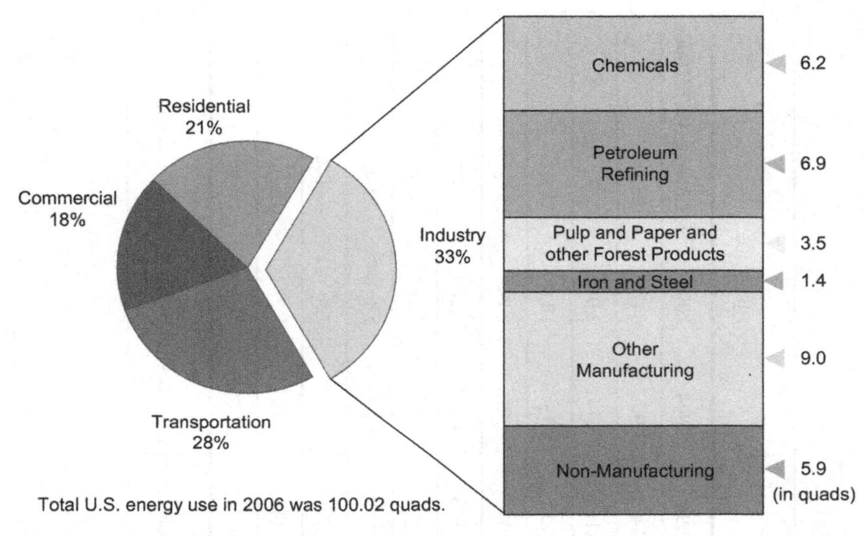

FIGURE 1 Energy Use in the US Industrial Sector in 2006 (Quadrillion Btu). *Source: EIA (2006)*

2002 had an embodied energy content of about 14 quads, far surpassing the 9 quads of embodied energy of U.S. exports (National Academies, 2009). Similar trends are occurring in Europe and Japan. The energy embodied in international supply chains is a contentious issue in discussions about carbon footprint metrics and responsibilities for addressing climate change.

U.S. manufacturing has undergone significant change in production and value added over the last several decades, modifying its strategies to improve market competitiveness and increase profit. On the one hand, the forest products industries have enlarged their production of pulp and paper by 38% and 66%, respectively, since 1985. This industry shows a clear strategy of specialization in industrial production with an orientation toward high value-added products, reducing the production of commodities with lower market profit. On the other hand, the iron and steel industries have shrunk their production by 35% and 33%, respectively. In spite of these swings in production, in general the manufacturing industry has sustained a similar overall level of energy consumption with only a slight reduction of 420 trillion BTU (or 1.9%) since 1978 (Table 1). The variation in trends across industries reflects shifts in composition in the economy, offshore movement of manufacturing, and advances in energy efficiency.

The U.S. manufacturing industry has more than 211,000 plants of which 76% are small firms (with 5 to 49 employees), 20% are medium-sized firms (50 to 249 employees), and only 4% correspond to large firms (more than 250 employees). With respect to energy consumption, the distribution has an inverse relationship, with large firms consuming 67% of the total industrial

TABLE 1 Total US Industrial Energy Use: 1978-2006
(Excluding Non-fuel Uses of Coal, Oil, and Natural Gas, in Trillion Btus)

Industry	1978	1985	1990	1995	2004	2006	Change 1978/2006
Wood Product Mfg. (321)	637.6	523.1	592.1	674.5	695.7	642.9	0.80%
Paper Mfg.(322)	2,384	2,662	3,161	3,168	3,141	2,902	22%
Printing and Related Support Activities (323)	161	147	195	219	233	183	13%
Petroleum and Coal Products Mfg. (324)	3,091	2,006	3,365	3,373	3,916	3,743	21%
Chemical Mfg. (325)	4,204	3,047	4,218	4,216	4,063	4,284	1.90%
Nonmetallic Mineral Product Mfg. (327)	1,617	1,165	1,289	1,235	1,322	1,466	−10%
Primary Metal Mfg. (331)	5,005	2,427	2,730	2,737	2,702	2,716	−46%
Fabricated Metal Product Mfg. (332)	664	576	645	747	718	708	6.60%
Other Manufacturing (339)	4,549	4,220	4,584	5,345	5,301	5,252	0%
Total (Manufacturing)	22,313	16,773	20,781	21,713	22,092	21,893	−1.90%

Note: NAICS codes are presented in parenthesis.
Source: U.S. Department of Energy, U.S. Energy Intensity Indicators, Trend Data, Industrial Sector, available at: http://intensityindicators.pnl.gov/trend_data.stm.

energy consumption, followed by medium-sized firms with 26%, and small firms with only 7% of the total industrial energy consumption (see Figure 2.)

The increase in production of some manufacturing industries, such as pulp and paper, chemicals, and cement has not been accompanied by a proportionate increase in energy consumption. As a result, many of these expanding industries have reduced their energy intensity (measured as total energy use per value of shipment). This improvement in energy productivity is explained by advances in production technologies and better operational practices, which were particularly important following the oil crises in the 1970s.

The petroleum and coal products manufacturing industry experienced a particularly significant improvement in energy intensity with a reduction of 60% in 2004 relative to 1977, followed by chemical manufacturing with a 42% reduction, plastic and rubber with 31%, nonmetallic minerals with 25%, and primary metals with 23%. Paper manufacturing was the only industry of this group that did not decrease in energy intensity (Figure 3).

Advances in engineering, materials, thermodynamics, sensors and controls, and information technologies, among others, offer the potential to transform industrial processes in response to emerging climate change policies. As the era of cheap energy comes to an end, successful manufacturers will increasingly focus on technological innovations that allow for order-of-magnitude

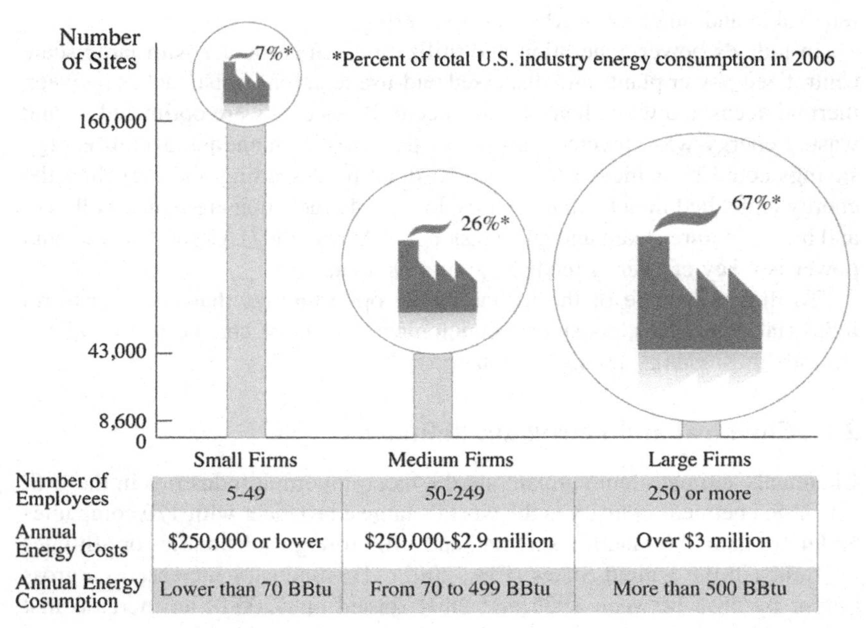

	Small Firms	Medium Firms	Large Firms
Number of Employees	5-49	50-249	250 or more
Annual Energy Costs	$250,000 or lower	$250,000-$2.9 million	Over $3 million
Annual Energy Cosumption	Lower than 70 BBtu	From 70 to 499 BBtu	More than 500 BBtu

FIGURE 2 U.S. Industrial Classification Respect to Energy Consumption. *Source: MECS 2006 and U.S. CENSUS Bureau office 2007*

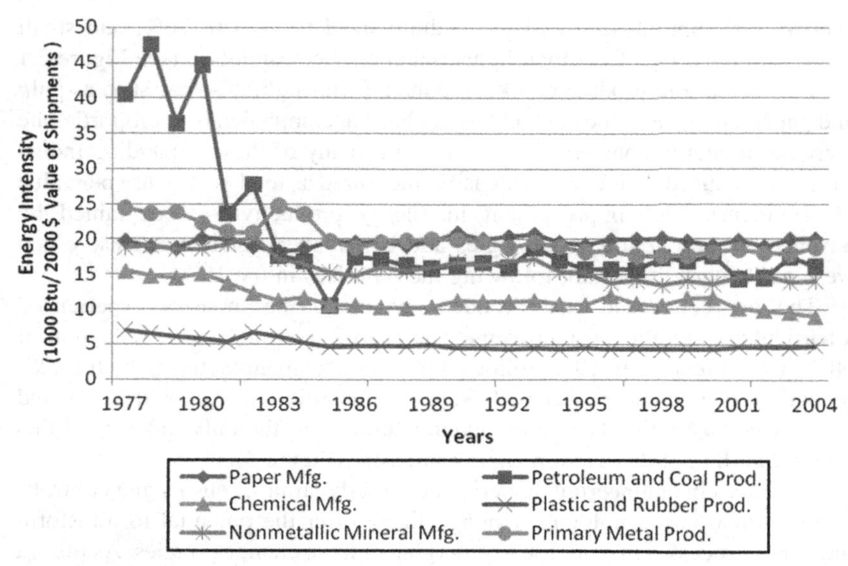

FIGURE 3 Changes in Energy Intensity in Six Key US Industries (1977-2004). *Source: DOE 2010, U.S. Energy Intensity Indicators. Trend data*

reductions in energy consumption and the substitution of fossil fuels for renewable and other low-carbon energy resources.

In today's power generation and utilization infrastructure with large-scale centralized power plants and dispersed end-use locations, mismatches between thermal needs and waste heat streams occur. If systems were optimized so that wasted energy was recycled into productive uses, tremendous overall energy savings could be achieved. This can be done by cascading and recycling the energy embodied in hot exhaust gases, low-grade fuels that are typically flared, and high-pressure steam and gas (Casten and Ayres, 2007). Combined heat and power is a key efficiency technology in this area.

To illustrate some of the technological opportunities that may transform industrial complexes, consider technological drivers of change in five of the nation's most energy-intensive industries.

2.1. Chemical and Petroleum Refining

Chemicals and petroleum are among the most important industries in the U.S. The U.S. chemical industry is the world's largest producer with 170 companies and more than 2800 facilities abroad and 1700 foreign subsidiaries or affiliates operating in the United States (EIA, 2001). This industry increased its gross output by 58% between 1985 and 2004 (based on $2000), an increase that occurred in conjunction with a 15% reduction in electricity consumption and a 14% drop in energy intensity (Figure 4). Thus, its drop in energy intensity

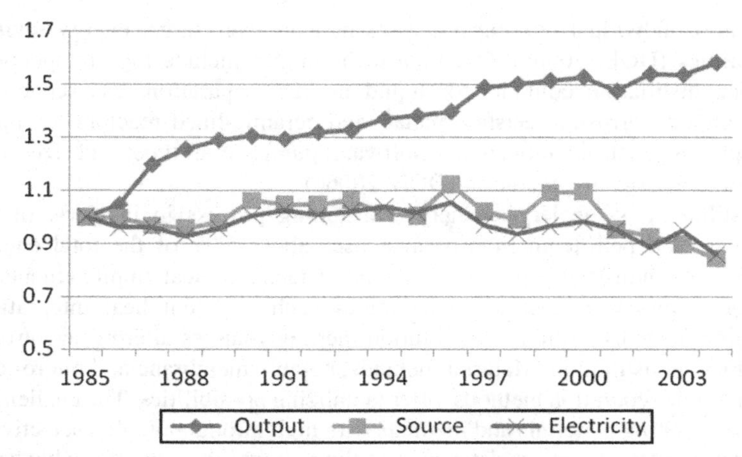

FIGURE 4 Chemical Products Energy Consumption and Intensity Indexes (1985=1). *Source: DOE 2010, U.S. Energy Intensity Indicators. Trend data*

following the Arab oil embargo of 1973–74 has been followed by declining energy intensity in more recent years.

The United States is also the largest producer of refined petroleum products in the world, with 25% of global production and 163 operating refineries. This industry's gross output increased by 27% between the years 1985 and 2004; at the same time it increased its electricity consumption by 40% and its energy intensity by 53% (Figure 5).

Benchmarking data indicate that most U.S. petroleum refineries can economically improve energy efficiency by 10–20% (LBNL, 2005), and

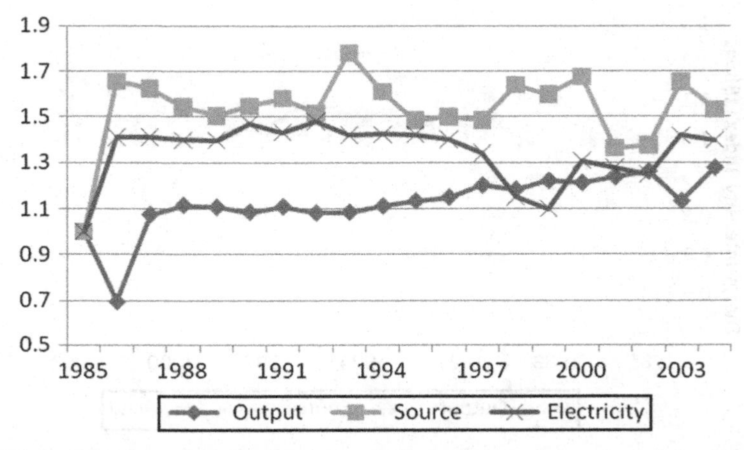

FIGURE 5 Petroleum and Coal Products Energy Consumption and Intensity Indexes (1985=1). *Source: DOE 2010, U.S. Energy Intensity Indicators. Trend data*

analysis of individual refining processes indicate even larger energy savings possibilities (DOE, 2006c). Common technologies include high-temperature reactors, distillation columns for liquid mixture separation, gas separation technologies, corrosion-resistant metal- and ceramic-lined reactors, sophisticated process control hardware and software, pumps of all types and sizes, and more efficient steam generation (DOE, 2006c).

Distillation is the largest energy-consuming process in industry. In the chemicals and petroleum industries, it uses about 53% of the total energy required for industrial separations. Potential technological improvements to distillation processes include technologies such as latent heat integration, multiple-effect distillation, and solution-thermodynamics-altering azeotropic or extractive distillation. Material methods, notably membrane and micro- and nano-particle separation methods, offer tantalizing possibilities. The challenges are in developing materials and methods with high throughput, high selectivity, low energy requirements, resistance to fouling, durability, and affordable costs (National Academies, 2009).

Membrane separation is the most widely applicable of all technologies for reducing energy of separation processes in the petroleum, chemical, and forest products industries (Nenoff et al., 2006; Banerjee et al., 2008). Zeolites are one of the kinds of materials to achieve separations that would not require direct heat. However the zeolite approach leaves the capturing material with the target material attached, so some removal process is required. Membranes may be made of organic materials for relatively low-temperature processes, inorganic materials such as ceramics for high temperature use, or a combination of the two. Membranes are currently used successfully to separate light hydrocarbons as well as hydrogen from gas streams; the separated light hydrocarbons have uses with values considerably higher than that of fuel.

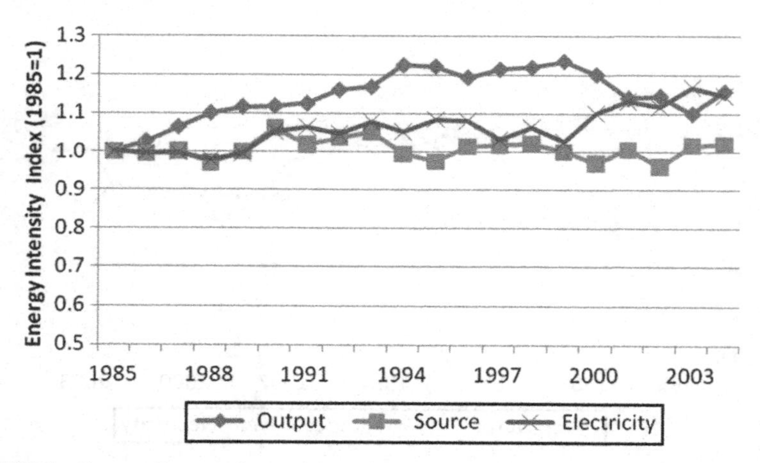

FIGURE 6 Energy Consumption and Intensity Indexes (1985=1). *Source: DOE 2010, U.S. Energy Intensity Indicators. Trend data*

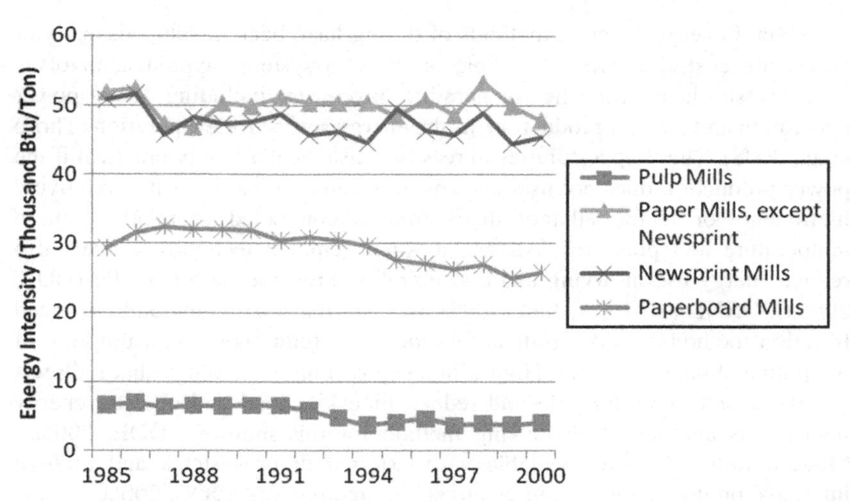

FIGURE 7 Pulp and Paper Industry Energy Intensity (1985-2000). *Source: DOE 2010, U.S. Energy Intensity Indicators. Trend data*

2.2. Pulp and Paper Industry

The U.S. pulp and paper industry is a global leader with 34% of the world's pulp production and 29% of the world's paper and paperboard production (Miller Freeman, 1998). The industry increased gross output between the years 1985 and 2004 by 15.6% with an increase in electricity consumption of 14.6% compared to 1985 with little change in energy intensity (Figure 6). Nevertheless, a much lower energy intensity results when the value of shipments is replaced by the total tonnage of production. Measured in this way, the energy intensity for pulp and paper decreases by 15%.

The principal products of this industry are pulp, paper, newsprint, and paperboard. Mills for each of these products have shown important improvements in energy use, especially the pulp mills decrease of 39% in energy intensity in the period 1985–2000, followed by paperboard mills with 23%, and paper mills and newsprint with 9% and 11%, respectively (Figure 7). Recycling also conserves a great deal of energy and represented roughly 40% of total paper production in the U.S. in 2005 (Houser et al., 2008). This percentage of recycling is lower than the 69% waste paper pulp in the U.K. (Confederation of Paper Industry, 2009). U.S. paper recycling has enough capacity to double of its current levels.[1]

1. Paper recycling has some limits on recovery due to loss of quality when paper is used for permanent records, destroyed, or contaminated. To improve the quality, virgin wood fibers are needed to replace those that are damaged. Taiwan has the maximum rate of paper recycling of 90% that could be considered as the current technical limit of recycling for the U.S. paper industry.

Several energy-efficient methods of drying have been developed, many of which are cost-effective today. One of these, a systems approach, involves using waste heat from heat-generating processes including from power generation and ethanol production, as the energy source for evaporation (Thorp et al., 2008). These opportunities to recycle waste heat are only practical if the power production does not use condensing turbines—that is, if it is relatively inefficient—or if the ethanol distillation is conducted at relatively high temperature and pressures. Advanced water removal technologies can also reduce energy use in drying and concentration processes substantially (DOE, 2005a). ORNL and BCA, Inc. (2005) estimate that membrane and advanced filtration methods could significantly reduce the total energy consumption of the pulp and paper industry. High-efficiency pulping technology that redirects green liquor to pretreat pulp and reduce lime kiln load and digester energy intensity is another energy-saving method for this industry (DOE, 2005a). Modern lime kilns are available with external dryer systems and modern internals, product coolers, and electrostatic precipitators (DOE, 2006c).

Kraft processing is a prominent way to produce wood pulp. In most Kraft mills today, the black liquor produced from delignifying wood chips is burned in a large recovery boiler. Because of its high water content, the combustion of black liquor is inefficient, and the possibility of electricity production from secondary steam production is limited by the steam's low pressures. Gasification of black liquor not only allows efficient combustion, but also enables the use of a gas turbine or combined cycle process with a high electrical efficiency, thereby offering the potential for increasing the production of electricity within pulp mills. The surplus of energy from the pulp process also allows for the possible production of useful heat, fuels, and chemicals—that is, the operation of "biorefineries" (Worrell et al., 2004, pp. 22–23).

There are many novel sensors for a wide range of applications. In the papermaking industry, for example a fiber optic sensor measures paper basis weight to improve wet-end control in papermaking and make paper of a uniform basis weight and higher quality. It minimizes energy requirements. Another noncontacting laser sensor measures shear strength and bending stiffness. By measuring the rate of propagation of ultrasonic shock waves in the paper, this device could save the U.S. paper industry approximately $200 million annually in energy costs.[2]

2.3. Iron and Steel

The primary metal industry is composed principally of iron and steel and aluminum production. This industry has shown an impressive reduction of 46% in energy consumption during the last 30 years. This reduction has been a result

2. See *www.physorg.com/news4221.html*

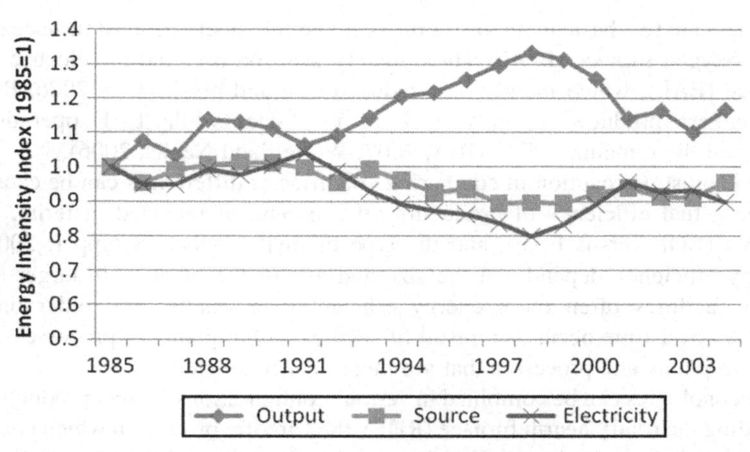

FIGURE 8 Primary Metal Energy Consumption and Intensity Indexes (1985=1). *Source: DOE 2010, U.S. Energy Intensity Indicators. Trend data*

of a 17% reduction in energy intensity—measured in terms of energy used per value of shipment—between 1985 and 2004, with an 11% reduction in electricity consumption (Figure 8). Recycling is widely utilized in this sector, with steel reaching rates of 83% in 2008. This contributes to declines in energy use in the sector (Steel Recycling Institute, 2009).

Figure 9 shows the 54% reduction in energy intensity in terms of the energy consumed per ton of iron and steel produced in the U.S. Given the nearly complete penetration of recycled resources, other advances will be needed to enable improvements of a similar magnitude in the future.

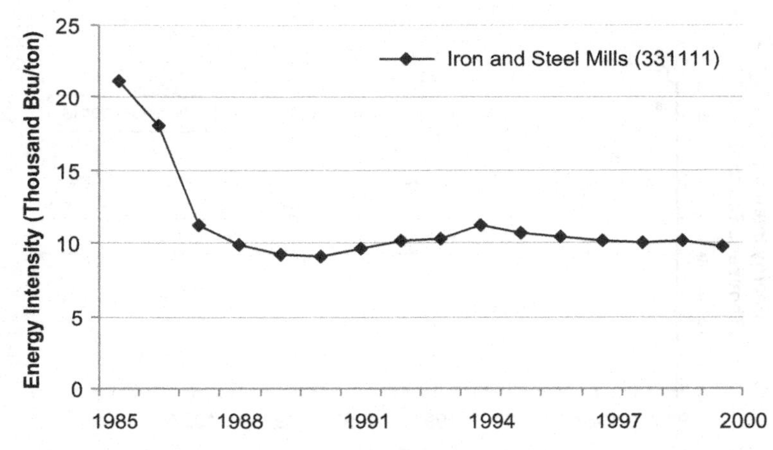

FIGURE 9 Iron and Steel Mills Energy Intensity (1985-2000). *Source: DOE 2010, U.S. Energy Intensity Indicators. Trend data*

There are two basic methods for producing crude steel: the blast furnace and basic oxygen furnace (BOF), which mainly use iron ore, and the electric arc furnace (EAF), which mainly uses reduced iron and pig iron. In 2006, BOF steelmakers produced roughly 43% of raw steel while EAF operations produced the remaining 57% (IEA, 2007; Worrell and Neelis, 2006).

One must use caution in comparing countries as differences can be caused by the actual efficiency of production, the amount of recycled material, the process (BOF versus EAF), and the type of final product (Schipper, 2004). Energy efficiency depends on the size and age of the plant, with larger and newer facilities often more energy efficient than smaller and older ones. Changes over time occur as a result of savings within plants or processes and shifts to plants and processes that are more energy efficient.

Technologies can be combined in various configurations in steel production, including the rotary hearth furnace (RHF), the Circofer process in which coal is charred and ore is partly metallized in a single first step and then completed in a bubbling second step, and the RHF with a submerged arc furnace; the energy consequences of these alternatives are unclear (Fruehan, 2008). Several revolutionary new steelmaking technologies are also under development, such as the use of hydrogen as an iron ore reductant or furnace fuel, and electrolytic or biometallurgical-based iron and steel production. Success with these could significantly reduce the carbon footprint of these industries.

2.4. Cement Industry

The U.S. cement industry consists of 39 companies that operate 118 cement plants in 38 states. While its production levels have grown since 1985, the industry's energy intensity declined by 35% between 1985 and 2000 (Figure 10).

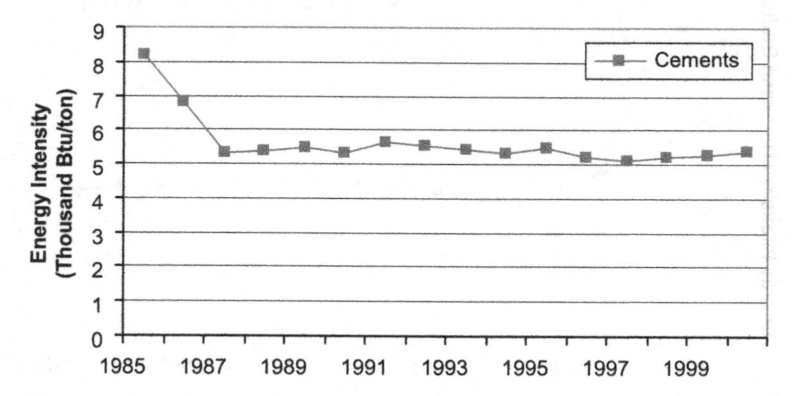

FIGURE 10 Cement Energy Intensity (1985-2000). *Source: DOE 2010, U.S. Energy Intensity Indicators. Trend data*

The cement manufacturing process involves three components: the mining and preparation of inputs; the chemical reactions that produce clinker; and the grinding of clinker with other additives to produce cement. The feed for older kilns is a slurry of inputs, the wet kiln process, while large new plants mix dry materials for introduction to the kiln. Energy use varies with the process and characteristics of the plant, but in general about 90% of the energy use, and all of the fuel use, occurs in the manufacture of clinker in the kiln. The chemical process that converts limestone to lime, produces roughly the same amount of carbon dioxide gas as that generated by the energy used in its production for coal-fired kilns. Technologies that allow production of cement with a lower per-ton share of clinker thus yield multiple benefits.

Upgrading a kiln from wet to dry, and from a long dry kiln to a preheater, precalciner kiln results in major energy efficiency gains but for a price that requires a payback period of at least ten years. Worrell et al. (2004) conclude that these upgrades are attractive only when an old kiln needs to be replaced. More incremental upgrades could yield commercially attractive benefits including advanced control systems, combustion improvements, indirect firing, and optimization of components such as the heat shell. While opportunities vary with specific plants, the combination of these activities appears to yield an improvement in energy use on the order of 10%. Recovering heat from the cooling stage also yields substantial savings. If the heat is used for power generation, it can save up to half of the electricity used in the clinker process. However, taking full advantage of the heat recovery savings may require other major upgrades (National Academies, 2009).

Changing the chemistry of cement to reduce the need for calcination can decrease the high share of clinker that characterizes U.S. production. Options for blended cements include fly ash and steel slag. Fly ash may be particularly promising as it is a coal combustion byproduct that can be reused in many different contexts, such as construction and pavement. Worrell et al. (2004) identify potential energy savings of up to 20% from deployment of blended cement technologies, and larger carbon dioxide emission reductions. Advanced technologies with potential to further improve energy efficiency and emissions include carbon capture and storage technology, fluidized bed kilns, advanced comminution technologies, and the substitution of mineral polymers for clinker (Worrell et al., 2004; Battelle, 2002).

3. POTENTIAL ENERGY SAVINGS IN ENERGY-INTENSIVE INDUSTRIES

Numerous studies have shown high energy-savings potential in energy-intensive U.S. industries. A recent study by the National Academies (2009) compiled these studies for five industries for 2020. The results are summarized in Figure 11.

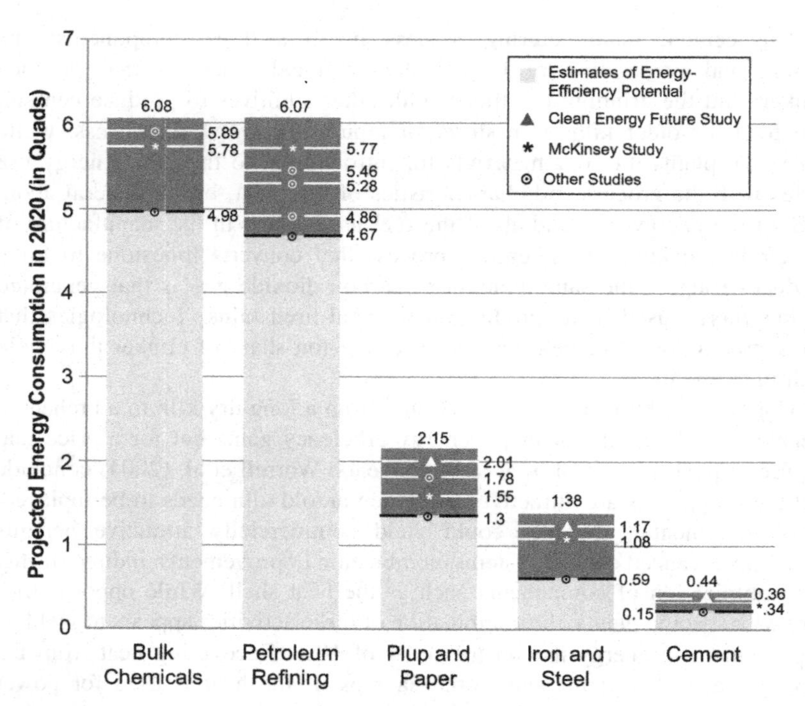

FIGURE 11 Potential for Improving Energy Efficiency in 5 Key Industries in 2020. *Source: Authors, based on National Academies (2009)*

The chemical manufacturing industry was analyzed by three studies. The estimates of energy savings in 2020 in the U.S. chemical industry are wide ranging from 3.1% savings estimated by NREL (2002), 5% of savings presented by McKinsey and Company (2008), and more than 18% of saving estimated by Energetics Inc. (2007).

The petroleum refining industry's energy savings in 2020 are presented in three recent studies. The lowest estimate of 5% of energy savings is provided by McKinsey and Company (2008). The intermediate range of savings, between 12% and 24%, was published in a study by LBNL (2005). The highest estimate is a range of 28–65% of savings published in a DOE (2006c) report.

The pulp and paper industry also represents a significant potential for energy savings through its process improvement. Estimates of achievable energy savings range from 6.1% in the *Scenarios for a Clean Energy Future* (CEF) study to 37% of energy savings estimated by the study by Jacobs and IPST (DOE 2006c).

The iron and steel industries also offer an important opportunity for energy savings. McKinsey and Company (2008) estimated 22% of energy savings potential in 2020. The AISI study (2005) provided a significantly higher level of energy savings potential of 58% of current energy use.

Finally, three studies analyzed energy savings potential in the cement industry. The lower estimation of energy savings potential is presented by the CEF Study (Brown and Levine, 2001) with 19% of saving in 2020, followed by McKinsey and Company (2008) with 23% savings. The highest potential of savings is presented in the study by Worrell et al. (2004), with 67% of energy saving in 2020.

By applying these percentage savings potentials to the AEO business-as-usual forecast of future industrial energy consumption in the U.S., it is possible to compare and contrast the studies in a common framework (Figure 11).

If similar efforts were implemented worldwide, particularly in the rapidly expanding economies of Brazil, Russia, India, and China—the BRIC countries—these energy savings could be multiplied several times. Of course, greenfield industrial complexes start with the advantage of more advanced equipment such as dry kilns, electric arc furnaces, and lime kilns with external dryer systems.

4. BARRIERS TO TECHNOLOGICAL INNOVATION IN INDUSTRY

Energy efficiency tends to thrive in a culture of innovation, where companies are committed to progressive thinking (McKinsey and Company, 2008, p. 8). The broader application of high-efficiency industrial technologies, on the other hand, is impeded by a range of technical, corporate, regulatory, and workforce barriers. These include:

- Technical risks
- Lack of specialized knowledge
- High transaction costs for obtaining reliable information
- Relatively high initial costs
- Lack of access to capital
- Unfavorable fiscal policies
- Unfavorable regulations
- External benefits and costs

Companies must consider the *technical risks* of adopting a new industrial technology. When energy costs are low, industry has little incentive to make investments in efficiency measures, particularly if there are uncertainties about the benefits and impacts of novel approaches can be significant. Small technology changes, particularly in large integrated process plants, can lead to major changes in process and product performance. In today's manufacturing environment with 24/7 operations, reliability and operational risks represent major concerns for industry. The need to keep a process running in a predictable fashion, for example, often overrides the inclination to replace equipment with a more efficient model. An historic example is provided by the American steel industry, where companies continued to build open hearth furnaces after World

War II, despite the demonstration of superior basic oxygen furnaces. The old technology was familiar and the new technology was considered to be a risk (National Academies, 2009). A more modern and streamlined version of the vetting process is used by the Dow Corporation, which has a group established to present energy-efficiency upgrades for a plant. These "tech centers" work with efficiency experts on staff to assess the quality and reliability of proposed plant upgrades. They then work with production managers and jointly make an implementation decision about proposed upgrades as described in Prindle (2010).

Lack of specialized knowledge of energy engineering and energy management is another impediment to adoption. Industrial managers can be overwhelmed by the numerous products and programs that tout energy efficiency, and without in-house energy experts, may find it risky to rely on third-party information to guide investments. For example, plant managers at the United Corporation Technologies (UTC) find it difficult to rely solely on facility experts and has created a special energy-focused team to work directly with its 300 facilities to identify savings opportunities (Prindle, 2010). To make optimal energy-efficiency decisions, plant managers must have working knowledge of a massive number of technologies (McKinsey and Company, 2008). External expertise is available, but manufacturers generally do not support third-party installers or consultants such as energy services companies (ESCOs) and utilities (CCCSTI, 2009; Prindle, 2010). Energy consulting firms often lack the industry-specific knowledge to provide accurate energy and operational cost assessments, and many industrial operations don't have in-house engineering resources to sort through or analyze the information.

This barrier is exacerbated by *high transaction costs for obtaining reliable information* (Worrell and Biermans, 2005). Researching new technologies and collecting other relevant information consumes time and resources, especially for small firms, and many industries prefer to expend human and financial capital on other investment priorities. Overall, corporate decision-makers are predisposed toward investments that result in more output. Although the reduction of costs through investments in efficiency may have the same impact as increases in productivity on overall profit, there is a tendency for investments to be focused on increasing revenue as opposed to decreasing costs. In some cases, industrial managers and decision-makers are simply not aware of energy efficiency opportunities and low-cost ways to implement them. In others, they don't believe they have enough time or money to research new technologies. In more progressive companies, divisions are established to root out these savings (Prindle, 2010).

Relatively high initial costs for industrial energy-efficiency improvements can be an impediment to investments. New energy-efficient technologies often have longer payback periods than traditional equipment and represent a greater financial risk since there is significant uncertainty about future energy prices. Senior managers also often postpone capital investment and refurbishment because they are uncertain about the longevity of their companies (McKinsey and Company, 2008, p. 9). The global economic downturn beginning in 2008 has exacerbated concerns about enduring profitability.

The *lack of access to capital* is one of the most significant barriers to energy efficiency improvements in industry. Projects to improve energy efficiency have to compete for financial and technical resources against projects that achieve other company goals and against more familiar technologies. A large share of capital goes toward meeting government standards for health, safety, security, and emissions; the remaining discretionary capital is then allocated to other goals such as product improvement, production expansion, and (finally) cost savings such as energy efficiency. Although, in theory, firms might be expected to borrow capital any time a profitable investment opportunity presents itself, in practice firms often ration capital—that is, they impose internal limits on capital investment (Canepa and Stoneman, 2004). As a result, companies impose high ROI requirements on efficiency investments (CCCSTI, 2009). In addition, if the technology involved is new to the market in question, even if it is well-demonstrated elsewhere, the problem of raising capital may be further complicated.

In the United States, existing *fiscal policies are often unfavorable to investments in end-use efficiency.* The current federal tax code discourages capital investments in general, as opposed to direct expensing of energy costs. More specifically, tax credits designed to encourage technology adoption are limited by alternative minimum tax rules, tax credit ceilings, and limited tax credit carryover to following years; these limitations prevent the credits from being used to their full potential by qualified companies. Furthermore, outdated tax depreciation rules require firms to depreciate energy efficiency investments over a longer period of time than many other investments (Brown and Chandler, 2008). Significant utility company interconnection fees, overly layered permitting processes, and lack of net-metering policies provide disincentives for manufacturing plants to capture waste energy for the generation of electricity in combined heat and power systems (CCCSTI, 2009). However, in response to increasing peak demand and growing strain on existing capacity, utilities are pursuing demand response and energy efficiency strategies with industry.

Existing *regulations can also be unfavorable* to industrial energy efficiency. EPA's New Source Review (NSR) Program can also hinder energy efficiency improvements at industrial facilities. As part of the 1977 Clean Air Act Amendments,[3] Congress established the NSR program and modified it in the 1990 Amendments, but exempted old coal plants and industrial facilities from the New Source Performance Standards (NSPS) to be set. NSPS standards are intended to promote use of the best air pollution control technologies, taking

3. P.L. 95–95; 91 Stat. 685.

into account the cost of such technology and any other non-air quality, health, and environmental impact and energy requirements. However, investment in an upgrade could trigger an NSR, and the threat of such a review has prevented many upgrades from occurring. NSR thus imposes pollution controls where they are least needed and artificially inflates the value of the dirtiest plants. Altogether, these effects have led some critics to question whether the NSR program and the NSPS have resulted in higher levels of pollution than would have occurred in the absence of regulation (Brown and Chandler, 2008; List, 2004).

External benefits and costs are difficult to value and inhibit reduction of greenhouse gas (GHG) emissions by industrial plant managers. In general, companies invest in emissions reduction or other environmental improvements only when the investments are offset by lower energy or raw material costs or other cost benefits. Suppliers, who typically introduce innovations to the industrial sector, are often reluctant to expend resources in developing GHG emissions-reducing technologies without an assured market. Policy uncertainty and the absence of an international climate agreement is also leading to competitiveness concerns and reduced cooperation across firms.

Given all of these inhibitors to reinventing industrial energy use, can energy and/or tax policies influence the future course of industrial innovation? Would restrictions on greenhouse gas emissions become a driver for change? Evidence from other countries is encouraging, as is the experience of some U.S. federal programs and individual state initiatives.

5. POLICY DRIVERS OF CHANGE

A variety of approaches have been utilized globally to promote industrial energy efficiency. This section describes some of the lessons learned by the Netherlands, Denmark, India, Japan, and China, and concludes with a summary of the policies utilized in the United States. While many of these nations have similar policies in place, their differences and points of success and policy innovation are highlighted here. The trajectory of these countries' energy intensity from 1980 through 2005 suggests an improvement in energy efficiency overall for each country, punctuated by periodic slippages (Figure 12). Between 1980 and 2005, China underwent a marked increase in energy efficiency. Still, even with this massive improvement, China today is only slightly more efficient than the United States was in 1980, and has recently undergone an increase in energy intensity. The graph suggests a more gradual improvement across the other five countries.

The Netherlands has taken a proactive stance on industrial energy efficiency, beginning with their Long Term Agreements on Energy Efficiency with industry beginning in 1992. These agreements were established through an understanding by industry that the government is closely observing energy consumption and will not initiate strong regulations so long as industry meets the targets (Nuijen and Booij, 2002). This program had a goal of increasing

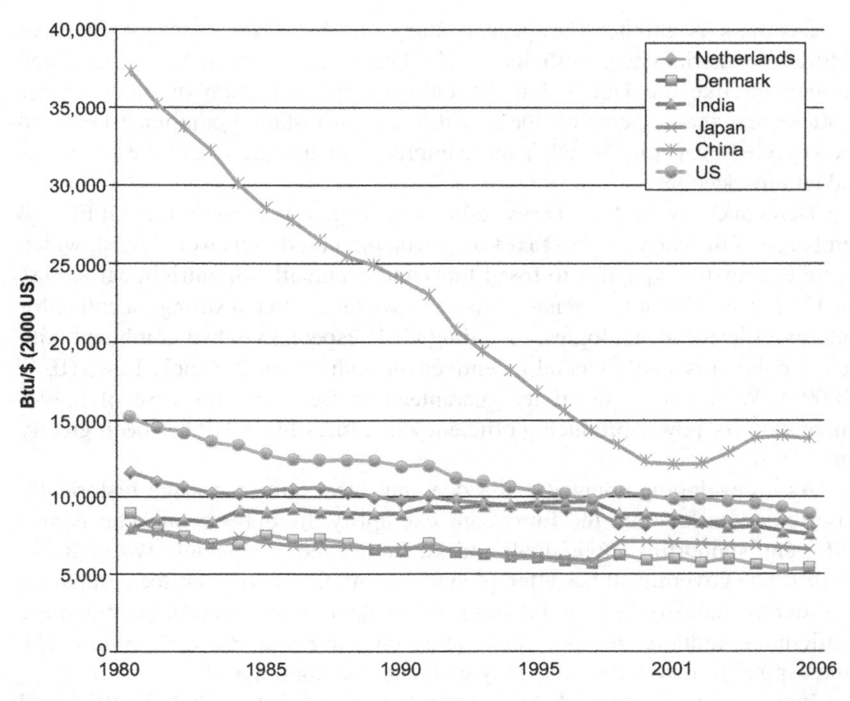

FIGURE 12 Energy Intensity Trends in Selected Countries: 1980-2006. *Source: Energy Information Administration, International Energy Annual*

energy efficiency by 20% over a 1989 baseline by 2000. The results were better than anticipated, achieving a 22% savings in affected industries, which represent 90% of industrial energy consumption in the Netherlands. The country experienced annual net savings of roughly €700 million annually, increasing the competitiveness of Dutch-produced goods in the global market.

The Netherlands established a second phase of the Long Term Agreements in 2000 to operate until 2012. In this phase, the most energy intensive industries will be benchmarked to comparable industries worldwide. The affected industries must be best in class in energy efficiency, and in return, the government will not implement additional stringent climate change policies. Curiously, analysis of the benchmarking mechanisms suggests that estimated energy savings will be smaller than under a continuation of the first phase of the Long Term Agreements (Phylipsen et al., 2002). This is due to a change in the policy from continued energy savings in the original Long Term Agreements to a benchmarking standard in the second phase. With the expiration of the second phase in 2012, it remains to be seen whether the initial increase in efficiency gains will be maintained over the entire period. Other industries remain covered under the goals of the Long Term Agreements.

Denmark is another European country that has taken extensive steps to address industrial energy efficiency. The Danish government has a negotiated agreement like the Dutch, but the unique implementation of other energy policies has made Denmark the world leader in installed combined heat and power (CHP) capacity, which is more impressive when the size of the country is taken into account.

Denmark's push for energy efficiency began following the OPEC oil embargo of the early 1970s. Taxes on petroleum based fuels were levied, which were eventually expanded to fossil fuels and eventually an outright carbon tax in 1992. The constant presence of these taxes has created a strong incentive for energy efficient technologies, including CHP, especially when combined with some regulatory and financial incentives through the Heat Supply Laws (IEA, 2009a). With grid connectivity guaranteed in Denmark, the ease of implementation for power producing efficiency measures like CHP has been greatly increased.

As a developing country, India does not have quite the same historically coordinated effort that the Europeans exemplify. Its industry makeup is also different, supporting more small and medium-sized companies (World Bank, 2008). The government has attempted to incentivize energy service companies to enter the industrial sector, but has had a difficult time doing so. Despite these difficulties, India is currently less energy intensive than the U.S. (Figure 11), and aspires to match the efficiency of Japan (Lamont, 2009).

India's newest approach to the problem is innovative. It has introduced an energy efficiency trading program designed to reduce energy intensity by 5% a year through certificate trading. It is expected this market will be worth $15 billion and will cover nine sectors by 2015 (Lamont, 2009). This approach is very similar to other markets for efficiency credits, but India mandates the reductions and the program is designed like a cap-and-trade program. This is a unique approach for a developing country, with the expected outcome of more rapid deployment of efficient technologies throughout the Indian economy.

The two oil crises of the 1970s also spurred the government of Japan to start actively pursuing industrial energy efficiency policies. By 1991, Japan had achieved a 35% improvement in energy efficiency, but started to see its energy intensity rise. Japan implemented a new set of policies in 1993 to further energy efficiency throughout industry and its economy in general. Tax credits for small and medium-sized industry were established, as were a large number of low-interest loans, which covered both the purchase of highly efficient equipment and cogeneration installations (Sato, 2000).

In 2006, Japan updated its efficiency goals in response to rising energy prices and the anticipation of increasing global energy demand. The New National Energy Strategy featured five focus areas for energy, including energy efficiency. With the Energy Conservation Frontrunner Plan, the goal of improving energy efficiency 30% by 2030 was established. To achieve this ambitious goal, Japan's Ministry of Economy, Trade and Industry mandates energy management plans for industry, the appointment of a certified energy manager for each business, and the introduction of benchmarking for industrial sectors (Energy Conservation Center, Japan, 2009). Future progress is expected to come from a number of bills addressing climate change currently working through the Japanese government, with Tokyo launching Asia's first mandatory carbon trading scheme in early April, 2010 (Soble, 2010)

From 1980 through 2000, China experienced a reduction in national energy intensity of 65% (Zhang, 2003). These reductions were the result of process and technological changes, as well as structural shifts throughout Chinese industry. Rapidly developing countries typically see an increase in energy intensity; China was able to buck this trend through a series of policy reforms allocating capital toward energy efficiency and developing energy service conservation and energy management centers, which act similarly to energy service companies (Wang et al., 1995; Sinton et al., 1999). China intended to continue this trend, with goals and mandates in the Energy Conservation Law of 1997 (ECL) and the 10th Five-Year Plan.

However, China has recently faced difficulties with these goals. The early 2000s saw energy consumption outpace GDP growth, and thus saw an *increase* in energy intensity for the first time in decades. Part of this increase was almost certainly driven by difficulties in implementation of the ECL itself, which required provincial energy plans that were slow to develop and difficult to enforce (Wang, 1999).

Noting the deteriorating conditions, the Chinese government announced a mandatory reduction in energy intensity of 20% by 2010 in late 2005. Initial responses were not sufficient to reverse the trend, inspiring new policies and strategies to meet the mandate (Lin et al., 2006). The ECL was revised, tax policy was modified for export products, tax credits for efficiency investments were granted, and numerous buildings and appliance policies came into effect, being adopted in the 11th Five-Year Plan. The Top-1000 Energy Consuming Enterprises program has promoted energy-efficiency throughout large-sized industry.

It is anticipated that these top energy-consuming businesses will contribute 25% of the overall efficiency gains required by the 11th Five-Year Plan, and additional businesses are being added to the program. The end-result of these policies has placed China on a path toward reaching its mandates and reducing energy intensity once again (Zhou et al., 2009). Even so, with highly energy consumptive industries including steel, cement, and so on, experiencing increasing demand for their products as the global economy recovers from the recent recession, continuing the progress may prove difficult, and increases in overall consumption are virtually guaranteed. (See also this book's Chapter 15, "Why China Matters," by Buijs.)

Just before the December 2009 Copenhagen Summit began, China announced a commitment to reduce the carbon intensity of its economy to 40–45% below 2005 by 2020. This will require a 4% reduction in GHG emissions each year from projected emissions increases, at the same time as China's economy could grow at an annual rate of 8% or higher. Achieving such a goal may involve expanding the scope of major efficiency improvements to China's smaller industrial facilities in addition to potentially imposing new regulations and continuing to close inefficient plants (Friedman, 2009). Others have estimated that the 40% goal represents the business-as-usual case for China, and will be easier to meet with faster economic growth. Both the 40% and 45% emissions trajectories still push global emissions beyond the IEA 450 ppm CO_2 scenario, so even if China is successful in achieving its own goals, the world would need greater efforts to stay below 450 ppm CO_2 (Seligsohn and Levin, 2010).

The policies pursued by different nations illustrate the variety of approaches used to promote industrial energy efficiency. In the United States, the implementation of federal activities is distributed among federal agencies, with more than a dozen involved in the administration of 72 currently funded and active deployment programs working on energy efficiency in industry (CCCSTI, 2009).

Reflecting the importance of informed decision-making, remedying a lack of specialized knowledge and addressing incomplete and imperfect knowledge barriers are important policy priorities in the U.S. context. As a result, "labeling and information dissemination" are the most common type of deployment program targeting industrial energy efficiency (Figure 13).

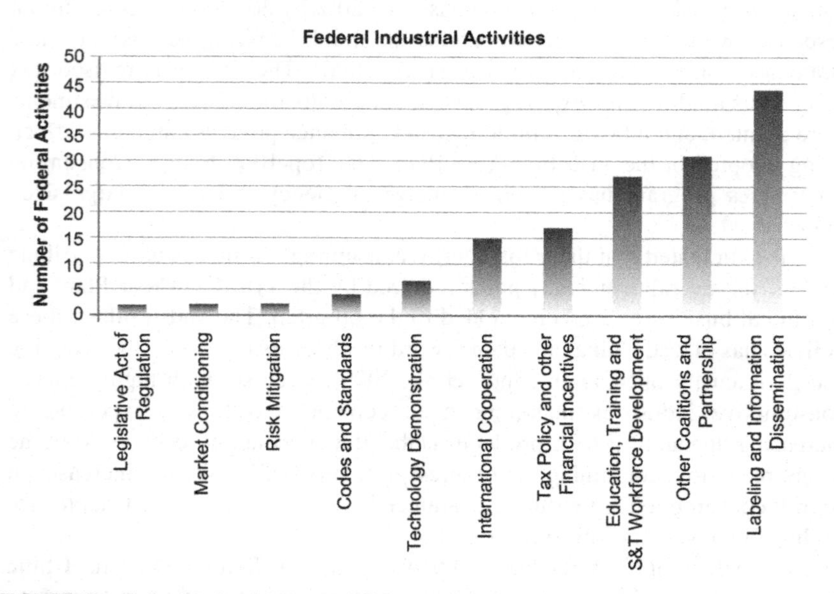

FIGURE 13 Federal Industrial Energy Efficiency Policies and Measures Operating in the United States, in 2009. *Source: CCCSTI Energetics Deployment Database, September 2009*

In the U.S., the focus has been significantly less driven by regulation. Instead, there are many public-private partnerships with industry. For example, programs like Save Energy Now, administered through the Department of Energy's Industrial Technologies Program, work with large industrial partners in energy-intensive industries to identify areas of significant efficiency gains. Save Energy Now recognizes industrial energy efficiency leaders and works through the supply chain as well.

The Industrial Technologies Program also works with small and medium-sized firms through the audits performed by the Industrial Assessment Centers at universities throughout the country. This program identifies cost-effective opportunities for energy efficiency throughout the firms' operations. Unfortunately, implementation of these recommendations was only 47% from program initiation in 1981 through 2007 (DOE, 2007), suggesting that significant benefits are not being captured.

Another public-private partnership in the U.S. couples the government with manufacturers to reduce energy intensity by 2.5% or more per year. This is done through energy management standards, which almost always include a comprehensive energy plan and an energy manager to oversee the implementation of the plan. This type of project ensures that equipment continues to operate as efficiently as possible, as energy use is constantly monitored.

Finally, the multiagency Climate Change Technology Program has begun to work recently on addressing barriers to industrial efficiency through crosscutting policy options. A workshop was held with a mix of academics and industry leaders to discuss the barriers to industrial energy efficiency and preferred policy responses. Some of the policy options being considered include establishing a national energy efficiency resource standard that qualifies CHP, enabling municipalities to establish clean energy property tax liens, superior energy performance standards, and making third-party financing available for industrial energy efficiency upgrades. All of these policies would represent a significant step forward in addressing significant financing, regulatory, and information barriers to industrial energy efficiency (Brown et al., 2011).

6. MANUFACTURING THE NEXT GENERATION OF GREEN TECHNOLOGIES

Most of the current dialogue focuses on new technology that lowers industry's energy use. In some cases, more important energy savings come from adapting the new technology for use in other sectors. For example, developing a new generation of fuel cells may lead to greater savings in motor vehicles. Other possibilities include "on-demand" manufacturing that applies ink-jet printing systems to three-dimensional fabrication, or new plastics that double as integrated photovoltaic systems (Laitner and Brown, 2005). This role of industry in the development of emerging technologies highlights even greater energy savings than might be apparent from looking at industry's own energy use patterns alone. With the growing focus on corporate sustainability, industry is adopting a much broader view of its energy and environmental responsibilities, extending its concern to issues surrounding the sustainability of the products and services it offers and including the sustainability of its chain of suppliers. Wal-Mart, for example, has included indicators of energy sustainability in metrics used to select product and service providers.[4] Accordingly, contractors with minimal environmental impacts are preferred.

Wal-Mart is not alone in this initiative, as many other corporations have taken voluntary action to reduce the GHG emissions of their operations. These efforts are not yet operating at the scale needed to address the challenges of climate change and energy (Southworth, 2009); however, they appear to be expanding as corporate commitments to sustainability grow, and as consumer and shareholders demand greater effort (Prindle, 2010).

Industry is often viewed as a recipient of new technologies to meet production demands. While many innovations are created at research hubs like top tier universities, industry is often a source of technological innovation as well. In the energy realm, many next-generation technologies in areas such as fuel cells, solid-state lighting, and biofuels, are being developed by industry alone and also in public-private partnerships. Industry is not just a recipient of new technologies, but in fact plays a key role in developing the next wave of energy technologies.

Fuel cells provide a useful example. Different sectors of industry are innovating to create different uses and applications of fuel cells. Honda was recently recognized for its innovations in the use of fuel cells in transportation vehicles, winning awards and having its FCX Clarity model named the 2009 world green car. Honda reports that this vehicle is 2–3 times more fuel-efficient than gasoline-based vehicles, and gets 1.5 times better fuel economy than a hybrid electric-gasoline vehicle (Honda, 2009).

However, personal transportation is not the market where fuel cells have really seen competitive advantage and uptake; that distinct honor resides with auxiliary power units, marine systems, and forklifts. In fact, a recent Department of Energy report found that 3 KW proton exchange membrane (PEM) fuel cell-powered forklifts currently have total system costs nearly half that of their conventional lead-acid battery counterparts (DOE, 2008). Industry is actively experimenting with at least six different fuel cell technologies and innovations continue in all areas (DOE, 2009). As sales continue to increase, it is expected the technology will continue down the learning curve and costs will decrease.

Another example where industry is leading through innovation is the search for super-efficient solid-state lighting. This area of innovation is generally in light emitting diode (LED) technology. LEDs are much more efficient generators of light than incandescent and fluorescent lighting technologies, and they

4. Jim Stanway, Wal-Mart, personal communication, 2007.

also have longer lifetimes. LEDs are useful in many applications, including traffic and street lighting, indoor lighting, and flat-screen displays. This varied application list results in companies from different sectors being involved in RD&D, including Sony and Philips. While the U.S. government enters into many public–private partnerships and provides assistance in overcoming barriers (such as product testing standards) to deployment, the variety of applications for solid-state lighting technologies have industry leading the way in innovation (Building Technologies Program, 2009).

Finally, industry is developing next-generation biofuels that are sustainably produced with a near net-zero carbon footprints. Some promising examples are cellulosic ethanol and algae-based biofuels. BP Biofuels has a number of partnerships for developing feedstocks and technology, representing over $2 billion in private investment between seven companies (Semans and deFontaine, 2009). These companies are working together to develop cellulosic ethanol while respecting the environmental, agricultural, and social impacts producing feedstocks can create (Scotti, 2009). ExxonMobil has teamed with renowned geneticist Craig Venter and his start-up, Synthetic Genomics, to develop genetically modified algae as a source of biofuels. An initial investment of $600 million has been made, and Exxon has publicly acknowledged it intends to invest billions more for commercialization and deployment once R&D is sufficiently advanced (Johnson, 2009).

Many of the new technologies are being developed in public–private partnerships, representing the shared interest of developing new, more efficient, and environmentally friendly next generation technologies. Fuel cells, solid-state lighting, and cellulosic and algal-based biofuels all represent significant advances in currently deployed technologies, but all still face significant barriers. The public-private interfaces in each of these areas help to overcome many of the economic barriers. The potential for increased efficiency and sustainability in the next generation of technologies stands to show that industry itself is a driver of innovation.

CONCLUSIONS

The dual goals of advancing energy efficiency at industrial plants and advancing product innovation for broader use are both critical to promoting the more productive consumption of energy in a resource-constrained world.

Developing and deploying more efficient technology is the key to reducing carbon intensity in industry. Advanced industrial technologies and best practices in energy management are already working to improve energy efficiency and lower GHG emissions. These efforts have helped the industrial sector diminish GHG emissions in some of the nation's most energy-intensive industrial facilities.

Still, barriers to broader application of technologies suitable for commercialization in this sector remain. As a result, independent studies using different approaches agree that the economic potential for improved energy efficiency in industry is large. Of the 34.3 quads of energy forecasted to be consumed by U.S. industry in 2020 (EIA, 2008a), 14–22% could be saved through the implementation of cost-effective energy-efficiency improvements (National Academies, 2009). Large mismatches abound between the thermal needs and waste heat streams of industrial facilities served by large-scale centralized power plants. If systems were optimized so that the vast majority of wasted energy was recycled into productive uses, industrial energy consumption could be cut tremendously.

Comparisons of the energy content of manufactured products across countries underscore the potential for U.S. industry to reduce its energy intensity. Japan and Korea, for instance, have particularly low levels of industrial energy intensity. Many energy-intensive industries have devoted considerable resources to increasing their energy efficiency. For many other industries, energy represents a small fraction of their costs and is not a priority. Until the chief executives of U.S. industry become a force for clean energy and environmental progress, the challenges of climate change and resource depletion cannot be adequately addressed.

Ultimately, we need to transform the vision of industry as a necessary evil exiled to remote locations to avoid contaminating pollution. Instead, imagine a future where concepts of industrial ecology are taken to an extreme, so that people will want these facilities and jobs in their communities. Because they are clean and green, people want to live close to industrial parks to reduce their commute to work, expand their commitment to community, and help make industry part of the climate solution. The public's imagination has been captivated by zero-energy buildings and cars that operate like pollution vacuum cleaners. Now we need a new vision of industry—factories-of-the-future with minimal resource requirements, that clean up our ecosystems, contribute to human health, produce valuable goods, promote innovation, and improve standards of living.

ACKNOWLEDGMENTS

This chapter's description of energy efficiency opportunities in energy-intensive industries in the U.S. draws from the findings and conclusions of Chapter 4 of the National Academies 2009 publication, *Real Prospect for Energy Efficiency in the United States.* Marilyn Brown was the lead author of that chapter, with significant contributions by Steve Berry (University of Chicago), Linda Cohen (University of California, Irvine), Alexander MacLaughlan (retired from E.I. du Pont de Nemours & Company), Maxine Savitz (retired from Honeywell, Inc.), and Madeline Woodruff (National Academies). We also wish to acknowledge the valuable review comments we received from Andre de Fontaine (Pew Center on Global Climate Change and now at the DOE Industrial Technology Program) and Perry Sioshansi (CEO and Founder of Menlo Energy Economics). Finally, we benefited from our ongoing dialogue with the U.S. Department of Energy's Climate Change Technology Program and the Industrial Technologies program. Any remaining errors are entirely the responsibility of the authors.

BIBLIOGRAPHY

American Iron and Steel Institute (AISI). (2005). *Saving one barrel of oil per ton*. AISI.

Banerjee, R., Phan, A., Wang, B., Knobler, C., Furukawa, H., O'Keeffe, M., et al. (2008). High-throughput synthesis of zeolitic imidazolate frameworks and application to CO_2 capture. *Science, 319*, 939–943.

Battelle (Battelle Memorial Institute). (2002). *Toward a Sustainable Cement Industry: Climate Change*. Substudy 8, independent study commissioned by the World Business Council for Sustainable Development. Battelle Memorial Institute.

Brown, M., Chandler, J., Lapsa, M., & Sovacool, B. (2008). *Carbon lock-in: Barriers to deploying climate change mitigation technologies*. Report TM-2007/124. Oak Ridge National Laboratory. *www.ornl.gov/sci/btc/pdfs/brown_doc7435_tm124_08.pdf*.

Brown, M., Jackson, R., Cox, M., Deitchman, B., Cortes, R., & Lapsa, M. (2011). *Making industry part of the climate solution*. Report Forthcoming. Oak Ridge National Laboratory.

Brown, M., Levine, M., Short, W., & Koomey, J. (2001). Scenarios for a clean energy future. *Energy Policy, 29*(14), 1179–1196.

Building Technologies Program. (2009). *Solid-state lighting research and development: Manufacturing roadmap*. U.S: Department of Energy.

Canepa, A., & Stoneman, P. (2004). Comparative international diffusion: Patterns, determinants and policies. *Economics of Innovation and New Technology, 13*(3), 279–298.

Casten, T. R., & Ayres, R. U. (2007). Are worldwide power systems economically and environmentally optimal? In M. Brown & B. Sovacool (Eds.), *Energy and American Society – 13 Myths* Springer.

CCCSTI. (2009). (Committee on Climate Change Science and Technology Integration). *Strategies for the commercialization and deployment of greenhouse gas intensity-reducing technologies and practices*.

Confederation of Paper Industry. (2009). *Recovery and recycling of paper and board fact sheets. www.paper.org.uk/information/factsheets/recovery_and_recycling.pdf*

Cowart, R. (2001). *Efficient reliability: The critical role of demand-side resources in power systems and markets*. Report to the National Association of Regulatory Utility Commissioners by the Regulatory Assistance Project. *www.raponline.org/Pubs/General/EffReli.pdf*

DOE (Department of Energy). (2000). *Energy and environmental profile of the U.S. chemical industry*. Prepared by Energetics Incorporated. DOE. *www1.eere.energy.gov/industry/chemicals/tools_profile.html*

DOE. (2005a). *Energy and environmental profile of the U.S. pulp and paper industry*. DOE, Industrial Technologies Program.

DOE. (2006b). *Energy bandwidth for petroleum refining processes*. Prepared by Energetics Incorporated. DOE. *www1.eere.energy.gov/industry/petroleum_refining/bandwidth.html*

DOE. (2006). Pulp and paper industry energy bandwidth study. *Prepared by Jacobs Engineering Group and the Institute of Paper Science and Technology for the American Institute of Chemical Engineers and DOE. www1.eere.energy.gov/industry/forest/bandwidth.html*.

DOE. (2007). *IAC database*. Industrial Technologies Program.

DOE. (2008). *Early markets: Fuel cells for material handling equipment*. Department of Energy Hydrogen Program.

DOE. (2009). 2007 *fuel cell technologies market report*. Department of Energy, Energy Efficiency and Renewable Energy.

DOE. (2010). *U.S. energy intensity indicators: Trend data www1.eere.energy.gov/ba/pba/intensityindicators/trend_data*

EIA. (2001). *Chemical analysis brief. www.eia.doe.gov/emeu/mecs/iab/chemicals/*

EIA. (2008a). *Annual energy outlook 2008*. DOE.

EIA. (2008b). *International energy annual* 2006. DOE.

Energetics Incorporated. (2007)

Energetics Incorporated. (2009). *Inventory of federal activities that promote deployment and commercialization of GHG-intensity reducing technologies and practices.*

Energy Conservation Center, Japan. (2009). *National strategies and plans. www.asiaeec-col.eccj.or.jp/nsp/index.html*

Friedman, L. (2009). *China, U.S. give Copenhagen negotiators some targets*. The New York: Times. *www.nytimes.com/cwire/2009/11/30/30climatewire-china-us-give-copenhagen-negotiators-some-ta-73618.html?pagewanted=1* 30 November 2009.

Fruehan, R. J. (2008). Future steelmaking processes. *Materials Science and Engineering Department*. Carnegie Mellon University. *www.osti.gov/bridge/servlets/purl/840930-yRKP7T/webviewable/840930.PDF.*

Honda. (2009). *Honda wins 2009 Grove Medal for FCX Clarity fuel cell technology. http://www.honda.ie/contentv3/index.cfm?fuseaction=page&pageID=17397*

Houser, T., Bradley, R., Childs, B., Werksman, J., Heilmayr, R. (2008). *Leveling the carbon playing field.* World Resources Institute.

IEA (International Energy Agency). (2003). *Energy to 2050: Scenarios for a sustainable future*. IEA.

IEA. (2007). *Tracking industrial energy efficiency and CO2 emissions*. IEA.

IEA. (2009a). *Cogeneration and district energy: Sustainable energy technologies for today and tomorrow*. IEA. *www.iea.org/files/CHPbrochure09.pdf*

IEA. (2009b). *IEA scoreboard 2009*. IEA.

IEA. (2009c). *World energy outlook*. IEA.

Jacobs Engineering and Ipst Doe. (2006c). *Pulp and paper industry energy bandwidth study.* Prepared by Jacobs Engineering Group and the Institute of Paper Science and Technology for the American Institute of Chemical Engineers and DOE. *www1.eere.energy.gov/industry/forest/bandwidth.html*

Johnson, K. (2009). Biofuels bonanza: Exxon, Venter to team up on algae. *Environmental Capital—Wall Street Journal* blog. *http://blogs.wsj.com/environmentalcapital/2009/07/14/biofuels-bonanza-exxon-venter-to-team-up-on-algae/*

Laitner, J. A., & Brown, M. A. (2005). *Emerging industrial innovations to create new energy efficient technologies*. In: *Proceedings of the Summer Study on Energy Efficiency in Industry.* Washington, DC: American Council for an Energy-Efficient Economy.

Lamont, J. (2009). Industrial to launch energy-efficiency trading. *Financial Times.*

LBNL (Lawrence Berkeley National Laboratory). (2005). *Energy efficiency improvement and cost saving opportunities for petroleum refineries, an ENERGY STAR guide for energy and plant managers*. LBNL-57260-Revision. Prepared by C. Galitsky, S. Chang, E. Worrell, & E. Masanet. Berkeley, Calif.: LBNL.

Lin, J., Zhou, N., Levine, M. D., & Fridley, D. (2006). *Achieving China's target for energy intensity reduction in 2010: An exploration of recent trends and possible future scenarios.* Lawrence Berkeley: National Laboratory.

McKinsey and Company. (2008). *The untapped energy efficiency opportunity of the U.S. industrial sector: Details of research, 2008*. New York: McKinsey and Company.

Miller Freeman. (1998). *Pulp and Paper North American Fact Book 1998—1999*. San Francisco: Miller Freeman.

Mullins, O. C., & Berry, R. S. (1984). Minimization of entropy production in distillation. *J. Phys. Chem., 88*, 723—728.

National Academies. (2009). *Real prospect for energy efficiency in the United States*. Washington, DC: The National Academies Press.

Nenoff, T. M., Ulutagay-Kartin, M., Bennett, R., Johnson, K., Gray, G., Anderson, T., et al. (2006). *Novel modified zeolites for energy-efficient hydrocarbon separations. SAND2006–6891*. Albuquerque, NM: Sandia National Laboratories.

NRC (National Research Council). (2000). *Materials technologies for the process industries of the future*. Washington, DC: National Academy Press.

NREL (National Renewable Energy Laboratory). (2002). *Chemical industry of the future: Resources and tools for energy efficiency and cost reduction now*. DOE/GO-102002–1529; NREL/CD-840-30969. *www.nrel.gov/docs/fy03osti/30969.pdf*

Nuijen, W., & Booij, M. (2002). Experiences with long term agreements on energy efficiency and an outlook to policy for the next 10 years. *Netherlands Agency from Energy and the Environment*.

Phylipsen, D., Blok, K., Worrell, E., & de Beer, J. (2002). Benchmarking the energy efficiency of Dutch industry: An assessment of the expected effect on energy consumption and CO_2 emissions. *Energy Policy, 30*.

Prindle, W. (2010). *From shop floor to top floor: Best business practices in energy efficiency*. Washington, DC: Pew Center.

Sato, A. (2000). Promotion of energy efficiency investments in Japan. In R. Wahnschafft (Ed.), *Promotion of Energy Efficiency in Industry and Financing of Investments*. United Nations.

Schipper, L. (2004). International comparisons of energy end use: Benefits and risks. In C. J. Cleveland (Ed.), *Encyclopedia of Energy: Vol. 3*. (pp. 1–27) Amsterdam: Elsevier.

Scotti, R. (2009). *BP biofuels: A growing alternative*. BP.

Seligsohn, D., & Levin, K. (2010). *China's carbon intensity goal: A guide for the perplexed*. World Resources Institute.

Semans, T., & deFontaine, A. (2009). Innovating through alliance: A case study of the DuPont–BP partnership on biofuels. *Pew Center on Global Climate Change*.

Sinton, J., Levine, M., Fridley, D., Yang, F., & Lin, J. (1999). *Status report on energy efficiency policy and programs in China*. Lawrence Berkeley: National Laboratory.

Soble, J. (2010). Tokyo starts scheme for carbon trading. *Financial Times*, 8 April 2010. *www.ft.com/cms/s/0/fd971baa-42a4-11df-91d6-00144feabdc0.html*

Southworth, K. (2009). Corporate voluntary action: A valuable but incomplete solution to climate change and energy security challenges. *Policy and Society, 27*(4), 329–350.

Steel Recycling Institute. (2009). *US steel recycling rate hits all-time high*. SRI.

Thorp IV, B.A., & Murdock-Thorp, L.D. (2008). Compelling case for integrated biorefineries. Paper presented at the 2008 PAPERCON Conference, May 4-7, 2008, Dallas, Texas. Norcross, Ga.: Technical Association of the Pulp and Paper Industry.

Wang, A. (1999). *A comparative analysis of the 1997 energy conservation law of China and the implementing regulations of Shandong*. Zhejiang, and Shanghai: Natural Resources Defense Council.

Wang, Q., Sinton, J. E., & Levine, M. D. (1995). *China's energy conservation policies and their implementation*. Lawrence Berkeley: National Laboratory.

World Bank. (2008). India: Energy efficiency lending focuses on industry clusters. *http://go.worldbank.org/RWBUAW0RM0*

Worrell, E., & Biermans, G. (2005). Move over! Stock turnover, retrofit, and industrial energy efficiency. *Energy Policy, 33*, 949–962.

Worrell, E., & Galitsky, C. (2004). *Energy efficiency improvement and cost saving opportunities for cement making—an ENERGY STAR (R) guide for energy and plant managers. LBNL-54036*. Lawrence Berkeley: National Laboratory.

Worrell, E., Galitsky, C., & Price, L. (2008). *Energy efficiency improvement opportunities for the cement industry.* Lawrence Berkeley: National Laboratory.

Worrell, E., & Neelis, M. (2006). *Worlds best practice energy intensity values for selected industrial sectors. LBNL-62806.* Lawrence Berkeley: National Laboratory.

Worrell, E., Price, L., & Galitsky, C. (2004). *Emerging energy-efficient technologies in industry: Case studies of selected technologies. LBNL-54828.* Lawrence Berkeley: National Laboratory.

Worrell, E., Price, L., Martin, N., Hendriks, C., & Meida, L. O. (2001). Carbon dioxide emissions from the global cement industry. *Annual Review of Energy and the Environment, 26*, 303–329.

Zhang, Z. (2003). Why did the energy intensity fall in China's industrial sector in the 1990s? The relative importance of structural change and intensity change. *Energy Economics, 25*(6), 625–638.

Chapter 13

Prospects for Renewable Energy

Douglas Arent,* Paul Denholm,* Easan Drury,* Rachel Gelman,* Maureen Hand,* Chuck Kutscher,* Margaret Mann,* Mark Mehos,* and Alison Wise†

*National Renewable Energy Laboratory (NREL), Golden, Colorado, †Ecotech Institute

Chapter Outline

1. INTRODUCTION

With the growing concern regarding climate change worldwide, renewable energy (RE) technologies have become increasingly important during the past decade. This chapter examines availability, markets, and the technical potential of RE resources in meeting energy demand in a redefined energy economy. These new energy challenges include energy security, environmental integrity, climate change, and economic prosperity. This chapter will look at how far renewables have come during the past decades and their potential to provide a larger portion of our energy needs in the future. Section 2 examines the current status of renewables capacity as well as technology investment; it includes trends by country and by technology as well as the impact of the global economic situation. Section 3 outlines market and technology trends for resources such as hydropower, wind, solar, geothermal, and bioenergy; it also provides market projections. Section 4 looks more to the future with an analysis of the prospects—and challenges—for selected RE resources and technologies.

2. OVERALL STATUS OF RENEWABLE ENERGY CAPACITY AND INVESTMENT

2.1. Capacity Growth

In 2008, renewables (new renewables, excluding large hydropower and traditional biomass) contributed about 3.4% (depending on the calculation rules applied) to global energy generation demand. Including large hydropower and biomass, renewable generation increases to about 24% (REN21, 2009; UNEP, 2009). In 2008, new renewables represented about 229 gigawatts (GW) (excluding large hydro and biomass) of the electricity generating capacity, which is nearly 5% of total global power capacity (about 4700 GW). In 2008, global wind capacity was 53% (121 GW) of global renewable capacity (excluding large hydro and biomass), small hydropower was 37% (85 GW), grid-connected solar photovoltaics (PV) was 6% (13 GW), and geothermal was 4% (10 GW). As a percent of installed capacity, renewables increased by about 75% between 2000 and 2008 (Figure 1 and Table 1)—the share of new renewables in global electricity production is also increasing. In 2008, the highest installed capacities were found in China (76 GW), the United States (40 GW), Germany (34 GW), Spain (22 GW), and India (13 GW). Further, RE accounted for 5 GW of new electric generating capacity, and 4.2% of installed generating capacity in 2008 (REN21, 2009).

For renewable fuels, market growth has been similarly strong. Globally, biodiesel production has expanded more than six-fold, from 555 million gallons in 2004, to 3200 million gallons in 2008 and bioethanol from approximately 11,000 million gallons in 2004 to 17,300 in 2008 (DOE 2009). The growth in RE markets is dominated by a few countries, namely Germany, Spain, the United States, India, Brazil, China, and Japan. Argentina also contributes significantly to biodiesel production.

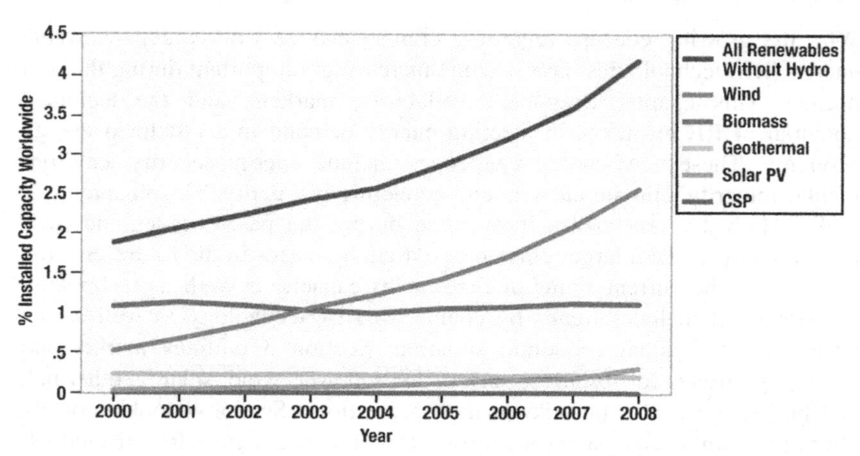

FIGURE 1 Renewables as a Percent of Total Installed Capacity Worldwide. *Source: DOE (2009).*

TABLE 1 Renewable Electric Power Capacity as of 2008

Technology	World Total	Developing Countries	EU-25	China	United States	Germany	Spain	India	Japan
Gigawatts Wind power	121.0	24.0	65.0	12.2	25.2	23.9	16.8	9.6	1.9
Small hydropower	85.0	65.0	12.0	60.0	3.0	1.7	1.8	2.0	3.5
Biomass power	52.0	25.0	15.0	3.6	8.0	3.0	0.4	1.5	>0.1
Solar photovoltaic-grid	13.0	>0.1	9.5	>0.1	0.7	5.4	3.3	~0.0	0.5
Geothermal power	10.0	4.8	0.8	~0.0	3.0	0.0	0.0	0.0	0.5
Solar thermal power-CSP	0.5	0.0	0.1	0.0	0.4	0.0	0.1	0.0	0.0
Ocean (tidal) power	0.3	0.0	0.3	0.0	0.0	0.0	0.0	0.0	0.0
Total renewable power capacity (excluding large hydro)	207.0	119.0	96.0	76.0	40.0	34.0	22.0	13.0	8.0
For comparison Large hydropower	860.0								
Total electric power capacity	4700.0								

Note: Small amounts, on the order of a few megawatts, are designated by "~0." Biomass power, large hydropower, and total electric power capacity are approximate. Global estimate for 2007 total renewable power capacity is 240 GW.
Source: REN21 (2009).

2.2. Investment

In recent years, the world has experienced major developments in research, demonstration, and deployment of new RE technologies. In 2008, investors devoted more than $155 billion globally to new RE capacity, manufacturing plants, and research and development (R&D)—this was an increase of 5% from 2007 (UNEP, 2009).

2.2.1. *Renewable Energy Investment Trends*

Annual investments in new renewable capacity increased from $60 billion in 2005 to $93 billion in 2006, $148 billion in 2007, and to $155 billion in 2008 (UNEP, 2009).[1] In 1999, this figure was about $10 billion. In 2008, these investments were dominated by wind power, followed by solar PV and biofuels. According to more aggressive estimations, venture capital (VC) investment in green energy technologies exceeded $4.1 billion in the third quarter of 2008 (Figures 2 and 3), far exceeding any previous quarter on record (the first quarter of 2008 was $2.2 billion and the second quarter of 2008 was $3.2 billion). Investment in solar technology led the VC charge in the third quarter with more than $1.5 billion invested in 26 VC financing rounds (Greentech, 2008). Investment levels are on track to reach $450 billion a year by 2012 and $600 billion a year in 2020 (UNEP, 2008). However, due to the economic crisis of 2008 and beyond, these expectations are probably no longer valid.

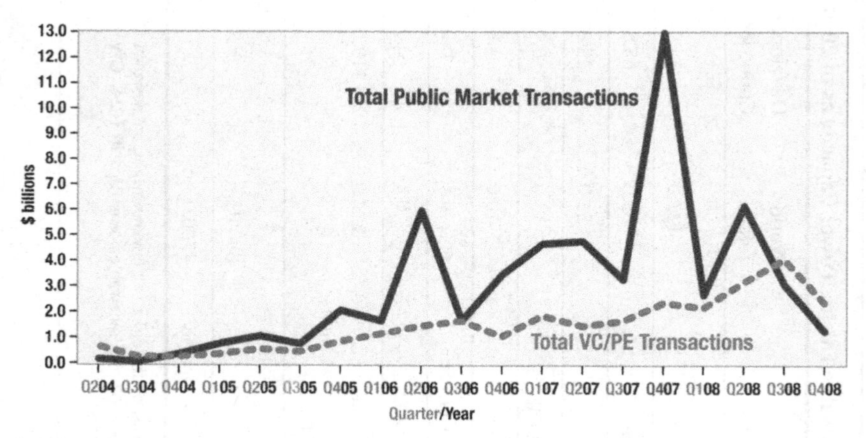

FIGURE 2 Public Market Transactions versus Venture Capital (VC)/private Equity (PE) Transactions between 2004 and 2008. *Source: NEF (2009).*

1. According to REN21, the investment figures are a more conservative $39 billion in 2005, $63 billion in 2006, $104 billion in 2007, and $120 billion in 2008.

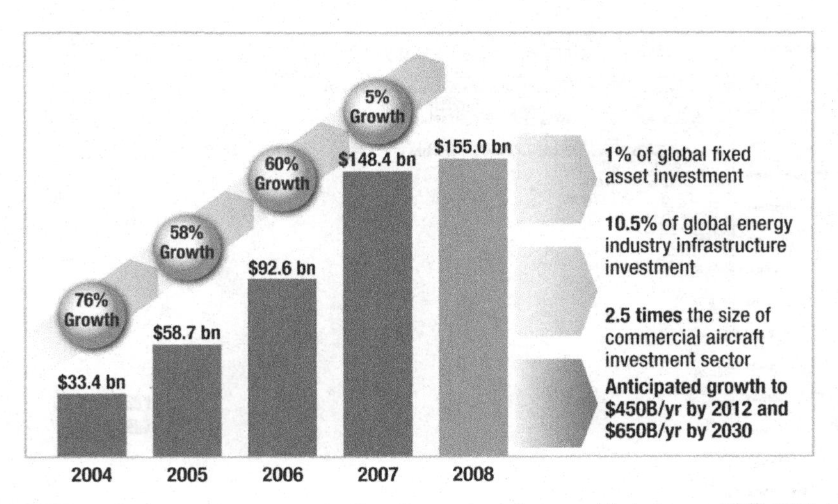

FIGURE 3 Global New Investments in Clean Energy (mainly renewables). *Source: NEF (2008a).*

2.2.2. Investment Trends by Country

Figure 4 shows investment by country between 2007 and 2008. In 2007, the United States led the world globally in RE investments, followed by Spain and China. In 2008, investments decreased globally, but the United States still maintained its lead, followed by Spain, Brazil, and China.

2.2.3. Investment Trends by Technology

In 2008, only solar and geothermal experienced significant growth in terms of VC and private equity (PE) investments. Solar investments grew 49% from 2007 levels to reach $33.4 billion, and geothermal grew 149% to reach $2.2 billion. Wind remained relatively stable, growing only 1% to $51.8 billion in investments in 2008. Compared to 2005/2006 levels, investments in renewable technologies are up more than tenfold.

Other technologies experienced a decrease in funding from VC and PE investors (Figure 5). Biomass power dropped 25% from 2007 to $7.9 billion in 2008, and biofuels fell 9% to $16.9 billion. There was a significant decrease in funding for efficiency technologies, with a drop of 33% from 2007 levels to reach $1.8 billion.

2.2.4. Impact of Global Financial and Economic Situation on Investment

As of the end of 2008 and early 2009, it is hard to determine the impact that the global financial and economic crisis of 2008/2009 (and beyond) will have on the future of investment in RE technologies. Early-stage investment seems to be shielded from most of the negative impact of the downward trend in the

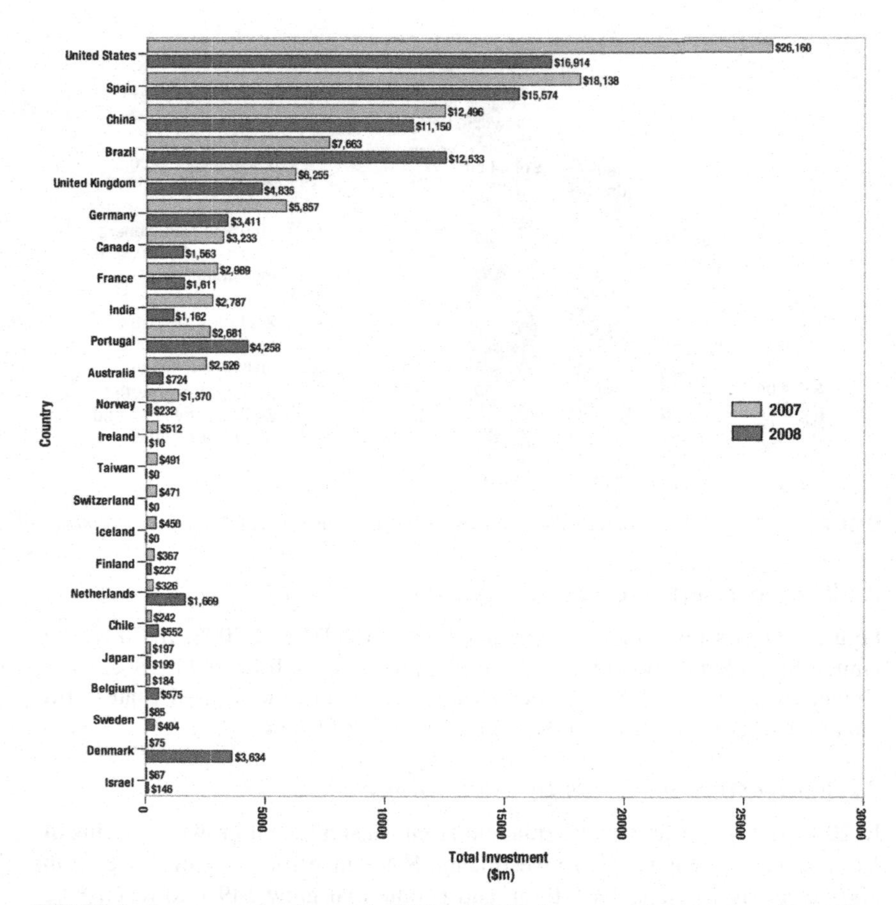

FIGURE 4 Investments by Country in 2007 and 2008: Investment Includes Asset Financing, Public Market Investment, and Venture Capital/Private Equity. *Source: NEF (2008a) and UNEP (2008).*

financial markets, primarily because this capital is sourced from third parties who are not directly impacted by the credit market. It is the large-scale deployment projects that may find the most difficulty securing financing options going forward in the near term, given that available debt has become more scarce (Greentech, 2008; REW, 2008; WSJ, 2008). Because the supply of money available to lend has contracted, the cost of borrowing is expected to be substantially higher. This could impede the incorporation of debt financing for RE equipment, increase construction costs, and affect project development as a whole. While this negative effect is significant, it is considered less of an issue than the reduced tax equity investments.

The market for tax equity is sizeably smaller than recent "boom" years in the financial and RE industries, falling precipitously in October 2008 with the

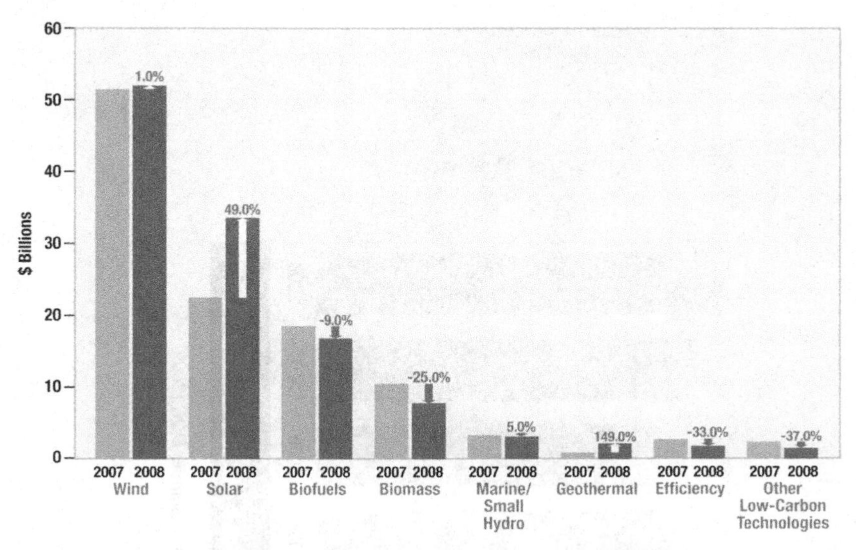

FIGURE 5 Investment by Technology in 2007 and 2008. *Source: NEF (2009) and UNEP (2009).*

U.S. financial crisis. Most, if not all, of the handful of traditional tax equity investors saw a massive loss in profitability—and some no longer exist. Replenishing these sources of investment will likely prove to be a difficult challenge for near-term financing of RE projects. Attracting new tax equity investors is possible, but the project's tax equity returns will need to increase. Wind projects may be particularly impacted by the reduction of tax equity due to their larger project size and shorter tax incentive extensions.

The impact of the financial crisis on RE financing also depends on the time horizon, often based on 40-years. Most immediate-term projects that obtained financing prior to October 2008 are being honored and moving forward. However, the near-term development of RE projects seems to be most impacted by the crisis, and there are wide-ranging estimates regarding the extent to which near-term projects are in jeopardy. On a longer-term basis, however, the outlook for RE financing and project development is positive because the financing issues are mitigated, perhaps through further government assistance or an eventual economic rebound. As of early 2010, market conditions appear to be stabilizing, with substantial investments being placed in technology companies, project finance, and strategic acquisitions. Post-Copenhagen pledges by some 100 countries are anticipated to provide substantial longer-term policy environments.

3. MARKET AND TECHNOLOGY TRENDS

This section covers trends by specific technology.

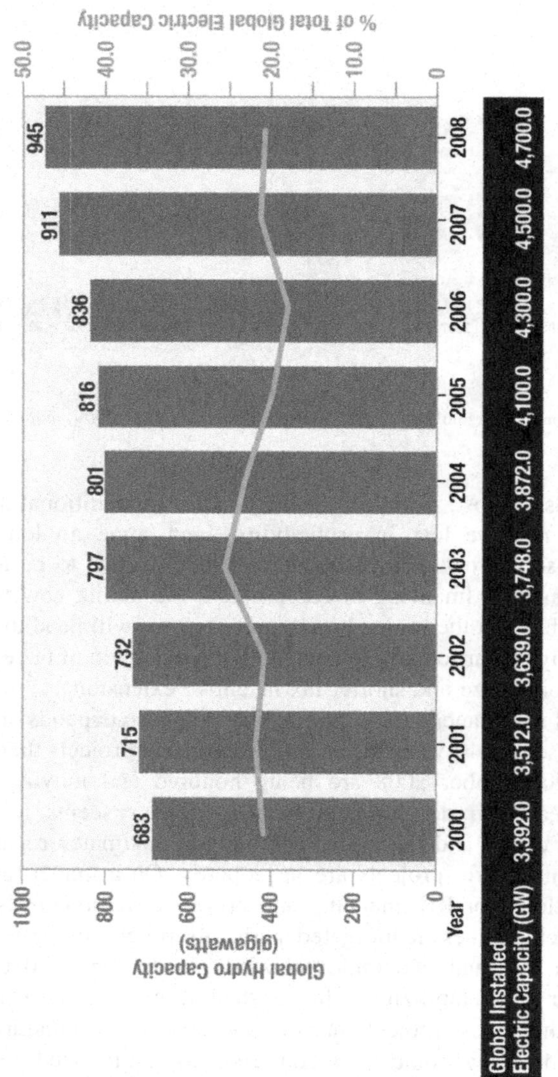

FIGURE 6 Development of Globally Installed Hydropower. *Source: DOE (2009).*

3.1 Growth of Hydropower Generation

In 2008, hydropower represented 20% of global electric capacity. Figure 6 shows the development of installed hydropower between 2000 and 2008.

3.2. Growth of Wind Power Generation

In 2008, more than 70 countries had wind power. In 2008, the wind capacity grew by about 29%, and the total installed capacity reached about 121 GW (Figure 7). Offshore wind power—being more costly and creating new maintenance concerns—grew in recent years with a few hundred megawatts annually (REN21, 2009; UNEP, 2009).

Worldwide, the wind industry has seen enormous growth in both generation and installed capacity in the past two decades. In 2006/2007, the wind power industry was experiencing supply chain difficulties due to booming demand. The wind industries in China and India continued to grow.

Germany, Spain, and Denmark have employed public policy mechanisms for the period represented in the figure. Germany's consistent policy approach has led to strong consistent growth and created the largest wind market worldwide in terms of capacity installed. Spain has experienced major growth since the appropriate policy was passed in 1997. Denmark's wind industry experienced steady growth through the 1990s; though new growth has tapered in the past few years as market saturation and land constraints have been reached. Although surpassed in capacity by Germany in 1997, the United States also has a strong growth curve for wind (Figure 8), driven largely by the production tax credit (PTC). Since its establishment in 1992, the PTC has experienced a series of short-term extensions, but it was allowed to lapse in three years: 1999, 2001, and 2003. The impact of the tax credit expiration is clear in the figure, as growth in capacity stalled in 2000, 2002, and 2004. As an additional comparison, the Chinese wind market is also considered in this analysis. China's wind market has grown sharply since 2000, with average annual growth rate (2000–2007) of about 43%. Although the country does not have a national market-oriented policy, it does have government-centered policies that support wind energy.

The boom-and-bust cycle of wind development due to the short-lived nature of the PTC is apparent in Figure 9. The PTC expirations reduced the annual growth rates in the United States to less than 10% for the year following the expiration and have been disastrous to planning efforts.

In 2008, Europe had 55% of the global share of wind capacity (about 66 GW of total 121 GW). Germany and Spain—two European Union (EU) member states with aggressive feed-in tariff (FIT) policies—account for more than half of the EU-27 installed wind capacity. In 2008, Germany had 23.9 GW of installed capacity, Spain followed behind with an installed capacity of 16.8 GW, and Italy had a capacity of 3.7 GW (Figure 10).

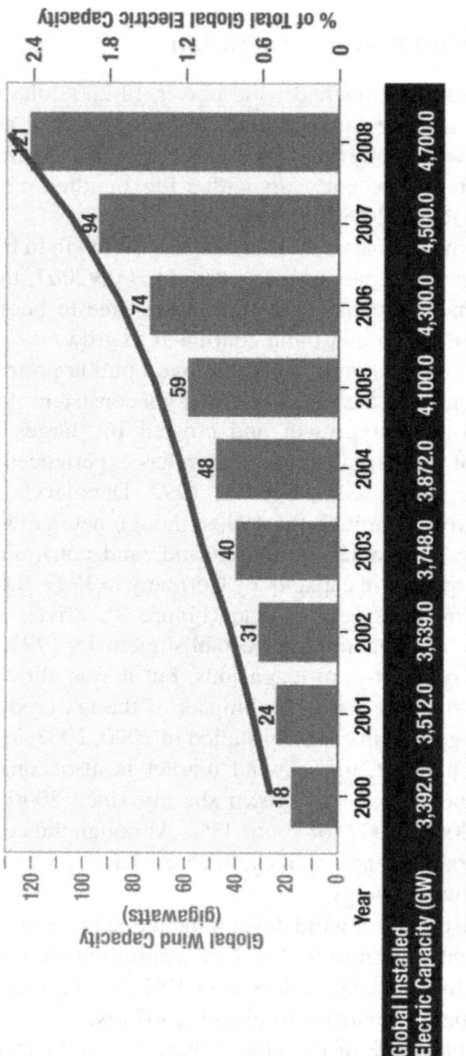

FIGURE 7 Global Growth of Wind Generation between 2000 and 2008. *Source: DOE (2009).*

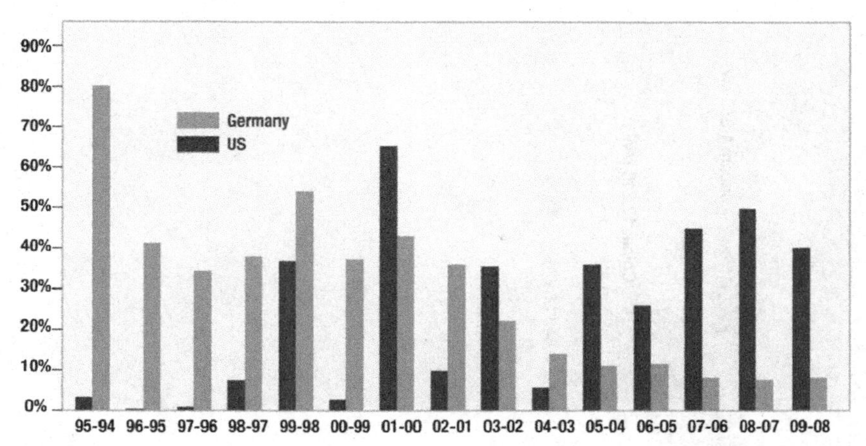

FIGURE 8 Wind Markets in the United States and Germany. *Source: AWEA (2010), GWEC (2010).*

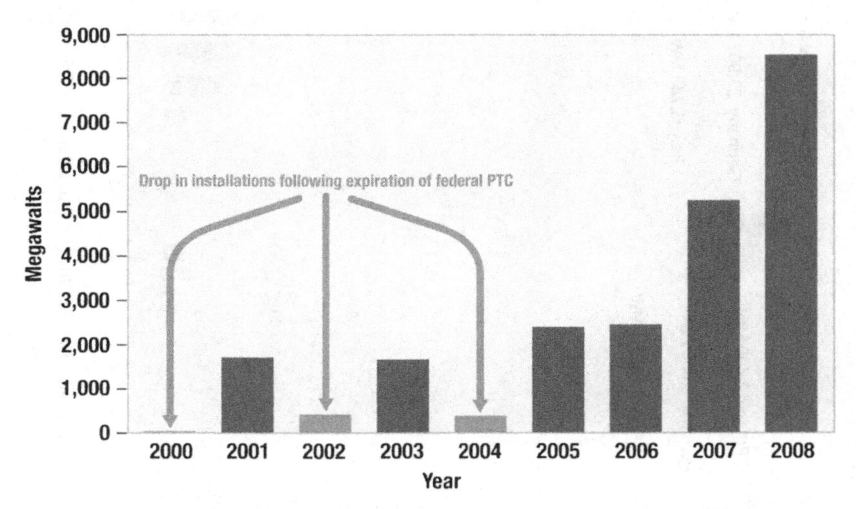

FIGURE 9 U.S. Annual Wind Capacity Additions. *Source: DOE (2009).*

3.3. Growth of Solar Power Generation

Solar technologies generally include direct electric power generation via solar cells or PV, concentrating thermal solar power (CSP), solar hot water (SHW), passive solar, and many smaller direct uses such as solar cook stoves. Within PV, multiple technologies are now available, including crystalline silicon, concentrating PV, and thin films comprised of many different technologies ranging from amorphous silicon, to copper indium gallium selenide, to organic compounds.

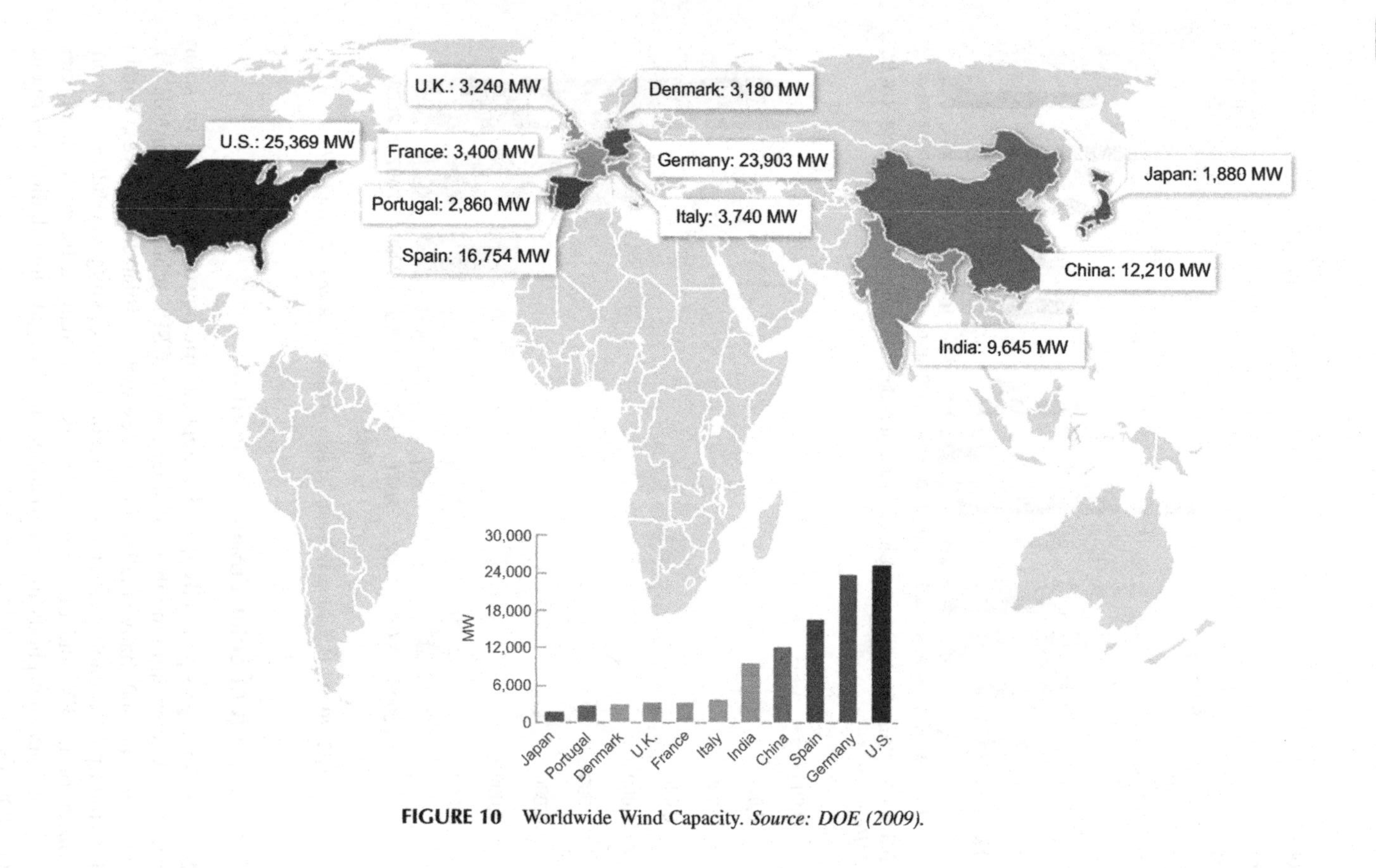

FIGURE 10 Worldwide Wind Capacity. *Source: DOE (2009).*

In 2008, solar PV production was 6.9 GW, up from 3.7 GW in 2007 and 2.5 GW in 2006. In recent years, solar PV market growth was centered in Germany, Japan, Spain, Italy, and South Korea; and in California and New Jersey in the United States. The installed capacity grew about 50% in 2006, 2007, and 2008, with a total installed capacity at the end of 2008 of nearly 16 GW (including off-grid installations) (REN21, 2009).

The development of solar thermal electricity (STE)—or CSP—stagnated from the early 1990s through 2004 (Figure 11). New initiatives were developed in Israel, Portugal, Spain, and the United States. In 2006/2007, Arizona completed a 1 MW trough plant and Nevada built a 64 MW solar thermal plant. Spain also completed a central receiver plant. Egypt, Mexico, and Morocco (with the support of the World Bank) developed three integrated CSP/combined-cycle plants—the CSP component of these plants is 20–30 MW. The total installed CSP capacity at the end of 2008 was 0.5 GW (REN21, 2009; World Bank, 2009).

In the United States, some 424 MW of CSP are operational in 2009, with more than 8,000 MW with signed power purchase agreements. Spain has 435 MW of commercial CSP generation; CSP projects with about 2,000 MW have provisional registration. Parabolic trough plants are expected to dominate this growth in coming years although several substantial tower and dish-engine projects are under development.

Almost 70 CSP electricity projects are in the planning phase, mainly in North Africa, Spain, and the USA. With the support of the current tariff, Spain completed the 11-MW PS10 power tower plant in 2007 and the 20-MW PS20 tower in 2009. In addition, Spain's 50-MW parabolic trough system with 7.5 hours of storage, Andasol 1, came on line at the end of 2008. Several additional plants have come on line in 2009 and 2010, giving Spain a total capacity of over 400 MW.

Rooftop solar collectors provide hot water to more than 60 million households worldwide, mostly in China. The global installed capacity in 2006 was about 105 GWh, an increase of 19% from 2005. As of the end of 2007, the installed capacity was about 126 GWh, an increase of 19% again (REN21, 2008; UNEP, 2008; REN21, 2009).

Four countries dominate in quantity of solar PVs installed. Figure 12 illustrates the increasing levels of capacity experienced in these countries: Germany, Spain, Japan, and the United States. From 2005–2007, Germany again showed the highest growth rate in PV capacity, followed by Spain. The outlook for solar PV is projected to shift from a supply-constrained market to one that is led by demand. The crystalline silicon module price is expected to fall to about $2.4/W in 2009; this decrease in price would be more rapid than the decline in incentives, which would help create longer-term expansion of the industry as demand strengthens in response to such low module prices. While investment declined in 2008, the trend is expected to turn toward increased investment in "downstream" companies such as developers, installers, and inverter manufacturers (NEF, 2008b). Top solar PV manufacturers in 2008 were Q-cells (Germany), First Solar (U.S.), and Suntech (China) (Prometheus Institute, 2009).

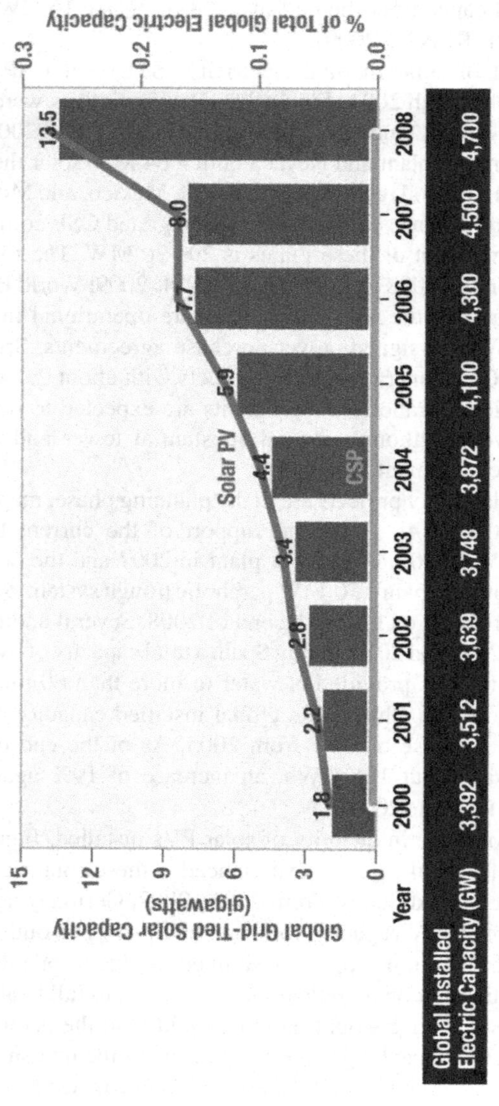

FIGURE 11 Global Growth of Solar Generation. *Source: DOE (2009).*

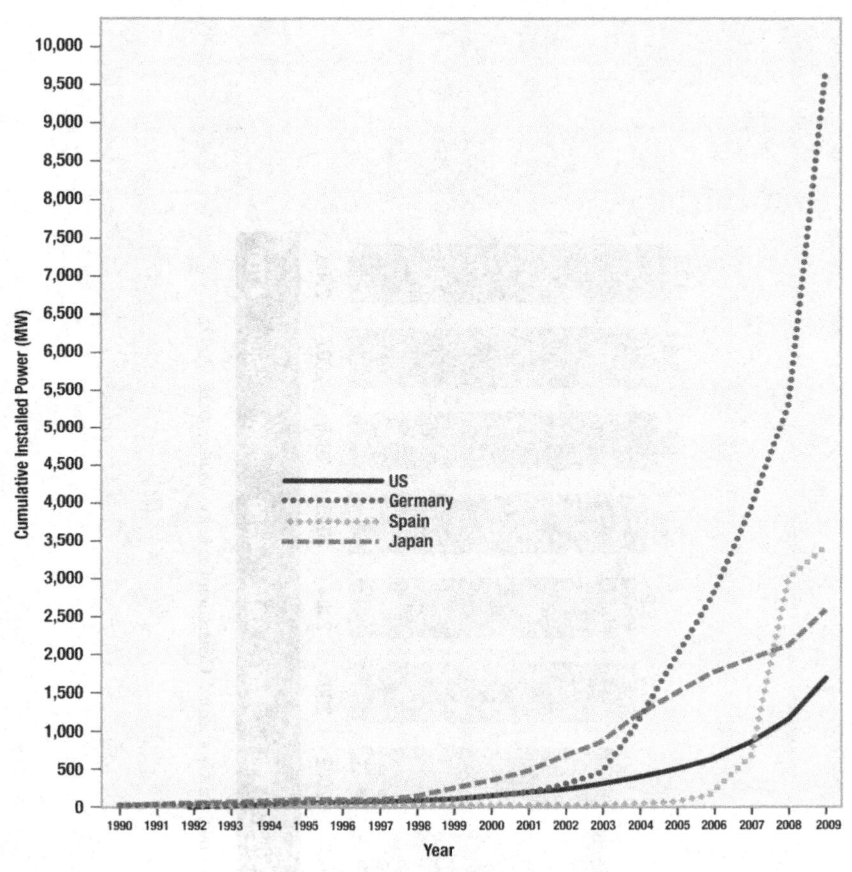

FIGURE 12 Trends in Photovoltaic Applications. *Source: IEA (2008b), SEIA 2010.*

3.4. Growth of Geothermal Power Generation

About 80 countries have geothermal plants, with Iceland in the lead, generating about 26% of its electricity from six large geothermal power plants (Orkustofnun, 2007). The heating capacity has increased by about 30%–40% annually in recent years. Geothermal heat pumps are a rapidly growing market, with more than 2 million heat pumps used in more than 30 countries. Most of the geothermal power capacity can be found in Italy, Japan, New Zealand, and the United States. By the end of 2008, the global installed electricity generation capacity was 10.0 GW (Figure 13), with an addition of approximately 400 MW during the year (REN21, 2009). Growth trends for geothermal power have been lower than other renewable power technologies due to significantly different location-specific resource base, technology development, and investment risk profile. However, technology development in enhanced

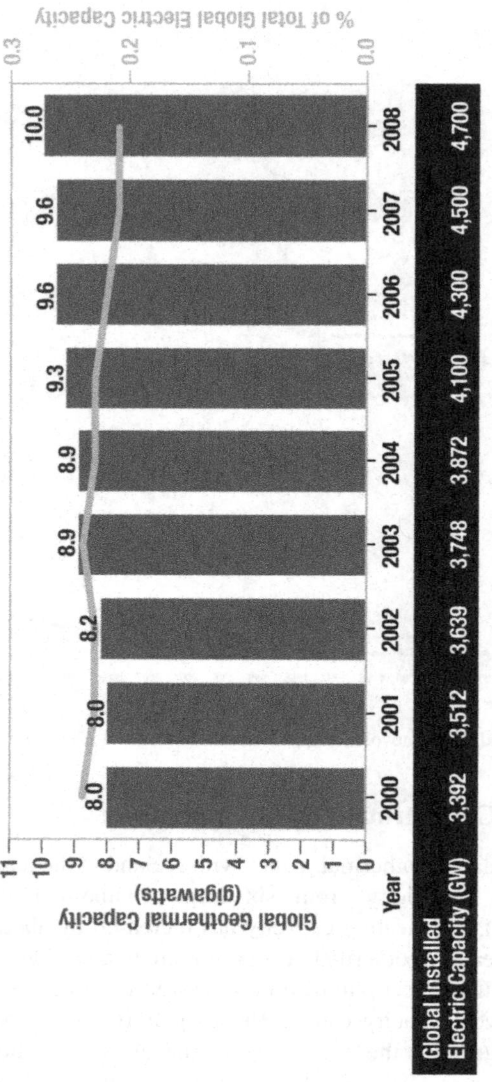

FIGURE 13 Global Growth of Geothermal Generation Capacity. *Source: DOE (2009).*

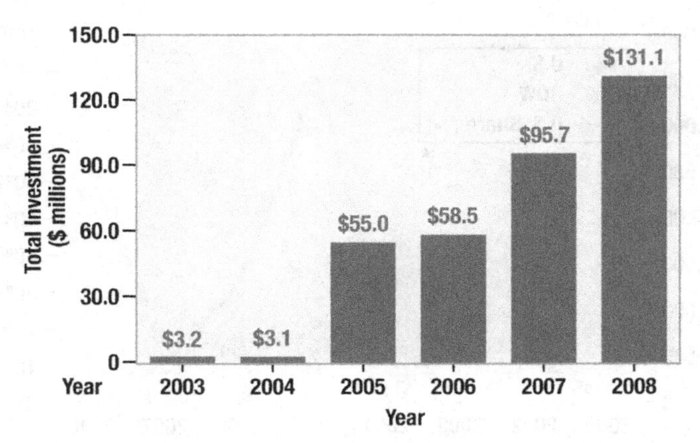

FIGURE 14 Enhanced Geothermal Systems (EGS) Investments. *Source: NEF (2008a).*

geothermal systems (EGS) offers the promise for significant global expansion of this power source (MIT, 2006), as described below.

The market for geothermal energy continues to grow, with investments being made and projects being built in both traditional geothermal and emerging geothermal technologies. There are two main focus areas for geothermal market development: hydrothermal technology and EGS technology (see Section 4 for more on the technologies).

EGS as a geothermal technology is particularly attractive from an investor perspective, especially because traditional geothermal technology projects are essentially capped for expansion. Based on external interviews, existing hydrothermal projects are perceived to be relatively small in generation capacity when compared to other base-load technologies—to an investor, these hydrothermal projects represent a declining asset value over time. Figure 14 shows that global investment activity in the technology during the period 2003–2008 was trending positively, with investments increasing from about $3 million in 2003 to $131 million in 2008.

Figure 15 shows investment trends in hydrothermal, with U.S. geothermal firms representing a significant share of the global total since 2004. Even when counting the enormous impact of the Philippine National Oil Company-Energy Development Corporation (PNOC-EDC) privatization in 2007, the U.S. share of total investments averaged 30% for 2000–2008. U.S. geothermal firms received more than $2 billion during that period, averaging about $400 million per year since 2004.

3.5. Growth of Bioenergy

Bioenergy remains one of the world's most used energy sources globally. However, most of this use is categorized as "traditional biomass" in

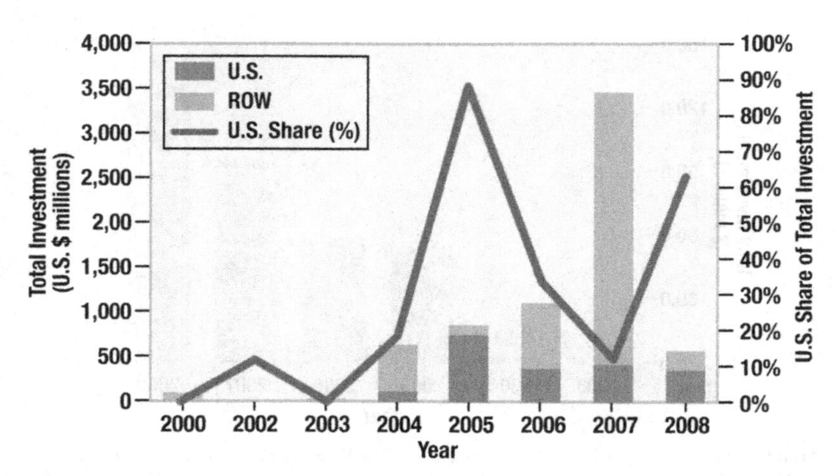

FIGURE 15 Hydrothermal Investment in the United States and the Rest of the World – ROW. *Source: NEF (2008a).*

developing countries. Modern bioenergy has two primary subuses: biopower and biofuel.

For biofuels, the United States has become the dominant ethanol producer (corn-based for blending gasoline), producing 18 billion liters in 2006, 24 billion liters in 2007 (RFA, 2009), and 34 billion liters in 2008 (REN21, 2009). However, ethanol production in Brazil increased to almost 19 billion liters in 2007 and 25 billion liters in 2008. Biodiesel production has increased in recent years at 20%–100% the annual rates, particularly in Germany, France, Brazil, Argentina, and the United States. Meanwhile, the impact of the production and consumption of biofuels on food prices, biodiversity, water consumption, and the mitigation of greenhouse gases (GHGs) is strongly debated, especially in Europe and the United States. Global biofuel production has more than quadrupled from 4.8 billion gallons[2] in 2000 to about 24 billion in 2009, but still accounts for less than 3% of the global transportation fuel supply (Figure 16). About 90% of production is concentrated in the United States, Brazil, and the EU (Figure 17). Production could become more dispersed if development programs in other countries, such as Malaysia and China, are successful. The leading feedstocks for producing biofuels include corn, sugar, and vegetable oils.

The installed capacity of biopower (biomass used to generate electricity) grew at a rate of about 4% a year between 2000 and 2008; the capacity increased from about 37 GW in 2000 to 52 GW in 2008 (Figure 18). The worldwide biopower sector is expected to increase by another 21 GW in the next five years to reach a cumulative installed capacity of 71 GW by 2012,

2. 1 gallon = 3.79 liters.

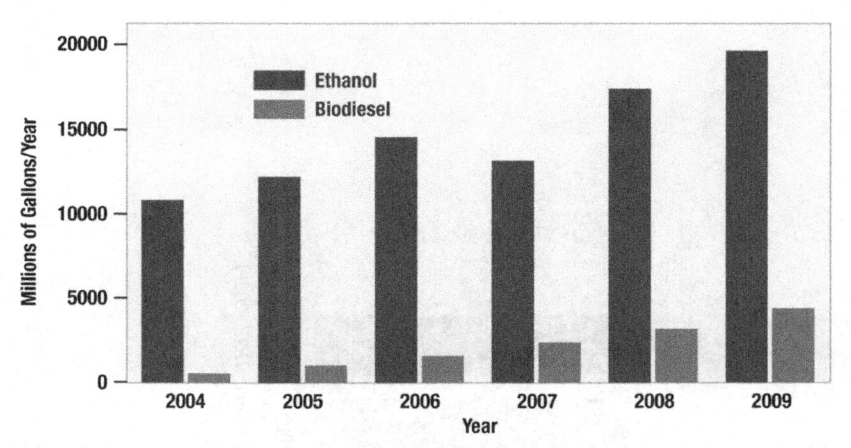

FIGURE 16 Global Production of Liquid Biofuels between 2000 and 2008. *Source: RFA 2010, REN21 2010*

suggesting a growth rate of about 9% a year (NEF, 2008a). The growth in the worldwide biopower energy market is particularly driven by the United States, Brazil, and Germany. However, by 2012, Spain and India are expected to surpass the leading nations in annual installed capacity—Spain is expected to add 8 GW and India is expected to add 4 GW annually, becoming the fastest-growing biopower nations in the world (NEF, 2008a).

3.5.1. Bioenergy Market Development

Bioenergy is poised to create a growing impact on the global energy infrastructure, provided that technology innovation and distribution issues continue to be addressed. Observers have predicted that the biofuels industry has a 10-year window of opportunity to evolve into a global, interdependent energy

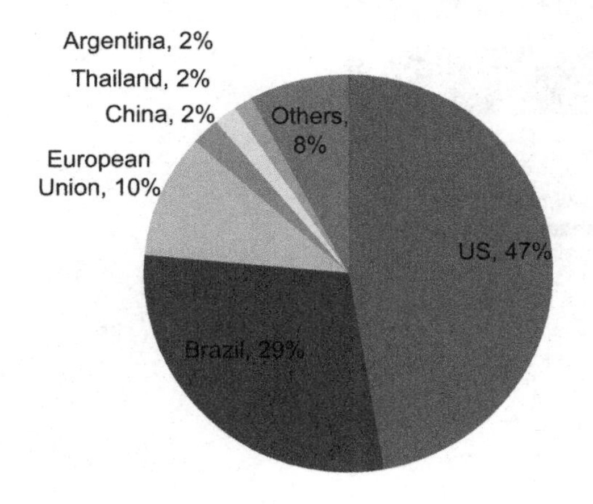

FIGURE 17 Global Biofuel Production by Country in 2009. *Source: RFA 2010, REN21 2010*

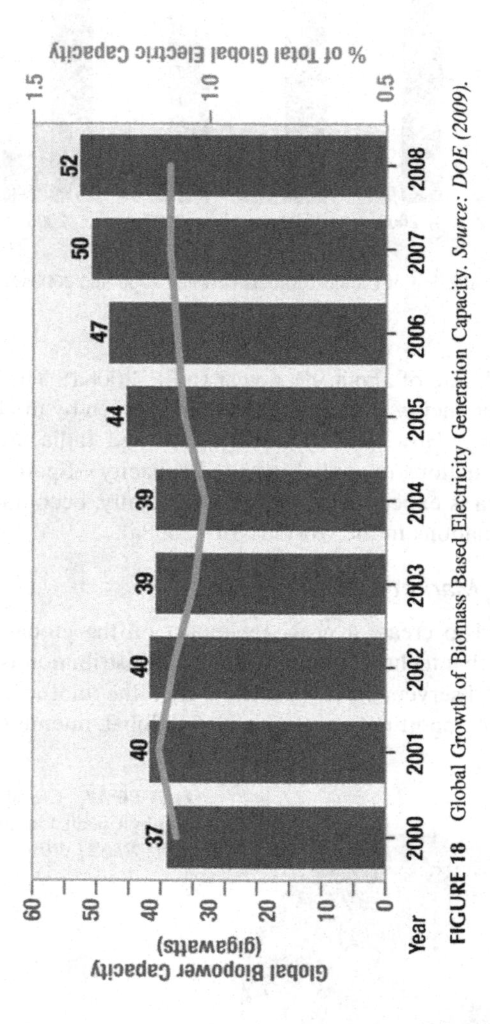

FIGURE 18 Global Growth of Biomass Based Electricity Generation Capacity. *Source: DOE (2009).*

system (Accenture, 2008). The growth of this industry will depend on the fluctuation of traditional fuel prices, potential instability of existing fossil fuel resources and distribution, and the technological innovation needed to adapt the worldwide vehicle and transport fleet to new fuels. Future trends could include the use of first-generation biofuels and "next-generation" biofuels side by side in the energy market, with fuels sourced from the more challenging cellulosic feedstocks entering after several years.

3.5.2. Biopower

Market data on biomass power capacity and generation is sparse: The European Commission (EC, 2008a; 2008b) does not distinguish between electricity generation from biomass and heat production from biomass in its statistics. Germany has the most extensive data collected on biomass electricity generation and capacity of any country. Germany provides differentiated tariffs for each type of biomass electricity generation, which has led to market development in all areas. At the end of 2008, the globally installed biopower capacity was estimated at 52 GW, mostly in developing countries (25 GW) and EU-27 (15 GW) (REN21, 2009) (Figure 19).

3.5.3. Biofuels

New investments in biofuels reached $16.9 billion in 2008 (UNEP, 2009) (Figure 20). For 2008, investors announced more than $6 billion for ethanol production facilities in Brazil, Canada, France, Spain, and the United States. In 2007, about 50 billion liters of ethanol fuel were produced globally (RFA, 2008), and in 2008 about 67 billion liters (REN21, 2009).

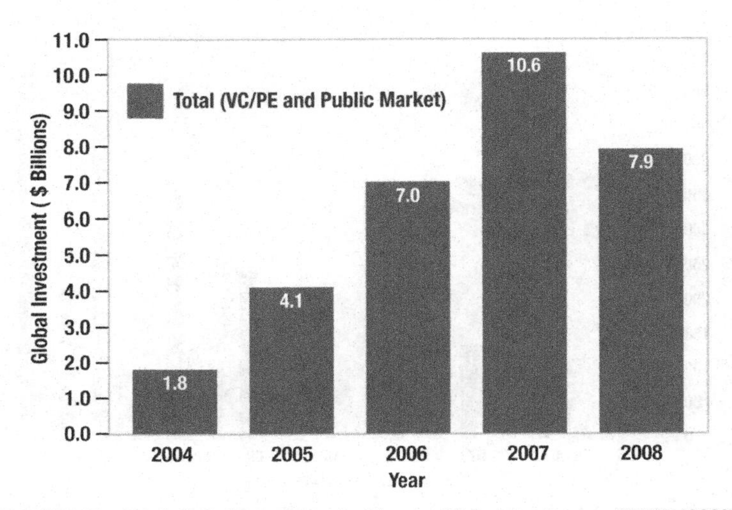

FIGURE 19 Total Global Investment in Biomass Electricity. *Source: UNEP (2009).*

Figure 20 highlights investment trends for technologies that appear to hold the most market promise for "next-generation" applications. From October 2006 to December 2008, there was $915 million privately invested in this new generation of biofuels (NEF, 2009). Approximately $220 million of this has gone to researching algae as a potential feedstock, as companies look to alternative feedstocks to address controversy regarding energy versus food markets. In this period, investments were made in biobutanol companies totaling $154 million; biobutanol can be used in the existing vehicle fleet and can be distributed in the existing pipeline infrastructure (NEF, 2009a). Finally, $539 million was invested in enzymatic hydrolysis; this technology is attractive because of its compatibility with "first-generation" starch and sugar hydrolysis and fermentation.

The term "next generation" in bioenergy is widely accepted for a technology that uses nonfood feedstocks and new ways of converting the power stored in plant-based carbohydrates to usable energy. Figure 21 captures the essence of this by categorizing three primary nonfood feedstocks, with six conversion technology pathways, and multiple fuel and product outputs. A fourth important fuel output category is "fungible" fuels, which are fully compatible with existing petroleum refining, distribution, and use value chains. These advances could widen the market for biofuels considerably. New feedstocks could also dramatically reduce the cost of fuel production and reduce land-use requirements and environmental emissions.

3.5.4. Bioenergy and Food versus Fuel Debate

The perception that land and feedstocks for bioenergy may be competing with their potential use for food can have negative ramifications on continued government support for the technology. Renewable energy RE production

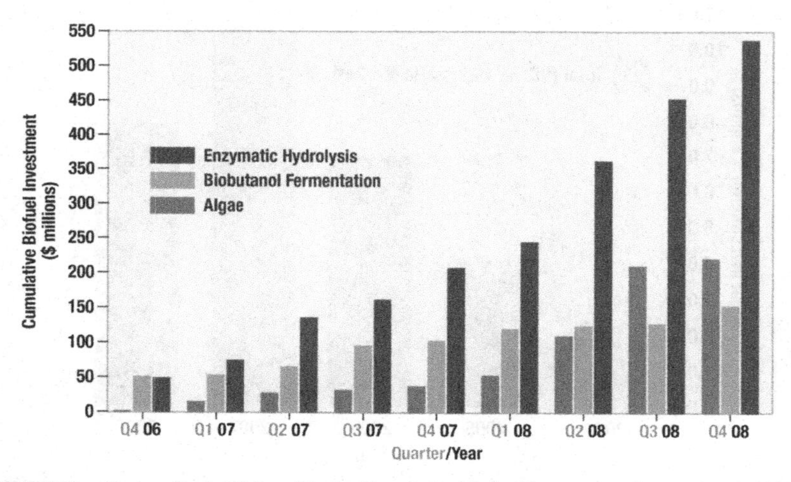

FIGURE 20 Venture Capital/Private Equity Cumulative Biofuel Investment. *Source: NEF (2009a).*

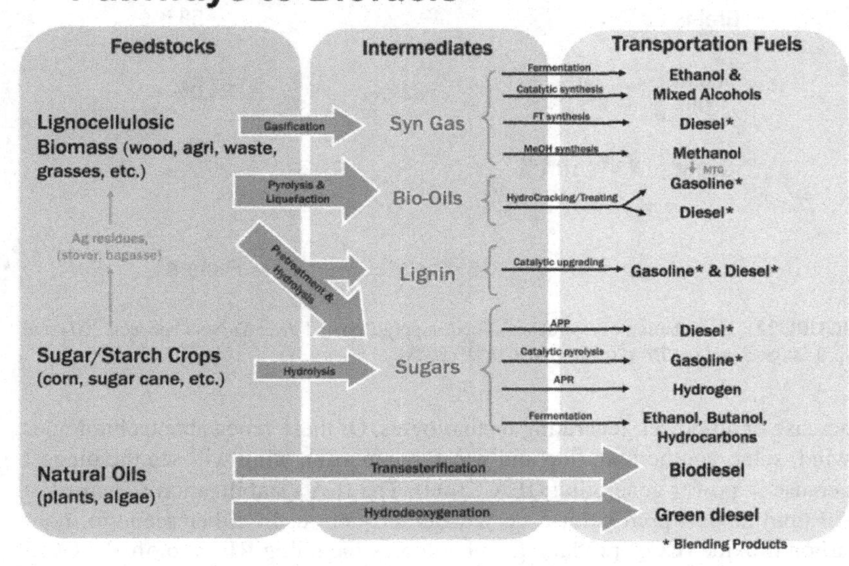

FIGURE 21 Pathways to Biofuels. *Source: NREL*

continues to rely on public policy support; as of late 2008, subsidies for renewables globally are greater than $20 billion per year, a majority of which are allocated for biofuels. While perception is key, analyses conducted by New Energy Finance, the Department of Energy (DOE), and the Food and Agriculture Organization (FAO) (NEF, 2008a; Karsner, 2008; Glauber 2008) note that biofuel production has not been the dominant factor in the steady increase in food prices from 2004 to 2008. The authors conclude that, while biofuel production has been one driver of food-price inflation, more significant drivers are the increase of input costs, changes in consumption habits, and increase in global population.[3] The decrease of food prices after mid-2008 supports this finding. Figure 22 shows the estimated fraction that biofuels contribute to the supply and demand price drivers for grains, food oils, and sugar.

3.6. Market Projections: Potential versus Pragmatic Penetration

RE use for electricity and for fuel is expected to grow in the coming decades. Globally, under business-as-usual assumptions, the share of electricity generation from nonhydro renewable sources is projected by International Energy Agency (IEA) to increase from 2.5% in 2007 to 8.6% in 2030, the fastest rate of

3. For the first time in decades, population growth has not been matched by an increase in agricultural yields, particularly grains.

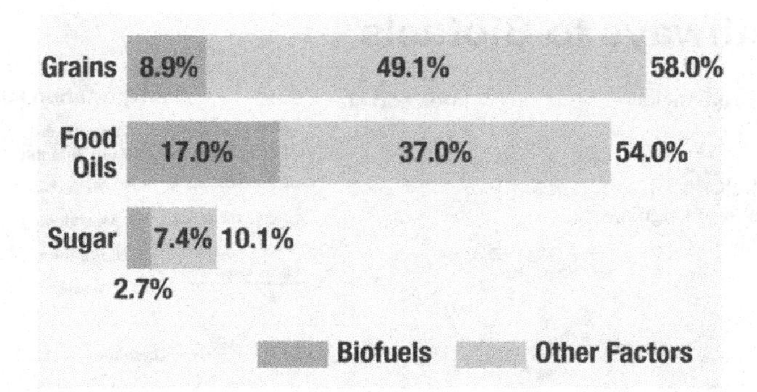

FIGURE 22 Contribution of Biofuels to Food Supply/Demand Price Drivers between 2004 and April 2008. *Source: NEF (2008a) and UDOP (2008).*

increase of all power generating technologies. Of these renewable technologies (wind, solar, geothermal, tide and wave, bioenergy), wind will see the biggest increase in power generation (IEA, 2009). The IEA's stabilization scenarios at 550 ppm or 450 ppm paint a very different picture. In either scenario, non-carbon dioxide (CO_2) producing technologies including RETs, biofuels, CCS, and nuclear contribute significantly more to the world's energy mix. These are only two of many possible futures for RE technologies. Further, the U.S. policy environment (highlighted by numerous renewable electric standard or RES bills and the Waxman-Markey comprehensive climate/RES/energy efficiency standard bill, as well as investments under the American Reinvestment and Recovery Act), are poised to significantly accelerate near- to mid-term deployment of low-carbon technologies in the United States. An example of the potential use of RETs in the United States under a climate scenario indicates more than tripling the contributions of RETs to U.S. energy supply by 2030 (UCS, 2009). Additionally, China, and numerous other countries have recently announced increased goals for the use of RETs.

Other reports, such as the National Academy of Sciences, have completed U.S.-centric assessments indicating possible contributions of more than 20% by 2030 (NAS, 2010). Other reports—based on scenario frameworks, spanning from advanced technology, policy, fossil fuel disruption, and so on—indicate a broad range of potential contributions from as low as today's 3% to more than 50% by 2050.[4] We should recognize, as described below, that any market projection reflects complex interactions among a set of assumptions related to technology, policy, and markets, whereas the "absolute" or technical potential, as described below is significantly higher and not a constraint.

4. Many such reports exist and are too numerous to catalog here. For example, see 20% Wind by 2030 at *www.eere.energy.gov/windandhydro/pdfs/41869.pdf* or Energy [R]evolution 2050 at *www.energyblueprint.info/*.

It is more difficult to make market projections about the use of bioenergy and renewable fuels, because many of the assumptions about market penetration are dependent on first-generation technologies, which limit the applicability of market uptake. "Next-generation" renewable fuels—specifically biofuels—present a significant opportunity for accelerated market penetration but remain a largely unknown entity for estimating market penetration. A 2008 research report by the consulting group Accenture estimates that biofuels could make up 10% to 15% of the global market in the next 10 to 20 years—but reaching that level will be particularly challenging (Accenture, 2008). These are consistent with previous technical estimates (IEA Bioenergy, 2009), although it is recognized that technical estimates vary considerably due to assumptions and methodologies.

3.6.1. Resource Potential

The U.S. is fortunate to have significant renewable resources. Figure 23 shows the combined resources with their geographic distributions. Their superposition illustrates the complexity of matching resource with energy demand, which involves a complex mix of spatially-optimized technologies. A vast quantity of energy can be supplied by renewables, but careful technology development, policy planning, and market adoption measures will be required. Clearly, the resource potential, particularly for solar and wind, is enormous compared to today's, and projected, power requirements. Well-known issues of variability have begun to gain significant attention, and do not appear to be major barriers for market shares up to ~30%. (UWIG Library, 2009). Moving toward a power system with a majority of renewable generation is the subject of current investigation, particularly with increased emphasis on systems level solutions, including the introduction of IT-enabled power management, advanced forecasting, adaptive and shiftable loads, balancing area enlargement, additional transmission infrastructure, and in very high penetration scenarios technology advances in areas such as storage. The combination of these mulitiple enabling capabilities will likely open up new opportunties for renewables to have an even larger role in the power system.

3.6.2. Pragmatic Penetration

There are many perspectives on the key drivers to market growth in the future. Those that have been largely accepted fall into three broad categories.

3.6.3. Infrastructure Evolution

For electric-sector technologies, access to and ability to build new transmission has been frequently identified as a major barrier to expanding the role of RETs. Issues related to siting, permitting, financing, cost allocation, and repayments, as well as access and integration costs for the system operator and system utilities have been identified. As we look forward, infrastructure evolution refers to the

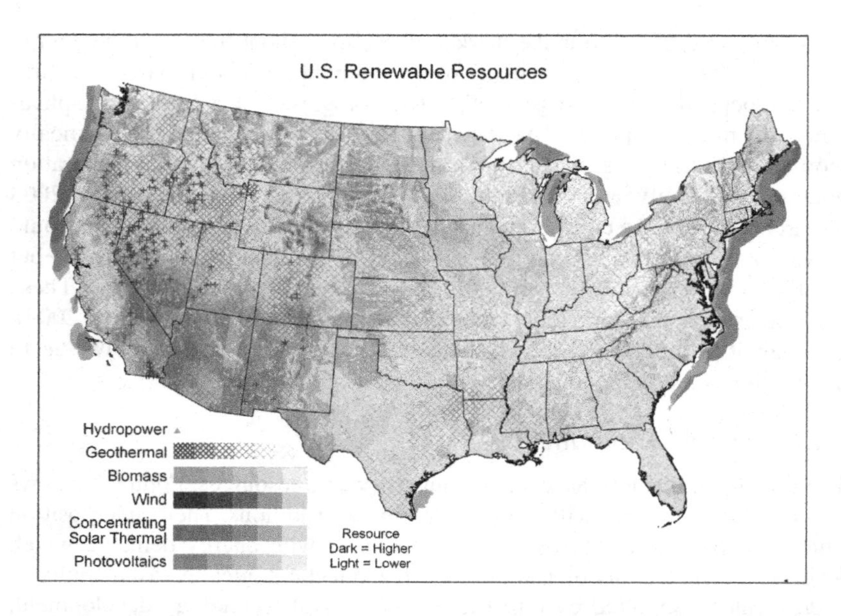

Resource	Solar PV/CSP)	Wind	Geothermal	Water Power	Biopower
Theoretical Potential	206,000 GW (PV) 11,100GW (CSP)	8,000 GW (onshore) 2,200 GW (offshore to 50 nm)	39 GW (conventional) 520 GW (EGS) 4 GW (co-produced)	140 GW	78 GW

FIGURE 23 Renewable Resources in the Continguous United States. *Source: NREL (2009) – see data notes in References section.*

continued development of the so-called "smart grid." An "intelligent" generation, transmission, and distribution system for energy will enable more renewables to engage with the energy infrastructure. Without this intelligence, it will be more difficult for the existing grid to incorporate larger amounts of intermittent RE. This situation can be compared to the interaction between automobiles, roads, and traffic lights. If we were to introduce increasing numbers of vehicles to our roads with no signals to direct the flow of traffic, the whole system would collapse quickly into accidents and blockages. It is this intelligence that enables vehicles to move through the system relatively seamlessly. The roads need to be accessible by the vehicles being introduced to the system, so more roads will need to be built to accommodate these new vehicles.

Similarly, power generated from sustainable sources such as wind and solar needs to be generated and introduced into the grid. The variability of these resources is greatly reduced with an increase in the number of distinct power generation sites at different geographic locations, and can be accommodated at a systems level with a combination of advanced forecasting, load shifting, demand response (e.g. "smart appliances" and interruptible loads). Reserve requirements and fast ramping generation will likely be needed, which could also be addressed by pumped hydro or compressed air energy storage, or through innovation in storage capabilities such as batteries, that will have to compete economically with options such natural gas combined cycle. Strong wind and sun resources may not be located near existing transmission and distribution lines, and these distant sites may be economic to build, but will require new transmission. Energy from these sources needs to be matched intelligently to the needs of the end user to integrate these renewable resources. These "smart" innovations are anticipated to allow for more dynamic control of loads, which enables power system operators to more nimbly match load and supply at higher penetrations (much greater than 20%) of renewables.

Similarly, infrastructure innovation and evolution may be needed for some, but not all alternative fuels. Hydrogen, for example, will require the build-out of hydrogen generation, storage, and distribution stations. Similarly, ethanol, as it penetrates beyond a blending agent, requires the build-out of both flex fuel fleets as well as high blend (e.g., 85% or more) ethanol distribution. This is not necessarily the case for "fungible biofuels" such as "green diesel" or bio-based gasoline, or JP-8 via Fisher Tropsche processes.

3.6.4. Financing Innovations

Financing innovations address up-front costs for renewable projects and long-term value. Because the global economic system has not successfully monetized the "value" of RE (or negative externalities such as GHG emissions), innovative financing measures attempt to rebalance the equation by mitigating up-front capital costs—this levels the playing field when compared to competing extractive energy projects. For example, the property-assessed clean energy (PACE) model assesses energy efficiency and renewable projects, which helps the regional public sector provide financing vehicles that are tied to the property where that project takes place. This allows the up-front capital costs of the project to be paid back over time through property taxes. This financing model is enacted through specific policy legislation (at this point, usually by amending existing legislation). Similarly, FITs are financial transactions mandated by specific policy, where RE generation must be purchased by the existing utility at a specified rate. This requirement enables the value proposition of renewable project developers to become more robust, which minimizes perceived risks and assures financial returns based on the high capital costs and long recovery period.

3.6.5. Policy Motivators

Finally, policies that require specific levels of renewables to be developed—such as a renewable portfolio standard (RPS) or a renewable fuel standard (RFS)—may continue to drive the market for renewable technologies. Such quantitative target policies are implemented as part of a broader portfolio of policies, including FITs, tax incentives, performance and service standards, net metering, and many others.

4. TECHNOLOGY/RESOURCE PROSPECTS

This section discusses the future prospects for RE technologies, including both cost trends and technology trends.

4.1. Overall Cost Trends

As depicted in Figure 24, RE technologies have experienced considerable cost reductions during the past decades. There is considerable literature on learning and cost reductions (Grübler, 2003; Nemet, 2006), indicating the importance of R&D as well as market growth. As these technologies continue to expand their market growth, as described above, additional cost reductions are anticipated.

The future cost trends show a narrower distribution of possible future costs. But, from today's perspective, the range of potential costs should remain rather broad. This view is based on uncertainties of technology innovation and market adaptation. That said, however, the historical learning trends of the RE technologies have shown robust temporal trends for the past 30 years. As technologies mature, however, diminishing marginal learning may be anticipated.

The costs of RE options must, of course, be considered within the context of competing technologies and the operational policy environment. For example, policies such as RPSs, FITs, and solar set-asides and direct subsidies for PV are mechanisms that have been employed (along with many others) to require and provide net positive investment environments for renewable technologies. These exist during a period when competing fossil fuel generation technologies, such as coal and gas turbines, may be lower cost on the margin (given fuel prices, cost of capital, etc.), and in which carbon externalities are not accounted for in the economic evaluation. Many of these more mature power generation technologies have experienced very little, if any, cost reductions during the past decades (Nemet, 2006).

This section reviews a few select technology trends in detail.

4.2. Wind Technologies

During the past 25 years, average wind turbine ratings have grown almost linearly with time since the introduction of 50 kW turbines in the early 1980s. Current commercial machines are rated at 1.5 MW to 3 MW for land-based turbines, and offshore turbines as large as 5 MW are being deployed with larger machines on the drawing boards of several manufacturers. During the past 25 years, wind turbine designers have predicted that the current generation of turbines had grown as large as they would ever be. However, with each new

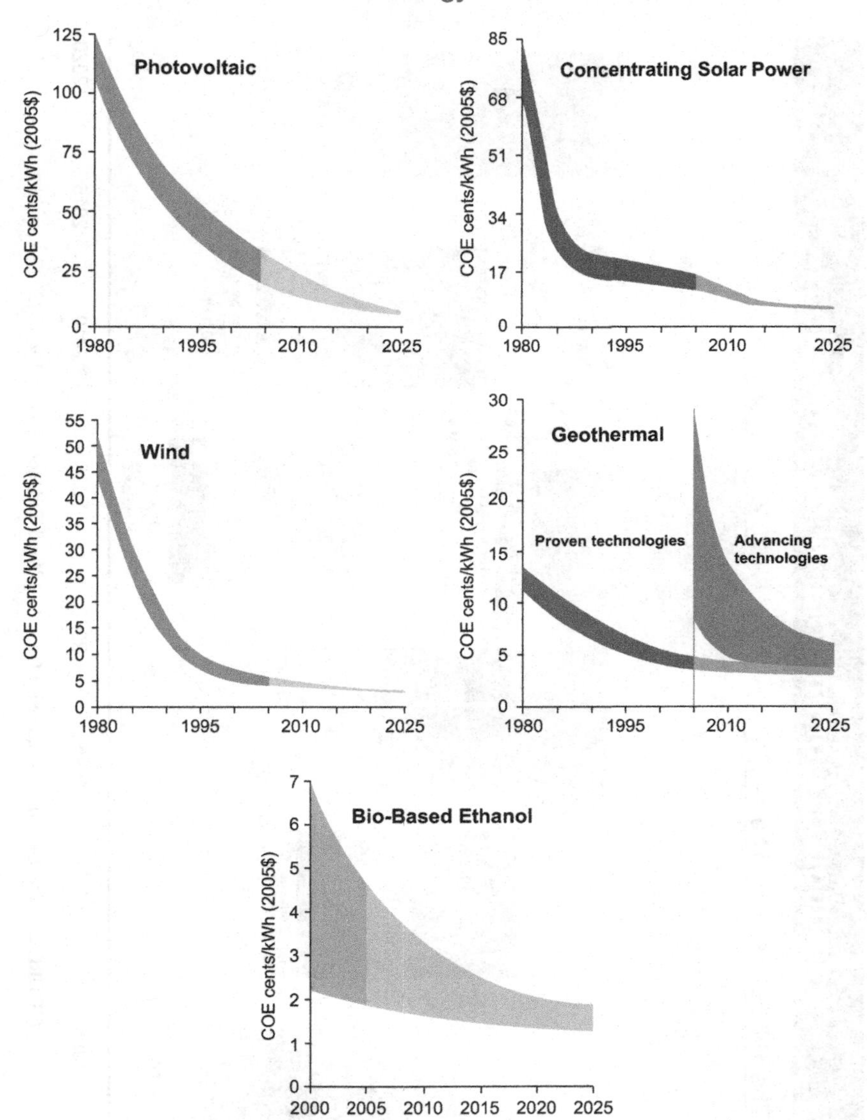

FIGURE 24 Renewable Energy Cost Trends (DRAFT). Note: These Graphs are Reflections of Historical Cost Trends NOT Precise Annual Historical Data. *Source: NREL (2005).*

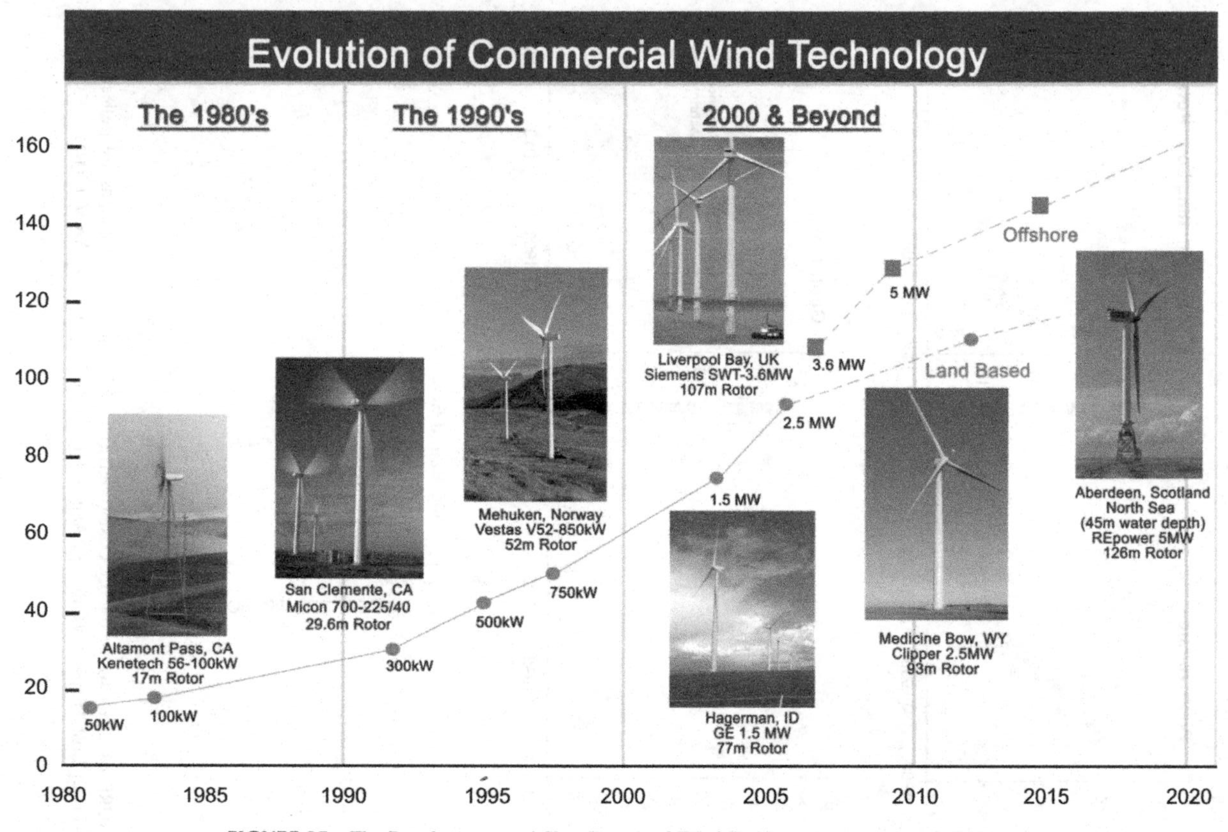

FIGURE 25 The Development and Size Growth of Wind Turbines. *Source: Musial (2007).*

generation of wind turbines, the size has increased along a linear curve and resulted in a reduction in life-cycle cost of energy with each increase in size. This impressive evolution of wind turbine technology is illustrated in Figure 25.

The long-term drive to develop larger turbines stems from a desire to take advantage of wind shear (wind speed increases with height above the ground) by placing rotors in the higher, much more energetic winds at greater elevations above ground. This is a major reason that the capacity factor of wind turbines has increased over time. However, there are constraints to this continued growth; in general, it costs more to build a larger turbine. The primary argument for a size limit for wind turbines is based on the "square-cube law." Roughly stated, it says that "as a wind turbine rotor increases in size, its energy output increases as the rotor-swept area (the diameter squared), while the volume of material, and therefore its mass and cost, increases as the cube of the diameter." In other words, at some size the cost for a larger turbine will grow faster than the resulting energy output and revenue, making scaling a losing economic game. Engineers have successfully skirted this law by changing the design rules with increasing size and removing material or by using material more efficiently to trim weight and cost. Figure 26 illustrates how successive generations of larger blade design have moved off the cubic weight growth curve (green line) to keep weight down and reduce blade material costs and beat the "square-cube" law (movement to blue curve, and more recently to weight/size ratios indicated by the red dots).

Land transportation constraints can also pose limiting factors to on-shore wind turbine growth. Cost-effective transportation can only be achieved by remaining within standard over-the-road trailer dimensions of about 4.1 m high

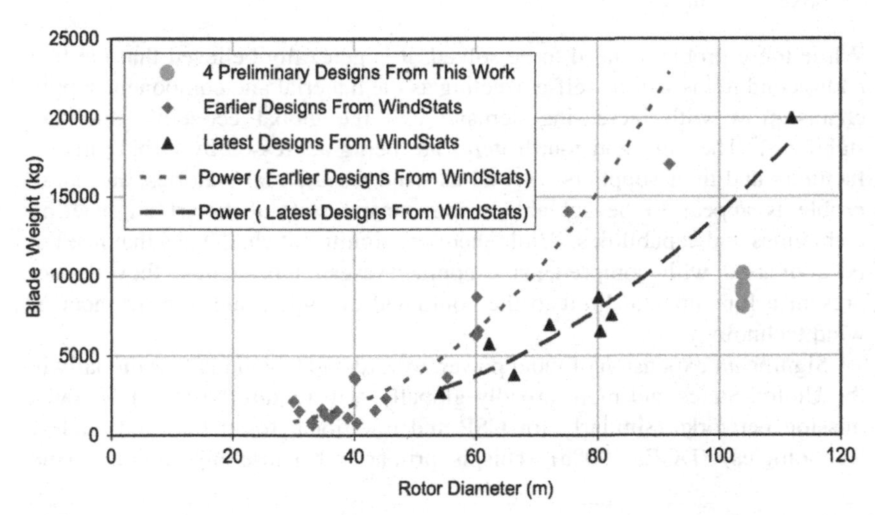

FIGURE 26 Reduced Growth in Blade Weight Due to the Introduction of New Technology. *Source: Griffin (2001).*

by 2.6 m wide depending on the road classification and the country. Rail transportation is often even more dimensionally limited. Offshore wind logistics are less constrained by size, but at this early stage of market development are constrained by availability of appropriately configured mobile support ships.

If advanced R&D provides better design methods, as well as new materials and manufacturing methods that allow the entire turbine to scale as the diameter squared, it would be possible to continue to innovate around the square-cube law and clever engineering may provide new transportation solutions.

4.2.1. Near-Term Land-Based Challenges

During the past few years, the price of land-based wind-generated electricity has been increasing after a 20-year downward price trend. Currently, the unsubsidized cost of wind energy sold to utilities in the United States is about 5 to 9 U.S. cents/kWh depending on the wind regime and many other financial factors as documented by Wiser and Bolinger (2007). Some of the reasons for this recent price increase have been attributed to the following near-term issues:

- A significant rise in material costs such as steel and copper, as well as transportation fuels, during the past several years.
- Fluctuating exchange rates in the global marketplace that increases the cost of a few critical components that can currently be manufactured only in a few countries.
- Transportation and installation cost increases due to the larger turbine size and larger crane requirements.
- Reliability problems with some major turbine components, such as gearboxes and blades.

While these problems need to be solved, it is generally believed that the first and second items will be self-correcting as the material and component supply chains grow with increasing demand and the global economic situation stabilizes.[5] The third and fourth items are being addressed by turbine manufacturers and their suppliers, as well as wind research laboratories, and these problems appear to be solvable using current state-of-the-art engineering techniques and capabilities. While there are significant challenges that need to be addressed with comprehensive applied research programs, they do not present a fundamental limit to the continued evolution and improvement of wind technology.

Significant expansion of wind power, toward and beyond 20% nationally in the United States and more broadly globally, will require build out of transmission networks (similarly for CSP and traditional fossil fuel and nuclear technologies) (DOE, 2008a). This is primarily because high-quality wind

5. There are many who believe higher commodity prices are here to stay as global demand for steel, concrete, copper, and fuel (particularly in China) remains high.

resources are often not located near demand centers. Integration of wind at these higher penetration levels has been studied in detail for Europe and many areas of the United States. In general, these studies indicate technical feasibility for wind penetration ranging from 20%–40%, and detail the transmission requirements, ancillary services, and operational issues to be addressed (UWIG, 2009).

4.2.2. Potential Future Land-Based Turbine Technology Improvements

Both the EU (2008) and the U.S. DOE (2008) have identified a broad array of wind energy R&D activities that have the potential to significantly improve the cost and performance of wind technology, which are generally quite similar in perspective (also see WindPACT study, 1999). The U.S. DOE report titled *20% Wind Energy by 2030* (DOE, 2008a) summarizes the range of potential impacts on energy production and capital costs from a number of these advances—these ranges are shown in Table 2. Although not all of these potential improvements may be achieved, there is sufficient potential to warrant continued R&D. The most likely scenario, as shown in Table 2, is a sizeable increase in energy production with a modest drop in capital cost (compared to 2002 levels, which are the baseline for the estimates in Table 2).

In summary, current thinking in the United States and Europe indicates that no "big breakthrough" is on the horizon for land-based wind technology. However, many evolutionary R&D steps executed with technical skill can cumulatively bring about a 30%–40% improvement in the cost-effectiveness of wind technology during the next two decades. This continued learning (cost reduction with increased market use) compares favorably to more mature technologies such as fossil fuel power. Technology improvement as described above will help wind become more competitive with conventional generation technologies. Cost reductions can also be achieved through good practice in siting and improving O&M strategies.

4.2.3. Offshore Wind Turbine Technology Status

Worldwide offshore wind energy resources are abundant, indigenous, and broadly dispersed and many of the most expensive and highly constrained electric load centers are also located on or near a coastline. At the end of 2007, the wind industry had developed 25 offshore projects in European waters, many of them large-scale and fully commercial with a total capacity of around 1100 MW. The European Wind Energy Association (2007; 2008) foresees offshore wind installations growing to between 20 and 40 GW depending on the policies implemented, while the U.S. *20% Wind Energy by 2030* report (DOE, 2008a) indicates that approximately 50 GW could be offshore. The current estimate for offshore wind energy is in the range of 10 to 15 U.S. cents/kWh depending on the wind conditions, water depth, and installation complexity. There is less

TABLE 2 Areas of Potential Technology Improvement

Technical Area	Potential Advances	Increments from Baseline (Best/Expected/Least, Percent)	
		Annual Energy Production (%)	Turbine Capital Cost (%)
Advanced Tower Concepts	* Taller towers in difficult locations * New materials and/or processes * Advanced structures/foundations * Self-erecting, initial, or for service	+11/+11/+11	+8/+12/+20
Advanced (Enlarged) Rotors	* Advanced materials * Improved structural-aero design * Active controls * Passive controls * Higher tip speed/lower acoustics	+35/+25/+10	−6/−3/+3
Reduced Energy Losses and Improved Availability	* Reduced blade soiling losses * Damage tolerant sensors * Robust control systems * Prognostic maintenance	+7/+5/0	0/0/0

Advanced Drive Trains (Gearboxes, Generators, and Power Electronics)	* Fewer gear stages or direct drive * Medium-/low-speed generators * Distributed gearbox topologies * Permanent-magnet generators * Medium-voltage equipment * Advanced gear tooth profiles * New circuit topologies * New semiconductor devices * New materials (GaAs, SiC)	+8/+4/0	−11/−6/+1
Manufacturing Learning	* Sustained, incremental design and process improvements * Large-scale manufacturing * Reduced design loads	0/0/0	−27/−13/−3
Totals		+61/+45/+21	−36/−10/+21

Note: Since the 2002 baseline for this analysis, there has been a sizeable improvement in capacity factor—from just over 30% to almost 35%—while capital costs have increased due to large increases in commodity costs and a drop in the value of the dollar. Therefore, working from a 2006 baseline, we can expect a more modest increase in capacity factor. The 10% capital cost reduction is still possible, although beginning from a higher 2008 starting point, because commodity prices may not drop back to 2002 levels.

Source: DOE (2008a).

operational experience with offshore wind, so the long-term operations and maintenance costs are still in question.

The typical shallow water baseline offshore wind turbine is essentially a marinized version of the standard land-based turbine with some system redesigns to account for ocean conditions (Musial, 2007). These modifications include structural upgrades to the tower to address the added loading from waves, pressurized nacelles and environmental controls to prevent corrosive sea air from degrading critical drive train and electrical components, and personnel access platforms to facilitate maintenance and provide emergency shelter. Offshore turbines must have corrosion protection systems at the sea interface, and high-grade marine coatings on most exterior components. For marine navigational safety, turbine arrays are equipped with warning lights, vivid markers on tower bases, and fog signals. To minimize expensive servicing, offshore turbines may be equipped with enhanced condition monitoring systems, automatic bearing lubrication systems, onboard service cranes, and oil temperature regulation systems, all of which exceed the standard for land-based designs. Lightning protection is mandatory for both land-based and offshore systems. The major portion of the turbines' nacelle covers and towers are painted light blue or grey colors to minimize their visual impact especially at long distances.

Today's offshore turbines range from 2 to 5 MW and are typically represented by architectures that comprise a three-bladed horizontal-axis upwind rotor, nominally 80 m to 126 m in diameter. Tip speeds of offshore turbines are typically higher than land-based turbines at 80 m/s or more. The drivetrain topology consists of a modular three stage hybrid planetary-helical gearbox that steps up to generator speeds between 1000 and 1800 rpm, which is generally run with variable speed torque control, although direct drive generators may prove to be a viable alternative. Tower heights offshore are lower than land-based turbines because wind shear profiles are less steep, tempering the energy capture gains sought with increased elevation.

The offshore foundation system differs most substantially from land-based turbines. The baseline offshore technology is deployed in arrays using monopiles at water depths of about 20 m. Monopiles are large steel tubes with wall thickness of up to 60 mm and diameters of 6 m. The embedment depth will vary with soil type but a typical North Sea installation will require a pile that is embedded 25 m to 30 m below the mudline and that extends above the surface where a transition piece with a flange to fasten the tower is leveled and grouted on. The monopile foundation requires a special class of installation equipment for driving the pile into the seabed and lifting the turbine and tower in place. Mobilization of the infrastructure and logistical support for a large offshore wind farm is a significant portion of the system cost. The wind turbines are arranged in arrays that take advantage of the measured prevailing wind conditions at the site. Turbine spacing is chosen to minimize aggregate power plant power losses, interior plant turbulence, and the cost of cabling between turbines just as land-based wind farms do, except water depth presents a siting obstacle just as rough terrain does on land.

4.2.4. Potential Future of Offshore Turbine Technology Improvements

There are many engineering challenges for offshore wind technology associated with the greatly increased scale and the added complexity of the offshore wind and ocean environment. The European *UpWind* research project (Jensen, 2007) envisions the turbines growing in scale to 8–10 MW and having rotor diameters greater than 120 m, which is a challenge to design, build, install, and operate at sea. The *UpWind* project has been established to address this multitude of engineering challenges and the project addresses the following technical areas:

- Aerodynamics and aeroelasticity
- Rotor structure and materials
- Foundations and support structure
- Control systems
- Remote sensing
- Condition monitoring
- Flow
- Electrical grid
- Management

The *UpWind* project also addresses the systems integration topics of: integrated system design and standards, metrology, training and education, innovative rotor systems, electricity transmission and conversion, smart rotor blades, and system up-scaling, through the use of a matrix organizational structure. In general, offshore wind technology is expecting to gain improvements similar to those envisioned for land-based wind turbines, but at a much larger machine scale and in a more hostile operating environment. Clearly, to reach the turbine rotor sizes planned, the researchers must continue to beat the "square-cube law" and design larger and lighter rotors, as illustrated in Figure 27.

4.3. Solar Technologies

The Earth receives more energy from the sun in one hour than the global population uses each year. Not only does solar energy have a larger technical potential than any other RE resource, it is readily available in every inhabited environment. However, capturing this large but diffuse energy resource will require overcoming both technical and economic challenges.

As highlighted earlier in the chapter, in recent years, solar PV market growth was centered in Germany, Japan, Spain, Italy, and South Korea; and in California and New Jersey in the United States. The installed capacity

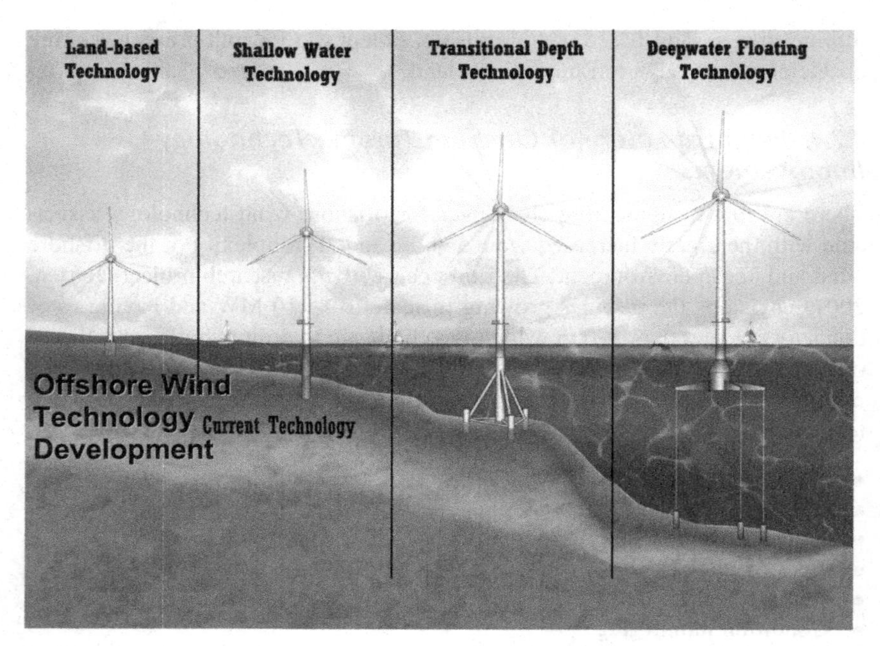

FIGURE 27 These Pathways Show How Wind Turbine Technology Evolves from Land-Based Turbines to Shallow Water Offshore and Then Into Deeper Waters. *Source: Musial (2007).*

grew about 50% in 2006, 2007, and 2008, with a total installed capacity at the end of 2008 of nearly 16 GW (including off-grid installations) (REN21, 2009).

Solar technologies have the potential to significantly reduce global carbon emissions. However, reaching solar penetration levels high enough to effectively reduce emissions will require significant growth in the global PV and CSP markets, which will need to be driven by continued price reductions and efficiency gains. Going beyond this level will bring additional challenges such as integrating intermittent PV generation into the electricity grid and building additional transmission capacity linking CSP in areas with a good solar resource to load centers (Denholm et al., 2009).

4.3.1. Solar Cost Trends

PV modules have experienced significant efficiency improvements during the past few decades (from fundamental design aspects and also from process improvements), and have undergone unprecedented market growth during recent years. This growth has been supported by government subsidies and other financial incentives. Module prices in 2007 and 2008 were nearly $4/Watt-peak (Wp); module prices are at this time still higher than the often cited $1/Wp needed for grid parity. However, reductions in supply chain constraints and increased manufacturing output (particularly of thin films) have led to a significant drop in wholesale prices. One manufacturer, FirstSolar, announced in late 2009 that they had achieved less than $1/W manufacturing costs. This does not include balance of systems, installation, or financing costs.

Although PV prices increased in the past several years, prices have been steadily falling during the past few decades (Figure 27). The primary force driving PV prices has been the market—supply and demand factors, including complex dynamics on the supply side, which are now being addressed through new silicon supply and increased manufacturing throughput for thin-film technologies. Although prices have increased in recent years, PV prices started falling again after silicon prices eased and demand weakened from Germany and Spain in 2009. In Figure 28, the green curve represents nominal prices while the blue curve represents the prices in 2007 dollars (this does not include thin-film technologies).

PV module prices have been steadily falling during the past few decades, primarily due to technology improvements—using lower cost feedstocks, efficiency increases, thinner solar cells, reduced technical losses, increased manufacturing throughput, and so on. The price of raw silicon has already started to decrease due to increased supply, a trend expected to continue through 2015. Table 3 shows the spot market prices for polysilicon feedstocks, and manufactured wafers and solar cells. If the polysilicon price continues to fall to $50/kg, solar module costs may be reduced to less than $3/Wp. Depending on PV demand, this decrease in silicon price could be used to increase profit margin or to lower PV system prices.

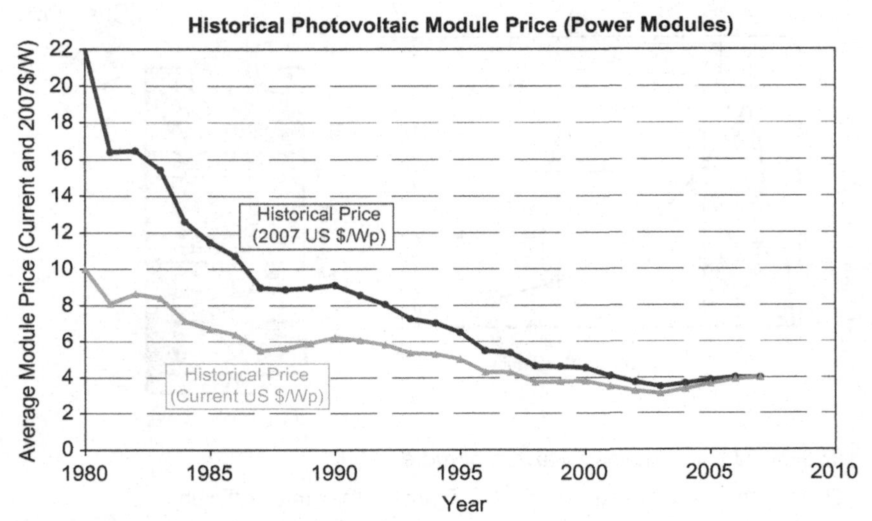

FIGURE 28 Photovoltaic Module Prices. *Source: Navigant Consulting (2008).*

TABLE 3 Polysilicon Spot Market Price Drop

Polycrystalline Silicon	2007 Price	2008 Price (Sept. 2008)
Materials	$400–430/kg	$350/kg for solar grade
Wafers	$11–12/in.	$10-10.50/in.
Cells	$3.4–3.8/W	$3.2-3.4/W

Source: Taipei and Hwang (2008).

Solar industry analysts predict that with new polysilicon feedstock manufacturing plants coming online in 2008, supply bottlenecks will decrease, possibly to the point where supply outpaces demand. If this happens, polysilicon "oversupply" could cause a sharp decline in silicon prices. However, sudden price drops in silicon may be self-correcting as new PV manufacturing technologies continue to reduce PV costs, increasing demand for both PV and silicon (Price, 2010). A comprehensive report by the International Energy Agency (IEA 2010) indicates investment costs for concentrating solar power (CSP) have the potential to be reduced by 30% to 40% in the next decade.

4.3.2. Solar Power Research Prospects

PV cell development continues to bring new advances in cell design, new materials systems, and new record efficiencies. A broad technology portfolio

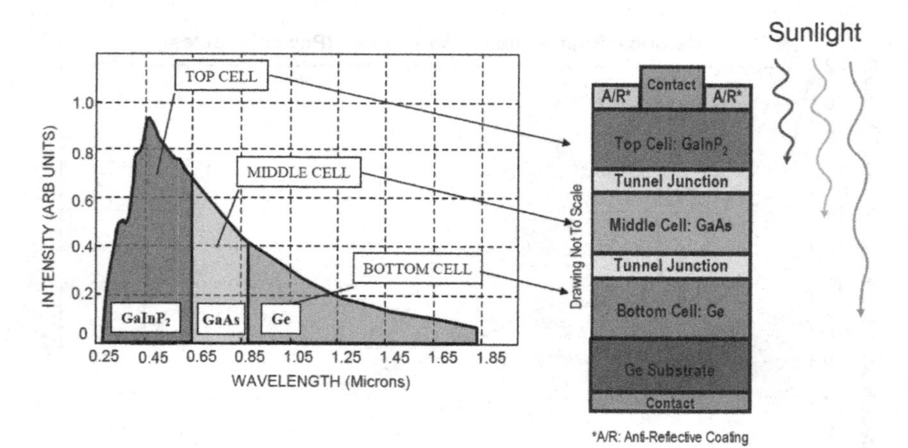

FIGURE 29 A Diagram of a ~41% Efficient Multiple Junction Concentrator Cell (Developed by Spectrolab/Boeing/NREL). *Source: Geisz (2007).*

includes PV R&D for next-generation technologies focused on increasing efficiency (multijunction), reducing costs (organic, dye-sensitized), or both (quantum dots, etc.). For example, multijunction single crystal cells (typically for concentrator or space systems) have been demonstrated in the lab with more than 40% efficiency (Kurtz, 2009). A typical design is shown in Figure 29, which involves more than 20 thin layers and extreme control during deposition and processing.

Thin-film organic cells are now approaching 8% efficiency, with prospects for inexpensive manufacturing at scale. These new materials will compete with CdTe (cadmium telluride) and CIGS (copper indium gallium selenide) thin films, which are commercial today and gaining economies-of-scale advantages (Figure 30).

Even more revolutionary concepts for new solar PVs are under investigation, stemming from growing interest and capabilities in nanostructures and quantum physics. For example, Nozik (2008a) reported quantum efficiencies greater than 1 for the creation of multiple excitons from a single photon in properly sized quantum dots (Figure 31). Theoretical calculations of electrical conversion efficiencies up to 80% may be possible (Nozik, 2008b).

These novel approaches to solar conversion offer improvements to the overall cost-effectiveness of solar electricity. Other approaches, as briefly mentioned above, include large-scale manufacturing of thin films at moderate conversion efficiencies, as well as systems approaches that include concentration, combined electrical and thermal output, and integration into building

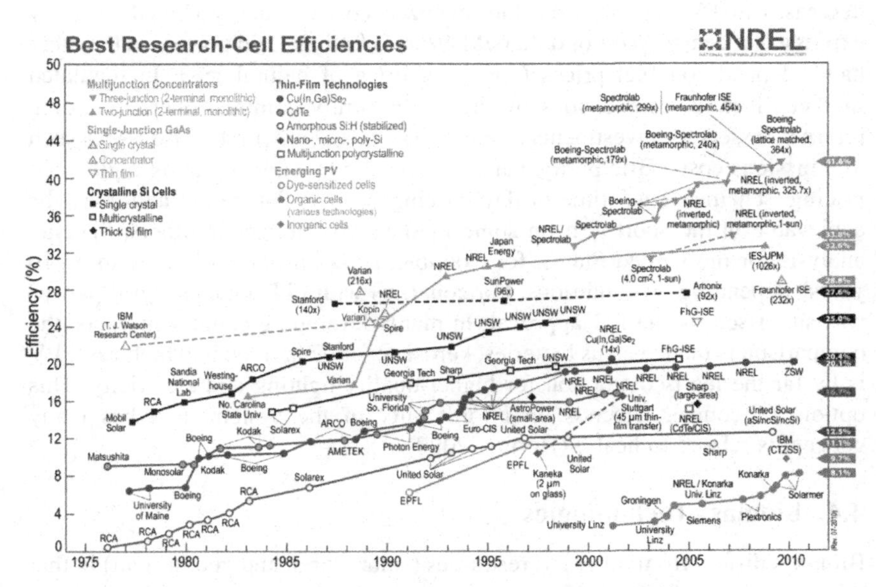

FIGURE 30 PV Cell Efficiency. *Source: Kurtz 2009*

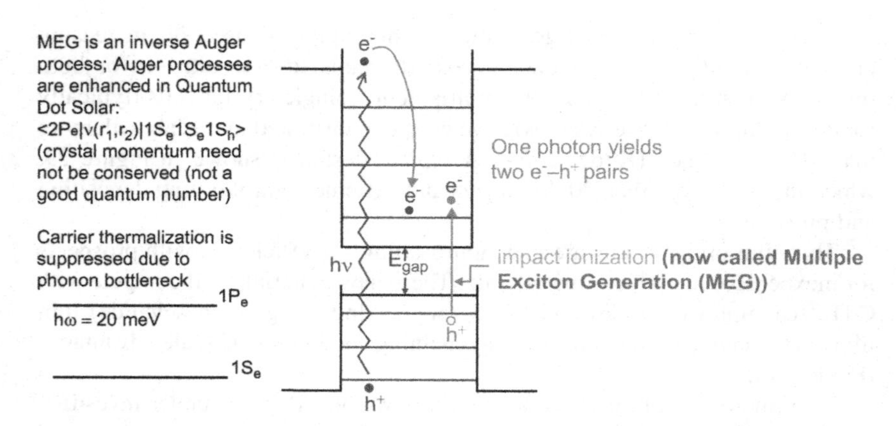

FIGURE 31 Enhanced Photovoltaic Efficiency in Quantum Dot Solar Cells by Multiple Exciton Generation. *Source: Nozik (2008a).*

materials such as façade glass or roofing. The multitude of approaches shows that a number of solutions may be effective, and will be adaptable for the range of market conditions across both developed- and developing-country applications.

A classical argument for solar PV (and other) technology has been "grid parity." This argument is based on the delivered price per kW/hr, which is a function of capital price, financing cost and duration, and O&M costs compared to the price of "grid-delivered" power. Clearly, with continued decreases in PV capital costs, the levelized cost of energy (LCOE, usually expressed in cents/kW/hr or dollars/MW/hr) is falling. Grid power, on the other hand, depends on fuel prices (e.g., spot price of natural gas). In regulated markets, it involves decisions by the public utility commissioners regarding return on assets for investments in transmission, and other rate-case issues such as emissions costs. Grid parity, per se, is specific to a given pricing region and pricing scheme (e.g., time-of-day pricing versus flat rate), and may be achievable in the short term in some markets, and longer in others. Specific analysis for the United States, for example, shows a range of "now to 5–10 years" depending on conditions. For countries with FIT policies, "grid parity" was surpassed by the FIT approval. In many developing country markets, the comparison is made versus batteries, kerosene, or open fires; in which case, PV is by far the least-cost option for higher-quality lighting and electricity. This option, of course, depends on affordability of the system, and has many cobenefits related to health (Horton, 2009).

4.4. Biomass Technologies

Biomass differs from other RE resources (solar, wind, and geothermal) in that it has a non-zero fuel cost. The total biomass cost, amounting to $20–$60 per ton, stems from planting, management, harvesting, collecting, and transportation. Because of the transportation cost, biomass power plants are typically located within 50 miles of available resources. This limits plant capacity to about 50 MW (although more efficient integrated gasification combined cycle or IGCC plants could be twice that size), with the average plant producing about 20 MW. Because the biomass resource is well-dispersed, there are many opportunities to locate new plants near existing transmission lines (Kutscher, 2009a).

The American Solar Energy Society's Tackling Climate Change study (ASES, 2007) concluded that biomass power could provide 110,000 MW of electric power in the U.S. by 2030, enough to provide about one-third of the electricity currently generated by coal. The main hold-up in deployment is that its electricity cost (about 8 cents per kilowatt-hour) is not currently competitive with coal. However, biomass is much cheaper than new nuclear power plants. And if CO_2 is eventually priced at about $30 per ton of CO_2 or more, the cost of biomass power will be on a par with that of coal.

Biomass power production offers many benefits. Capturing biomass waste products reduces needed landfill capacity. Collection of forest residues reduces fire danger. Perhaps most important, biomass power tackles climate change in a number of ways. It can help shift agricultural emissions from methane to CO_2, which, pound for pound, would reduce the global warming impact by a factor of 25 (Kutscher, 2009a). Studies have shown that the GHG emissions associated with burning forest residues are much lower than if those same residues are allowed to decay on the forest floor. And biomass power plants can provide baseload power, thus displacing coal plants, the worst carbon emitters in our nation's electric system.

Biomass also has the potential to fight climate change in a major way that other renewable resources cannot. Biomass power using appropriate feedstocks is usually considered to produce little to no net carbon emissions, exclusive of harvesting and transport impacts. (Although biomass emits CO_2 when it is burned, it removes about a near-equivalent amount of CO_2 from the air as it grows.) But if carbon capture and storage technologies were used in conjunction with high-efficiency biomass-IGCC plants, biomass power could potentially become carbon-negative and help decrease atmospheric CO_2 (Kutscher, 2009a).

4.5. Geothermal Technologies

Geothermal power plants are similar to coal plants in that they generate electricity by producing a high-temperature, high-pressure vapor that passes through a turbine, which spins an electric generator (Kutscher, 2009b). There are a few ideal places in the world where dry (superheated) steam exists near the surface. After drilling, that steam can simply be routed directly to a steam turbine. Direct steam plants are used in Northern

California at The Geysers, the world's largest geothermal power plant complex. More commonly, geothermal wells tap pressurized hot water. If the temperature of that water is higher than about 350°F (or about 175°C), it can be rapidly boiled in a low-pressure flash tank where a fraction of the water becomes steam. The steam is then routed to a steam turbine.

For lower temperature resources, the hot geothermal fluid is passed through heat exchangers, where it boils a secondary fluid having a lower boiling point than water, like pentane or isobutane. The resulting vapor spins a specially designed turbine.

These are called binary-cycle plants. While all geothermal power plants have very low emissions, binary-cycle plants have virtually none, because all of the geothermal fluid is returned to the ground. Binary-cycle plants also have the advantage that they do not result in any water draw-down from a geothermal reservoir, although they can decrease reservoir temperature over time.

The best geothermal resources, so-called hydrothermal resources, have three qualities: high near-surface temperatures (preferably at least 240°F, or 116°C; the higher the better), fluid content in the form of pressurized water or steam, and permeable rock. All of the existing geothermal power plants in the world are sited on hydrothermal resources.

But these resources are limited. Worse, we don't even have a good handle on how much we have. A 2008 U.S. Geological Survey (USGS) study estimated that the United States has power production potential from identified hydrothermal resources of between 4000 MW (95% confidence level) and 13,000 MW (5% confidence level). When the USGS attempts to include potentially undiscovered resources, the range jumps to 11,000–90,000 MW (with a mean of about 40,000 MW). This is a big range, and this resource uncertainty is a key question mark for the future exploitation of geothermal resources (Kutscher, 2009b).

The resource picture gets brighter if we ease our expectations for the three qualities of good geothermal resources. If we are willing to drill down 3 to 10 kilometers, there are wide areas throughout the Western United States with adequate temperature. And although those areas tend to be dry and low in permeability, we can potentially inject water at high pressure and fracture the rock, as is often done to enhance oil and gas recovery. This is the concept referred to as EGS. When the same USGS study estimated EGS resources, the range of power production potential jumped to 350,000–720,000 MW with a mean of about 500,000 MW, which would be sufficient power to provide virtually all the annual kilowatt hours of electricity needed in the United States.

EGS has been researched in France and Australia and continues to be a topic of intense investigation, given the ubiquity and size of the global resource.

A 2006 Massachusetts Institute of Technology (MIT) report, *The Future of Geothermal Energy* (MIT, 2006), was quite optimistic about EGS and helped generate considerable interest. But while the EGS resource is extremely large, exploiting it in a cost-effective way is no trivial matter. For one thing, drilling costs go up exponentially with depth. And there is the challenge of properly fracturing the rock without inducing seismicity. The concept involves an injection well where high-pressure water is pumped down into the hot rock, creating permeability. The water then picks up heat as it flows to one or more nearby production wells. But fracturing the rock in a way that allows the flowing water to communicate with a large volume of hot rock without significant loss needs to be demonstrated. Water loss is not only an obvious problem for arid regions typical of EGS resources, but it also represents wasted pumping power.

The MIT study concluded that to be economical, an EGS project must obtain a continuous production well water flow rate of 80 kilograms per second at 200°C (about 390°F). Limited water availability and high drilling costs may result in shallower wells and lower temperatures, supporting binary-cycle plants. Thus far, the highest achieved flow rate at an experimental EGS site has been 25 kg per second.

A 2008 DOE study, *An Evaluation of Enhanced Geothermal Systems Technology* (DOE, 2008b), evaluated the MIT analysis and described the significant amount of R&D needed to create and sustain a viable EGS reservoir. The DOE has had limited funding to tackle this. That changed recently when the Obama administration announced that $80 million to support EGS R&D will be provided from the American Reinvestment and Recovery Act. Even this is a small amount compared to what will be needed. But the potential for EGS to produce large quantities of baseload power justifies the expenditure and the risk (Kutscher, 2009b).

CONCLUSIONS

Stabilizing the atmospheric concentration of anthropogenic GHGs will require decarbonization of the global economy during the next century. Even with reasonable expectations for nuclear power and capture and storage of CO_2 from remaining fossil fuel use, this goal implies massive expansion of renewable, CO_2-free, sources of energy. Further, RE systems offer tremendous opportunity for distributed energy for rural communities, not only providing cleaner, more sustainable energy, but also—and perhaps more important—means for income generation, education, and reduction of health impacts of nonsustainable energy supplies. Such "distributed energy systems" must be optimized to the local resource and demand conditions. For example, optimal systems may include PV, wind, and pico/mini or small hydro. Complementary installations of solar thermal systems may be very attractive in markets ranging from home systems to industrial process heat.

And the process is already under way. Global wind capacity doubled from about 50 GW in 2005 to more than 120 GW in 2008; worldwide shipments of solar PVs increased from about 200 MW in 1999 to nearly 6.9 GW in 2008—according to some estimates, it may reach 60 GW by 2015; and global biofuel production tripled from 4.8 billion gallons in 2000 to about 24 billion in 2009. Yet, even with this rapid growth, these three sources today account for only about 3% of global primary energy use. Continued, even accelerated, expansion of RE technologies is poised to contribute to global, state, local, and individual goals of reliable, sustainable energy. Under predictable policies that enable a "levelized playing field" (inclusive of environmental externalities and cobenefits), RE technologies will continue to grow at a double-digit pace. The benefits of renewables are multiple: local and global environmental benefits, health and education cobenefits, energy security and balance of trade for importing countries, and the support for local employment. Given the breadth of benefits, renewables are likely to continue exceptionally strong growth; and, if climate change scenarios are indicative, will help the world stabilize and reduce global GHG emissions to an acceptable level to avoid drastic adaptation requirements in the next century.

Multiple global models have analyzed possible future scenarios (Edenhofer, 2010; EMF, 2009). These scenarios, while highly dependent on the model and modeling assumptions, indicate orders of magnitude increases in the use of RE over the next decades, which represent an addressable market of billions to trillions worldwide.

Based on today's available assessments/studies, can we answer the question: Will renewables grow to play a significant role in our energy future at reasonable cost? The answer, of course, "depends" on continued technology progress and stable policy environments that create sustainable, profitable business opportunities. Another question remains: Can renewables meet more than 20% of global energy needs by 2050? Technically, the answer is yes"; but operationally, again, answer depends on commitments to achieve the attributes that RE technologies offer, and the competitive positioning relative to other options. To grow from 3.4% of global energy today to a significant (and some studies indicate technical feasibility of a majority) share of US and global energy supply will require substantial investment and a willingness from general consumers, business, and government and a concerted effort to build out the necessary infrastructure and adapt systems to accommodate the necessary changes in operations. Renewable energy potential is both ubiquitous and well more than 200x the total U.S. energy consumption today. Realizing this potential represents both a significant opportunity and challenge, and would offer significant portfolio of benefits addressing energy security, economic prosperity and environmental stewardship.

BIBLIOGRAPHY

Accenture. (2008). *Biofuels' time of transition: Achieving high performance in a world of increasing fuel diversity.* Accenture. *http://newsroom.accenture.com/article_display.cfm?article_id=4747. http://www.accenture.com/Global/Services/By_Industry/Energy/R_and_I/BiofuelsStudyPartTwo.htm*

American Solar Energy Society (ASES). (2007). *Tackling climate change in the U.S.* Chuck Kutscher (Ed.). American Solar Energy Society, January 2007. *www.ases.org/images/stories/file/ASES/climate_change.pdf.*

American Wind Energy Association (AWEA). (2010). *AWEA Year End 2009 Market Report.* http://www.awea.org/publications/reports/4Q09.pdf

Denholm, P., Drury, E., Margolis, R., & Mehos, M. (2009). Solar energy: The largest energy resource. In F. P. Sioshansi (Ed.), *Generating Electricity in a Carbon-Constrained World.* Elsevier.

Department of Energy (DOE). (2009). 2008 renewable energy data book. Office of Energy Efficiency and Renewable Energy (EERE). *www1.eere.energy.gov/maps_data/pdfs/eere_databook.pdf*

Department of Energy (DOE). (2010). "2009 Renewable Energy Data Book." Published by the Office of Energy Efficiency and Renewable Energy (EERE), August 2010. Accessed August 2010 at http://www.nrel.gov/docs/fy10osti/48178.pdf

Department of Energy (DOE). (2008a). *20% wind energy by 2030: Increasing wind energy's contribution to the U.S. electricity supply.* In: *NREL Report No. NREL/SR-500−41869; DOE/GO-102008-2567.* Golden, CO: National Renewable Energy Laboratory. *www.eere.energy.gov/windandhydro.*

Department of Energy (DOE). (2008b). *An evaluation of enhanced geothermal systems technology.* Office of Energy Efficiency and Renewable Energy (EERE). *www1.eere.energy.gov/geothermal/pdfs/evaluation_egs_Tech_2008.pdf.*

Edenhofer, O. (2010). The economics of low stabilization. *The Energy Journal, 1*(1), page numbers.

Energy Modeling Forum (EMF). (2009). *http://emf.stanford.edu/*

European Commission (EC). (2008a). EUROSTAT 2008 *http://epp.eurostat.ec.europa.eu/portal/page?_pageid=1090, 30070682, 1090_33076576&_dad=portal&_schema=PORTAL.* http://epp.eurostat.ec.europa.eu/portal/page/portal/eurostat/home/

European Commission (EC). (2008b). Second strategic energy review − securing our energy future. *European Commission. http://ec.europa.eu/energy/strategies/2008/2008_11_ser2_en.htm.*

European Wind Energy Association (EWEA). (2007). *Delivering offshore wind power in Europe. www.ewea.org/fileadmin/ewea_documents/images/publications/offshore_report/ewea-offshore_report.pdf.*

European Wind Energy Association (EWEA). (2008). *Wind energy statistics. www.ewea.org/index.php?id=180.*

Geisz, J. F., Kurtz, S., Wanlass, M. W., Ward, J. S., Duda, A., Friedman, D. J., et al. (2007). *High-efficiency GaInP/GaAs/InGaAs triple-junction solar cells grown inverted with a metamorphic bottom junction.* NREL Report No. JA- 520−41602.

Glauber, J. (2008). Food and Agriculture Organization (FAO). Statement before the Committee on Energy and Natural Resources, United States Senate, June 12.

Global Wind Energy Council (GWEC). (2010). Global Wind 2009 Report. *http://www.gwec.net/fileadmin/documents/Publications/Global_Wind_2007_report/GWEC_Global_Wind_2009_Report_LOWRES_15th.%20Apr.pdf.*

Greentech Media. (2008). The venture power report, 2006−2008. Greentech Media. *http://www.greentechmedia.com/articles/read/the-2008-greentech-market-taxonomy-342/ and http://www.greentechmedia.com/articles/category/finance-and-vc/*

Griffin, D. A. (2001). *WindPACT turbine design scaling studies technical area 1 – composite blades for 80- to 120-Meter Rotor.* National Renewable Energy Laboratory, Golden, CO. NREL Report No. SR-500-29492, p 44. *www.nrel.gov/publications/*, *http://www.nrel.gov/docs/fy01osti/29492.pdf*

Grübler, A. (2003). *Technology and global change*. Cambridge University Press.

Horton, R. (Ed.). (2009). *Health and climate change*. The Lancet, November. *www.thelancet.com/*. *http://www.thelancet.com/journals/lancet/article/PIIS0140-6736(09)61994-2/fulltext#article_upsell*.

International Energy Agency (IEA). (2010). "Technology Roadmap–Concentrating Solar Power" Published by the International Energy Agency, 2010. Accessed November 2010 at *http://www.iea.org/papers/2010/csp_roadmap.pdf*.

International Energy Agency (IEA). (2008a). *World energy outlook 2008*. *www.worldenergyoutlook.org/2008.asp*.

International Energy Agency (IEA). (2008b). *Trends in photovoltaic applications: Survey report of selected IEA countries between 1992 and 2007*. International Energy Agency. *www.iea-pvps.org/products/download/rep1_17.pdf*.

International Energy Agency (IEA). (2009). *World energy outlook 2009*. *www.worldenergyoutlook.org/docs/weo2009/WEO2009_es_english.pdf*.

International Energy Agency (IEA) Bioenergy. (2009). Bioenergy – a sustainable and reliable energy source. *A review of status and prospects*. IEA Bioenergy: ExCo: 2009:05 37.

Jensen, P. H. (2007). Upwind: Wind energy research project under the 6. framework programme. In L. Sønderberg Petersen & H. Larsen (Eds.), *Energy solutions for sustainable development. Proceedings. Risø international energy conference 2007, Risø (DK), 22–24 May 2007*. Risø National Laboratory (DK), Systems Analysis Department; Risø National Laboratory (DK), Information Service Department. Risø-R-1608(EN), pp. 135–139. *www.risoe.dtu.dk/rispubl/reports/ris-r-1608.pdf*.

Karsner, A. (2008). Department of Energy (DOE). Statement before the Committee on Energy and Natural Resources, United States Senate. June 12.

Kazmerski, L. (2010). Updated and maintained for the National Renewable Energy Laboratory (NREL).

Kurtz, S. (2009). *Opportunities and challenges for development of a mature concentrating photovoltaic power industry (revision)*. NREL Report No. TP-520-43208, p. 27.

Kutscher, C. (2009a). The biomass solution. *Solar Today*, November/December. American Solar Energy Society (ASES).

Kutscher, C. (2009b). Can geothermal power replace coal? *Solar Today*, September/October. American Solar Energy Society (ASES).

Massachusetts Institute of Technology (MIT). (2006). *The future of geothermal energy – impact of enhanced geothermal systems (EGS) on the United States in the 21st century*. MIT.

Musial, W. (2007). Offshore wind electricity: A viable energy option for coastal USA. *Marine Technology Society Journal, 41*(3), 32–43. National Renewable Energy Laboratory, Golden, CO. NREL Report NO. NREL/JA-500-41338. *www.nrel.gov/publications/ http://www.ingentaconnect.com/content/mts/mtsj/2007/00000041/00000003/art00005?token=005218f6d876228341333c4a2f7a40386f3847462823743b624f6d62222c227e372530332976539a8e*

National Academy of Sciences (NAS). (2010). *Electricity from renewable resources: Status, prospects and impediments*. Washington, DC: National Academy of Sciences.

National Renewable Energy Laboratory (NREL). (2005). *Renewable energy cost trends*. *www.nrel.gov/analysis/docs/cost_curves_2005.ppt*.

National Renewable Energy Laboratory (NREL). (2009). Renewable resources in the continguous United States, along with theoretical potential for individual technologies: solar,[6] wind,[7] geothermal,[8] hydroelectric,[9] biopower.[10] *http://www.nrel.gov/gis/docs/resource_maps_200905.ppt.*

Navigant Consulting. (2008). *Photovoltaic manufacturer shipments & competitive analysis 2007/2008. www.navigantconsulting.com/downloads/PV_Manuf_Report08_US_EG.pdf.*

Nemet, G. F. (2006). Beyond the learning curve: Factors influencing cost reductions in photovoltaics. *Energy Policy, 34*(17), 3218–3232.

New Energy Finance (NEF). (2008a). News website. *www.bnef.com*

New Energy Finance (NEF). (2008b). *PV market outlook.*

New Energy Finance (NEF). (2009). Web site published by New Energy Finance. Accessed December 2009 at *www.newenergyfinance.com.* http://bnef.com/

Nozik, A. J. (2008a). Multiple exciton generation in semiconductor quantum dots and novel molecules: Applications to third generation solar photon conversion. *Proceedings of the 17th IEEE International Symposium on the Applications of Ferroelectrics* (ISAF, 2008), 23–28 February, Santa Fe, New Mexico. Piscataway, NJ: Institute of Electrical and Electronics Engineers, Inc. (IEEE) Vol. 1, pp. 1–2; NREL Report No. CP-5A0-45095; doi:10.1109/ISAF.2008.4693953.

Nozik, A. J. (2008b). Multiple exciton generation in semiconductor quantum dots. *Chemical Physics Letters, 457*, 3–11, NREL Report No. JA-100-42889, doi:10.1016/j.cplett.2008.03.094.

---

6. Solar: Estimates of potentials for both PV and CSP are assumed to be unconstrained by grid limitations such as lack of storage or transmission capacity. For PV, the solar resource potential (NREL, 2003b) was restricted by excluding federal and sensitive lands as sites for major collector installations; allowing installations only where land surface slopes are less than 5% excluding agricultural land used for food production (both farmland and rangeland); and requiring a minimum insolation value of 6 kilowatt-hours per square meter per day (kWh/m^2/day). The CSP resource is restricted based on insolation values, to areas in the southwestern United States. Site selection constraints reduce the land areas that can be used for CSP by precluding access to federal and sensitive lands, land with a surface slope less than 1% slope, major urban areas and features, and parcels less than 1 km^2 in area. The remaining area determined the technical potential for CSP, assuming 50 MW/ km^2

7. Wind: 2008 DOE report, *20% Wind by 2030*, p.8. The nation has more than 8000 GW of available land-based wind resources, plus 2200 GW of offshore wind class 5 and better between 0 and 50 nm from shore, based on NREL's most recent offshore resource estimates (offshore modeled data for TX, LA, GA, New England, and Great Lakes; offshore portions of onshore modeled resource datasets; and estimates by NREL's wind resource assessment group for remaining areas). Potential capacity estimated assuming 5 MW/km^2.

8. Geothermal resource shown is temperature at a depth of 6.5 km (from Southern Methodist University Laboratory, 2004). High temperature areas are most favorable for early development using EGS technology, while low temperature areas are possible resources in the future. Electric energy producing potential from EGS for entire U.S. is ~520 GW (USGS, 2008. *http://energy.usgs.gov/flash/geothermal_slideshow.swf*). Individual spots identified by a dark + symbol indicate high temperature anomalies where identified conventional hydrothermal resources are located. Electric power potential from these sites is 9 GW, with another 30 GW of undiscovered hydrothermal resources estimated (USGS, 2008). Electric power potential from using high temperature fluids coproduced with oil and gas, primarily concentrated in Texas and Louisiana Gulf Region, not shown.

9. Long-term hydroelectric potential is drawn directly from a GIS-based study by INL: *Feasibility assessment of the water energy resources of the United States for new low power and small hydro classes of hydroelectric plants*, January 2006.

10. Biopower potential estimated based on available residue data produced in the report *Geographic perspective on the current biomass resource availability in the United States* (NREL, 2004), assuming conversion efficiencies between 30% and 35% depending on the residue type.

Orkustofnun. (2007). Energy statistics in Iceland. Orkustofnun National Energy Authority. *www.os.is/Apps/WebObjects/Orkustofnun.woa/swdocument/20644/Energy_Statistics_2007.pdf http://www.nea.is/.*

Price, S. (2010). *2008 solar technologies market report, p. 131*. NREL Report No. TP-6A2-46025; DOE/GO-102010-2867.

Prometheus Institute. (2009). 25th Annual Data Collection Results: PV Production Explodes in 2008. *PV News, 28*(4).

Renewable Energy Policy Network for the 21st Century. (REN21). (2008). *Renewables Global Status Report 2007. www.ren21.net/pdf/RE2007_Global_Status_Report.pdf. http://www.ren21.net/Portals/97/documents/GSR/RE2007_Global_Status_Report.pdf.*

Renewable Energy Policy Network for the 21st Century (REN21). (2009). Renewables global status report – 2009 update. *www.ren21.net/pdf/RE2007_Global_Status_Report.pdf. http://www.ren21.net/Portals/97/documents/GSR/RE_GSR_2009_Update.pdf.*

Renewable Energy World (REW). (2008). *2008 wrap-up: What will a financially disastrous year mean for renewables? www.renewableenergyworld.com/rea/news/story?id=54382. http://www.renewableenergyworld.com/rea/news/article/2008/12/2008-wrap-up-what-will-a-financially-disastrous-year-mean-for-renewables-54382.*

Renewable Fuels Association (RFA). (2008). Changing the climate: Ethanol industry outlook 2008. *www.ethanolrfa.org/objects/pdf/outlook/RFA_Outlook_2008.pdf. http://www.ethanolrfa.org/page/-/objects/pdf/outlook/RFA_Outlook_2008.pdf?nocdn=1.*

Renewable Fuels Association (RFA). (2009). *Growing innovation: 2009 ethanol industry outlook. www.ethanolrfa.org/objects/pdf/outlook/RFA_Outlook_2009.pdf. http://www.ethanolrfa.org/page/-/objects/pdf/outlook/RFA_Outlook_2009.pdf?nocdn=1.*

Renewable Fuels Association (RFA). (2010). "Ethanol Industry Statistics". Accessed October 2010 at *http://www.ethanolrfa.org/pages/statistics*

Solar Energy Industries Association (SEIA). (2010). US Solar Industry Year in Review 2009. *http://seia.org/galleries/default-file/2009%20Solar%20Industry%20Year%20in%20Review.pdf.*

Taipei, N., & Hwang, A. (2008). Solar spot market prices fall. *Digitimes*. October.

União dos Produtores de Bioenergia (UDOP). (2008). *www.udop.com.br/index.php?cod=1043131&item=noticias. http://www.udop.com.br/.*

Union of Concerned Scientists (UCS). (2009). *Climate 2030: A national blueprint for a clean energy economy, (May).*

United Nations Energy Program (UNEP). (2008). *Global trends in sustainable energy investment 2008. http://sefi.unep.org/english/globaltrends.html. http://sefi.unep.org/fileadmin/media/sefi/docs/publications/Global_Trends_2008.pdf.*

United Nations Energy Program (UNEP). (2009). *Global trends in sustainable energy investment 2009.* New Energy Finance/United Nations Energy. *http://sefi.unep.org/fileadmin/media/sefi/docs/publications/UNEP_SEFI_Global_Trends_Report_2009_f.pdf.*

United States Department of Agriculture (USDA). (2007). *The future of biofuels: A global perspective. www.ers.usda.gov/AmberWaves/November07/Features/Biofuels.htm.*

Utility Wind Integration Group (UWIG) Library. (2009). *www.uwig.org/opimpactsdocs.html.*

Wall Street Journal (WSJ). (2008). Clean energy confronts messy reality. *Wall Street Journal online. http://online.wsj.com/article/SB122714114743842743.html?mod=googlenews_wsj.*

WindPACT – The Wind Partnerships for Advanced Component Technology. (1999). Various projects on advanced wind technology. *www.nrel.gov/wind/advanced_technology.html.*

Wiser, R., Bolinger, M. (2007). *Annual report on U.S. wind power installation, cost, and performance trends*: 2007. Department of Energy, Energy Efficiency and Renewable Energy; DOE/GO-102008-2590. *www.osti.gov/bridge. http://www.nrel.gov/docs/fy08osti/43025.pdf.*

World Bank. (2009). *State and trends of the carbon market.* World Bank Institute. *http://wbcarbonfinance.org/docs/State___Trends_of_the_Carbon_Market_2009-FINAL_26_May09.pdf. http://siteresources.worldbank.org/EXTCARBONFINANCE/Resources/State_and_Trends_of_the_Carbon_Market_2009-FINALb.pdf.*

Chapter 14

Heating Systems When Little Heating Is Needed

Klaas Bauermann, and Christoph Weber
University of Duisburg-Essen

Chapter Outline

1. INTRODUCTION

The threat of Global Warming has pushed substantial reductions in CO_2 emissions to the top of the agenda for politicians and decision-makers around the world.

On the road to a more efficient and sustainable future, the building sector will have to play a major role in reducing energy consumption and lowering CO_2 emissions, given its substantial contribution to today's emissions. Considerable potentials for emission reductions have been repeatedly identified in the building sector (e.g. McKinsey, 2007). Yet, buildings are long-lived and complex, durable goods, and many factors influence energy use within buildings, including heating, cooling, lighting, and other purposes. This means that while advanced technology may be incorporated in *new* buildings, the much larger *existing* stock poses significant challenges. A key driver for lowering CO_2 emissions from buildings would be notably improved insulation. However, other parameters like the size of the buildings and the heating system installed also directly influence the CO_2 emission. Consequently, a number of attributes of the building sector will have to be taken into account if the emission reduction potentials of buildings are to be effectively tapped.

Once construction is completed, buildings remain unchanged for some decades. This has two important implications: First, during construction, building design should be optimized to lower the energy demand for heating and cooling. Even more attention has to be devoted to increasing the energy efficiency of the existing building stock. Due to their longevity and incremental modifications, the building stock cannot be compared to cars, for example. Forty years is a long time-span for cars, yet it may be less than half of the lifetime for a typical European building. In this context a key question is whether the slow process of renewal and remodeling will be enough to achieve a low-carbon building stock if the right measures are applied. A related question is how will the existing heating infrastructure, notably gas and district heating systems, be affected by substantial changes in the buildings' energy demand.

This chapter examines different pathways toward a low-carbon building stock. Based on German regulations and the built environment, five different scenarios are examined and their potentials to lower heating-related emissions are analyzed. The chapter's main focus is to determine if the measures and standards set by the government are sufficient to tap the full CO_2 reduction potential of the building stock or if other measures are more promising and need to be implemented. The chapter is organized into four sections. The introduction is followed by an overview of energy consumption and emissions in the building sector. The current long-term policy targets at the European level and in Germany are also briefly reviewed and the development of key driving factors for building-related emissions is analyzed. Section 3 examines future evolution of the building sector and the effect of recent developments using Germany as a case study. In Section 4, a number of alternative pathways for carbon emission reduction are presented and the impact of politics in the building sector discussed. The last section examines the implications of these developments for a utility offering district heating, followed by conclusions.

2. ENERGY AND CARBON CHALLENGES IN THE BUILDINGS SECTOR

To assess possible routes toward a low-carbon future, first a clear view of the starting point concerning energy consumption and CO_2 emissions within the building sector is needed. This section examines relevant existing European and German regulations as well as the potential evolution of these regulations over time. The main drivers of energy use within the building sector are examined using a simple model linking CO_2, energy use, and the building stock.

2.1. Relevance of the Buildings Sector

Since the initial climate conference in Rio de Janeiro, politicians have been discussing the climate challenge. Yet, despite all the proclamations, global CO_2

emissions have kept rising. Even within Europe, CO_2 emissions have risen since 1990, reaching an all-time high in 2006. Since then, European emissions have declined by 2.2% to 5478 million tons in 2008.[1]

For the year 2020, the European Union has adopted the so-called 20-20-20 targets, requiring a reduction of greenhouse gas (GHG) emissions by at least 20% below 1990 levels, a 20% share in energy consumption to be provided from renewable sources, and a 20% improvement in energy efficiency.[2] Yet these targets can only be considered as a first step toward a low-carbon future. According to the Intergovernmental Panel on Climate Change (IPCC), GHG concentration in the atmosphere should not exceed 450 ppm_{CO2eq} to limit global warming to a maximum of + 2°C.[3] This is commonly believed to require a cut in global emissions by 50%−60% by 2050, requiring a cut in industrialized countries by at least 80% compared to 1990 levels, given the higher per capita level of emissions in these countries. However, no binding policies have been adopted at the European level to achieve this objective. Only the U.K. has officially adopted this goal in its Climate Change Act.[4] In Germany, a parliamentary inquiry commission referred to this target in 1994,[5] yet the objective has not become legally binding to date. This may be interpreted as an indication that radical emission cuts, be it in buildings or elsewhere, are not as easily achievable in practice as state-of-the art technology would suggest. Or rather, since the necessary technologies exist, political priorities seem to be different.[6]

Power plants and factories with smokestacks and cars, trucks, and buses with their exhaust pipes constantly remind the public that CO_2 is being emitted, even if it is not always the case. Consequently, these obvious polluters get the blame and are frequently the target of environmental activists. Meanwhile the society neglects the substantial emission contribution of houses and their heating and cooling systems. Clearly, to address the issue of climate change more attention has to be paid to the building sector. It uses a lot of energy and is therefore responsible for the associated emissions.

Germany has already achieved substantial CO_2 emission reductions over the past two decades (Figure 1). Yet, these have, to a large extent, been the immediate consequence of the reunification, especially in the energy sector and within the industry. Buildings account for roughly 30% of the current German GHG emissions (Figure 2). Space and water heating are responsible for the

---

1. Department of Commerce (2009), energy data.

2. *http://ec.europa.eu/environment/climat/climate_action.htm.*

3. IPCC, 2005, p. 57.

4. *http://ec.europa.eu/environment/climat/climate_action.htm.*

5. Enquete Kommission (1994).

6. This book's Chapter 6 by Bollino and Polinori discusses the political dimensions of climate change.

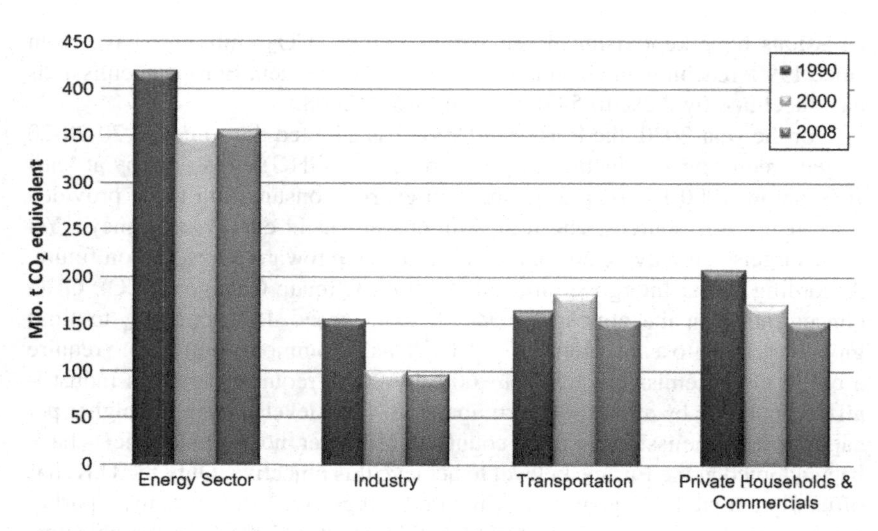

FIGURE 1 German CO_2 Emissions 2002 and 2010. *Source: UBA (2010).*

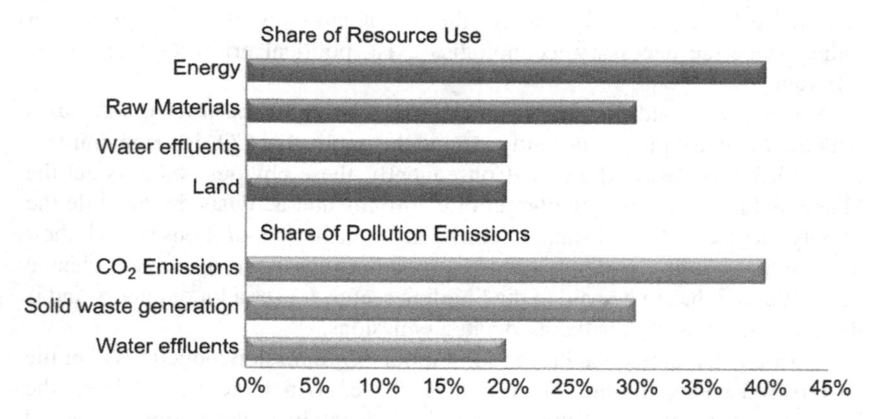

FIGURE 2 Environmental Impact of Buildings. *Source: UNEP (2006).*

major share of emissions in private households and the service sector, yet other indirect emissions including those associated with electricity production and district heating, as well as air-conditioning, must also be considered. In the case of Germany, CO_2 emissions from the energy sector are the biggest, followed by households, the commercial sector,[7] and the transportation sector. Emissions in all these sectors are higher than those in industry.

7. Commercials: In Germany this sector comprises trade, commerce, and services.

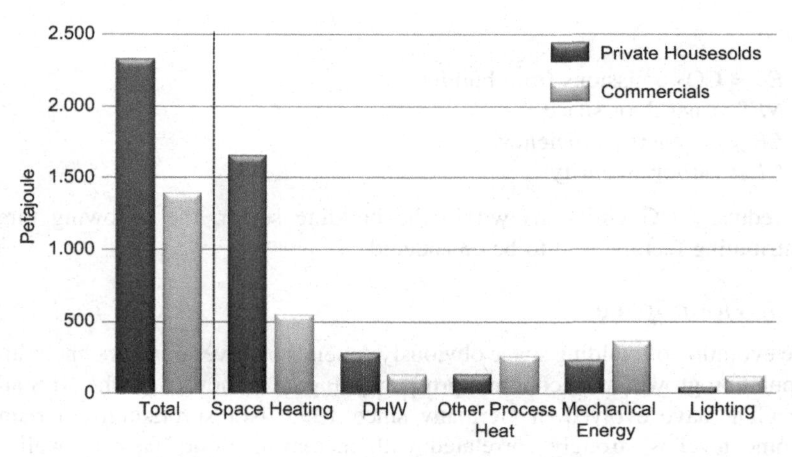

FIGURE 3 Final Energy Consumption in Germany (2007). *Source: Department of Commerce (2008).*

On a global scale, the environmental impact of the building sector is even more important, accounting for 40% of global CO_2 emissions.

In the case of Germany, space heating and domestic hot water (DHW) dominate energy use in the household as well as in the commercial sector, accounting for 83% and 48% of the total energy consumed, respectively (Figure 3).

Heating is by far the most energy-intensive activity in buildings in Germany, as in other Northern European countries. Hence, future measures have to focus on reducing the energy needed to heat buildings, both residential and nonresidential. Clearly, the energy requirements of buildings vary depending on climate and other variables, with space heating as the dominant energy user in colder climates and cooling in warmer climates. In all cases, better insulation can reduce both the heating demand in cold and the cooling demand in hot climates.

2.2. Building Stock, Energy Use, and CO_2 Emissions

The CO_2 emissions associated with energy use in buildings are affected by three key parameters:

- The total usable building space, measured as net floor space (NFS).
- The energy efficiency of the building: the quantity of energy used per unit of useful building space.
- The carbon intensity of the fuels used.
- This can be represented as:

$$E_b = NFS * EE_{BS,t} * CI$$

where:

E_b = CO_2 emissions from buildings
NFS = net floor space
$EE_{BS,t}$ = energy efficiency
CI = carbon intensity

To reduce GHG emissions within the building sector, the following three contributing factors have to be considered.

2.2.1. Floor Space

The evolution of building space obviously depends on several factors, including population growth and economic prosperity. Figure 4 shows how the NFS and the GDP have evolved in Germany since 1990. Not surprisingly, a rising income level is strongly correlated with increasing floor space as well as higher comfort levels, as described in this book's Chapter 9 by Gray and Zarnikau—a common feature in developing and developed economies. Economic growth and job creation is also associated with increased demand for office space, reinforcing the per capita energy consumption. In contrast, economic recessions are often first noticed as a downturn in the construction sector.

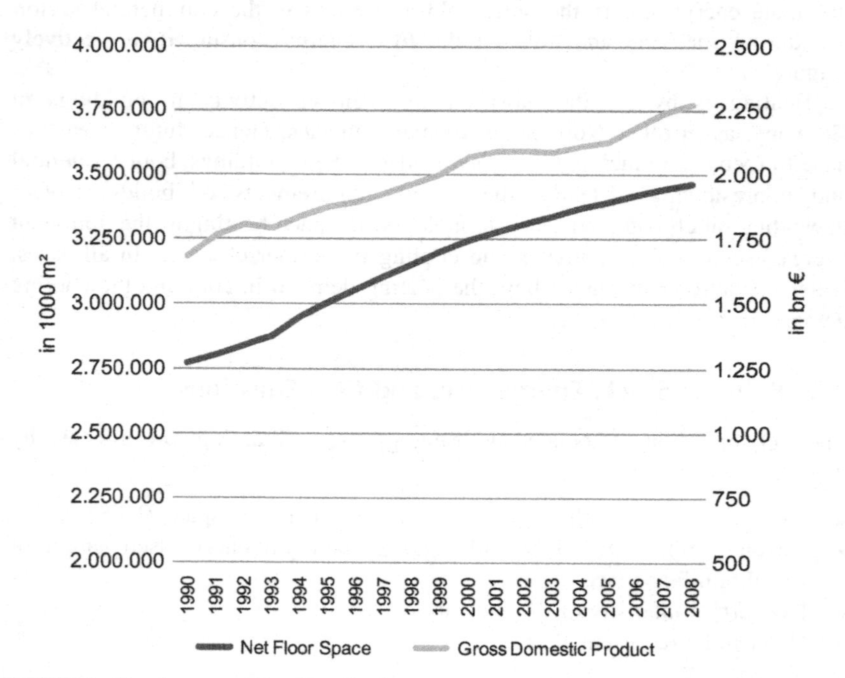

FIGURE 4 Development of Dwelling Stock (net floor space) and GDP in Germany (1990-2008). *Source: Own calculation based on Federal Statistical Office (2010).*

This simple economic observation may be formalized by computing the elasticity of floor space demand as a function of GDP per capita for Germany. For the years 1990–2008, an elasticity of 0.70 is observed. If the elasticity calculation is done separately for the two periods 1990–2000 and 2001–2008, the elasticity is found to decline from 0.78 in the first period to 0.62 in the second. After the German Reunification in 1990, Eastern Germany was in the process of catching up to the economic and social standards of Western Germany. Aside from the positive connection between GDP and dwelling size, other demographic and social factors have to be considered, and these factors are multifaceted. An increasing share of single households—common in many Western countries—also affects the evolution of the building stock. Overall, the NFS in Germany has increased by around 25% since 1990, while GDP per capita has increased by around 30%.

2.2.2. Energy Efficiency Regulations

Germany has implemented regulations to limit the energy requirements of buildings since the first oil crisis. The first of these regulations was the 1977 Heat Insulation Ordinance, followed by two revised versions in 1982 and 1995. Their aim was to reduce heating demand of new buildings by measures focusing on the building shell or specifying the requirements for the heating systems. Since 2002, these two types of ordinances have been combined in the German regulation for energy saving in buildings and building systems, known as EnEV. In the latest version of the EnEV, issued in 2009, the "Directive 2002/91/EC of the European Parliament and of the Council on the Energy Performance of Buildings" has been incorporated and the energy efficiency requirements for new buildings have been tightened by another 30%. The EnEV has thus become the key instrument to address the energy efficiency of buildings. The impact of these mandatory energy requirements on new buildings is shown in Figure 5, illustrating that such regulations do have a strong and direct effect on the energy demand of buildings.

However energy demand regulations typically do not affect the existing building stock, which accounts for the vast majority of energy consumption and emissions. Therefore the scope of the EnEV has already been extended to cover existing buildings that are substantially redeveloped,[8] that is, if refurbishment affects more than 10% of the surface of a building component.[9] Hence the regulatory framework now covers new constructions and in some cases redeveloped buildings. And with the next tightening of the regulations, scheduled for 2012, a particular focus on the existing buildings is expected. This is of particular relevance given the longevity of the existing stock. In Germany,

8. EnEV (2009), § 9.

9. This is only applicable to components that are part of the building envelope like outside walls, roof, windows, etc.

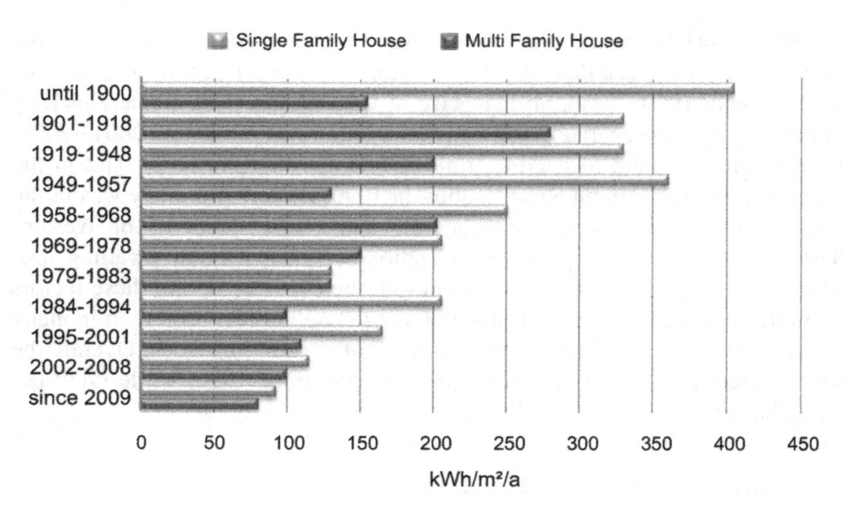

FIGURE 5 Measured Average Energy Consumption for Heating and DHW per Year of Construction. *Source: Fraunhofer-Institut (2007) own amendments.*

demolition and new construction activities are at a very low level, as in many other developed countries. In 2008, for example, 39,000[10] dwellings or only 0.1% of the existing building stock were torn down, while 175,900[11] or 0.45% were newly constructed. Moreover, demolition and construction rates have shown a declining trend in recent years.

On the one hand, the high average lifespan of buildings implies long-lasting effects for energy efficiency achieved through building codes. On the other hand, the long lifespan reduces the opportunities for energetic retrofitting. This is notably true for the exterior walls, which have a lifespan between 40 and 80 years, depending on construction type and quality.[12] Available information about energetic retrofitting indicates rates between 1.3%[13] and 2.2%[14] of the existing building stock per year. To raise these retrofit rates, the government has introduced subsidies and low-interest loans for retrofitting activities, differentiated according to the intensity of the measures. By providing high incentives to those that redevelop their buildings to new built standards and less subsidies for those insulating less, these polices aim not only at raising retrofitting frequencies but also at convincing building owners to do as much as possible.[15] Yet today,

10. Federal Statistical Office (2008).

11. Federal Statistical Office (2009a).

12. Blesl (2002), p. 18.

13. Department of Commerce (2007) EEAP, p. 53.

14. Federal Ministry of Transport (2007), p. 7.

15. *http://www.kfw-foerderbank.de/DE_Home/Service/Other_Languages/Summary_PB_Englisch.pdf.*

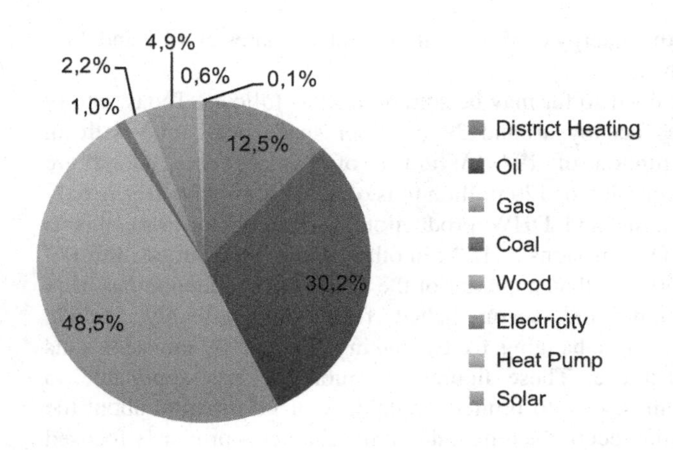

FIGURE 6 Share of Heating Systems in the Residential Building Stock (2008). *Source: BDEW & Federal Statistical Office (2008).*

after years of promoting energy efficiency, the primary energy demand for heating and DHW of the German building stock still averages about 194 $kWh_{PE}/m^2/a$[16] as compared to 237 $kWh_{PE}/m^2/a$ in 1990. Hence the improvement in average energy efficiency of about 18% has not been sufficient to compensate for the increase in dwelling space of about 25% over the same period.

2.2.3. Carbon Intensity

The carbon intensity—the average amount of emissions produced per kWh by heating systems—mainly depends on the fuel mix and the heating systems used. It is influenced by the electricity generation mix in the case of electric heating and whether or not district heating is coal-based. The distribution of heating systems and technologies in the German building stock is shown in Figure 6.

The corresponding carbon intensity is 0.210 kg CO_2/kWh_{PE} including upstream emissions in electricity and district heating production. This is 19% lower than the 1990 value of 0.259 kg CO_2/kWh_{PE}.

Before the EnEV came into effect, the Heating System Ordinance regulated the installation of heating systems and their efficiency. Today the Renewable Energy Sources Act additionally prescribes the use of renewable energy in every newly built house, be it residential or nonresidential and in case of major refurbishments.[17] Yet given the replacement frequencies of heating systems and the number of new buildings constructed every year, the existing building stock will still largely determine the carbon intensity in the years to come. The mixture of heating systems today shows that gas and oil are still the most

16. Own calculation based on data of BMWi (2008).

17. Federal Environment Ministry (2008), p. 4.

frequently used heating energy carriers, with respective shares of 48% and 30% of all heating systems.

The findings obtained so far may be summarized as follows: Today, nearly 82 million Germans occupy around 39 m^2 floor space per capita with an approximate consumption of 194 $kWh/m^2/a$ of primary energy and are responsible for the emission of 129 million tons of CO_2 every year. This is only from residential heating and DHW production. Compared to 1990, this is a reduction in the CO_2 emissions of 18%; in other words, in about one third of the time span to 2050, less than a quarter of the aimed 80% reduction has been achieved. Furthermore, the accomplished reductions probably include a substantial share of low-hanging fruits, making future CO_2 emission cuts more difficult to achieve. These figures are more or less applicable to nonresidential buildings. Due to limited availability of information about the nonresidential building sector, the remainder of the chapter is primarily focused on the residential sector.

3. ASSESSING POSSIBLE ROUTES FOR LOW-CARBON FUTURES

Given the interdependences between the building stock, its energy use and the resulting CO_2 emissions discussed above, the relevant question is how emissions may evolve in the future and what reduction targets are achievable. This section discusses the potential development of the three main drivers in detail to identify the possible future pathways for building-related carbon emissions. First, the building stock is considered to get a clearer picture of the growth in dwelling space as well as energy efficiency developments. This is followed by a more detailed description of the age structure of the German building stock to identify future potentials and to evaluate what has to be done to lower the CO_2 footprint of buildings. Next, a similar analysis is done for heating systems. To identify possible changes, the use of renewables in the heating market and the shift away from fossil fuels is considered.

3.1. Growth of Floor Space and Building Stock Evolution

3.1.1. Floor Space

The NFS is a main driver for the energy consumption in buildings. In Germany, the average growth in residential and nonresidential building space has been 1.24% per annum since the Reunification in 1990. This has been accompanied by an average GDP increase of 1.47% per annum.[18] Besides the income elasticity of the dwelling size, demographic changes play a crucial role in the evolution of the building stock. An expected decline in German population in

18. Own calculation based on data of the Federal Statistical Office (2008; 2010) and the IMF (2009).

the coming decades may decrease the upward pressure on building space and thus contribute to decreasing energy use. But this should not be considered as a matter of fact. Multiple contrary effects, for example, increased use of electric energy for various appliances in households and offices—notably for air-conditioning—may offset any reduction of energy consumption induced by a declining population. Nevertheless it is useful to link the total *NFS* to the population, *P*, using the floor space per capita, FS_{pc}, as proportionality factor:

$$NFS = FS_{PC} * P$$

where:

FS_{PC} = floor space per capita

P = population

It is unlikely that with a rising GDP the majority of people will end up living in a 400 m^2 flat in 2050. Yet a continuation of the slight upward trend observed during the last decade seems realistic.

Another observable trend is the increasing share of single households—a common feature of many developed countries with aging populations. The number of one-person households is estimated to increase by 11% by 2025 compared to the 2005 level, while the number of four-person households is expected to decrease by 24% over the same period.

The decline of the average household size is not only a sign of a changing lifestyle. A rise in expected lifetime leads to more elderly people living predominantly in one- and two-person households. Moreover, a lower birth rate, an increasing share of couples keeping two households, and a higher job-related mobility contribute to the decline in average household size and a higher per-capita dwelling size.[19]

These developments are correlated to ongoing social, demographic, and behavioral changes. Clearly, their impact on CO_2 emissions has to be taken into consideration. Equally obvious is the fact that government policies play a relatively minor influence on these sociodemographic processes.

In the case of Germany, growing wealth and changing lifestyles, including the increasing proportion of single households, will also lead to a growing NFS. The growth rate is likely to decline over time due to the overall decline in population, reaching zero in the 2040s. Yet extrapolating the current GDP per capita growth rate and income elasticity of dwelling space leads to a NFS in 2050, which is 23% higher than the 2010 level.

3.1.2 Energy Efficiency

The average energy efficiency of the building sector $EE_{BS,t}$ is influenced by refurbishments of existing buildings, demolition of old buildings, and

19. Federal Statistical Office (2007), press release # 518.

construction of new buildings. Since retrofitting activities may vary in the efficiency level achieved, this may be summarized by the following equation:

$$EE_{BS,t} = \frac{1}{NFS_t}((NFS_{t-1} - DM_t) * EE_{BS,t-1} + NC_t * EE_{NC,t} - REF_{I,t} * ES_{REF\ I,t} - REF_{II,t} * ES_{REF\ II,t})$$

$$ES_{REF\ I,t} = (EE_{BS,t} - EE_{RI,t})$$

$$ES_{REF\ II} = (EE_{BS,t} - EE_{R\ II,t})$$

Where:

EE = average energetic standard in $kWh_{PE}/m^2/a$
DM = demolition—NFS in buildings being demolished
EE_{BS} = energetic standard of the building stock in $kWh/m^2/a$
NC = new construction—NFS in buildings been newly constructed
EE_{NC} = energetic standard of the new construction
REF_I = redeveloped buildings—NFS in buildings being completely redeveloped
ES_{REF} = energy savings through redevelopment in $kWh_{PE}/m^2/a$
EE_{RI} = energetic standard of completely redeveloped buildings
REF_{II} = redeveloped buildings—NFS in buildings being partly redeveloped
$EE_{R\ II}$ = energetic standard of partly redeveloped buildings

New construction as well as refurbishment of existing buildings lead to an improvement of the average insulation standard, that is, reduces the overall energy consumption. Furthermore, demolition of old inefficient buildings may contribute to improvements in energy efficiency.

A general improvement in building insulation is one of the key factors to lower emissions. Policies and the society will have to take adequate measures to tap these huge potentials efficiently. Although new construction and demolition of existing buildings may have a role to play, the main factor to improve insulation standards is to redevelop existing buildings.

Provisions taken for new buildings at first sight look comprehensive (cf. Figure 5), but still energy efficiency measures lag far behind official ambitions. So, many observers expect that CO_2 reduction targets in the building sector will not be achieved, neither in 2020 nor in 2050, if no additional measures are implemented.[20]

To achieve substantial emission reductions, building codes have to be tightened further. Thereby the full range of technological advances has to be considered, including zero-energy and passive houses. Technologies are existent but ordinary new buildings are still too energy intensive to achieve

20. IWU (2007), p. 34.

ambitious CO_2 reduction goals. The energy efficiency of a typical building has improved with the latest building standard (cf. Figure 5) and another revision of the building code, including a further reduction of primary energy consumption by about 30%, is planned for the year 2012. But still the current consumption level of around 100 $kWh/m^2/a$ is too high. Currently, the German Reconstruction Loan Corporation (KfW), as a public bank and financial agency, also supports low-energy houses. The so-called KfW-70 and KfW-55 configurations describe buildings that consume respectively 30% and 45% less than a building constructed according to the current building standard (labeled as KfW-100). Technology is available today to build passive houses or even zero-energy houses in the near future. The European Commission recently released a proposal to tighten requirements of the directive on the energy performance of buildings. Nearly zero net energy buildings are proposed to become the standard for new constructions from 2020 on.[21] It is, however, not clear whether these proposals will become legally binding. One available technical solution is the German Passivhaus concept for buildings requiring around 15 $kWh/m^2/a$ for heating. But this concept is far from being generally applied today and even further away from being implemented in any building code. A realistic assumption is that building standards will require comparable insulation levels from the year 2035 onwards.

Furthermore, given the historical structure of the German building stock, initiatives to raise retrofitting rates need to be taken. In Germany, only 37% of the available NFS was built after 1979. The remaining 63% were built before any requirements to regulate energy consumption.[22]

Consequently, the retrofitting rate for existing buildings must be increased substantially. Following the aforementioned IWU analysis, the frequency of renovation has to double to achieve the ambitious German CO_2 reduction targets in 2050.[23] However, private house owners often fear the costs of redeveloping, being unable to calculate the benefits, while property owners are worried that they will not be able to pass on such costs to their tenants, due to legal restrictions. Therefore, it is vital to fill information gaps and to overcome any other existing barriers. Financial incentives offered need better promotion and coordination. Still, there is uncertainty as to how to achieve this. Politicians are worried that the problem may only be solved by implementing mandatory retrofitting standards, which might violate property rights or significantly increase rental charges.

---

21. This book's Chapter 17 by Rajkovich et al. describes similar measures proposed for California, starting in 2020.

22. WDV (2009), p. 15.

23. IWU (2007) p. 34.

3.2 Heating System Evolution

The third main factor for building CO_2 emissions, the carbon intensity, depends on the type of heating systems and energy sources used in the buildings and the corresponding emissions per kWh of heat. Regulations since 1977 focus on the energy demand (kWh/m^2/a) of buildings and emissions are, so far, not targeted directly. In general, the carbon intensity may be written as a function of the heating system distribution using a vector equation:

$$CI = HSD^T * \theta_{HS}$$

where:

HSD = heating system distribution
θ_{HD} = emission coefficients for heating systems

The carbon intensity describes the average CO_2 emissions per unit of primary energy consumed for heating purposes. There are large differences in the carbon footprint of heating systems between the very few old houses with coal fired furnaces and new buildings with heat pumps or biogas firing. Among the newly installed heating systems, the shares of the various energy carriers vary from year to year, depending on market prices and legal dispositions. Substantial differences are thereby observed between new buildings and those undergoing a heating system renewal.

Ordinances can influence the choice of heating systems for new buildings, but the influence on the building stock is weak. Since January 2009, the Renewable Energy Heating Act has forced owners of newly erected buildings in Germany to use renewable energies for heating. Being part of the integrated energy and climate program of the German government, this act is intended to raise the share of renewable energies in space heating significantly. The obligation covers all types of buildings and all type of owners. Various forms of renewables, like solar thermal energy, biogas, or geothermal energy may be used individually or in combination, yet the requirements differ depending on the type of renewable energy. In the case of solar energy, a minimum of 15% of the heating energy has to be solar. When using gaseous biomass, at least 30% of the heating energy has to be from the renewable source; in the case of liquid or solid biomass or geothermal energy, the minimum share is 50%. Instead of using renewable energies, building owners may also take alternative climate change mitigation measures such as improving the insulation, connecting to district heating, or using heat from combined heat and power.[24]

Figure 7 shows the distribution of heating systems in the building stock and compares to the installations in new buildings during the first nine months of 2009.

24. Federal Environment Ministry (2008), p. 3.

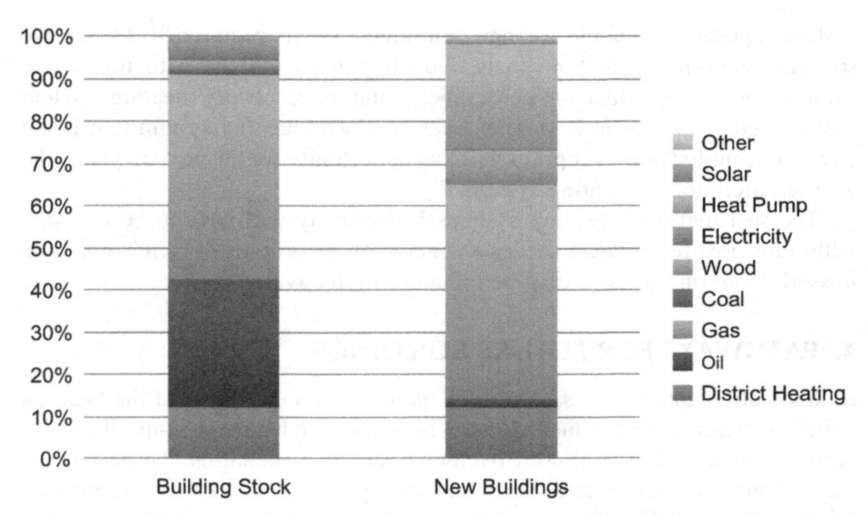

FIGURE 7 Distribution of Heating Systems Installed in the Building Stock and in New Buildings (built between Jan – Sep. 2009). *Source: AGEB & Federal Statistical Office (2009).*

The obvious differences are at least partly attributable to the new regulation. The shift away from oil has been observed over the past two decades and amplified by the recent price increases. Gas systems are still dominating the market of newly installed heating systems since the combination of a gas condensing boilers with solar thermal collectors suffices to fulfill the regulatory obligation. Still many people avoid a heating system purely based on renewables, but nevertheless heat pumps and biomass-firing gain market shares. The general interest in heating energy demand and its costs has increased over the last few years, and homeowners are increasingly shifting away from fossil fuels.

Policies aim at reaching a 14% share of renewable energies in the total heating energy supply by 2020.[25] The new regulation is intended to a decrease the emission intensity of new buildings but does not reach far enough to tackle emissions of the existing buildings. An estimated 70% of the heating systems in Germany were installed between 10 and 24 years ago, 20% are even older than 24 years.[26] Assuming an average lifetime of 20 to 30 years, the majority of heating systems will have to be renewed soon. The Renewable Energy Sources Act fails to consider these systems.

A more far-reaching legislation has been implemented by the German federal state of Baden-Wuerttemberg, which has passed its own Renewable Energy Sources Act. It requires that at least 10% of the produced heat has to come from renewable energy sources. This is mandatory for every heating

25. Renewable Energy Heating Act (2008), § 1, para. 2.

26. BDH (2008).

system replacement, while for new buildings 20% is required.[27] To achieve stringent emission reduction goals, federal politics will have to follow the example given by Baden-Wuerttemberg and cover every heating system replacement. The reason is simple: Every replaced heating system that is not environmentally friendly represents a lost opportunity for the next 25 years, the average lifetime of a heating system.

The fact that most heating systems in Germany will have to be replaced within the next few years provides a window of opportunity, which is not to be missed if carbon intensity is to be influenced effectively.

4. PATHWAYS FOR FUTURE BUILDINGS

In this section, alternative scenarios of future carbon emissions of the German building sector are examined. First a base case or business-as-usual (BAU) scenario is presented, followed by four alternative scenarios. These include technological advances and additional policy measures as well as possible lifestyle changes. The impacts of these different pathways on CO_2 emissions are compared to identify the most promising measures for achieving significant emission reductions.

4.1. BAU Scenario

The base case or BAU scenario is based on recent regulations and announced changes for the near future and describes a likely evolution of the CO_2 emissions though 2050.

The population forecast published by the Federal Statistical Office in 2006 is used as the basis of population projections. Low birth rates, a maturing population, and less immigration imply an expected 16% decline in German population by 2050 compared to 2008. For GDP, a continuation of the average per capita GDP growth observed between 1990 and 2008 is assumed. These assumptions result in an increase in per capita floor space, which in turn determines the rate of new construction (Table 1). The stipulated development of insulation standards and refurbishment rates are indicated in Table 1. A renovation rate of 1.5% p.a., split equally between full and partial energetic retrofitting, is assumed; a rather moderate increase in renovation frequency compared to the present situation. Also the heating system distribution is assumed to evolve rather slowly, as shown in Table 1 and Figure 8.

This base case scenario captures the effect of existing and expected measures on the CO_2 emissions from the building sector through 2050 in Germany (Figure 8).

In this case, the CO_2 emission reductions turn out to be insufficient to achieve the 80% reduction target for 2050. Cutting emissions to 46.9 Mt in

27. EWärmeG BW (2009), § 4, para. 1.

TABLE 1 Exogenous Parameters, Scenario Variables, and Related CO_2 Emissions for the BAU Scenario

	2008	2015	2020	2025	2030	2035	2040	2045	2050
Exogenous Parameters									
GDP growth rate p. a.	1,47%	1,47%	1,47%	1,47%	1,47%	1,47%	1,47%	1,47%	1,47%
Demolition rate p. a.	0.10%	0.10%	0.20%	0.30%	0.45%	0.43%	0.43%	0.60%	0.60%
Refurbishment rate p. a. total (of which full retrofit)	1.50% (0.75%)	1.50% (0.75%)	1.50% (0.75%)	1.50% (0.75%)	1.50% (0.75%)	1.50% (0.75%)	1.50% (0.75%)	1.50% (0.75%)	1.50% (0.75%)
Primary Energy Consumption in kWh/m²/a									
New construction	100	80	80	60	60	40	40	20	20
Redeveloped building	100	80	80	60	60	40	40	20	20
Partly redeveloped building	140	120	120	100	100	80	80	60	60
Endogenous Variables BAU Scenario									
New construction rate	0.81%	0.78%	0.74%	0.82%	0.80%	0.75%	0.59%	0.69%	0.10%
Net floor space per capita in m^2	38.8	41.5	43.4	45.4	47.6	49.8	52.1	54.6	57.1
Energy efficiency of the building stock in $kWh/m^2/a$	194	185	171	157	141	125	109	96	83
Carbon intensity in kg CO_2/kWh primary energy	0.210	0.205	0.201	0.194	0.187	0.178	0.169	0.157	0.144
Emission Development BAU Scenario									
CO_2 emissions in Mt	129.3	127.3	119.8	109.0	96.6	83.4	70.6	58.1	46.9

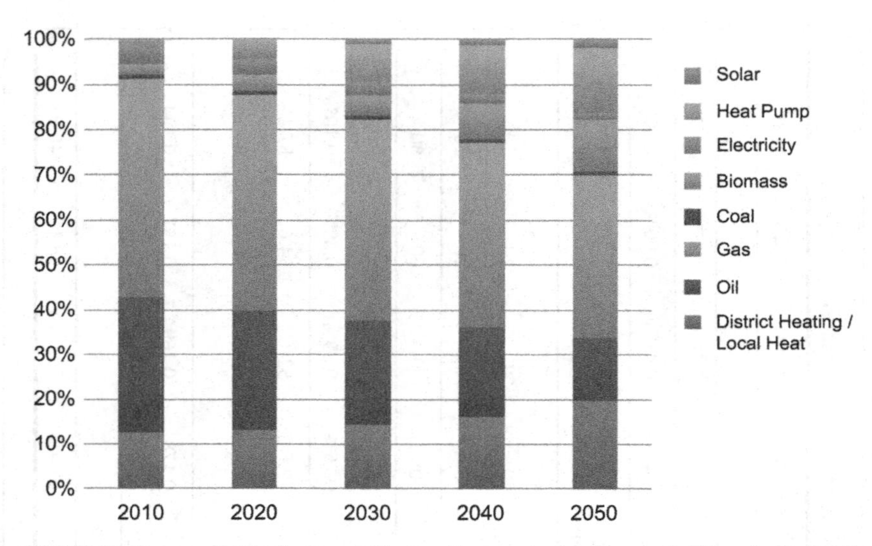

FIGURE 8 Forecasted Distribution of Heating Systems in the German Building Stock (BAU scenario). *Source: Own calculation.*

2050 corresponds to a reduction of 70% compared to 1990. Moreover, under the BAU scenario, per capita emissions merely decrease by 64%.

4.2. Scenarios for Low-Carbon Futures

The preceding discussion indicates that more ambitious policies are needed to achieve the 80% target. To examine the effect of alternative policy measures, four alternative future scenarios, further described in Table 2, are examined:

1. The first scenario assumes policies promoting higher retrofit rates in existing buildings.
2. The second scenario examines alternative lifestyle changes.
3. The third scenario combines the first and the second one.
4. The fourth scenario focuses on heating system changes.

The "higher retrofit" scenario assumes increased refurbishment as a consequence of subsidies, information campaigns, and higher cost awareness resulting in a frequency of 2.6% energetic renovation per year split into 2% p.a. in full and 0.6% in partial energetic building retrofits. Given the long lifetime of building facades of 40 years and more, this may be considered as an upper boundary for achievable renovation rates.[28] This means that 84,500,000 m^2 are renovated, instead of 48,800,000 m^2 in the BAU scenario in 2010, resulting in

28. Various sources suggest that such a rate would be sufficient to tap the CO_2 emission reduction potentials of buildings, cf. notably Department of Commerce (2007), p. 53.

TABLE 2 Alternative Scenarios – Changes in Building Retrofit, Dwelling Space, and Heating Systems in the Four Low Carbon Scenarios

Scenario	Higher Retrofit	Lifestyle	Higher Retrofit & Lifestyle	Alternative Heating System
Building retrofit	Increased rate cf. Table 3	As in BAU	Increased rate cf. Table 3	As in BAU
Increase in dwelling space	As in BAU	Reduced rate cf. Table 3	Reduced rate cf. Table 3	As in BAU
Heating system	As in BAU	As in BAU	As in BAU	Modified cf. Figure 9

an additional 11 $kWh_{PE}/m^2/a$ reduction in the average energy demand through 2050, roughly 13% below the BAU scenario.

Despite these improvements compared to the BAU scenario, overall CO_2 abatement actually increases by four percentage points, resulting in a 74% overall reduction by 2050. Hence, even with a higher assumed rate of retrofitting frequency, the desired 80% target will *not* be met. While a 74% reduction in emissions is not far off the mark, it must be pointed out that the assumptions made to achieve it are unlikely to be achieved in the foreseeable future. Furthermore, building codes are mandatory only for new buildings, while redevelopment is not covered in most cases. Moreover, the population decline in Germany is likely to put a downward pressure on new construction rates in the future, thus reducing the effect of the EnEV. This suggests that if future emissions in the building sector are to be seriously addressed through improved insulation, the policy has to focus on the *existing* building stock.

The "lifestyle" scenario assumes that within the residential building sector total available dwelling space remains constant from 2010 onwards, resulting in a 19% increase in per capita floor space by 2050, given the shrinking population (Table 3). In this case, GHG emissions are reduced by 73% by 2050, suggesting that relying on possible lifestyle and preference changes to achieve ambitious goals will be helpful but not sufficient. One effect of this scenario is that new building construction is considerably slowed down, also implying slower pace of efficiency improvements.

The prior two scenarios show that even if Germany slows down the trend toward bigger homes and more living space or reaches rates of renovation and retrofitting that are far beyond current levels, the target of 80% CO_2 emission reduction by 2050 will not be met. Consequently, a combination of these two scenarios is considered. The "higher retrofit and lifestyle" scenario combines the energy efficiency improvements associated with a high rate of retrofits (i.e., 2.6% p.a.) with the assumption of constant floor space after 2010.

TABLE 3 Modified Exogenous Parameters and Resulting CO_2 Emissions for the "Higher Retrofit" and "Lifestyle" Scenarios

	2008	2015	2020	2025	2030	2035	2040	2045	2050
Exogenous Parameters "Higher Retrofit" Scenario									
Refurbishment rate p.a. total (of which full retrofit)	2.60% (2.00%)	2.60% (2.00%)	2.60% (2.00%)	2.60% (2.00%)	2.60% (2.00%)	2.60% (2.00%)	2.60% (2.00%)	2.60% (2.00%)	2.60% (2.00%)
Endogenous Variables "Higher Retrofit" Scenario									
Energy efficiency of the building stock in $kWh/m^2/a$	194	179	162	145	127	111	96	84	72
Emission Development "Higher Retrofit" Scenario									
CO_2 emissions in Mt^2	129.3	123.7	113.0	100.4	87.0	74.2	61.9	50.9	40.8
Exogenous Parameter "Lifestyle" Scenario									
Net floor space per capita in m^2	39.6	40.1	40.5	41.3	42.0	43.1	44.2	45.6	47.2
Endogenous Variable "Lifestyle" Scenario									
Energy efficiency of the building stock in $kWh/m^2/a$	194	188	177	164	149	133	116	103	90
Emission Development "Lifestyle" Scenario									
CO_2 emissions in Mt^2	129.3	124.5	114.8	103.1	90.3	76.6	63.7	52.2	42.1

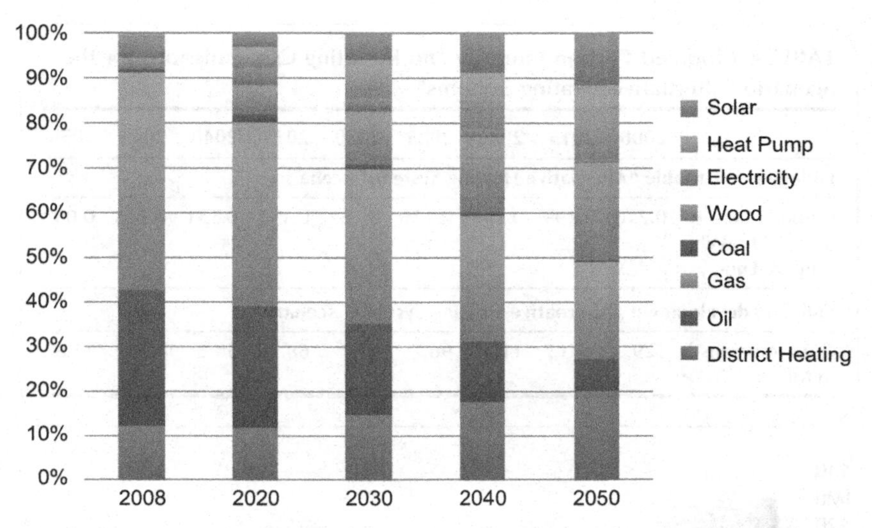

FIGURE 9 Distribution of Heating Systems in the "Alternative Heating Systems" Scenario Until 2050. *Source: Own calculation based on BMU Prognos (2009).*

Surprisingly, the combination of these two scenarios does not have a multiplicative impact. The simple explanation is that while the limitation of floor space slows down the CO_2 emissions, it also reduces the scope of available efficiency improvements, resulting in overall CO_2 reduction of 76.4% by 2050, still short of the 80% target.

The final scenario, "alternative heating systems," examines the impact of a faster turnover in the heating systems. It assumes a mixture of a policy measures for the German Federal Environment Ministry and a more conservative forecast of the Swiss-based Prognos institute[29] to achieve an alternative evolution of the heating system, as illustrated in Figure 9.

Assuming a smaller proportion of oil and gas fired heating systems and a significantly higher proportion of solar, biomass, and heat pump systems compared to the BAU scenario (Figure 8) results in a carbon intensity (Table 4) considerably below that of the BAU scenario (Table 1) and achieves a 79.7% reduction in carbon emissions by 2050 relative to 1990 levels, virtually meeting the desired 80% target.

The preceding discussion demonstrates the range of policy options that could impact GHG emissions associated with building heating systems and illustrates the challenges in trying to achieve an 80% reduction by 2050 in view of the fact that building owners frequently stay with the same energy carrier when replacing their heating system. Overall the difference between the BAU

29. BMU (2009), p. 53, and Prognos AG (2009), p.56.

TABLE 4 Modified Carbon Intensity and Resulting CO_2 Emissions for the Scenario "Alternative Heating Systems"

	2008	2015	2020	2025	2030	2035	2040	2045	2050
Endogenous variable "Alternative Heating Systems" scenario									
Carbon Intensity in kg CO_2/kWh Primary Energy	0.210	0.199	0.189	0.176	0.162	0.147	0.131	0.115	0.098
Emission development "Alternative Heating Systems" scenario									
CO_2 emissions in Mt^2	129.3	123.7	112.9	98.7	83.7	68.7	54.8	42.6	31.9

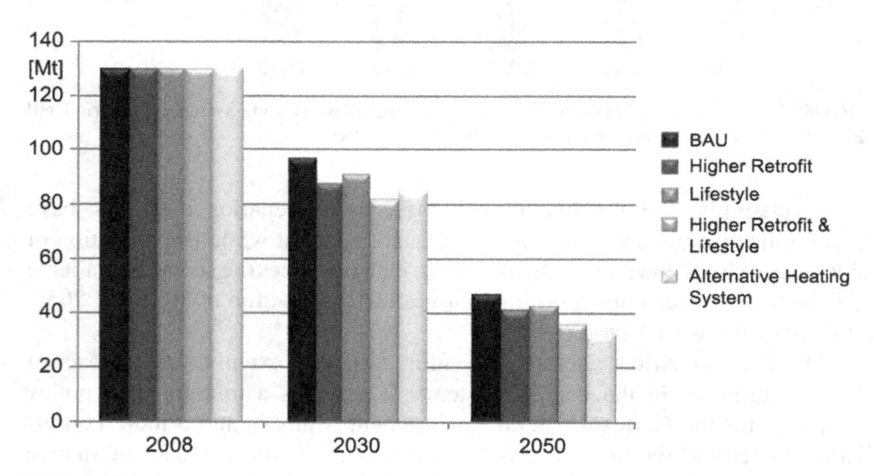

FIGURE 10 Comparison of Total CO_2 Emissions Over all Scenarios Until 2050. *Source: Own calculation.*

scenario and the best performing "alternative heating systems" scenario in 2050 is roughly 33% or 15 Mt of CO_2 (Figure 10).

5. IMPLICATIONS FOR GRID INFRASTRUCTURES AND HEAT SUPPLIERS

Germany has a long tradition of district heating. The first European district heating plant was erected in Dresden in 1900. In the early twentieth century, district heating systems were developed in cities such as Hamburg, Frankfurt, and Berlin. After World War II, district heating systems became popular notably in Eastern Germany where it became the standard for apartment dwellings, especially in typical Soviet-era tower-blocks, which served as heat sinks to local power plants.

An increasing number of these buildings are now vacant or are being torn down, resulting in a gradual decline in demand for district heating. In many cases, the existing suppliers are confronted with the question of whether upgrading of existing heating infrastructure is economical given a steady decline in demand.

This challenge becomes even more pressing in the coming years as buildings are better insulated, thus requiring less heating. Simultaneously, the aging grid infrastructure will require substantial replacement investments. Existing operators must assess the business case for upgrading these systems in view of falling demand and profitability relative to original investment several decades ago.

Analysis of selected urban areas indicates that supplying heating energy to city centers will remain an attractive option even with decreasing heat demand. In suburban areas with lower population density, the situation may be less attractive. Under a scenario such as "higher retrofit" described earlier, reinvestments in existing districting heating systems might not make good business sense in suburban areas. In all cases, heat suppliers are likely to postpone as far as possible any reinvestment to save costs in the short run, to gain a better sense of the rate of decline in heating demand.

In general, an advantageous strategy for utilities with existing district heating systems may consist of successively giving up branches of their distribution system with low heat demand located at the end of the network and to operate the existing system without making major replacements for as long as possible. Even making compensation payments to abandoned customers may be an advantageous option in the presence of a rapidly declining heat demand, if the alternative is an even more costly renewal of the aging network.

Overall, the analysis indicates that gradual improvements in building insulation constitute a threat for heat suppliers in the longer run. Residential areas, especially, that were once economically supplied when the district heating systems were originally installed become less economical with the passage of time and given the need for replacement investments. Not surprisingly, new residential areas with high insulation standards and superior building codes are already unattractive for heat suppliers. Moreover, existing areas undergoing energetic retrofit may also become unattractive in the future as the demand for heating drops. Policies concentrating on lowering energy demand per floor space present a direct threat to gas and heat suppliers, gradually leading to the demise of the post-war model of centralized heat supply. City centers and other densely populated urban areas together with industrial customers are likely to be the only viable outlets for district heating and gas distribution.

CONCLUSIONS

The preceding discussion leads to the chapter's four main conclusions:

1. First, since buildings account for roughly 40% of all energy use globally, more attention must be paid to reducing the energy consumption of the building sector—and associated GHG emissions.

2. Second, improved standards for building energy efficiency, building insulation, and improved heating systems typically focus on new construction, leaving the significantly larger share of existing buildings essentially unaffected.
3. Third, the most effective means of reducing building related GHG emissions in colder climates is to change the way buildings are heated. Since the average lifetime of heating systems is around 25 years, or roughly half the lifetime of typical building shells, this offers a much more effective way to reduce energy use in the building sector.
4. Fourth, as buildings get better insulated and their heating demand decreases, the business case for existing district heating systems and gas infrastructure will become less compelling, resulting in their eventual abandonment in all but highly concentrated areas.

Clearly, in a future where buildings are much better insulated and use superior technologies for heating, there will be diminished need for heating, and this offers a significant opportunity for reducing GHG emissions.

BIBLIOGRAPHY

AG Energiebilanzen e.V. (AGEB). (2009a). *Energieverbrauch in Deutschland – Daten für das 1.-4- Quartal*. Berlin.

AG Energiebilanzen e.V. (2009b). *Energiebilanz der Bundesrepublik Deutschland. www.ag-energiebilanzen.de/viewpage.php?idpage=63*.

AG Energiebilanzen e.V. (2009c). *Auswertungstabellen zur Energiebilanz für die Bundesrepublik Deutschland. www.ag-energiebilanzen.de/viewpage.php?idpage=139*.

Blesl, M. (2002). *Räumlich hoch aufgelöste Modellierung leitungsgebundener Energieversorgungssysteme zur Deckung des Niedertemperaturwärmebedarfs*. Stuttgart: Institut fur Energiewirtschaft und Rationelle Energieanwendung (IER).

Born, R., et al. (2007). *Energieeinsparung durch Verbesserung des Wärmeschutzes und Modernisierung der Heizungsanlage für 31 Musterhäuser der Gebäudetypologie*. Darmstadt: Institut Wohnen und Umwelt GmbH (IWU).

Bundesgesetzblatt. (2008). Teil 1 Nr. 36 *Gesetz zur Förderung Erneuerbarer Energien im Wärmebereich* (Erneuerbare-Energien-Wärmegesetz – EEWärmeG). Berlin.

Bundesgesetzblatt. (2009). Teil 1 Nr. 23 *Verordnung zur Änderung der Energieeinsparverordnung (EnEV)*. Berlin.

Bundesverband der Energie- und Wasserwirtschaft e.V.(BDEW). (2010). Beheizungsstruktur des Wohnugsbestandes http://www.bdew.de/bdew.nsf/id/DE_Beheizungsstruktur_des_gesamten_Wohnungsbestandes/$file/10%2007%2016%20Beheizungsstruktur%20im%20Bestand%201975-2009p.pdf.

Bundesverband der Energie- und Wasserwirtschaft e.V.(BDEW). (2010). Beheizungsstruktur im Bestand 2008. *www.bdew.de/bdew.nsf/id/DE_Beheizungsstruktur_im_Wohnungsbestand/$file/09%2011%2013%20Beheizungsstruktur%20im%20Wohnungsbestand.pdf*.

Department of Commerce. (2007). *Nationaler Energieeffizienz-Aktionsplan (EEAP) der Bundesrepublik Deutschland gemäß EU-Richtlinie über Endenergieeffizienz und Energiedienstleistungen. (2006/32/EG)*. Berlin: Öffentlichkeitsarbeit.

Department of Commerce. (2007). *Bericht zur Umsetzung der in der Kabinettsklausur am 23./24.08.2007 in Meseberg beschlossenen Eckpunkte für ein Integriertes Energie- und Klimaprogramm.* Berlin.

Department of Commerce. (2010) *http://www.bmwi.de/BMWi/Navigation/Energie/Statistik-und-Prognosen/energiedaten.html, did=176660*.html.

Diefenbach, N., et al. (2005). *Beiträge der EnEV und des KfW-CO2-Gebäudesanierungsprogramms zum Nationalen Klimaschutzprogramm.* Darmstadt: Institut Wohnen und Umwelt.

Eichener, V. (2009). *Die für 2009 geplante Verschärfung der EnEV ist nicht wirtschaftlich verträglich.* Ernst & Sohn Mauerwerk 13(2009), Heft 1.

Enquete Kommission. (1994). *Mehr Zukunft für die Erde – Nachhaltige Energiepolitik für dauerhaften Klimaschutz.* Bonn: Bundesanzeiger Verlagsgesellschaft mbH.

Erdmenger, C. et al. (2007). *Klimaschutz in Deutschland: 40%-Senkung der CO2-Emissionen bis 2020 gegenüber 1990.* Federal Environment Agency (UBA).

Fachverband Wärmedämm-Verbundsysteme e.V. (WDVSysteme). (2009). Energiesparkompass 2009. Baden-Baden.

Federal Environment Agency (UBA). (1990). *http://www.umweltbundesamt.de/emissionen/publikationen.htm.*

Federal Environment Agency. (2010). *National trend tables for the German atmospheric emission reporting 1990–2008.* Dessau: Umweltbundesamt.

Federal Environment Ministry (BMU). (2006). *Nationaler Allokationsplan 2008–2012 für die Bundesrepublik Deutschland.* Berlin.

Federal Environment Ministry (BMU). (2008). *Weiterentwicklung der Ausbaustrategie Erneuerbarer Energien Leitstudie 2008.* Berlin: Referat Öffentlichkeitsarbeit.

Federal Environment Ministry (BMU). (2009). *Langfristszenarien und Strategien für den Ausbau erneuerbarer Energien in Deutschland Leitszenario 2009.* Berlin: Referat Öffentlichkeitsarbeit.

Federal Environment Ministry. *The renewable energies heat act in brief.* (2008) *www.bmu.de/english/renewable_energy/downloads/doc/42193.php.*

Federal Ministry of Health. (2003). *Nachhaltigkeit in der Finanzierung der sozialen Sicherungssysteme – Bericht der Kommission.* Berlin: Öffentlichkeitsarbeit.

Federal Ministry of Transport. (2007). *CO_2 Gebäudereport 2007.* Berlin: Referat Öffentlichkeitsarbeit.

Federal Ministry of Transport. (2009). *Beurteilung energetischer Anforderungen an Nichtwohngebäude in Zusammenhang mit der Fortschreibung der EnEV.* Berlin.

Federal Statistical Office. (2006). *Bevölkerung Deutschlands bis 2050.* Wiesbaden: Pressestelle.

Federal Statistical Office. (2007). Press Release Nr.518. *www.destatis.de/jetspeed/portal/cms/Sites/destatis/Internet/DE/Presse/pm/2007/12/PD07__518__12421, templateId=renderPrint.psml.*

Federal Statistical Office. (2008). Wiesbaden: Wirtschaft und Statistik.

Federal Statistical Office. (2009a). *Bauen und Wohnen Baugenehmigungen von Wohn- und Nichtwohngebäuden u.a. nach Bauherren Lage Reihen z.T. ab 1980.* Wiesbaden.

Federal Statistical Office. (2009b). *Bauen und Wohnen Bestand an Wohnungen.* Wiesbaden.

Federal Statistical Office. (2009c). *Baugenehmigungen/Baufertigstellungen nach Gebäudeart – Lange Reihen z.T. ab 1960.* Wiesbaden.

Federal Statistical Office. (2009d). *Gebäude und Wohnungen Bestand an Wohnungen und Wohngebäuden Abgang von Wohnungen und Wohngebäuden Lage Reihen ab 1969–2008.* Wiesbaden.

Herring, H. (2009). National building stocks: addressing energy consumption or decarbonization? *Building Research & Information, 37*(2), 192–195.

Intergovernmental Panel on Climate Change (IPCC). (2005). *Carbon dioxide capture and storage.* Cambridge University Press.

International Monetary Fund (IMF). (2009). *www.imf.org/external/pubs/ft/weo/2009/01/weodata/download.aspx*

Kosten und Potenziale der Vermeidung von Treibhausgasemissionen in Deutschland – Sektorperspektive Gebäude, McKinsey & Company. Inc., 2007

Loga, T., et al. (2007). *Querschnittsbericht Energieeffizienz im Wohngebäudebestand Techniken, Potenziale, Kosten und Wirtschaftlichkeit*. Darmstadt: Institut Wohnen und Umwelt (IWU).

Lomas, K. J. (2009). Decarbonizing national housing stocks: strategies, barriers and measurement. *Building Research & Information, 37*(2), 187–191.

Power, A. (2008). Does demolition or refurbishment of old and inefficient homes help to increase our environmental, social and economic viability? *Energy Policy, 36*, 4487–4501.

Prognos, A. G., & Öko-Institut. (2009). *Modell Deutschland Klimaschutz bis 2050: Vom Ziel her denken*. Basel/Berlin: World Wide Fund For Nature (WWF).

Rabenstein, D. (2008). Auf dem Weg zu einer CO2-freien Wärmeversorgung – Probleme mit den Wegweiser? *Ernst & Sohn Bauphysik*, 30(1).

Ravetz, J. (2008). State of the stock – what do we know about existing buildings and their future prospects? *Energy Policy, 36*, 4462–4470.

Royal Institution of Chartered Surveyors (RICS). (2008). Breaking the vicious circle of blame – making the business case for sustainable buildings. *RICS EU*. Brussels: Public Affairs.

Schuler, A., Weber, C., & Fahl, U. (1997). Energy consumption for space heating of West-German households: Empirical evidence, scenario projections and policy implications. *Energy Policy, 28*, 877–894.

Schwarz, N., & Sommer, B. (2009). *Auswirkungen des demographischen Wandels – Daten der amtlichen Statistik*. Wiesbaden: Federal Statistical Office.

United Nations Environment Programme, Sustainable Building & Construction Initiative. SBCI Secretariat. http://www.unepsbci.org/SBCIRessources/Brochures/documents/UNEP_SBCI___Sustainable_Building__Construction_Initiative/SBCI_Broch_2.pdf.

Weber, C. (1999). *Konsumentenverhalten und Umwelt – eine empirische Untersuchung am Beispiel von Energienutzung und Emissionen*. Frankfurt am Main: Europäischer Verlag der Wissenschaften.

Part III

Case Studies

Chapter 15

Why China Matters

Bram Buijs
Clingendael International Energy Programme

Chapter Outline

1. INTRODUCTION[1]

There is hardly any field where the impact of China and its rapid development is not felt, and energy and climate change are no exception. In a formidable feat the world's most populous country has followed smaller Asian neighbors in a high-speed developmental trajectory, lifting millions out of poverty in the process. Already it has become hard to recall to memory that as the second half of the twentieth century started, China was an almost completely agrarian country. Since the reforms and "opening up" policy introduced by Deng Xiaoping in the late 1970s, China's gross domestic product (GDP) has increased more than tenfold as it maintained annual growth rates of around 10% on average. The introduction of market principles combined with an abundant supply of cheap labor triggered a growing influx of foreign investment and unleashed an enormous economic activity domestically. As the first decade of the twenty-first century has come to a close, the Chinese economy still shows little signs of slowing down.

The success of China's development, however, comes with consequences not only for China itself but for the world at large. China's economic growth up to

1. This chapter is based on the Clingendael Energy Paper, *China, Copenhagen and Beyond*, Clingendael International Energy Programme, September 2009. Available online at: *www.clingendael.nl/publications/2009/20090900_ciep_report_buijs_china_copenhagen_beyond.pdf*.

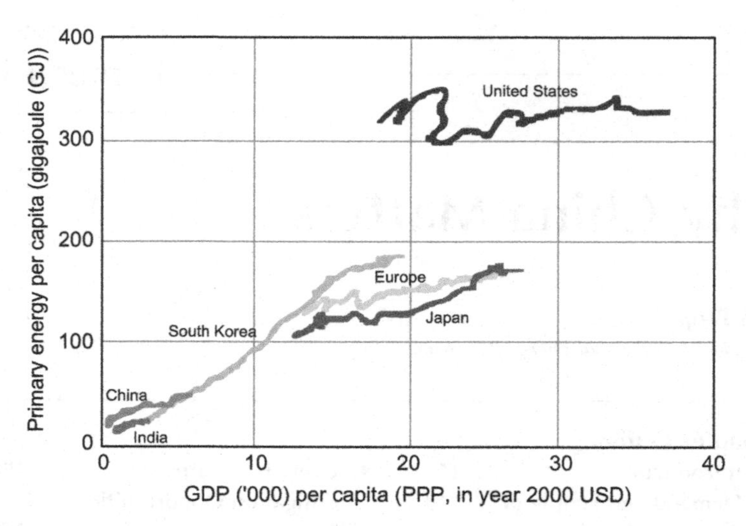

FIGURE 1 Developmental Trajectories (per capita primary energy vs. per capita GDP). *Source: World Energy Council, World Energy and Climate Policy: 2009 Assessment, 2009, p. 60. Adapted from Shell Energy Scenarios to 2050. Available online at: http://www.worldenergy.org/documents/report_final_3.pdf.*

now has followed a traditional developmental pattern: Rapid industrialization has been driving economic growth and the energy system underlying this growth is fueled predominantly by fossil energy resources, especially coal. Yet if we compare the growth patterns that are characteristic of the development path that was followed by other Asian countries such as Japan and South Korea, the implications of China's path will be enormous in terms of energy use and greenhouse gas (GHG) emissions (Figure 1). Simple calculations show that energy consumption in China at OECD levels is hardly imaginable: With per capita oil consumption levels similar to the United States, China would require all of today's global oil production of roughly 85 million barrels per day. In terms of electricity, its current per capita usage of 2.4 MWh per year stands at less than one-third of the OECD average (IEA, 2009a, p. 51). Even if China manages to improve the environmental performance of its carbon-intensive power sector, it will remain extremely difficult to offset the consequences of electricity demand growth on emissions. To illustrate the point: If China were to achieve the same per capita electricity consumption level as Germany *with the same level of carbon dioxide (CO_2) emissions per kWh as Germany,* its total emissions from power and heat generation would nearly double compared to current levels.[2]

2. This would imply almost halving China's current emissions level per kWh which stand at 777 gCO_2/kWh compared to 412 gCO_2/kWh in Germany. IEA. (2009). *CO_2 emissions from fuel combustion highlights (2009 Edition)*, pp. 101. Available online at: *www.iea.org/co2highlights/CO_2highlights.pdf.*

In this sense, China is the key protagonist illustrating the fundamental energy sustainability dilemma: If the whole world would have the same energy consumption patterns as the richest few, neither fossil energy production nor the climate would be able to bear the consequences. China with its population of 1.3 billion—almost one-fifth of the current global population—is probably the first country that faces this challenge directly with respect to its own development. The acknowledgment of this reality by the Chinese government is in fact driving much of the progressive policy that has been implemented in the recent past. China's leadership recognizes that China will need to find a different developmental model that will allow continued growth without becoming restrained by scarcity of energy resources or energy-related environmental issues[3] (Jiang, 2008; CAS, 2007).

To address energy security concerns and the long-term challenge of securing sufficient energy resources for development, China's energy policy includes a strong focus on energy efficiency, energy conservation, and the promotion of renewable energy sources. Although China's rising energy consumption has an increasing impact on the global availability of energy resources, the first and foremost consequences will be felt in the attempts to address the challenge of climate change. According to proposed stabilization schemes to limit the global temperature increase to 2°C, global emissions should peak no later than 2020. As the world's largest emitter of GHGs, China's contribution to attain such a target will be critical. However, as this chapter will show, China's developmental path will need to drastically change course if the required reductions in both energy consumption and emissions are to be achieved.

It is important to point out that China is not only important for climate change because it has become the world's largest emitter, but also because it holds some of the world's largest potential for climate change mitigation. Since China is still in the midst of its development, very significant opportunities to change future energy consumption and emissions levels exist. This holds especially true for sectors that are going through rapid expansion at the moment, such as power generation, housing, and transportation. The deployment of low-carbon and energy-efficient technologies in these sectors could have a significant impact on the future levels of energy demand and emissions. Yet to avoid the lock-in of carbon-intensive technologies, the speed of

3. Jiang Zemin [president of the People's Republic of China from 1993 to 2003]. (2008). 对中国能源问题的思考 [Reflections on energy issues in China], *Journal of Shanghai Jiaotong University, 42*(3), 257–8, 263–4: "To meet the ever increasing energy demand by one billion plus people in the course of building a moderately prosperous and modern society in an all-round way, China will build the world's largest energy supply and consumption system in the coming 10 to 20 years. Therefore, the urgent task before us is to blaze a new path in energy development with Chinese characteristics, in order to achieve the nation's strategic goal of modernization with a minimal cost of energy resources and impact on the environment." Also see (CAS, 2007).

implementation is crucial: Delaying strong action for a few years or more will mean the largest abatement potential will have been lost. There are also large gains to be had in terms of improving energy efficiency and energy conservation. In many sectors including industry, power generation, and housing, the difference with developed country standards and "best available technologies" is still considerable and closing this gap would contribute significantly to reining in China's demand for energy and related emissions. In some cases technological leapfrogging can actually be cost-effective, as the deployment of state-of-the-art power plants and advanced industrial production techniques in China's economy demonstrates. Yet for the implementation of energy-efficient and low-carbon technologies that are still under development and not economically competitive, overcoming cost barriers in China is as much of a challenge as it is in the developed world.

The crux of the matter regarding a true transition to a more sustainable energy system in China is the relative abundance of coal. When considering the projected levels of energy consumption—even under relatively energy-efficient scenarios—an enormous expansion of energy supply will be necessary. Since China has some of the world's largest coal reserves and it remains one of the cheapest sources of energy, coal will likely maintain a central position in its energy system. Although constraints on the supply of coal domestically might mean that it will have to look out for another staple fuel in the long run, the reserves are so vast that China can remain self-sufficient for a long time still. Unlike with other fossil fuels like oil and gas, energy security concerns will run counter to a big shift away from coal. The large-scale implementation of carbon capture and storage (CCS) technologies, that might allow the continued use of coal while reducing the carbon emissions, is fraught with difficulties and will most likely carry significant economic cost. This means that it will be very hard to turn away from coal and achieve the transformation to a low-carbon energy system in China on the short term, which is required by the stabilization scenarios.

This chapter will discuss the challenges outlined above. The following two sections aim to provide a fundamental understanding of China's developmental stage, its energy system, and its policy measures on energy and climate change. The fourth section will focus on China's growth patterns and opportunities for emissions mitigation in several key areas including industrial energy demand, the power sector, energy efficiency of buildings and the transportation sector. Several quantitative scenarios on China's future development are examined in Section 5, followed by conclusions.

2. DEMAND AND SUPPLY OF ENERGY IN CHINA

This section provides an overview of China's energy use, first discussing its energy consumption in the context of overall development followed by a discussion of the supply side of the energy system.

2.1. China's Development and Energy Consumption

Although in some aspects China already appears to be a fully developed country, it is important to realize that China as a whole is still a country in transition with an uneven level of development.

Average per capita income levels, although having increased significantly in the course of the past few decades, still stand at US$6,600 per year when measured at purchasing power parity terms compared to US$46,400 in the United States. As of 2009, more than half of China's population still lives in rural areas and almost 40% is employed in the agricultural sector, even though that sector only contributes little more than one-tenth to the GDP (see Box 1; and CIA, 2010).

Indicative of China's uneven developmental stage is the fact that energy consumption is dominated by industry, which accounts for almost 60% of the final energy consumption and 75% of electricity demand. The industrial sector, which accounts for about half of China's GDP, has a large share of energy-intensive sectors such as iron and steel, cement, chemicals, aluminum, other nonferrous metals, and pulp and paper. The main driver for these industries is

BOX 1 People's Republic of China – Key Statistics Including Global Rankings

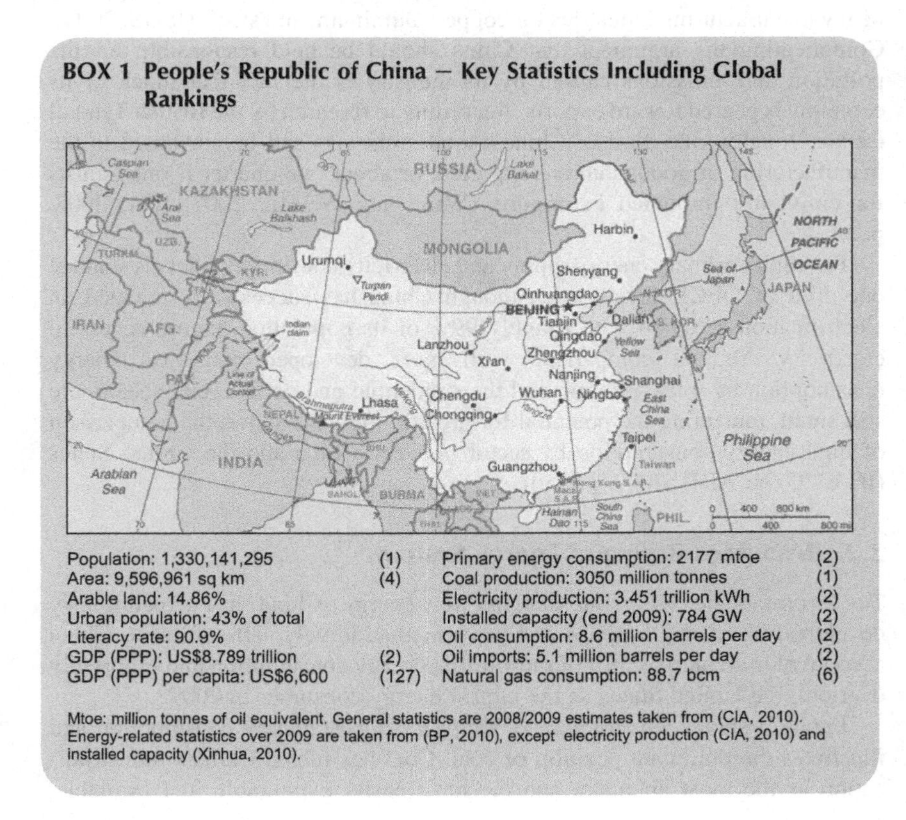

Population: 1,330,141,295	(1)	Primary energy consumption: 2177 mtoe	(2)
Area: 9,596,961 sq km	(4)	Coal production: 3050 million tonnes	(1)
Arable land: 14.86%		Electricity production: 3.451 trillion kWh	(2)
Urban population: 43% of total		Installed capacity (end 2009): 784 GW	(2)
Literacy rate: 90.9%		Oil consumption: 8.6 million barrels per day	(2)
GDP (PPP): US$8.789 trillion	(2)	Oil imports: 5.1 million barrels per day	(2)
GDP (PPP) per capita: US$6,600	(127)	Natural gas consumption: 88.7 bcm	(6)

Mtoe: million tonnes of oil equivalent. General statistics are 2008/2009 estimates taken from (CIA, 2010). Energy-related statistics over 2009 are taken from (BP, 2010), except electricity production (CIA, 2010) and installed capacity (Xinhua, 2010).

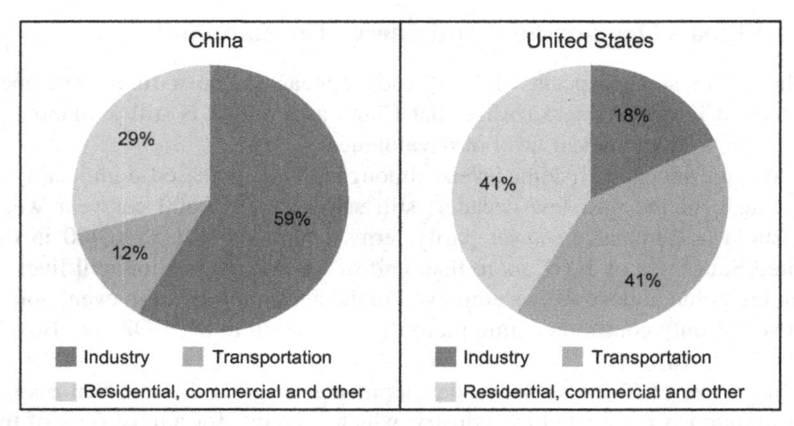

FIGURE 2 Final energy demand by sector in China and the United States. *Source: APERC,* APEC Energy Overview 2008, *March 2009.*

the enormous infrastructure development taking place in China, increasing demand for cement and construction materials, and demand from both light and heavy manufacturing industries for copper, aluminum, and steel (USGS, 2004). Complicating the argument that China should be held responsible for the pollution and emissions caused by its industry is the fact that much of its economy is geared toward exports. According to research by the British Tyndall Centre, roughly one-third of China's total emissions can be attributed to the manufacturing of goods that are exported, or about one-quarter if one adjusts for emissions embodied in imports (Wang and Watson, 2007; IEA, 2008, p. 387).

Per capita primary energy supply and electricity consumption levels stand at one-third of OECD levels, even though China has successfully pursued an electrification program with roughly 99% of its population having access to electricity. Yet, in comparison with more developed countries, energy consumption by transportation and the residential and commercial sectors are still small, indicating the potential for growth. Figure 2 shows the comparison of final energy consumption by sector between China and the United States (IEA, 2009a; ADB, 2009, p. 150).

2.2. Resources Base and Energy System

To accommodate its rising demand for energy, China has expeditiously developed its energy supplies and has remained largely self-sufficient. China accounts for about one-fifth of the world's energy consumption and reportedly overtook the United States as the largest energy consumer in 2009.

Figure 3 shows the structure of China's primary energy consumption and illustrates the dominant position of coal. Coal has fueled China's industrialization as the most abundant and the most easily exploitable fuel available.

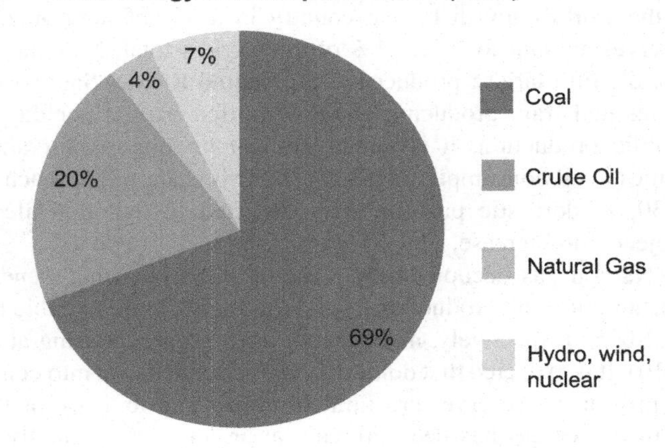

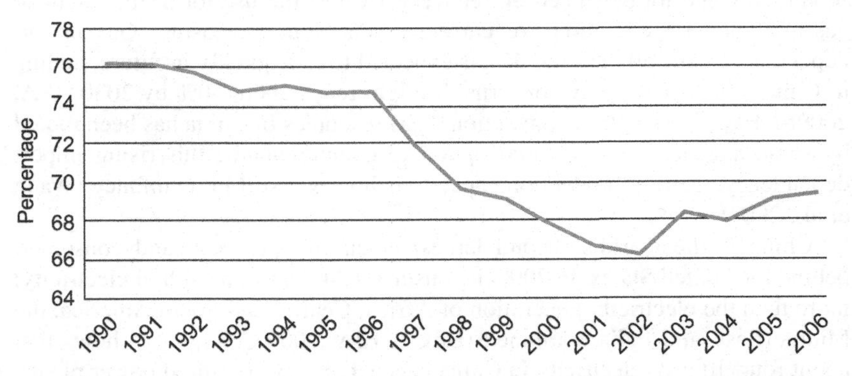

FIGURE 3 China's fuel mix structure and the share of coal 1990-2006. *Source: China Statistical Yearbook (2009).*

China holds the world's third-largest coal reserves, behind the United States and Russia, and the reserves amount to 13.9% of the world's total (BP, 2010). China has emerged as both the world's largest consumer as well as producer of coal. It accounts for about 45% of global coal demand and production and produces more than twice as much as the United States, which ranks second. Moreover, production is increasing at an astonishing rate, doubling between 2001 and 2008 and growing faster than primary energy demand growth in recent years. Although energy policy has aimed to limit the use of coal due to the negative environmental impact, this has had only limited effect up to the present. Figure 3 shows how in past decades the share of coal steadily declined until around 2002, after which it increased again to just under 70%, due to a sudden boom in industrial energy demand.

The second most important fuel in China's energy mix is oil. Even though it is the world's fourth largest country in terms of land area, China's proven reserves amount to only 1.1% of the world's total. Nonetheless, China is the world's fifth largest producer of oil, behind Russia, Saudi Arabia, the United States and Iran, producing about 4 million barrels per day. In spite of this prolific production, it is already reliant on imports for almost 60% of its domestic oil consumption (BP, 2010). This share might reach 80% or more by 2030, as domestic production is expected to flatten while consumption is projected to increase.

Natural gas occupies only a minor share in China's energy system. Gas consumption and production levels stood at 88.7 billion cubic meters (bcm) and 85.2 bcm, respectively, in 2009, but have been increasing at a rapid rate (BP, 2010). It is expected that domestic production will run into constraints however, as proven gas reserves are limited, amounting to 1.3% of the world's total. Demand for gas has been growing at around 10% annually since 2000 (BP, 2010). Gas is used mainly by the petrochemical industry, for fertilizer production, and for enhanced oil recovery. Residential use for heating and for gas-fired power generation are currently small but increasing. Gas import dependency, currently around 5%, is expected to rise quickly in future, adding to China's energy security concerns. It might reach about 50% by 2030 (IEA, 2009b). Exploitation of unconventional gas resources in China has been touted by some analysts as a potential option to counterbalance this rising import dependency. Current production capacity is however still in its infancy (Wang et al., 2009).

China is the world's second largest electricity producer and consumer, behind the United States. In 2008, it consumed 3451 billion kWh of electricity: more than the electricity generation of Africa, Central and South America, the Middle East, and India combined (Box 1; EIA, 2009).[4] Figure 4 shows that about four-fifths of electricity in China is generated by coal-fired power plants. Hydropower is the only other significant source of power generation, contributing 15%, while nuclear power, oil- and gas-fired power plants occupy only minor shares. The large role for hydropower reflects that China holds the world's largest hydropower resources and has emerged as the world's largest producer of hydroelectricity. As hydroelectricity still is by far the world's most important source of renewable energy—accounting for four-fifths of all "renewable" electricity at a global level—this makes China the world's largest producer of renewable energy as well (REN21, 2010). However, the

4. The recent estimate of Chinese electricity consumption (of 3451 billion kWh in 2008) is taken from *CIA World Factbook*, 2010. The international comparison however is based on (preliminary) figures for 2006 from U.S. Energy Information Administration (EIA), Annual Energy Review 2008, June 2009: *www.eia.doe.gov/aer/pdf/aer.pdf*, p. 337. Net electricity generation (in billion kWh) of China is 2717.5, Africa 546.8, the Middle East 641.4, Central and South America 951.0 and India 703.3.

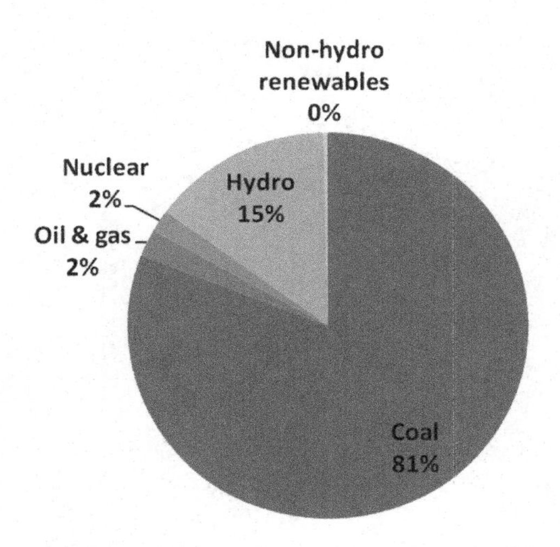

FIGURE 4 China's power generation fuel mix. *Source: IEA, WEO2009 (2007 data).*

contribution of nonhydro renewable energy sources in China, such as wind, biomass, and solar, is still rather marginal and accounts for less than 1% of the electricity supply (IEA, 2009b).

3. ENERGY AND CLIMATE CHANGE POLICY

Chinese energy policy has been driven mainly by energy security concerns that favored the development of domestic energy resources. Although energy security remains of paramount importance, concerns about the environmental impact of excessive coal use and perceived economic opportunities in new energy technologies have become important factors as well, as further described in this book's Chapter 2 by Felder et al. This section discusses various aspects of Chinese energy policy and the implications for its policy on climate change.

3.1. Energy Policy

Traditionally, there has been a strong emphasis on energy efficiency and energy conservation in Chinese energy policy, which has led to the remarkable growth pattern observed in the two decades from 1980 to 2000. During this period GDP quadrupled while energy consumption merely doubled, which signifies quite an impressive feat for an industrializing country.

China has reiterated the goal of quadrupling GDP while only doubling energy consumption for the period 2000 to 2020, but with a surge in energy demand in the first decade of the new millennium, reaching this objective has become practically impossible (Figure 5). Nonetheless, China is continuing to strongly promote energy efficiency and set a 20% reduction target in energy

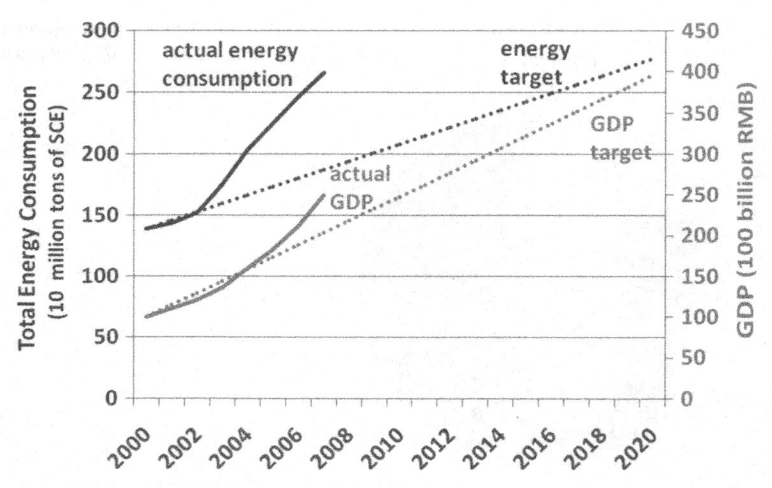

FIGURE 5 Chinese energy demand and GDP growth compared to 2020 targets. *Source:* China Statistical Yearbook 2009, *SCE denotes (Chinese) standard coal equivalent. Based upon: Mark D. Levine, Nan Zhou, Lynn Price, 'The Greening of the Middle Kingdom: The Story of Energy Efficiency in China', Lawrence Berkeley National Laboratory, May 2009, p.22.*

intensity, i.e., energy consumption per unit of GDP, into its 11th Five-Year Plan covering 2005 to 2010. It has been estimated that the emissions savings resulting from this policy are around 1.5 billion tons of CO_2, almost five times as much as the absolute amount set by Europe (EU-15) under its Kyoto Protocol commitment (LBNL, 2007).

To implement its energy efficiency policy China has initiated quite effective programs such as the *Top-1000 Enterprises Program*, which targets the one thousand largest industrial energy consumers that together account for almost half of China's energy demand and two-thirds of its industrial energy demand. Through semivoluntary targets and energy audits at these companies, large improvements in energy efficiency are sought and implemented (Price et al., 2008; and this book's Chapter 12 by Brown et al.). Fuel efficiency standards for cars have been raised following European requirements, and are already stricter than in the United States. In terms of raising efficiency in power plants and industry, China is pursuing a policy to close down small inefficient units and factories. According to government estimates, 55.5 GW of capacity has been closed down in the period 2006–2010 (Reuters, 2009). Energy efficiency labeling has been made mandatory for many consumer products and appliances. Furthermore, a new building code standard has been introduced to achieve a 50% saving standard from previous requirements (WRI, 2009; APERC, 2008, p. 109).

Chinese policy aimed at the power sector has several aspects. Concerning coal-fired power plants, which make up the bulk of power generation capacity, energy policy is directed at modernization of the fleet. Overall energy conversion efficiency of Chinese power plants is estimated be about 33.8% on average, about 6%–7% lower than coal-fired power plants in developed countries. It can be as high as 45%–47% for new state-of-the-art ultra-supercritical power plants, and China is introducing such plants at a significant scale (CIEP, 2009, p. 68).

A second objective is to limit coal-fired power generation due to environmental concerns over air pollution and acid rain. To this end, China is strongly promoting the use of renewable energy and nuclear power. A goal has been set to increase the share of nonfossil energy sources from 7.5% to 15% of primary energy consumption by 2020. An additional incentive is the aim of establishing a strong domestic industry in wind, solar, and nuclear energy (CIEP, 2009, pp. 75–77).

The expansion of hydropower is still the mainstay of renewable energy development and government policies have strongly encouraged the development of both large-scale hydropower projects, such as the 22.5 GW Three Gorges Dam, and small-scale hydropower in rural areas. More recently, nonhydro renewables are stimulated by various policies including *feed-in tariffs* and a *renewable energy portfolio standard* (RPS) for grid and power companies (Martinot and Li, 2007). For wind energy this has resulted in a spectacular growth with total installed capacity doubling four years in a row. China emerged as the largest growth market for wind turbines in 2009.[5] Similarly, China's solar energy industry has been growing rapidly. In contrast to the Chinese wind industry, it has been almost completely directed at the export market. In very little time Chinese solar cell and panel manufacturers have gained significant global market shares. In its *Renewable Energy Medium and Long Term Development Plan* the government announced targets for 2010 and 2020 (Table 1), but some of these have already been exceeded. The 30 GW target for wind by 2020, set in 2007, has already been surpassed. It is likely to be revised upward to 100 GW or even 150 GW. For solar power, the target has been set at 1.8 GW for 2020, although some officials have signaled this could be put much higher at 10–20 GW. This would be quite an ambitious goal considering that there was not even 1 GW of grid-connected solar power installed in China in 2009 and about 21 GW of solar-power installed worldwide as of 2009 (REN21, 2010). The deployment of solar hot water has been more widespread: China holds 70% of all global solar hot water capacity and continued growth is strongly supported by the government (REN21, 2010, p. 12).

The Chinese definition of nonfossil energy also includes nuclear power, which currently plays only a minor role in electricity supply (Figure 4). At present, 13 reactors are in operation with a combined capacity of about 10 GW.

5. Installed capacity of wind power doubled four years in a row: from 1.3 GW in 2005 to 2.7 GW in 2006, 5.9 GW in 2007, 12.2 GW in 2008, and 25.1 GW in 2009 (GWEC, 2009, p. 27).

TABLE 1 China's Renewable Energy Targets for 2020

RE source	Total potential	2005	Target 2010	Target 2020
Hydropower	400 GW (540)	117 GW	190 GW	300 GW
Biomass				
biomass power	-	-	– 5.5 GW	– 30 GW
biomass pellets			– 1m tonnes	– 50m tonnes
biogas			– 19bn m^3	– 44bn m^3
bio-ethanol			– 2m tonnes	– 10m tonnes
bio-diesel			– 0.2m tonnes	– 2m tonnes
Wind power	300 GW onshore, 700 GW offshore	1.26 GW	5 GW onshore, 200 MW offshore	29 GW onshore, 1 GW offshore *(under revision)*
Solar power PV		70 MW	0.3 GW	1.8 GW
Solar thermal		80m m^2	150m m^2 (30 Mtce)	300 m^2 (60 Mtce)
Geothermal power			4 Mtce	12 Mtce
Tidal power			-	100 MW

Source: National Development and Research Commission, *Medium and Long Term Development Plan for Renewable Energy in China* (draft), September 2007.

However, the contribution of nuclear power is set to increase fast, as more than one-third of all nuclear power plants under construction worldwide are being built in China. Official targets aim to expand the nuclear power capacity to 40 GW by 2020, but this target is quite likely to be revised upwards to 60 GW or even 70 GW. Policy documents indicate that China aims to build up a domestic nuclear industry and strong technology transfer conditions were included in contracts awarded to foreign suppliers of nuclear technology (CIEP, 2009; World Nuclear Association, 2010).

3.2. Policy on Climate Change

China unveiled an explicit comprehensive policy on climate change with the launch of its *National Climate Change Programme* in 2007 (NDRC, 2007a). Measures that were included in the document can largely be interpreted as cobenefits arising from China's energy policy (Table 2). Yet even if these policies are partly driven by energy security, and environmental and economic considerations, they already yield significant benefits for climate change stabilization objectives.

TABLE 2 Chinese Estimates of Avoided Emissions, due to Mitigation Measures in *China's National Climate Change Programme*

Emissions avoided by 2010 (Mt of CO_2e)	Measure
550	Implement various energy conservation programmes
500	Continue to expand hydropower for electricity generation
200	Develop coal-bed methane (CBM) and coal-mine methane (CMM)
110	Upgrade thermal power generation: develop (ultra)-supercritical units, combined-cycle units, heat/power cogeneration, heat/power/coal gas multiple supply units
60	Utilize wind, solar, geothermal, and tidal energy
50	Increase forest rate to 20% and enhance carbon sinks
50	Continue to promote nuclear energy
30	Promote bio-energy for power generation and fuels
1550	**TOTAL**

Source: National Development and Reform Commission, *China's National Climate Change Programme*, June 2007.

However, this does not hold for all measures. Gas-fired power plants, for instance, which emit approximately half the amount of CO_2 per kWh, are also set to increase to 70 GW but are not overly promoted due to concerns over the forecasted rise in gas imports (CIEP, 2009, p. 67). China's system of energy pricing is another aspect that is hampering a drive for energy-efficiency and demand reduction. The price of electricity in particular is tightly controlled and kept relatively low. As a consequence, there is little economic incentive for energy-efficiency improvements. Additionally, China is developing coal liquefaction technologies with energy security as the main driver. However, due to the polluting production process, the carbon footprint of coal-based petroleum is much higher than conventional petroleum.

To truly implement policies and measures that will meet climate change objectives on the longer term, it will be necessary to take a step beyond energy security driven policies. The most crucial aspect of that will be to turn away from coal as a main fuel in China's fuel mix or to ensure the widespread deployment of CCS technologies. The former would run counter to energy security and economic concerns, while the latter carries significant costs. China is pursuing advanced coal technologies, including integrated gasification combined cycle (IGCC) power plants and CCS, as demonstrated by the

"GreenGen" project near Tianjin. However, some kind of carbon price or international financial assistance program would be necessary to incentivize the implementation of CCS as long as a direct economic rationale is lacking (Morse et al., 2009).

The adoption of carbon intensity targets in the run-up to the COP-15 summit in Copenhagen in December 2009 should be seen in this light. China announced the goal of lowering the carbon intensity of its economy by 40%–45% by 2020 compared to 2005 levels as a voluntary initiative. This should be regarded as a significant step, as it will also turn the focus to improving emissions monitoring and achieving emissions reductions within China's energy system, on top of improving energy efficiency. However, as illustrated in Section 5, the target aligns with improvements already following from recent energy policy measures and in itself it will not be sufficient to drastically change China's developmental path.

4. GROWTH PATTERNS AND MITIGATION OPPORTUNITIES

This section discusses several areas that can be considered of special importance given their impact on energy and emissions and the potential for mitigation measures.

China's demand for energy is driven mainly by its growing economy and increasing levels of prosperity. Population growth is less of a factor as the growth rate has slowed considerably due to China's one-child policy and now stands at 0.655% annually. This is less than half of India, which is expected to overtake China as the world's most populous country by 2025 (UN, 2008).[6]

The sectors that are the most important in China's current energy consumption system are industry and the power generation sector. They also play a vital role in China's GHG emissions: combined, the two sectors account for practically all coal consumption within China, and coal combustion causes about 75% of China's total CO_2 emissions. These emissions in turn account for about four-fifths of China's total GHG emissions.

Addressing these two sectors will be crucial to curtail rising emissions in China. Both reducing the energy intensity and carbon intensity of further growth will be necessary to make China's energy future more sustainable. While for the power sector there are technological options available to decarbonize the supply of energy, for industry this will remain a significant challenge as described in this book's Chapter 12 by Brown et al. For both sectors there are also significant gains to be had from energy-efficiency and demand reduction. However, other sectors that still contribute less to energy demand and emissions will also be important for pre-empting future growth.

6. According to these same "median variant" UN projections, China's population might already peak in 2030 at almost 1.5 billion and fall to 1.4 billion by 2050. In its high growth rate variant, however, China's population would increase further to 1.6 billion people by 2050.

4.1. Economic Growth and Industrial Energy Demand

Economic growth forecasts for China remain quite robust and the expansion of economic activity will be the main driving factor for an increasing demand for energy. When considering the ratio between the primary energy supply and GDP, China's economy is still more than four times as energy-intensive as the OECD average. The main reason for this is the predominance of energy-intensive industries in China's economy, mentioned in Section 2.1. China is a major global producer in many heavy industries: it accounts for 33% of the world's production of aluminum, 49% of cement, 51% of pig iron, and 38% of raw steel (USGS, 2010). A second contributing factor is the fact that the energy efficiency of the production processes in these industries are still significantly below the standard of best available technologies. This holds especially true for energy-intensive heavy industries such as steel, copper, aluminum, ammonia, plate glass, and cement, where energy intensity levels are still 25%–60% higher than the advanced international levels (APERC, 2008)[7]. As a consequence, improving energy efficiency in these sectors can contribute significantly to reducing energy consumption levels. However, in general these energy-efficiency and energy-conservation gains cannot be expected to offset significantly higher demand caused by expanding production in these sectors (IEA, 2009c). In the end, industrial restructuring and moving toward a more service-oriented economy will be essential for lowering China's future energy demand (Wang and Watson, 2008 and 2009).

4.2. The Power Sector and Electricity Demand

At a global level, the existing power generation sector already determines much of our future carbon emissions. Three-quarters of all generated electricity in 2020—and more than half in 2030—is estimated to come from power plants in operation today, according to the IEA (IEA, 2008, p. 12). In that sense, China and the massive expansion of its power sector offers a unique opportunity to influence the future energy system and level of emissions.

The power sector in China has been going through a phase of frenzied growth as consumption of electricity has been soaring. Demand has been increasing with growth rates between 9% and 15% in the past decade and installed capacity more than doubled in size since 2000. To illustrate the scale of this expansion, for every year in the three-year period 2005–2008, the equivalent of the whole power sector of the United Kingdom has been added in

7. APERC, *APERC Energy Overview 2007*, 2008, p. 46. APERC, *Understanding Energy in China*, 2008, pp. 101–103. Potential energy efficiency gains in energy resource consumption per unit of output for various industries: coal-fired power (17%), steel (18%), copper smelting (56%), aluminium (38%), ammonia (25%), cement (14%), plate glass (44%), and paper and paper products (120%).

China. Although the economic and financial crisis caused a drop in electricity consumption in the beginning of 2009, the total yearly electricity consumption in 2009 grew 6% and installed capacity reached 784 GW (MIT, 2008; CBS, 2009; Xinhua, 2010).

The previous section on Chinese energy policy indicated that China is striving to limit the role of coal in its power system. However, despite increasing investment in nuclear power and renewable energy sources, the majority of the growth in electricity supply is still coal-fired (Figure 6). China has been building the equivalent of several 500 MW coal-fired power plants per week (MIT, 2008). Considering that one 500 MW coal-fired power plant emits about 3 million tonnes of CO_2 per year, large amounts of CO_2 emissions are being locked-in (MIT, 2007, p. ix). Apart from the impediments to CCS that were already discussed, offsetting the coal-related emissions would require a tremendous upscaling of the technology; today's largest CCS activities at Sleipner (Norway) In Salah (Algeria) and Weyburn (U.S./Canada) store less than 5 million tons of CO_2 per year in total.

Table 3 shows that coal would still account for 58% of all installed capacity if China would meet all its ambitious targets by 2020 and electricity demand would follow the relatively conservative projection of the IEA. Given the fact that the renewable energy sources do not generate electricity according to their full capacity due to intermittency, the share of coal in terms of total generated electricity would even be higher.

In the medium term, the development of renewables in China will also face several impediments. One very significant challenge is that the development of hydropower in China will run into natural constraints. According to China's *Medium and Long Term Development Plan for Renewable Energy* (NDRC, 2007b), the total economically feasible potential for hydropower is estimated to be 400 GW, with a technically feasible upper limit of 540 GW. This means that

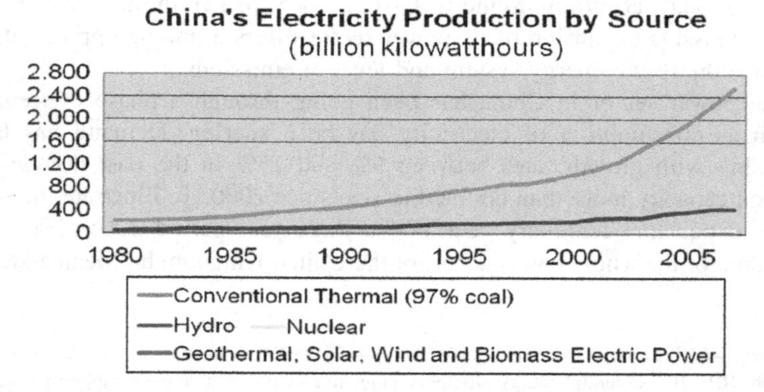

FIGURE 6 Development of Chinese electricity production by source 1980-2007. *Source: Energy Information Administration (EIA), U.S. Department of Energy,* International Energy Data, *Net Electric Power Generation. Most Recent Annual Estimates (2008).*

TABLE 3 Overview of Chinese Power Generation Capacity by 2020, Under the Assumption that Suggested Policy Objectives will be Achieved, in Gigawatt

Energy type	Capacity (2006)[x]	Shares	Capacity (2020)	Shares
Hydro	132	21%	300	21%
Wind	3	0%	100	7%
Gas	14	2%	70	5%
Nuclear	7	1%	70	5%
Biomass	2	0%	30	2%
Oil	16	3%	20[x]	1.4%
Solar	0	0%	10	0.7%
Total non-coal	*174*	*28%*	*600*	*42%*
Coal	449	72%	818	58%
Total capacity	623	100%	1418[x]	100%

Source: CIEP, 2009, p. 63. Based upon various government targets, additional projections (marked by [x]) taken from the Reference Scenario in International Energy Agency, *World Energy Outlook 2008*, p. 531.

already nearly half of the total potential economically viable hydropower reserves have been utilized, and this will grow to three-quarters if China achieves its target of establishing 300 GW of hydropower by 2020 (Table 3).

Regarding the development of wind and solar energy, China has a large potential but significant challenges exist as well. The resource potential for wind power is estimated at 1000 GW, of which about 300 GW is onshore and 700 GW offshore (Table 1; NDRC, 2007b). The best onshore resources are located in northern and western provinces (Figure 7). A consequence is that considerable transmission capacity is needed to transport the wind energy to urban demand centers. Problems with connecting far-off wind farms to the grid and the intermittency of the electricity supply are impeding China's wind energy expansion already. It is estimated that one-third of all wind farms in China are not connected to the grid. For developing solar energy resources, similar problems can be expected, as China's western provinces such as Tibet, Xinjiang, and Qinghai are the most promising but located far away from consumption centers (Figure 8). Offshore wind resources have the advantage that they are located close to the densely populated coastal regions, but investment costs are still relatively high. The deployment of offshore wind in

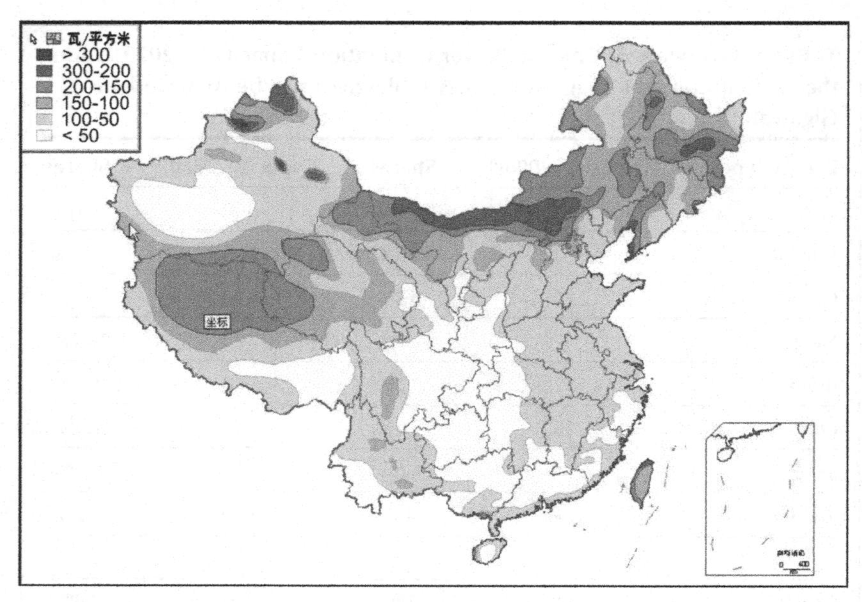

FIGURE 7 Map indicating regional spread of wind energy resources in China. *Source: Asia Pacific Energy Research Centre (APERC) (2004).* New and Renewable Energy Overview in the APEC Region.

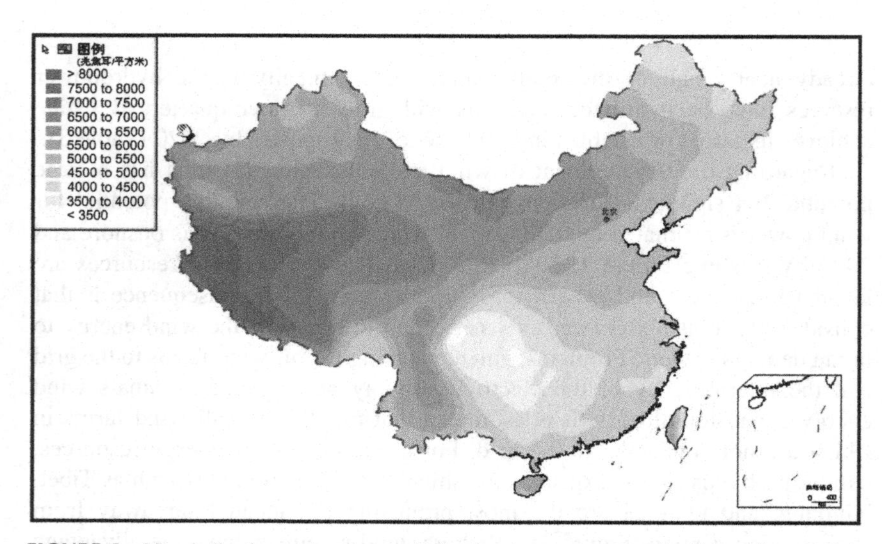

FIGURE 8 Map indicating regional spread of solar energy resources in China. *Source: Asia Pacific Energy Research Centre (APERC), 2004. New and Renewable Energy Overview in the APEC Region.*

China just started with a 100 MW wind farm off the coast near Shanghai, but a rapid growth of offshore wind farms is expected.

As far as biomass power is concerned, there is a potential for using large volumes of agricultural waste products, but developing and improving energy grasses and second-generation energy crops will be required to greatly expand biomass as a source of electricity. Currently biomass-fueled power capacity stands at only 2 GW and growth will need to accelerate rapidly to meet the 2020 target of 30 GW (Table 3).

Nuclear energy has the advantage that uranium and other reactor fuels are relatively abundant globally and it can play a serious role in reducing carbon emissions from China's power sector. The cost-effectiveness and high load-factor once in operation make it a viable option that could have a large impact. The main impediments are capacity constraints concerning the construction of nuclear power plants at a massive scale and potential environmental concerns over safety and nuclear waste.

4.3. Urbanization and the Energy Efficiency of Buildings

Urbanization is playing a key role in the growth of energy demand in several ways. First, urban per capita energy consumption levels are much higher than in rural areas. Second, the construction of housing and infrastructure is driving much of the demand for products of China's energy-intensive industries such as cement, steel, and other building materials. Lastly, the residential and commercial buildings constructed now will lay the foundation for future energy consumption levels to a great extent (LBNL, 2008a and 2008b).

The fundamentals underlying China's urbanization imply the trend is by no means exhausted. More than half of China's population still lives in rural areas and China's urban population is steadily increasing at the rate of 15–20 million annually. From 2000 to 2006, China's urban population expanded by 26% from 459 to 577 million and it is expected to surpass the 1 billion mark between 2025 and 2030. This means China could have almost 110 cities with over 1 million people by 2015, growing to more than 220 cities by 2025 (McKinsey, 2009b).

China is currently engaged in an unprecedented housing boom to accommodate the growing urban population. It accounts for about half of all building construction taking place worldwide, with about 2 billion square meters of floor-space added every single year.

While the construction sector contributes to current industrial energy use by creating demand for China's energy-intensive industries, it can be considered even more important for mitigation purposes in the long run. In developed countries, buildings on average account for 30%–40% of the total energy consumption via heating, cooling, lighting, and other appliances. New building techniques, passive design concepts, and smart metering technologies can however significantly lower this energy demand, as discussed in this book's Chapter 9 by Gray and Zarnikau, Chapter 10 by Ehrhardt-Martinez et al., and Chapter 14 by Bauermann and Weber.

Implementation of such features in China could have a lasting major effect on future levels of energy consumption and emissions.

4.4. Transportation

Private car ownership in China is still low by international standards as only 3 people out of 100 own a car, compared to developed countries where the figure stands at roughly 50 per 100 people in European countries and exceeds 76 per 100 in the United States. As a consequence, transportation energy demand has been relatively small, up to now. Figure 2 showed that transportation in China accounts for only 12% of the total final energy use, compared to 41% in the United States.

This, however, is set to change rapidly as sales volumes of cars in China have been skyrocketing, taking over the United States as the largest automobile market in 2009. A growing number of middle-class citizens can now afford to own cars with significant implications for energy use and carbon emissions. Analysts expect car ownership to surge fivefold by 2020 to reach about 15 cars per 100 residents. Aviation is also experiencing rapid growth, with Beijing-Shanghai already ranking as the world's sixth busiest route, with slightly more than 4 million passengers per year (China Daily, 2009; UNEP, 2008).

On a global level, emissions from transportation amount to almost one-quarter of energy-related CO_2 emissions and are projected to continue to rise swiftly. The sector remains one of the hardest to address in moving toward a more sustainable system in terms of energy and emissions.[8] This also holds for China, but as a sector still in expansion there is still scope for influencing the development.

Several options are being pursued. First, the growth in private car ownership is being discouraged with restrictions being in place in China's most populous cities. Second, large investments are being made in public transport, both within cities as well as in-between cities. Metro line construction is progressing at an enormous pace and a high-speed railway network is being developed between the largest cities (MIT, 2010). There are also experiments with other concepts such as a bus rapid transit system that will be introduced in Guangzhou.

The most commonly identified alternatives to petroleum-based transportation are biofuels, electric and/or hybrid vehicles, or hydrogen-based vehicles. Since China has limited arable land suited for agriculture (Box 1), expanding biofuels at a large-scale does not seem to be a promising option, at least not until suitable second-generation biofuels become available. The development of hydrogen-based vehicles might be an option, depending on breakthroughs in technology and costs.

8. For this reason, the IEA projects that the most incremental investment for achieving their 450 Scenario will be needed in the transport sector: about three times as much as for buildings (ranking second) and more than three times as much as for the power generation sector (ranking third). Mentioned is estimated cumulative investment for the period 2010–2020. IEA, *World Energy Outlook* 2009, p. 263, Fig. 7.2.

Electric cars can be considered as the most promising alternative for China, as this would align well with the objectives of reducing its oil import dependency and building up an internationally competitive car industry. There are several factors which make China well-positioned to take a lead in the development of electric cars: China's domestic car industry has focused largely on small efficient cars and China also has a strong base in battery manufacturing industry. The Chinese firm BYD, originally a battery manufacturer that moved into the electric car sector, is one prime example that caught quite a lot of attention and might be an indication of future developments. The ability of the Chinese government to issue strong centralized policy could be a major advantage in enforcing necessary standardization and infrastructure adjustments for electric vehicles. The potential impact of government measures is illustrated by the ban on gasoline scooters that has been issued as an air pollution reduction measure in major cities such as Shanghai and Beijing, which led to a rapid and near-complete shift toward electric bicycles and scooters.

However, for electric cars to really take off in China, or anywhere else, large improvements in cost and convenience will need to be made. The government has been running several promotion schemes that give consumers subsidies for buying an electric car, but prices are still relatively high and domestic sales volumes have been negligible up to now. Furthermore, although the introduction of electric cars would have an immediate effect in reducing oil demand, it would have little impact on emissions if the electricity used for transportation would still be generated by a highly carbon-intensive power sector.

5. FUTURE SCENARIOS

China's GHG emissions have been growing prodigiously, together with its energy consumption. The massive population but low per capita income and energy consumption levels clearly point to the staggering potential growth that might still take place.

Although the overall long term trend is inexorably upward, it has proven to be rather difficult to establish a "business-as-usual" (BAU) scenario for China. Previous projections have seriously underestimated the growth rates for both energy and emissions. In particular the surge in primary energy consumption and emissions which took place in the years following 2002, due to a sudden boom in heavy industry, was unforeseen by most analysts[9] (LBNL, 2008c).

Taking these considerations into account, several scenarios that illustrate BAU and alternative growth trajectories for China will be examined.

9. The U.S. Energy Information Administration in its *International Energy Outlook 2004,* for instance, projected that it would take China until beyond 2025 to overtake the United States as the largest emitter of CO_2. In fact, this already happened in 2007. As a consequence, emissions trajectories for China have been significantly revised upwards in more recent projections. Estimates for Chinese emissions in 2030 by the IEA, to take another illustrative example, were increased by 70% between its *World Energy Outlook* of 2004 and 2007 (LBNL, 2008c).

5.1. Business-As-Usual Scenarios

According to China's own statistics, total GHG emissions increased from 4.060 million tons of CO_2 equivalent (Mt CO_2e) in 1994 to 6.100 in 2004: an increase of about 50% in one decade (NDRC, 2007a, p. 6). Total Chinese GHG emissions for 2007 have been estimated at 7.6 Gt CO_2e, of which 6.1 Gt are energy-related CO_2 emissions.[10]

Table 4 shows that most BAU scenarios see China's emissions more or less doubling by 2030, taking into account that energy consumption and emissions have shown a significant acceleration since 2002. Energy-related emissions of CO_2 are expected to exceed 11 Gt by 2030, with total GHG emissions reaching 14.5 Gt according to McKinsey. The underlying assumption of economic growth is a very significant factor in making these projections. The McKinsey baseline scenario assumes an overall average GDP growth rate of 7.8% over the projection period.[11] In its *World Energy Outlook 2007*, the IEA also included a High Growth scenario which assumed Chinese average GDP growth over the

TABLE 4 Several Business-As-Usual Emissions Scenarios for China to 2030

Emissions (Gt CO_2e)	Base year emissions	2020	2030	Average growth
IEA, WEO2007, Reference Scenario	5.1 (2005)		11.4	3.3%
IEA, WEO2007, High Growth Scenario	5.1 (2005)		14.1	4.2%
EIA, IEO2009, Reference Scenario	6.0 (2006)	9.4	11.7	2.8%
EIA, IEO2009, High Ec. Growth Scenario	6.0 (2006)	9.9	12.9	3.2%
IEA, WEO2009, Reference Scenario	6.1 (2007)	9.6	11.6	2.8%
McKinsey, *China's Green Revolution* (2009), baseline scenario	7.6 (2007)*		14.5*	2.8%

Average growth denotes average annual growth rate of emissions over the period base year-2030.
(*): McKinsey emissions statistics are based on overall GHG emissions (i.e. all GHG gases), while IEA and EIA emissions statistics only consider energy-related CO_2 emissions. As a global average, energy-related CO_2 emissions represent 64% of total GHG emissions; for China this percentage lies higher at approx. 80%.
Sources: McKinsey, *China's Green Revolution*, 2009, pp. 22; IEA, WEO2007, pp. 389-402 (High Growth Scenario), 596-599 (Alternative Policy and Reference Scenario); EIA, EIO2009, pp. 131, 148; IEA, WEO2009, pp. 199-200, 210, 623, 647.

10. 7.6 Gt overall GHG in 2007 taken from McKinsey, *Green Revolution* (2009a), p. 29. See p. 22 for what is included in this estimate (CO_2, CH_4, N_2O, and carbon sinks). 6.071 Mt of en.rel. CO_2 emissions taken from IEA, *World Energy Outlook* (2009), p. 647.

11. Dropping from 9.9% between 2005 and 2010 to 8.2% between 2010 and 2020, and 6.5% between 2020 and 2030 (McKinsey, 2009a, p. 32).

period 2005–2030 would be 7.5% instead of the 6% annual growth used for the *WEO2007* Reference scenario. This would lead Chinese energy-related CO_2 emissions to exceed 14 Gt by 2030, pushing the total of all GHG emissions even higher (Table 4; IEA, 2007, pp. 389, 401). A few observations are worth noting:

- First, the official goal of quadrupling GDP between 2000 and 2020 while only doubling energy demand in the same period implies an average GDP growth rate of 7.2% and a primary energy demand growth of 3.5% annually up to 2020. If such a trend would continue until 2030, it would mean that China will be more in line with the High Growth scenario rather than the Reference scenario of the *WEO2007*.
- Second, the baseline scenarios already expect a significant decline in emissions growth compared to the past two decades. Annual growth rate of overall GHG emissions stood at 4.7% for the period 1990–2007, whereas most reference projections for 2030 see this slowing down to around 3% for 2005–2030 (Table 4; McKinsey, 2009a, p. 29). This reflects that baseline projections have already been adjusted to incorporate the effects of China's energy strong policy on energy efficiency, renewables, and nuclear power of the past years. The baseline scenario of McKinsey, for instance, projects an average energy intensity reduction of 17% in every five-year period between 2005 and 2030 (McKinsey, 2009a, p. 37). This nearly equals the much-touted 20% energy intensity reduction target that China issued for its 11th Five-Year Plan (2005–2010).
- Third, a similar observation holds with respect to China's carbon intensity target that was mentioned in Section 3. The *World Energy Outlook 2009* reference scenario already assumes a 37% decline in carbon intensity of China's economy, measured in CO_2 emissions per unit of GDP, over the period 2007–2020. This almost equals China's carbon intensity reduction target of 40%–45% by 2020 compared to 2005 levels that was announced in the run-up to the climate treaty negotiations at the Copenhagen COP-15 meeting. Similarly, the reference projections of the U.S. Energy Information Administration (EIA) that were published in May 2009 already projected a drop of 44% in carbon intensity for China's economy between 2006 and 2020 (IEA, 2009, p. 183; EIA, 2009, p. 148).

These observations point out that the BAU scenarios for China might be considered already reasonably "ambitious" and that there is also a risk that they underestimate the growth of Chinese energy consumption and emissions.

5.2. Alternative Scenarios

Several quantitative scenarios have been developed in other studies that indicate how energy consumption and emissions from China would develop if

China were to take a different course from what can be inferred from recent trends. Since many of China's progressive policies of recent years have already been incorporated in the BAU scenarios, this means a significant deviation from what has been implemented up until this point.

TABLE 5 Three 'Alternative' Emissions Scenarios for China to 2030

Emissions (Gt CO_2e)	Base year	2020	2030	Average growth
Alternative Scenario 1: IEA, WEO2009, *450 Scenario for China*	6.0 (2007)	8.4	7.0	0.7%
Extra details: Over the period 2007-2030, - CO_2 intensity of the vehicle fleet drops from 235 gCO_2/km to 90 gCO_2/km. - Power CO_2 intensity drops from 922 gCO_2/kWh to 448 gCO_2/kWh. - Share coal in power generation drops from 81% to 50%. Power sector in 2030 includes the following (approx.): 770 GW coal without CCS; 110 GW nuclear 100 GW of gas without CCS; 120 GW other RES 370 GW hydro; 250 GW wind 40 GW of coal and gas with CCS.				
Alternative Scenario 2: IEA, WEO2007, *Alternative Policy Scenario for China*	5.1 (2005)		8.9	2.3%
Extra details: - Coal supplies 64% of electricity by 2030. - Energy demand increases with 90% in 2030 compared to 2005, but is 15% lower than the reference scenario. Structural changes in the economy account for more than 40% of the total energy savings.				
Alternative Scenario 3: McKinsey, China's Green Revolution (2009), *abatement scenario*	7.6* (2007)		7.8*	0.1%
Extra details: Power sector in 2030 has total capacity of 2122 GW and includes the following: 550 GW coal (25% with CCS); 144 GW gas 317 GW (large) hydropower; 380 GW wind 120 GW (small) hydropower; 380 GW solar 182 GW nuclear; 48 GW other				

Average growth denotes average annual growth rate of emissions over the period base year-2030.
(*): McKinsey emissions statistics are based on overall GHG emissions (i.e. all GHG gases), while IEA and EIA emissions statistics consider energy-related CO_2 emissions. See note at Table 4.
Sources: McKinsey, *China's Green Revolution*, 2009, pp. 22; IEA, WEO2007, pp. 389-402 (High Growth Scenario), 596-599 (Alternative Policy and Reference Scenario); EIA, EIO2009, pp. 131, 148; IEA, WEO2009, pp. 199-200, 210, 623, 647.

We evaluate the findings of three different scenarios. One has been developed by using back-casting (i.e. calculating backwards from a "desired" outcome): the *IEA 450 Scenario* of the *World Energy Outlook 2009.*[12] Two more scenarios assess how special measures or policies might work out and impact on China's energy and emissions characteristics: the *Alternative Policy Scenario* of the IEA's *World Energy Outlook 2007* and the full abatement scenario in McKinsey's *China's Green Revolution* study (2009a).

Table 5 summarizes the emissions trajectories and key statistics of the different scenarios. Some background information of the scenarios is briefly sketched in the three subsections below, after which we will turn to the implications for climate change stabilization goals.

5.2.1. Alternative Scenario 1: IEA 450 Scenario—China (WEO, 2009)

The *450 Scenario* by the IEA envisions a future in which global emissions would be reduced to allow for a stabilization of GHGs in the atmosphere at 450 ppm, equivalent to a 50% chance of limiting the temperature increase to 2°C. China plays an essential role in this scenario as it contributes 37% and 33% to the required global primary energy demand and emissions reductions by 2030, respectively.

At the end of the projection period in 2030, China would occupy almost 27% of global energy-related emissions. Per capita emissions of CO_2 would be 4.8 tonnes of CO_2 per capita by 2030—roughly the same as in 2007—after peaking in 2020 at 5.9 tonnes per capita.

While there is no detailed description of how China would achieve this scenario, both energy demand reduction and decarbonization of the energy consumption are critical. If we focus on the year 2030, energy demand would need to be reduced by almost one-quarter compared to the reference scenario, and the average carbon content of electricity would need to drop with 49% from the expected 922 to 448 gCO_2/kWh. To this end, the share of coal-fired electricity generation in the fuel mix would need to decline from 81% to 50%, while the absolute amount of electricity supplied by nuclear power should increase more than 15-fold (by 2030). Renewables would also need to experience an astonishing growth: Electricity generated by hydro should increase fourfold; wind 70-fold; and other renewable energy sources 230-fold. What this would entail in terms of generating capacity by 2030 can be seen in Table 5. Considering transport, the CO_2 intensity of China's car fleet would need to be reduced by 57% from 235 grams of CO_2 per kilometer (gCO_2/km) to 90 gCO_2/km. To achieve this by 2030, 10% of all vehicles would need to be electric and 8.1% should run on biofuels. In terms of energy security consequences, China's gas imports would drop with 22% compared to the reference scenario, and oil

12. Another excellent back-casting analysis of Chinese emissions has also been made by the Tyndall Centre, that incorporates four different future scenarios (Wang and Watson, 2008; 2009).

imports would be limited to 11 million barrels per day by 2030 (IEA, 2009b, pp. 216–218).

5.2.2. Alternative Scenario 2: IEA Alternative Policy Scenario (WEO, 2007)

In this scenario, the IEA evaluated how several policy measures might work out that have been considered but not fully implemented by the Chinese government. This includes the strong promotion of natural gas over other fossil fuels, reforming the pricing system, the introduction of fuel taxes, and shifting the economy away from energy-intensive industry. Moreover, the scenario differs from the reference scenario (of WEO, 2007) in that it assumes a very effective enforcement of all related policy measures. Energy demand in 2030 would be 15% lower than its reference projection, but would nonetheless increase by 90% compared to 2005. Emissions would decrease by 2.6 Gt, equivalent to 22.5% compared to BAU: More than half of these reductions would be achieved through changes in the power sector, although coal would still supply 64% of all electricity in 2030. Structural changes in the economy would account for more than 40% of the total energy savings and also coal demand would fall with almost one quarter mostly due to less electricity demand. Emissions would stabilize soon after 2020 at 9 Gt CO_2 per year.

5.2.3. Alternative Scenario 3: McKinsey, China's Green Revolution (2009)

According to the McKinsey report, China's overall GHG emissions could almost be cut in half by 2030 compared to the baseline scenario, achieving a reduction of 6.7 Gt of CO_2e. The largest reductions (3.8 Gt) would take place in the power sector, where emissions would drop 70% compared to the baseline trajectory. The share of coal in the electricity generation fuel mix would decline to 34% by 2030, by vigorously promoting renewables, natural gas, and nuclear energy. CCS would need to be implemented to limit emissions from coal-fired power plants. Other important areas for abatement are emission-intensive industries (2.1 Gt), buildings and appliances (1.6 Gt), and road transportation (0.6 Gt).

As the power sector would contribute the most to reducing emissions, it is worthwhile to review the structure of the power generating capacity that would be required by 2030, which is summarized in Table 5, and compare this to our discussions in Section 3 and Section 4.2.

The costs of implementing this scenario are estimated at 150–200 billion euros per year over the period 2010–2030, on top of baseline investment figures. One-third of these investments are estimated to have positive economic returns, one-third will have slight to moderate economic cost, and the final one-third will have substantial costs associated with them.

Technology is critical in McKinsey's abatement scenario, as it requires new technologies that save on energy demand and emissions to be introduced across all sectors. This holds especially for the largest growth sectors identified in the baseline scenario: power generation, road transport, buildings, and appliances. In the coal sector, the most important new technologies to be implemented are *integrated gasification combined cycle* (IGCC) coal plants (implemented at 100 GW in the abatement scenario) and *carbon capture and storage* (to be implemented on one quarter of all coal-fired power plants by 2030), on top of more highly efficient ultra-supercritical plants already incorporated in the reference scenario. In the power sector, there will need to be more nuclear power, more wind (particularly offshore), more solar power, and more cofiring with bioenergy and (bio)power from switch grass and municipal solid waste. Hybrid and electric vehicles play a significant role in transportation apart from advanced fuel efficiency improvements in ordinary internal combustion engines. New technologies should be implemented in industry, especially in energy-intensive sectors such as steel and cement production. In terms of buildings, much can be gained by upgrading building codes and introducing passive design elements that have high energy savings for new buildings.

Finally, one crucial observation from the McKinsey report concerns the importance of the speed of implementation of mitigation measures. Since China's expansion in various energy-consuming sectors is taking place at such a rapid pace, it is estimated that after a five-year delay of starting full-scale of implementation of all options, 30% of the abatement potential would be lost. A ten-year delay would increase this to 60%.

5.3. Implications for Emissions Stabilization Scenarios

Stabilization scenarios have been developed by the Intergovernmental Panel on Climate Change (IPCC) of the United Nations as a suggested course of action to address climate change. Most attention has focused on limiting a global temperature increase to 2°C, which has been confirmed as an important threshold to limit ecological damage arising from climate change.

The objective of limiting a temperature increase to a maximum of 2°C is considered to be equivalent to stabilizing concentration levels of GHGs in the atmosphere at 450 parts per million (ppm), as this would give a 50% chance of keeping the temperature rise below that level. To improve the chances of not surpassing that threshold, even lower stabilization levels would be required.

As the concentration level of GHGs is determined by the cumulative total that is emitted over a certain period of time, various emissions trajectories are possible that would lead to a certain stabilization level. However, postponing emissions reductions will require a faster decline and steeper reductions later on.

According to the *Fourth Assessment Report* of the IPCC, stabilization scenarios in the range of 445 ppm–490 ppm would require global emissions to peak between 2000 and 2015 and decrease between 50% and 85% by 2050 compared to 2000 levels (IPCC, 2007a).

The burden-sharing of such global emissions reductions has been the subject of much discussion, but one of the most prominent suggestions has been for developed countries to reduce emissions by 25% to 40% by 2020 compared to 1990 levels, while asking developing countries to improve on their BAU projections of emissions by 15%–30%. In 2050, developed countries should have reduced emissions by 80%–95% compared to 1990 levels, allowing for more "carbon space" for developed countries as long as global emissions are reduced by 50% (IPCC, 2007a).

On the basis of the modeling results by the IPCC, the IEA has developed its own *450 Scenario* that sees global GHG emissions peaking in 2020 at 44 Gt of CO_2e and declining to 21 Gt in 2050. In terms of energy-related emissions this would mean a peak before 2020 at 30.9 Gt and a decline to 26.4 Gt in 2030 and 15 Gt in 2050. In this scenario, global energy related CO_2 emissions would have to decrease by about 1.5% per year in the period 2020–2050. Other suggested scenarios are even more strict: The trajectory suggested by the United Nations Environment Programme includes a global emissions target of 44 billion tonnes of CO_2e for 2020 and 16 billion tonnes of CO_2e by 2050 (UNEP, 2010).

The reference scenario projections for China, which were presented earlier, render such stabilization scenarios impossible or, at best, extremely unlikely. If China's energy-related CO_2 emissions were to continue to grow along the BAU trajectories indicated by the IEA and EIA (Table 4), they would already take up more than 40% of the annual global budget required for limiting a temperature increase to 2°C by 2030. The McKinsey baseline scenario, which also takes non-energy related GHGs into account, shows that Chinese emissions might reach 14.5 Gt by 2030 while still being on the increase. This would be rather hard to integrate with the UNEP target for global emissions in 2050 standing at 16 Gt (UNEP, 2010).

Of the three "alternative" quantitative scenarios that we examined, some still allow China's emissions to be incorporated into a 2°C stabilization goal. Of course, this is self-explanatory in the case of the *450 Scenario* by the IEA for China, which has been designed exactly to meet this objective. In this scenario, Chinese energy-related emissions would reach 7.0 Gt by 2030. The McKinsey abatement scenario projects *total* GHG emissions to be 7.8 Gt by 2030, which would still be in the range of what is required in the IEA's *450 Scenario*. However, a comparison between the required power sector structure for 2030 by McKinsey (Table 5) and the evaluation presented in Sections 3 and 4.2 indicate the enormous challenge that this scenario represents and questions the feasibility of such a transformation. The IEA *Alternative Policy* scenario shows that even with more stringent policy Chinese emissions might very well be above the required level for stabilization, projecting 2030 energy-related emissions at 8.9 Gt.

CONCLUSIONS

As a country still in the midst of its development, there are significant opportunities for China to develop along a more sustainable pathway than many countries that industrialized before it. China has a chance to put strong policy in place that can impact the energy system and energy-consuming infrastructure that is being laid out and that will determine future energy consumption and emissions levels up to a large extent. This holds especially true for several sectors that are undergoing a rapid phase of expansion, such as power generation, housing, and transportation. There are also large potential gains to be had by increasing the level of energy efficiency throughout China's economy and society.

China's current progressive policies are already having significant positive effects in reducing energy demand and energy-related emissions, as they stimulate energy efficiency and conservation, renewable energy sources, and nuclear power. The goal of establishing strong domestic industries in the field of advanced coal utilization, wind energy, solar energy, nuclear energy, and electric cars greatly contributes to a transition to a more sustainable energy system. China's role as a major global manufacturing center combined with these progressive industrial policies make China well-positioned to develop and deploy low-carbon and energy-efficient technologies. However, Chinese energy policy does not have sustainability as its main priority and there are serious impediments to achieving a sustainable energy system that will need to be addressed. As the cheapest, most abundant, and most carbon-intensive fuel available, the future role of coal in China's energy system is the crux of the matter regarding China's drive for sustainability.

Our analysis of several quantitative scenarios of China's future development showed that the demand for energy and especially energy-related emissions show an enormous increase, even if recent progressive policy measures—including China's carbon intensity targets—are taken into account. It can be concluded that BAU trajectories are impossible to reconcile with climate change stabilization scenarios that would limit a global temperature increase to 2°C. In fact, the most critical observation is that even with extreme measures and vigorous implementation, these global stabilization scenarios will be difficult to meet.

While current policies are not sufficient to achieve a sustainable energy system in the short-term span required in order to mitigate climate change, there are significant technical, economic, and political barriers that will hamper a strengthening of current policy measures. The main priority of China is economic development and the increase of welfare for its population. This means there are limits to implementing measures such as raising energy prices, limiting industrial energy demand, supporting renewable energy to expand even beyond current high growth rates, and developing and deploying CSS techniques to reduce emissions.

The necessity of a sustainable energy future for China should not be just a concern for China itself, however, as it will impact on the energy and climate future of the world at large. Given the global repercussions, the world will need to think about how to encourage an energy transition in China. Technologies, learning experiences, and best practices suggested in the other chapters of this book can hopefully provide inspiration and a contribution to this challenge.

BIBLIOGRAPHY

Asia Pacific Energy Research Centre (APERC). (2004). *New and renewable energy overview in the APEC region.*

Asia Pacific Energy Research Centre (APERC). (2008). *Understanding energy in China.*

Asian Development Bank (ADB). (2009). *Energy outlook of DMCs in East Asia.*

BP. (2010). *Statistical review of world energy.*

China National Bureau of Statistics (CBS). (2009). *China statistical yearbook 2008.* China Statistics Press.

China Daily. (2009). China car boom could last a few years: analysts. 9 July 2009. *www.chinadaily.com.cn/china/2009-07/09/content_8401265.htm.*

Chinese Academy of Sciences (CAS). (2007). Jointly addressing the challenge: Developing a sustainable energy system. 22 October 2007. (Announcing the release of the report: Chinese Academy of Sciences, *Addressing the Challenge: Developing a Sustainable Energy System*, October 2007.) http://english.cas.cn/Ne/CASE/200710/t20071022_18287.shtml.

Cho, J.-M., & Giannini-Spohn, S. (2007). Environmental and health threats from cement production in China. *A China Environmental Health Research Brief*, Wilson Center. *www.wilsoncenter.org/index.cfm?topic_id=1421*&fuseaction=topics.item&news_id=274782.

CIA. (2010). *CIA World Factbook – Country Profile China.* https://www.cia.gov/library/publications/the-world-factbook/geos/ch.html.

Clingendael International Energy Programme (CIEP). (2009). *Bram Buijs. China, Copenhagen and beyond.* Clingendael Energy Paper. *www.clingendael.nl/publications/2009/20090900_ciep_report_buijs_china_copenhagen_beyond.pdf.*

Energy Information Administration (EIA). (2009). *Annual energy review 2008.* U.S. Department of Energy. *www.eia.doe.gov/aer/pdf/aer.pdf.*

Global Wind Energy Council (GWEC). (2009). *Global wind report 2009. www.gwec.net/fileadmin/documents/Publications/Global_Wind_2007_report/GWEC_Global_Wind_2009_Report_LOWRES_15th.%20Apr.pdf.*

International Energy Agency (IEA). (2007). *World energy outlook 2007.*

International Energy Agency (IEA). (2008). *World energy outlook 2008.*

International Energy Agency (IEA). (2009a). *Key world energy statistics 2009.*

International Energy Agency (IEA). (2009b). *World energy outlook 2009.*

International Energy Agency (IEA). (2009c). *Energy technology transitions for industry.*

IPCC. (2007a). *Fourth assessment report. Climate change 2007 – synthesis report, summary for policymakers. www.ipcc.ch/pdf/assessment-report/ar4/syr/ar4_syr_spm.pdf.*

IPCC. (2007b). *Climate change 2007: Mitigation. contribution of working group III to the fourth assessment report of the IPCC – summary for policymakers.*

Jiang, Z. (2008). Reflections on energy issues in China. *Journal of Shanghai Jiaotong University, 42*(3).

Lawrence Berkeley National Laboratory (LBNL). (2007). *Taking out 1 billion tons of CO_2: The magic of China's 11th Five-Year Plan?* Lawrence Berkeley National Laboratory Report (LBNL-757E). Authors: Lin, J., Zhou, N., Levine, M., & Fridley, D.

Lawrence Berkeley National Laboratory (LBNL). (2008a). *Current status and future scenarios of residential building energy consumption in China.* Lawrence Berkeley National Laboratory Report (LNBL-2416E). Authors: Zhou, N., Nishida, M., & Gao, W.

Lawrence Berkeley National Laboratory (LBNL). (2008b). *Estimating total energy consumption and emissions of China's commercial and office buildings.* Lawrence Berkeley National Laboratory Report (LNBL-248E). Authors: Fridley, D.G., Zheng, N., & Zhou, N.

Lawrence Berkeley National Laboratory (LBNL). (2008c). *Global carbon emissions in the coming decades: The case of China.* Authors: Levine, M.D., & Aden, N. T.

Martinot, E., & Li, J. (2007). Powering China's development: The role of renewable energy. *Worldwatch Special Report.*

McKinsey. (2009a). *China's green revolution.*

McKinsey. (2009b). *Preparing for China's urban billion. www.mckinsey.com/mgi/reports/pdfs/China_Urban_Billion/MGI_Preparing_for_Chinas_Urban_Billion.pdf.*

MIT. (2007). *The future of coal.*

MIT. (2008). *Greener plants, greyer skies? A report from the frontline of China's energy sector.* China Energy Group, MIT Industrial Performance Center. Authors: Steinfeld, Lester, & Cunningham.

MIT. (2010). China's high-speed-rail revolution, *MIT Techology Review. www.technologyreview.com/energy/24341/page1/.*

Morse, R. K., Rai, V., & Gang, H. (2009). The real drivers of carbon capture and storage in China and implications for climate policy. *Program on Energy and Sustainable Development.* Stanford University. Working Paper no. 88. *http://papers.ssrn.com/sol3/Delivery.cfm/SSRN_ID1485003_code1265440.pdf?abstractid=1463572&mirid=3.*

National Development and Reform Commission (NDRC). (2007a). *China's national climate change programme.*

National Development and Reform Commission (NDRC). (2007b). *Medium and long term development plan for renewable energy.*

Price, L., Wang, X., & Jiang, Y. (2008). *China's top-1000 energy-consuming enterprises program: Reducing energy consumption of the 1000 largest industrial enterprises in China.* Lawrence Berkeley National Laboratory Report (LBNL-519E).

REN21. (2010). *Renewables 2010 – global status report.*

Reuters. (2009). *China to shut more small power plants in 2010 – media. http://uk.reuters.com/article/idUKTOE5BR06A20091228.*

UN. (2008). *World population prospects: The 2008 revision.* Population Division of the Department of Economic and Social Affairs of the United Nations Secretariat. *http://esa.un.org/unpp.*

UNEP. (2008). *Most commonly used air routes, in million passengers a year. UNEP/Grid Arendal. Data source: ENAC Air Transport Database.* French Civil Aviation University. *http://maps.grida.no/go/graphic/most-commonly-used-air-routes-in-million-passengers-a-year.*

UNEP. (2010). *Climate pledges. www.unep.org/climatepledges/.*

U. S. Geological Survey (USGS). (2004). *China's growing appetite for minerals.* Open-File Report 2004-1374 (presentation). *http://pubs.usgs.gov/of/2004/1374/2004-1374.pdf.*

U. S. Geological Survey (USGS). (2010). *Mineral commodity summaries 2010. http://minerals.usgs.gov/minerals/pubs/mcs/2010/mcs2010.pdf.*

Wang, H., Wang, G., Liu, H., Zhao, Q., & Liu, D. (2009). *Development trend of unconventional gas resources in China.* Langfang Branch Institute of the Research Institute of Petroleum

Exploration and Development, International Gas Union (IGU) Working Paper. *www.igu.org/html/wgc2009/papers/docs/wgcFinal00144.pdf.*

Wang, T., & Watson, J. (2007). *Who owns China's carbon emissions?* Tyndall Briefing Note No. 23. Tyndall Centre for Climate Change Research.

Wang, T., & Watson, J. (2008). *Carbon emissions scenarios for China to 2100.* Tyndall Centre for Climate Change Research.

Wang, T., & Watson, J. (2009). *China's energy transition. Pathways for low carbon development.* Tyndall Centre for Climate Change Research.

World Nuclear Association. (2010). *Nuclear power in China. www.world-nuclear.org/info/inf63.html.*

World Resource Institute (WRI). (2009). Energy and climate policy action in China. *http://pdf.wri.org/factsheets/factsheet_china_policy_2009-11-03.pdf.*

Xinhua. (2010). *China's power consumption grows 6% in 2009. http://news.xinhuanet.com/english/2010-01/06/content_12763227.htm.*

Chapter 16

Swiss 2000-Watt Society: A Sustainable Energy Vision for the Future

Roland Stulz,* Stephan Tanner,† and René Sigg**

Novatlantis, Swiss Federal Institute of Technology (ETH), Zurich, Switzerland,
*†Intep – Integrated Planning LLC, Minneapolis, Minnesota, **Intep – Integrated Planning GmbH, Zurich, Switzerland*

Chapter Outline

1. INTRODUCTION

Policymakers in Switzerland are bridging a gap to create an energy future that is sustainable. The collective vision of a 2000-Watt Society, which requires sharing global energy and material resources more fairly and equitably, has already become an indispensable component of the Swiss national energy and climate policy. Based on public–private partnerships, a number of case studies are being implemented to assess and demonstrate the feasibility of the concept. This chapter chronicles the progress of this vision for society hitherto and presents the essential requirements for potentially expanding the idea beyond Swiss borders.

Section 2 presents the original vision that led to the concept of Swiss 2000-Watt Society and what it entails, including the institutional support for the implementation of the vision and the twin strategy of one tonne of carbon dioxide (CO_2) per capita. Section 3 describes the fundamentals of the 2000-Watt Society including its technical feasibility. Section 4 highlights the results of a number of ongoing projects where the feasibility of the concept are being empirically including a number of remaining hurdles and barriers to full implementation of the concept. Section 5 presents an assessment of the potential implications of adopting similar targets for Europe, North America, or possibly in a broader context, followed by the chapter's conclusions.

2. THE 2000-WATT VISION

This section describes the original vision that led to the concept of Swiss 2000-Watt Society, provides the context behind this vision and the closely intertwined carbon reduction strategy, followed by a description of the current state of progress and implementation.

2.1. The Vision

The Swiss 2000-Watt Society started as a simple vision that seeks to make the current high living standards of Western countries universally available to everyone and attempts to accomplish this feat in a sustainable fashion.

The first essential aspect of the 2000-Watt Society is that current high standards of living in industrialized countries can be maintained, including the provision of *energy services* needed to cover all basic human needs, amenities, services, and goods while recognizing that this requires a paradigm shift to an "intelligent lifestyle, otherwise the 2000-Watt Society will remain simply a vision."[1]

The other important feature of the vision is that the world's finite resources should be distributed in a fair and equitable fashion among the world's inhabitants. The implementation of this vision is predicated on the belief that there are sufficient raw materials and resources for all if the energy utilization efficiency in developed countries is improved by a factor of 3 to 4.

Another critical feature of the vision is to reduce our global reliance on fossil fuels to the extent that the emission of greenhouse gases (GHGs) is lowered to roughly 1 ton of CO_2 per capita per annum—compared to between 8 and 10 today[2]—consistent with the recommendations of the Intergovernmental Panel on Climate Change (IPCC). This requires an increased reliance on renewable energy resources as well as using existing resources more efficiently.[3]

1. See Novatlantis (editor): *Smarter Living*, 2005, p. 6.

2. This book's Chapter 7 by Moran offers a discussion of current per capita emission levels and how these must be reduced to meet climate stabilization targets.

3. It must be noted that many proponents of the Swiss 2000-Watt concept do not favor increased reliance on nuclear energy as a means of managing GHG emissions—but this is not critical to the main concept.

The vision was originally conceived in 2004 by an interdisciplinary group of scientists and researchers at ETH Zurich (ETHZ) who examined the pathway toward a sustainable future and published their finding in a White Paper (Jochem, 2004). The main result of their analysis was that prosperity and economic development in Western countries *could* be sustained with roughly one third of the energy resources currently consumed. Moreover, they concluded that similar high standards of living *could* be provided globally *and* sustainable use of resources is technically feasible. These ideas, which led to a sustainable path for the twenty-first century, are the foundation of the Swiss 2000-Watt Society and have been adopted as a guiding component of Swiss climate and energy policy.

It is envisioned that the 2000-Watt per capita target is reached in two phases (Figure 1). In the first phase, roughly covering the period between now and the year 2050, the current energy requirements of 6500 watts per capita—the present Swiss average—are to be reduced to 3500 watts per capita. Simultaneously, annual per capita emissions of CO_2 are cut by roughly one quarter, from around 9 to 2 tons per person.

During the second phase, extending to 2150, per capita energy consumption is further reduced to 2000 watts of equivalent capacity to cover all human requirements while maintaining high standards of living. Incidentally, this

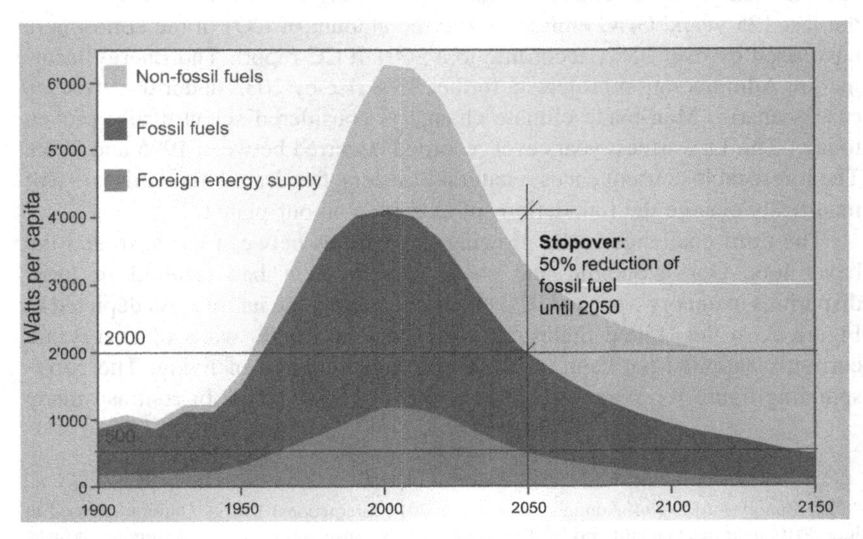

FIGURE 1 The sustainable path of the 2000-watt-society. The primary energy consumption level that sustains high standards of living in developed countries, currently around 6000 watts per capita for Switzerland, has to be reduced, together with declining reliance on fossil fuels. The figure depicts the potential path for development with roughly 50% reduction by 2050 towards the eventual goal of reaching 2000 watt per capita by 2150. All energy requirements are converted to equivalent electric capacity for ease of presentation. *Source: Novatlantis, 2005.*

happens to be what the Swiss population consumed in 1960s. Unlike then, however, fossil fuel resources are projected to make up only one quarter of total energy needs in the long term.

2.2. The Context

For 150 years, oil and other fossil fuel fuels have been the engine of the world's industrial development and their consumption has grown in tandem with economic growth. Currently over 80% of the energy that fuels our economies, especially in Western societies, is derived from fossil fuels.[4] The spiraling increases in growth and consumption have left many observers skeptical about the long-term sustainability of the business-as-usual scenario.

The future will be a time of great change and challenges. The first challenge may be the dwindling supplies of oil in accessible locations and at low cost. According to Fatih Birol, chief economist at the International Energy Agency (IEA),[5] we will have reached "peak oil" within a decade. Ironically, while the first half of the era of easy oil approaches its end, the number of people worldwide who wish to share in these dwindling resources is growing. There is apprehension about rising oil prices, growing global inequality, and increasing tension and conflict for control over dwindling natural resources.

The second challenge is that our over-reliance on fossil fuels is resulting in climate change as combustion of fuels is warming the Earth's atmosphere. In the last 125 years, GHG emissions and the amount of CO_2 in the atmosphere have risen by over 35%, according to a 2001 IPCC report. The Energy Information Administration projects a further 43% rise by 2035 under its reference case scenario. Man-made climate change is considered scientifically proven today.[6] The 11 warmest years ever recorded occurred between 1995 and 2006. The foreseeable consequences—natural disasters, floods, and food crises—will manifestly change the foundations of existence on our planet.

The third challenge is the growing inequalities between the have and the have nots. Our economic and social development has resulted in large disparities in energy use and GHG emissions among the nations. As depicted in Figure 2, in the United States the equivalent of 12,000 watts of energy are currently supplied per capita to sustain a high standard of living. The corresponding figure for Western Europe is 6000 to 8000 watts. In contrast, many

4. The Energy Information Administration, in its 2010 International Energy Outlook released in June 2010, indicates that currently 86% of global energy consumption is derived from fossil fuels. This percentage is projected to decline slightly to 80% by 2035 under the reference case, which assumes a continuation of business-as-usual practices and policies.

5. *The Independent*, 3 August 2009.

6. Lecture by ETH Professor Andreas Fischlin, coordinating and leading IPCC author, given in Zurich on August 28, 2009.

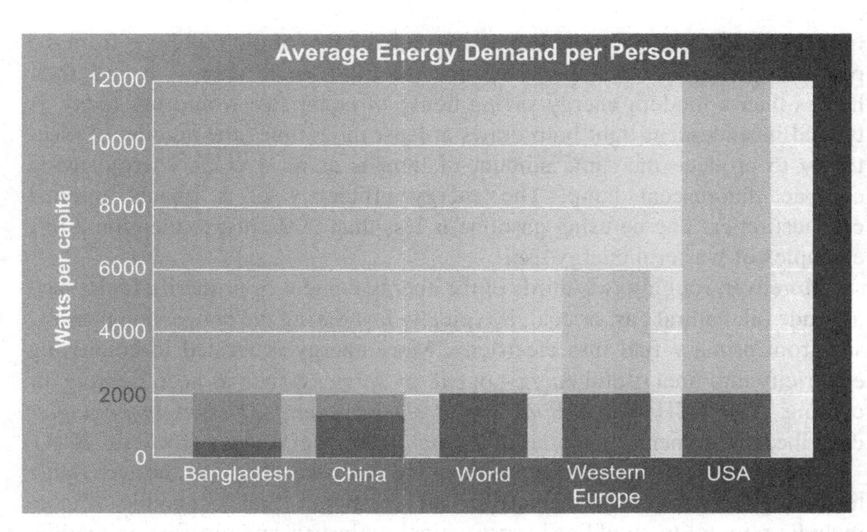

FIGURE 2 Current average energy demand per person in selected countries. The current global average is roughly 2000 Watts per capita. The green portion indicates countries where the current level is less than the global average, the yellow are countries exceeding the global average. *Source: Novatlantis, 2005.*

developing countries in Asia and Africa currently consume only several hundred watts per person, if that.

To address these challenges—and do it in a sustainable way—one needs to ask how much energy, on average, is available to meet the world's energy requirements and how much energy, on a per capita basis, is required to enjoy a lifestyle consistent with today's Western standards?[7]

ETH researchers, who originally posed these questions, estimate that 17,500-kilowatt hours (kWh) per capita per annum, which amounts to a continuous output of 2000 watts of installed capacity, is available for each global inhabitant if the global energy resources are evenly distributed. Moreover, they determined that there are sufficient amount of natural resources available worldwide to satisfy human needs if these resources are used in smarter ways, that is, more efficiently, sustainably, and equitably.[8]

Traditionally, fossil fuels have been used in great quantities and very inefficiently. In addition, in the everyday life of highly developed societies, the proportion of energy that is needlessly wasted, is inefficiently used, or

---

7. This book's Chapter 1 by Sioshansi also examines these issues.

8. The 2000 watts refer to the total energy consumed per person averaged over a year. The term watt for energy consumption may sound confusing but is in fact easy to understand. In physics the watt unit does indeed represent rated power. But energy consumption involves rated power as well. As a point of reference, 2000 watts is equivalent to energy consumption of 2000 joules per second, 172,800 kilojoules per day, or 48 kWh per day.

improperly utilized is simply staggering. Inhabitants of old buildings from the post-war period, for example, consume ten times more energy to heat their homes than a modern energy-saving house of equal size would use today. A typical incandescent light bulb draws at least three times the amount of electricity to produce the same amount of lumens as an average energy-saving compact fluorescent lamp. The energy efficiency of a typical internal combustion car engine using gasoline is less than 20%, just to mention a few examples of wasteful energy use.

Moreover, roughly two-thirds of the energy content of primarily fuels, such as crude oil, natural gas, or coal, is typically lost during the conversion process, say from primary fuel into electricity. More energy is wasted in converting electricity into final useful energy, or *energy services*, such as heat, lighting, or cooling. The ETH *White Book for R&D on energy-efficient technologies* describes the immense potential of *dormant* energy efficiency (Jochem, 2004).

The scientifically based conclusion of the study was that if the best available technologies were implemented in all consumer and economic sectors including mobility, building construction, industry, and energy extraction, energy utilization efficiency could be increased by a factor of five. Drawing from these findings, the 2000-Watt Society proposes to reduce primary energy demand by two thirds *without* jeopardizing citizens' prosperity, quality of life, or pace of economic development. But the realization of this vision requires more than mere technological fixes but also requires changes in our socio-economic systems as well as behavioral and lifestyle adjustments, topics repeated in other chapters of the book.

2.3. 2000-Watt Society and CO_2 Strategy

The vision formulated by ETH scientists for a 2000-Watt Society places energy efficiency at the center of the debate while striving to meet energy service needs *and* reduce GHG emissions at the same time. These intertwined objectives were initially presented to the public by the Energy Science Center (ESC) at ETHZ in February 2008.

Initially, much of the debate centered on how to reconcile between the two specified targets—2000 watts and the one tonne of CO_2 per person—on what basis they were calculated, and the consequences they would have for energy policy. Subsequently, a consensus has emerged that the two targets are complimentary in the sense that achieving one would contribute to the other. Moreover, it is broadly recognized that to achieve these targets requires adjustments in consumer behavior to reduce energy consumption *and* CO_2 emissions simultaneously.

Specifically, the goal to limit individual GHG emissions to one tonne per annum requires the application and implementation of a wide range of technologies and solutions supported by research and development (R&D) and scientific developments at ETH and elsewhere. The argument for a target of one tonne CO_2 is that it is absolutely paramount to diminish the climate effects of anthropogenic emissions and to massively reduce the use of fossil fuels to sustain high standards of living. ESC research outlines a dynamic road map toward a sustainable energy system in Switzerland to 2050 and beyond (Figure 1). This allows for scientifically based results to be integrated into public policy. Finally, the broad aims of the 2000-Watt Society are aligned with global efforts to address the climate challenge. The vision is to move toward a sustainable and equitable energy future where all human inhabitants benefit equally from the Earth's finite resources.

To further develop, promote, and implement this vision, the two ETH universities in Zurich and Lausanne and four affiliated Swiss research institutes established Novatlantis.[9] The aim is for knowledge and expertise from these research institutions to be used to achieve sustainable development. Another objective is to develop and apply technologies to conserve resources in collaboration with other partners in science and in the business community with supportive government policies.

2.4. A Success Story in the Making?

The success of the 2000-Watt Society, originally conceived as a *mere idea* can be measured by how quickly it has moved into the mainstream of the policy debate in Switzerland. The vision has since become an essential element of Swiss official energy and climate policy. The national energy authority, the Swiss Federal Office of Energy (SFOE), is engaged, for example, in establishing a special unit to implement the 2000-Watt Society.

Simultaneously, the Swiss Federal Institute of Technology continues to pursue the vision of the 2000-Watt Society as a platform for research and development. Moreover, the Novatlantis initiative, whose offices are adjacent to the ETH Competence Center for Energy and Mobility, has been placed in charge for developing a network for sustainability research that covers many aspects of everyday life across many sectors of the Swiss economy.

Most important, so far, Switzerland's two main cities of Basel and Zurich have embraced the 2000-Watt concept, while a number of other cities and cantons in the Swiss federal system are considering adopting similar initiatives. November 2008 may be regarded as a key milestone when the citizens of city of Zurich voted in favor of officially adopting the 2000-Watt vision in a public referendum, which is currently reflected in the city's municipal constitution.[10] Meanwhile, in Geneva, a coalition representing the business community, scientific organizations, and NGOs are in the process of lending broader political support to the aims of the 2000-Watt Society.

---

9. For further details about the mission of Novatlantis visit *www.novatlantis.ch*.

10. *www.stadt-zuerich.ch/content/gud/de/index/das_departement/strategie_und_politik/2000_watt_gesellschaft.html*.

3. THE WHITE BOOK

This section provides the context that led to the so-called *White Book* on sustainable energy use and an overview of the main areas of research with promising potential for sustainable future development.

3.1. Energy Consumption in Perspective

Energy consumption and its associated environmental side-effects pervade all aspects of our personal and public life to the point where we have become oblivious to it.[11] The normal state of affairs in today's developed countries is that energy is consumed everywhere at all times. Ten searches on the Internet, for example, require one-kilowatt hour (kWh) of electricity. A similar amount is required for a 30-watt energy saving lamp to burn uninterrupted for 36 hours. About 30 kWh of energy are used in the average home—for heating, hot water, and electrical appliances—every day. Drive 60 miles in a car and you consume the equivalent of 70 kWh.

In Switzerland, which is typical of other Western European countries, the two biggest components of energy consumption are buildings and transportation. Roughly 40%–50% of current energy use goes to provide energy services in buildings, with mobility consuming roughly 30%. The balance is used in the commercial and industrial sectors for the production of goods and delivery of services.

As numerous other studies have demonstrated and several chapters in the present volume point out, a number of options exist for reducing energy consumption and increasing efficiency within all sectors of the economy. What explains our failure to capture these cost-effective energy efficiency options are outlined in a few chapters in this book and will not be duplicated. Moreover, the over-consumption of natural resources is not merely limited to energy but extends to water, land use, and other mineral resources which are extracted at increasing rates from a limited pool and used excessively. Extraction, processing and consumption of most materials generate significant emissions during their life cycle and end up as trash if they are not recycled.

To achieve sustainability and avert climate change, it is imperative to use energy and other natural resources more intelligently, which requires a significant increase in energy utilization efficiency—in everyday life as well as in industrial processes and in the provision of public infrastructure. This suggests that the achievement of the 2000-Watt target must cover all aspects of energy needs in all economic sectors—in particular in the building sector, for provision of mobility, food, and communication services. Research summarized in the

11. This book's Chapter 1 by Sioshansi makes similar arguments, pointing out that the illusion of seemingly plentiful and historically cheap energy has resulted in over consumption and wasteful habits.

White Book of the 2000-Watt Society provides evidence of the untapped potential for efficiency and concludes that between 50% and 90% of today's energy consumption can be saved.[12]

3.2 Energy Efficiency Potential

The *White Book of the 2000-Watt Society* concludes that by using *existing* technology in Switzerland, it is possible to produce around 65% more energy services—such as heating and lighting in buildings, mobility, information and communication technologies, and so on—with a third of the energy currently consumed. There are reasons to believe that similar savings are available in other industrial countries.

Assuming optimized application of best available technologies across all consumption sectors can lead to an increase in energy efficiency by a factor of five, according to the *White Book* (Jochem, 2004). Some of the areas with the greatest potential for energy efficiency include the following:

- When managing building stock, up to 80% of resources can be saved. Key technologies include better insulation and windows; decentralized combined heating, cooling, and power systems; efficient low-temperature heating systems; integrated solar thermal energy and photovoltaics (PV); plus the use of near-surface geothermal energy.
- In the road transportation sector, energy efficiency can be increased by up to 50%. Private cars, for example, could halve their primary energy consumption per kilometer driven by substituting more energy-frugal engines and lighter vehicles. Significant improvements are also possible in air transport, railroads, shipping, and logistics. Ideas that influence the modal split among and between different transportation options in favor of the more efficient are equally important for reducing energy requirements.
- Considerable savings can also be made in industrial production processes. Despite their great variety, more intelligent processes and technical innovations may result in 30%−40% reduction in energy needs such as in heat production or chemical conversion processes.[13] There is major potential in the production of plastics, for example, if crude oil is replaced as the base material by biopolymers, which are produced synthetically from renewable raw materials.
- Material efficiency also involves great potential in reducing resource needs. Options include improved product design, recycling and the reuse of water and other materials, substitution of energy-intensive materials by low-energy equivalents, and structural change in usage strategies by means of sharing concepts or cascading use.

12. See also *www.novatlantis.ch* and *www.2000watt.ch*.

13. This book's Chapter 12 by Brown et al. covers industrial energy efficiency potentials.

- Increased reliance on information technology (IT) and electronics are key technologies, which can facilitate the adoption of more energy efficient options across many applications.

Innovative technologies are available in all areas or are currently being developed. However, the extent to which technically feasible efficiency increases can be implemented depends not only on the widespread use of these technologies but also on the investment and renewal cycles of many decision-makers. Technology on its own is not enough,[14] which is why it is crucial to optimize both individual product components as well as systematically analyze value added chains including product usage and waste disposal. Rebound effects of more efficient technology should also be considered in the analysis.

But technological fixes alone are not sufficient. They must be supported by policy, regulatory and fiscal instruments to be effectively applied, hence the need for appropriate incentives. In addition, consumer behavior and lifestyles play an important role as they ultimately determine how much energy is used for delivery of various energy services.

The *White Book* (Jochem, 2004) also examines the potential obstacles to the realization of the 2000-Watt vision by recognizing that any sustainable future energy pathway must consider two key factors:

- The range of potential technologies likely to be available.
- The investment strategies of various businesses, industries, or individuals in adopting the emerging technologies.

Within the building sector, for example, construction business has a long investment cycle but requires a fairly short payback period when considering new energy-saving technologies. Similarly, in private households, frequently economically justified efficiency strategies are not implemented due to non-economic factors such as lack of credible information on potential cost savings. Social and cultural aspirations play a major role in determining consumer behavior. A significant ratio of per capita consumption in affluent societies may be based on the desire to satisfy nonmaterial needs such as defining one's identity or status in the social context.[15]

Small- and medium-sized businesses often calculate their investments on the basis of amortization periods. Since energy costs usually make up a small fraction of their overall business expenditures, they often lack sufficient interest to pursue energy-efficient solutions or lack the necessary expertise. This applies especially to companies that are managed according to "lean management" principles and have to rely on technical expertise from external consultants.

14. This book's Chapter 11 by Prindle and Finlinson makes similar observations, including the importance of the human factor, operators, and consumer behavior.

15. This book's Chapter by 3 Bartiaux et al. describes the social and cultural aspects that define energy "needs."

Similarly, public institutions only rarely calculate their investments based on the entire life cycle of a product, system, or design. Instead, the provider with the least-cost up-front proposal usually gets the contract. Furthermore, a major proportion of municipal income comes from taxes or fees on volumetric consumption of electricity, gas, or water, reducing the motivation to lower consumption.

3.3. Key End-Use Technologies

The 2000-Watt vision places considerable attention on the provision of energy services to meet personal energy needs. It is recognized that the behavior of individuals is not only a sustainability indicator but that end users have considerable leverage in determining the widespread adoption of key energy-efficient technologies. Energy use in buildings and for mobility are among the most relevant areas with significant impact toward sustainable use of resources. Moreover, conservation of resources is only possible if fossil fuels are replaced by renewable energy sources.

The following provides a synopsis of the current progress on R&D in key areas with large potential impact on energy use, namely in buildings, for provision of mobility, and in energy supply.

3.3.1. Construction and Building Sector

The building sector can play a leading role toward sustainable use of resources. Consequently, living and working in a way that conserves resources and has a low carbon footprint must take center stage when planning for a sustainable future recognizing that roughly 80% of energy utilization within buildings is design-driven with only 20% attributable to occupant behavior. There are two promising developments in this sector:

- First, interest in low energy building standards has already been aroused with green buildings assuming a major role in the real estate market. One in every five new buildings in Switzerland is constructed with the Minergie[16] construction standard, meaning two-thirds lower heating requirements. Buildings certified to BREEAM in Britain and LEED standard in the U.S. have also gained in popularity in recent years.[17]
- Second, innovative technologies and processes for energy-efficient buildings with improved heat insulation and decentralized use of renewable energy sources have been successfully tested in many places. This means that the energy efficient building of the *future* is literally available *today*.

16. See also *www.minergie.ch*.

17. This book's Chapter 9 by Gray and Zarnikau and Chapter 17 by Rajkovich et al. discuss similar developments in the U.S.

> Numerous apartments, schools, and administrative buildings, which achieve the passive house or Minergie-P standard, representing 2 liters of heating oil equivalents per square meter of space, are no longer rarities. By comparison, the average Swiss house currently uses at least five times more.

Based on these promising developments, 2000-Watt compatible construction is planned in Switzerland based on the SIA Energy Efficiency Path standard.[18] This standard takes account of energy used in constructing the building, recognizing that construction materials embody energy resources and accounting for the fact that the choice of the building's location typically generates additional traffic, requiring a holistic approach to design. There is also growing interest in the concept of the energy self-sufficient house of the future.[19]

Given the relative size of existing real estate with its long life span, it is paramount that in addition to innovative planning concepts for *new* construction, major efforts must be direct toward energy-related renovation of *existing* buildings. The need for redevelopment is seen as significant while the willingness to act is limited.[20] There are at least two promising approaches to address this issue:

- First, further technical innovations are needed including in areas such as streamlining of retrofitting activities and the development of prefabricated renovation modules including cladding and roofing.
- Second, there is a need for policy schemes and financial incentives to encourage retrofitting and renovation of older buildings including improvements in socio-economic factors and provision of transparent market information, so the renewal rate can be improved.

3.3.2. Mobility

Recent progress in the development of more energy-efficient car propulsion systems has resulted in more efficient engines but these efforts have not translated into less fuel use in the transportation sector—which is a major consumer of fossil fuels and a key contributor to GHG emissions.

The average fuel consumption of a new car is still too high—averaging 7.4 liters of gasoline per 100 km in Switzerland in 2008—which means an emission of around 175 grams of CO_2 per km. The corresponding figures are much higher in many other countries. According to the U.S. Environmental Protection Agency (EPA), new cars sold in the U.S. in 2009 used *more* gasoline on

18. See also *www.sia.ch*.

19. In this book's Chapter 9, Gray and Zarnikau also discuss the concept of zero net energy homes of the future.

20. This book's Chapter 14 by Weber and Bauermann addresses similar issues in the German building sector context.

average than models sold in 1987. The main reason for this is the continuous increase in vehicle weight. One solution to correct this trend is to promote lightweight construction combined with minimum energy efficiency standards.

Beyond making individual vehicles lighter and more efficient—commendable strategies—one must consider more sustainable mobility including providing a range of appropriate transportation options suitable for different distances and needs. For short distances, for example, nonmotorized transportation is especially appropriate; medium distances should be covered by public transport if possible. Meanwhile long journeys, including intercontinental flights, may be limited to fossil fuel propelled aircraft. In the short term, the aim is to significantly lower CO_2 emissions for cars.

In urban and commuter transportation sectors, the proportion of electrically propelled cars is likely to increase. Light-duty vehicles with efficient propulsion are in demand in all sectors. Low-carbon fuels such as natural gas or biogas may make an additional contribution in reducing GHGs for mid-sized, commercial, and utility vehicles. The development of low-emission vehicles propelled by hydrogen and fuel cells is also underway.

However, it is not only important to develop efficient vehicles but also to focus on driving behavior. For example, it is possible to lower vehicle fuel use by up to 30% by adopting a fuel-saving driving style. Research specialists and practitioners are collaborating to implement superior mobility models. A good example is the Experimental Space Mobility, a pilot project in Basel through which cab drivers and the City Transport are offered the opportunity to test biogas vehicles.

3.3.3. Energy Supply

Renewable resources, including solar, geothermal, hydro, wind, and biomass, offer opportunities toward meeting the 2000-Watt target and are intended to substitute for fossil fuels.

Electricity consumption is likely to grow in the short- to medium-term due to increased electrification, a trend foreseen in the 2000-Watt pathway. Reasons for this are that in the building sector, fossil fuel heating systems are increasingly replaced by more efficient electricity-powered heat pumps. There will also be increasing sales of electricity-powered, climate-friendlier cars. Electrification, especially if generated from renewable resources, offers excellent opportunities to reduce GHG emissions.

A number of industrial countries with limited domestic energy resources are looking at renewable energy technologies to reduce their dependence on energy imports while reducing their GHG emissions. The European Union (EU), for example, has set a goal to increase the proportion of renewable energy to 20% by 2020. Governments in the EU, as well as in Switzerland, pay subsidies to develop the nascent solar and wind industries. Such strategies have multiple benefits including the development of low-carbon domestic energy supplies, reducing fossil fuel imports, reducing GHG emissions, and creating a more decentralized energy supply system.

The R&D need for improved effectiveness of these energy conversion and storage technologies cannot be underestimated at present. The spectrum of technologies yet to be developed includes fuel cells, thin film PV cells, development of smart grids, and direct current transmission, among others.

4. CURRENT PROGRESS

This section provides a synopsis of current efforts underway in several regions of Switzerland to demonstrate the feasibility of achieving the 2000-Watt vision and offers a status of progress to date.

Novatlantis, the ETH sustainability program, is in the process of putting together a network in Switzerland to implement the concept of 2000-Watt Society. As part of a public-private partnership, Switzerland's three largest cities of Basel, Zurich, and Geneva are involved in shortening the route from the laboratory to practical applications for highly efficient technologies and sustainable lifestyles. Public authorities, investors, businesses, the industry, and the scientific and design communities are also collaborating to demonstrate how low emission mobility, construction practices that conserve resources, and low-carbon sustainable urban development may look.

4.1. Basel Region

The city of Basel has been involved as a pilot study area since 2001. More recently, the local authorities have adopted the 2000-Watt vision and are currently striving to reduce energy consumption to one-third of the Swiss average while increasing the proportion of renewable energy to meet local energy needs. Basel's city parliament has formally endorsed these targets on two separate occasions.[21]

The first tangible step in accomplishing the city's vision is the implementation of a number of pilot and demonstration projects with public funding including a competition for renovation of office and residential buildings in compliance with the Minergie-P standard, where innovative proposals are sought.

At the same time, a number of experiments are underway to provide low-energy, low-carbon mobility in which efficient and low-emission vehicles are tested, including environmentally friendly fuels and drive technologies such as natural gas or biogas and hydrogen. Since 2009, the world's first street sweeper powered by a hydrogen fuel cell has been tested. Initial results indicate that the energy use of conventional diesel vehicles can be reduced by two-thirds. Further advantages of hydrogen-powered, fuel-cell vehicles are that there are no local noxious emissions, plus superior efficiency. The project also includes

21. See also *www.2000-watt.bs.ch*.

endurance testing of near zero emission vehicle (NZEV) catalytic converters for natural gas engines in collaboration with the local public transport agency and private taxi operators.

4.2. Zurich Region

In a public referendum conducted in November 2008, citizens of Zurich overwhelmingly approved the objectives of the 2000-Watt concept, which has been incorporated into the city's constitution. Over 75% of the citizens voted in favor of an ordinance to reduce per capita energy consumption by a factor of 3 and cut CO_2 emissions by a factor of 4–6 by 2050. While these targets are not identical to the 2000-Watt pathway, the city is nevertheless committed and has already embarked on a number of measures to implement the plan.

Among the specific steps taken thus far is the adoption of the Minergie standard as the minimum requirement both for renovations as well as new construction within the city boundaries. The city has also held design competitions in collaboration with real estate developers where the more ambitious Minergie-P-Eco standard is considered.

The scheduled expansion of the Triemli Hospital and the construction of a retirement facility are among the first projects in Switzerland that meet the 2000-Watt sustainability specifications. These buildings incorporate state-of-the-art energy-efficiency design with minimum energy input.[22] The buildings' remaining requirements for heating, cooling, and electricity are primarily covered from renewable energy resources including biomass, geothermal, and solar energy. The hospital's CO_2 emissions following the renovation are expected to be reduced from 6000 to around 800 tonnes per annum.

Zurich is also determined to convert its electricity supply to be climate-friendly and sustainable without resorting to new nuclear power stations. The city is expanding its reliance on renewable resources with the ultimate goal of supplying virtually all the city's electricity requirements from such resources within 60 years. Hydroelectric power will provide the predominant share, well over 30%, with PV supplying 10%, supplemented by biomass, wind, and deep level geothermal power. Currently, Zurich's administrative buildings are exclusively supplied with accredited eco-electricity.

4.3. Geneva Region

Like its sister cities, Geneva has also embraced the 2000-Watt vision partly because the local anti-nuclear lobby recognizes that this will obviate the need to build more nuclear power plants.

22. This book's Chapter 14 by Weber and Bauermann also describes buildings with minimal heating requirements.

The 2000-Watt pathway, however, is expected to increase electricity demand in the short- and medium-term due to increased electrification. This has prompted an electricity savings campaign targeting private households and property owners to reduce electricity consumption when feasible. The buildings at the UN headquarters in Geneva, for example, are cooled by water from Lake Geneva instead of relying on chillers.

To broaden support for the acceptance and implementation of the 2000-Watt concept in the Geneva region, a nonprofit and politically neutral association, Genève à 2000 Watts, has been founded providing a platform for exchanging ideas among the major players: the scientific and research community, city authorities, politicians, and professional and business associations.[23] The organization is proceeding to promote the 2000-Watt vision in a fashion analogous to Basel and Zurich.

4.4. National Vision for Switzerland and Beyond

The aims of the 2000-Watt Society are already recognized at a high political level in Switzerland. The state energy authority, the Swiss Federal Office of Energy (SFOE), is currently expanding its range of duties via a 2000-Watt Society Task Force. Additionally, SFOE is preparing an accreditation process geared to the objectives of 2000-Watt Society for cities and municipalities similar to the European Energy Award.

To adapt Swiss domestic energy supplies to the 2000-Watt goals by the mid-level 2050 goal, the energy research commission, which advises the Swiss government on energy and climate policy, has currently adopted the following core objectives:[24]

- Elimination of fossil fuels for provision of heating in buildings
- Reduction of energy consumption in buildings by half
- Increasing the use of biomass for energy supplies up to its net ecological production potential
- Reducing the average car fleet use of fossil fuels to 3 liters per 100 km

It must be noted that interest in the 2000-Watt vision has spread beyond Swiss borders. Consequently, Novatlantis has taken steps toward implementation of strategies for sustainable international development by engaging in collaborative efforts in other European countries, the Americas, and elsewhere. Among its notable initiatives to date is the establishment of the International Sustainable

23. The founding members, besides Novatlantis, include EIG (Engineers' School of Geneva), the Energy Center of ETH Lausanne, the University of Geneva, SIG (Industrial Services of Geneva), TPG (Public Transport of Geneva), WWF Geneva, FEDRE (European Foundation for the Sustainable Development of Regions), CGI Property (real estate agency), and OPI/CCSO (Service of Industrial Promotion).

24. *www.bfe.admin.ch/themen/00519/00520/index.html?lang=de&dossier_id=01495.*

Campus Network (ISCN), where 120 leading universities worldwide are involved. By organizing symposia and developing collaborative networks, Novatlantis is spreading its know-how while stimulating the exchange of ideas among researchers *and* making sustainability a study and research subject. Flagship projects include the Harvard Green Campus Initiative, the sustainability campaign at Stanford University, and the Science City-Campus at ETH in Zurich.

In the case of ETHZ campus, the 2000-Watt path serves as a guide for designing new buildings with the energy-efficient Minergie standard as well as reliance on climate-friendly energy resources. A dynamic underground tank system is intended to help reduce CO_2 emissions at the ETHZ campus by at least 50% by 2020. Geothermal heat exchangers supply the low-rated energy, which is converted with highly efficiency heat pumps and chillers into energy for heating or cooling.

5. BEYOND SWITZERLAND?

The underlying fundamentals of the 2000-Watt vision, namely concerns about a more equitable distribution of finite resources in a sustainable manner while averting climate change, are universal. Viewed in this context, the vision behind the 2000-Watt Society—or something akin to it with appropriate modifications—could be applied to other countries and regions. This section examines the potential ramifications of adopting goals and targets consistent with those envisioned by the 2000-Watt vision for other regions of the world.

5.1. Adopting the 2000-Watt Vision in Europe

Adopting the 2000-Watt principles within Europe is sensible given the similarities in political, cultural, and socio-economic status of development within Europe, certainly Western Europe. The EU countries to the north of Switzerland are particularly similar with regard to their current socio-economic status, high living standards, and the availability of energy-efficient products and services. When applying the 2000-Watt concept in Southern Europe, the warmer climate must be taken into consideration, especially in the Mediterranean region. Likewise, countries in Eastern Europe have rather different political structures and living standards.

A number of existing EU guidelines, including the 20% renewable target by 2020 and the proposal that requires new buildings to meet the Net Zero Energy House standard by 2019[25] are moving in the right direction, but must be further strengthened. Similarly, significant improvements in the energy efficiency of the transport sector are necessary to approach the 2000-Watt target.

25. According to the decision by the European Parliament in May 2010, amending the 2002 Energy Performance of Buildings Directive.

5.2. Adopting the 2000-Watt Concept in North America

According to recent figures, the total primary energy usage in the United States is equivalent to 12,000 watts per person,[26] double that of the average European usage. Moreover, the U.S. annual per capita emissions of GHGs are currently around 19 tons, also well above the European levels.[27] To be consistent with the 2000-Watt guidelines, these figures must be reduced to 2 tons per capita per annum by 2050 and 1 by 2150[28] requiring a Herculean effort, and can only be achieved by virtually abandoning the use of fossil fuels.

Special attention must be paid to the transport sector. While the centers of some U.S. and Canadian cities are densely packed and have public transportation, widespread suburban sprawl characterized by low density and long commuting distances offer difficult challenges for North America. To approach the 2000-Watt goals, a drastic switch from personal to public transportation within urban areas and between areas of high density must be implemented. Due to the regional structure and long distances, however, reducing motorized personal transportation will not be enough, which suggests switching to alternative types of propulsion and exclusive use of alternative energy sources.

In the build environment, there have been a number of individual initiatives including the pioneering efforts of people like Amory Lovins and his Rocky Mountain Institute or Harold Orr in Saskatchewan, Canada. More recent efforts include the U.S. Green Building Council, with its Leadership in Energy and Environmental Design, or LEED, certification program, which are incrementally moving in the right direction though fall short of meeting the 2000-Watt requirements.

Currently, there are efforts in the U.S. and Canada in adopting European building standards including the German Passive House and the Swiss Minergie standard that meet the 2000-Watt target. For example, das BioHaus, the first passive house certified building, was built in Bemidji, Minnesota at Concordia Language Villages.[29] This building, which is operating at 85%–90% below contemporary standard for school buildings, demonstrates that even under severe climatic conditions the 2000-Watt target can be met. A small-integrated PV system, yet to be added, would make das BioHaus a virtually zero net energy (ZNE) building.[30] If a building meeting the 2000-Watt standard can be built in the challenging climate of northern Minnesota within a reasonable cost today, then the concept should be possible in most other states as well.

---

26. Stadt Zürich: *Ein Kurswechsel mit Zukunft*, 2007, p. 1.

27. See Internet 00416: *http://mdgs.un.org/unsd/mdg/SeriesDetail.aspx?srid=751&crid=*

28. See Bébié, B., Gugerli, H., et al.: *Grundlagen für ein Umsetzungskonzept*, 2009, p. 10.

29. First certified Passive House Standard building in North America: das Waldsee Biohaus *http://waldseebiohaus.typepad.com/*

30. This book's Chapter 9 by Gray and Zarnikau and Chapter 17 by Rajkovich et al. describe the challenges to meet ZNE building standards.

5.3. Adopting the 2000-Watt Concept in Emerging and Developing Countries

The challenges facing many emerging and developing countries are similar to those of some Eastern European countries, which have less-developed infrastructures and economies. Moreover, for these countries, economic development and raising standards of living are pressing goals while environmental concerns and climate change are considered as luxuries. In this context, it is much more difficult to discuss sustainability or the 2000-Watt concept.

The most important point applicable to the developing countries is that they need *not* undergo the same development path pursued by the industrialized countries. Instead, they should strive to find a shortcut that allows them to attain higher living standards comparable to those of industrialized countries while avoiding the environmental side-effects associated with rapid development. These countries should seek sustainable development while avoiding high levels of energy consumption resulting from increased motorization and wasteful lifestyles based on frivolous consumption with its external costs and environmental degradation.

CONCLUSIONS

The 2000-Watt vision is focused on reducing primary energy demand in European countries by roughly two-thirds and on reducing CO_2 emissions by no less than 90% in the industrialized countries where the use of energy is disproportionately high.

Primary energy use and annual CO_2 emissions per person are the two key parameters under consideration. Achieving these objectives requires more rational use of energy as well as decarbonization, that is, less reliance on fossil fuels. The ambitious goals set by the 2000-Watt concept requires innovative solutions with cooperation and collaboration among the policymakers, businesses, the scientific and research community, and members of the society.

Since its inception, the 2000-Watt concept has started to take root in Switzerland, yet its objectives are universally appealing and can be adopted and modified elsewhere. To address longer-term global sustainability, to reach a more equitable distribution of finite resources, and to address the challenge of climate change requires coordinated action on a global level.

Based on the Swiss experience to date, state-of-the-art technology is essential but will not be sufficient to achieve the energy and climate goals of the 2000-Watt Society. Long-term success will require continuous reevaluation of our needs[31] and will necessitate raising public awareness and creating a sense of responsibility.

31. Gugerli, Dr. H., Referat "Erfolgreiche Umbau-Strategien", 2009.

Climate change is often invisible with its potential consequences removed both in terms of space and time.[32] Hence it is difficult for individuals to change their behavior and lifestyles—both considered necessary. Our socio-economic system may ultimately have to be altered based on a new economic paradigm, which isn't primarily based on continuous growth or increased consumption, production, and waste.

BIBLIOGRAPHY

Bébié, B., Gugerli, H., et al. (2009). *2000-Watt-Gesellschaft, Grundlagen für ein Umsetzungskonzept*. City of Zurich.

Boulouchos, K., et al. (Ed.). (2008). *Energiestrategie für die ETH Zürich*. Zurich: Energy Science Center (ESC) ETH.

Bretschger, L., Ramer, R., & Schwark, F. (2010). *How rich is the 2000-watt-society?* Zurich: CER ETH.

Bundesamt für Energie (BFE). (2007). *Die Energieperspektiven 2035*. Schlussberichte Band 1 bis Band 5, Swiss Federal Office for Energy.

Deutsche IPCC-Koordinierungsstelle (Ed.). (2008). Klimaänderung 2007. Synthesebericht.

Ekardt, Prof. Dr. F. (2009). "KlimawHandel – KonsumwEnde", lecture.

ETH-Rat. (Eds.), Schweizer Beitrag zur Energiezukunft. ETH-Domain, Zurich.

Gore, A. (2009). *Our choice: A plan to solve the climate crisis*. New York: Rodale Books.

Gugerli, Dr. H. (2009). "Erfolgreiche Umbau-Strategien", lecture.

Haefeli, U. (2008). *Verkehrspolitik und urbane Mobilität*. Stuttgart: Franz Steiner Verlag.

Jochem, E. (Ed.). (2004). *Steps towards a sustainable development. A white book for R & D of energy-efficient technologies*. Zurich: ETH.

Koschenz, M., & Pfeiffer, A. (2005). *Potenzial Wohngebäude*. Zurich: Faktor Verlag.

Novatlantis. (Ed.). (2010). *Leichter Leben, Auf dem Weg zu einer nachhaltigen Energiezukunft - am Beispiel der 2000-Watt-Gesellschaft*. Zurich: Novatlantis.

Stadt Zürich. (Ed.). (2007). *Ein Kurswechsel mit Zukunft*. City of Zurich.

Stadt Zürich. (2009). Bundesamt für Energie, Novatlantis. *Grundlagen für ein Umsetzungskonzept der 2000-Watt-Gesellschaft am Beispiel der Stadt Zürich*. LSP 4 – Nachhaltige Stadt Zürich – auf dem Weg zur 2000-Watt-Gesellschaft, City of Zurich.

VDI. (2009). *Mehr Wissen - weniger Ressourcen; Potenziale für eine ressourceneffiziente Wirtschaft*. Düsseldorf: Verein Deutscher Ingenieure.

32. Ekardt, Prof. Dr. F.: Vortrag "KlimawHandel – KonsumwEnde", 2009.

Chapter 17

Zeroing in on Zero Net Energy

Nicholas B. Rajkovich,[*] William C. Miller,[†] and Anna M. LaRue
[*]*University of Michigan,* [†]*Sentech, Inc.,*[1], [‡]*Resource Refocus LLC*

Chapter Outline

1. INTRODUCTION

A national effort is underway to achieve zero net energy (ZNE) in all new buildings by the year 2030 (Hawthorne, 2003). While no national mandate currently exists, federal legislation such as the Energy Policy Act of 2005 (EPAct 2005, Public Law No. 109-58) and the Energy Independence and Security Act of 2007 (EISAct 2007, Public Law No. 110-140) have authorized national initiatives that will develop and disseminate technologies, practices, and policies to move the U.S. market toward ZNE buildings.[2] One definition of ZNE is a building or community with "greatly reduced needs for energy through efficiency gains—60 to 70 percent less than conventional

1. The views and opinions expressed in this chapter are wholly those of the authors and do not necessarily reflect the official policy or position of the University of Michigan, Sentech, Inc., the U.S. Department of Energy, or Pacific Gas and Electric Company. William C. Miller, PhD is currently employed by Sentech, Inc. and provides support to the United States Department of Energy.

2. Proposed federal legislation might strengthen this mandate. For example S.3464 introduced by Senator Lugar (R-IN) would set national energy building standards in such a way that after the 2017 building code revisions, the Secretary of Energy would set subsequent revisions "...on a path to achieving net-zero-energy buildings." See Section 201 of the bill text available at *http://thomas.loc.gov/*.

practice—with the balance of energy needs supplied by renewable technologies" (DOE, 2008, p. 1–8). In a parallel effort, several state public utility commissions including those in California, Oregon, and Washington and energy offices such as the New York State Energy Research and Development Authority (NYSERDA) have initiated efforts to put the building sector on a path to ZNE. California continues to pursue ZNE as a key initiative to reducing overall electricity usage and greenhouse gas (GHG) emission.[3] The European Commission has also been discussing similar initiatives.

As described in this book's Chapter 9 by Gray and Zarnikau, advancing to ZNE will require significant changes in how we design, commission, and operate future buildings. As we consider how to transform the building sector, we will rely on the expertise and experience of local, regional, and statewide energy efficiency programs to assist in transforming the building sector. This chapter describes one such effort by a utility energy efficiency program, the Zero Net Energy Pilot Program developed by Pacific Gas and Electric Company (PG&E) for Northern California. The move to ZNE by PG&E is discussed as a reflection of the changing political environment and energy savings goals of the state of California, PG&E's utility regulators, and the utility itself.

To move to ZNE, it was important to address key gaps in existing energy efficiency programs. While a number of stakeholders are working to advance building technologies related to ZNE, the role for the California investor owned utilities (IOUs) was to demonstrate advanced technologies and encourage the market to transform as quickly as possible. Using the PG&E energy efficiency programs as a platform, it was envisioned that the movement toward ZNE would require important program changes. These include being involved very early in the schematic design phase of projects, accelerating the pace of technological development, supporting effective demonstration projects, and widespread education program development. These efforts were implemented to inform decision-making around energy savings related to location, infrastructure choices such as land use and transportation planning, and community-scale choices such as street orientation to optimize solar.

With the California Public Utilities Commission (CPUC) and the California IOUs now focusing on ZNE goals, it has become apparent that the models used for evaluating the energy savings and cost-effectiveness of energy efficiency programs are insufficiently comprehensive, resulting in an emphasis on short-term, often hardware-based savings. This misalignment will likely limit the success of long-term, strategic efforts in California, and if uncorrected, may be translated to other states and jurisdictions interested in similar goals. This chapter first explores a history of energy efficiency in California, outlines the

---

3. California biennially reanalyzes its energy system and revises its foundational energy policy. Final Commission Report, December 2009, CEC -100-2009-003-CMF recommends ZNE as its first policy recommendation. See page 227. That document also attributes almost 70% of the electricity used in California to residential and commercial building. See page 52.

process and ideas that led to the adoption of the state's Long-Term Energy Efficiency Strategic Plan, describes the development of PG&E's ZNE Pilot Program in response to that plan, and discusses how the conceptual frameworks underlying energy efficiency programs must adopt a wider view if we are to address the significant challenge posed by climate change.

2. A BRIEF HISTORICAL PERSPECTIVE ON ENERGY EFFICIENCY IN CALIFORNIA

Energy efficiency programs at California's IOUs, including PG&E, began over 30 years ago. While the California IOUs have offered energy efficiency programs for a significant period of time, program and utility objectives, funding levels, target energy savings, and customer segments of interest have repeatedly changed. Table 1 highlights key decisions by the CPUC affecting energy efficiency in California.[4]

The "decoupling," or separation, of utility revenues and sales, an important milestone in the regulated utility setting, occurred in the aftermath of the two substantial increases in world petroleum prices during the 1970s.[5] These price increases highlighted the difficulties of balancing stable and low rates, multi-year funding, and rate cycles with fuel procurement in unpredictable and volatile world markets. California's solution was to separate utilities fuel and nonfuel cost recovery and to set up a system to encourage cost-efficient operation of the utility itself. This highlighted those aspects of cost under the control of the utility, such as operating and capital costs while allowing uncontrollable costs, such as fuel costs, to be passed through, subject to a prudence review. For the costs over which the utility has control, the utility would be assured of the eventual recovery of the authorized amount of those costs, and could increase earnings—until the next general rate setting process—by conducting its business efficiently. Fuel costs were dealt with separately and these costs were passed through to ratepayers.[6]

This ratemaking mechanism removed any actual or perceived financial *disincentive* for the utility to participate in and promote energy efficiency. This is because without decoupling, sales below expectations imply lower revenues and profits. Since the objective of energy efficiency programs is to reduce sales below otherwise anticipated levels, the decoupled system assured the utility that it would receive authorized revenues regardless of the level of sales. If sales

4. As a point of reference, U.S. utilities currently spend approximately $3 billion on energy efficiency per annum, of which roughly $1 billion is spent in the state of California, making it the most aggressive in the country, by far.

5. This book's Chapter 8 by Long et al. also covers this important regulatory milestone.

6. The American Council for an Energy Efficient Economy summarizes decoupling at *www.aceee.org/pubs/u061.htm*, and with registration has reports on the topic. Many thorough expositions and analyses are available from the Regulatory Assistance Project at *www.raponline.org*.

TABLE 1 Key Dates and Decisions for California Investor Owned Utilities Energy Efficiency Programs

Issue:	Proceeding or Decision/ Date:
Separation of authorized utility revenues from variations in electric and natural gas sales in the late 1970's and early 1980's.	See the Electric Rate Adjustment Mechanism (ERAM) as detailed in CPUC Decision 93887, December 30, 1981.
The "Collaborative" Report of 1990 to the CPUC, which began a period of increased levels of funding, the treatment of efficiency as a resource comparable to supply-side resources, and the development of a comprehensive regulatory framework: cost-effectiveness screens, program approval procedures, after-implementation savings measurement requirements, and a utility risk/reward mechanism.	See CPUC Proceeding A.90-04-037. For a description of the collaborative process as applied to utility rate cases, see English et al., 1994.
A shift away from a resource perspective and toward "market transformation" during California's experiment with restructuring the electrical system supply-side in the late 1990's.	See CPUC Decision D.95-12-063, as modified by D.96-01-009 for the proposed competitive framework for the electric services industry.
The resurrection of resource energy efficiency following California's "crisis" in 2000-2002, including the adoption of explicit savings goals, increased program funding, the reintroduction of extensive measurement, and the restoration of utilities risk/reward mechanisms.	See, for example, CPUC D.02-03-056 (Selection of 2002 Programs), D.03-04-055 (Selection of 2003 Programs), D.03-12-060, and D.04-02-059 (Selection of 2004-2005 Programs).
The development of a "strategic plan" aligned with California's climate change legislation, and the beginning of a synthesis of resource and market transformation activities with achieving zero net buildings as a central goal and organizing principal.	See www.californiaenergyefficiency.com for a detailed description of the strategic plan process and a copy of the California Long-Term Energy Efficiency Strategic Plan, adopted by the CPUC in September of 2008.

were actually higher or lower than forecast—relative to the forecasts used in predicting future costs in rate-setting proceedings—leading to actual revenues being higher or lower than forecast, the utility would return any excess or recover any shortfall in future time periods. If energy efficiency itself caused a revenue shortfall, the revenue discrepancy would be made up in future periods. As a result of this mechanism, PG&E and the other California IOUs became financially indifferent to the effect of energy efficiency on their sales.[7]

During the 1980s the need for new generation, and the associated increase in fossil fuel consumption, led to an effort to stimulate expanded energy efficiency. Ralph Cavanagh of the Natural Resources Defense Council was pivotal in a multistakeholder, collaborative process from 1989 to 1992 that focused on energy efficiency policy (CPUC Proceeding A.90-04-037, 1990).[8] The "Collaborative" recommended that the CPUC adopt a revitalized regulatory framework treating energy efficiency as a *resource*, much like investments in supply-side options. This framework included increased program activity and funding, extensive measurement of energy efficiency savings and their persistence, and a shareholder risk/reward mechanism that shared the benefits of successful energy efficiency savings between ratepayers and the utility, referred to as a "shared-savings" scheme. Over the next four years the CPUC systematically implemented this program, first with a trial program, then with measurement approaches and interim risk/reward mechanisms, and finally with an adopted framework that incorporated comprehensive "resource" planning. This final program structure includes measurable savings programs, rigorous after-the-fact measurement of energy savings, and a shared-savings risk/reward mechanism to ensure utility management focus on program success (for example, see Vine et al., 2006).[9]

Critical details of this regulatory framework were the cost-effectiveness test the CPUC selected, the explicit construction of avoided costs to value energy savings, and the requirements on proposed portfolios to be cost-effective based on the savings measurable under CPUC measurement protocols. The key tests were the *total resource cost* (TRC) test, which measured cost-effectiveness from the perspective of all ratepayers or the region in which energy efficiency is implemented, and the administrators' cost test, which measures cost-effectiveness from the perspective of the administrators' costs and the benefits accruing to it. These criteria were applied to utility portfolios as a condition for approval.

It is important to note that in determining cost-effectiveness, all costs were included in the tests, but only the benefits arising from savings measurable from

7. As Long et al. point out (Chapter 8 of this volume), California's decoupling model is an anomaly among most states in the U.S., where utilities often have perverse incentives to increase sales rather than promoting EE.

8. See CPUC Proceeding D.05.01.055, pages 21–32.

9. Both ACEEE (*www.aceee.org*) and the Regulatory Assistance Project (*www.raponline.org*) have many presentations and papers on this and other related topics.

CPUC approved measurement protocols were included. Where "nonresource" programs—those without measurable savings, often long-term "strategic" programs—are a small part of this portfolio, this is not an important distinction. When there is interest in increasing the nonresource portion of the portfolio, cost-effectiveness can become an issue.

No sooner had this framework been established than California began to face the issues raised by the restructuring of its IOUs starting in 1996. While not central to the restructuring itself, energy efficiency for resource acquisition, like supply-side resource acquisition, was no longer regarded as an integral part of the electric distribution company's functions. Accordingly, the CPUC set out to find an independent administrator for energy efficiency activities focused on transforming the market for energy services and increasing energy efficiency. This exercise fared little better than the state's larger restructuring effort, and the effort to install an independent administrator for energy efficiency programs was not continued after the collapse of electric restructuring in 2000–2002.[11]

Subsequently, the acquisition of energy efficiency as a resource returned to the fore as one of the key instruments by which such crises could be prevented in the future. As the IOUs in California returned to the role of procuring electricity,[12] energy efficiency received increased emphasis and additional funding. The concept of state promoted market transformation[13] had now been permanently inserted into California's energy efficiency programs.

After the 2000–2002 crises had passed, the CPUC also systematically reviewed the policies that had been in place since the beginning of restructuring and established at PG&E an enlarged, resource-oriented energy efficiency portfolio. During the period from 2004 to 2006, the CPUC established energy efficiency savings goals (2004), affirmed the IOUs as energy efficiency administrators (2005), further increased funding and activity to support the higher goals (2005), and reinstituted a risk/reward mechanism (2006).[14]

At the close of 2006, PG&E was administering programs focusing on residential, commercial, industrial, agricultural, and institutional customers and covering all major energy using technologies and building systems, including lighting, heating, cooling and ventilation, motors, process, and residential and commercial new construction. PG&E offered customers information through

11. A reference describing California's restructuring experience is Sweeney, J. (2002). *The California electricity crisis*. Hoover Institution Press.

12. When the IOU's credit rating dropped in 2001, it was no longer able to transact on wholesale markets. The state of California temporarily assumed the function of procuring power on wholesale markets for retail customers.

13. The term goes back at least to 1994, see *www.aceee.org/pubs/e941.htm*, which describes it as the process of successive energy efficient innovations entering the market to the point of nonreversible saturation.

14. The California Public Utilities website provides further details of these decisions at *www.cpuc.ca.gov/PUC/energy/Energy+Efficiency/EE+Policy*.

channels ranging from TV spots to extensive design and engineering assistance; rebates to final customers or to producers, distributors, or retailers; and PG&E supported new and emerging technologies and advances in state and federal codes governing energy usage.

2.1. California's Long-Term Energy Efficiency Strategic Plan

Until 2006, longer term planning for energy efficiency had only occurred in the context of long-term energy planning. These plans were usually simply extensions of the one- to three-year plans approved by the CPUC. In 2007, the CPUC began the next logical step in moving from energy efficiency portfolios that were planned and implemented one to three years ahead to developing a "strategic plan" that would guide the trajectory and goals for energy efficiency programs through 2020 by attempting to focus on longer-term, more visionary possibilities for improving efficiency. Specifically, it used this vehicle to formally consider and approve the ZNE building goals needed to address climate challenges. The CPUC focused on this long-term strategy to align with California's implementation of Assembly Bill 32 (AB 32), the "California Global Warming Solutions Act of 2006," which requires the state's emissions of GHGs to be reduced to 1990 levels by 2020, and to set the strategic directions that could produce significant reductions in energy use, and therefore GHG emissions, over the coming decade. The goal and concept of ZNE was central to many of the findings of the strategic plan. Attaining these goals will cut the growth in California emissions significantly, as it pursues absolute declines through energy efficiency and other means. Currently, California expects that roughly 15% of its GHG reductions needed to meet the 2020 target will come from energy efficiency.[15]

As the keystone to this effort, the CPUC adopted the California Long-Term Energy Efficiency Strategic Plan (CEESP or Strategic Plan) in September 2008. The Strategic Plan sets forth a statewide roadmap to maximize the achievement of energy efficiency in California's electricity and natural gas sectors. The Plan also sets a number of specific goals for the years 2009 through 2030. Over 40 public workshops and meetings were convened by the CPUC and the IOUs leading to the adoption of the Strategic Plan (PUC, 2008).

With the CEESP, the state of California has created a roadmap for a scaling-up of statewide energy efficiency efforts. While the policies of the past three decades have been successful in raising public awareness of energy issues and laying the groundwork for large-scale efficiency efforts, the savings achieved through those policies have come through specific programs with targeted market impacts. The objective of the Strategic Plan is to sustain market transformation, moving California beyond its historic reliance on a near-term

15. For details on the AB 32 plan, including targets, see *www.arb.ca.gov/cc/facts/scoping_plan_fs.pdf*.

replacement of less efficient technology with more efficient technology and toward long-term, deeper savings achievable only through programs with broader, longer-term impact. This shift is in part to align with the goals of AB 32, but is also reflective of an evolution in program design as a result of a broader set of stakeholders engaged in energy efficiency planning.

Key to the Strategic Plan's success are four programmatic goals, widely viewed as ambitious, high-impact efforts. These goals, called the "Big, Bold Energy Efficiency Strategies" (BBEES), are listed in Table 2. The goals were selected not only for their potential impact, but also for their transparency and ability to galvanize market players. For each of these measures, the Strategic Plan provides recommendations for coordinated action among the state, its utilities, the private sector, and other market players. The recommendations take advantage of the wide variety of stakeholder expertise engaged in the strategic planning process. While specific energy savings have not been assigned to each of the goals, the structure of each utility-administered program has been redesigned to address meeting the BBEES.

The implementation of the Strategic Plan requires action on many fronts. In the near term, the CPUC has committed to leading the effort to coordinate necessary implementation actions. In addition the CPUC has authorized energy efficiency programs totaling $3.1 billion for the 2010–2012 timeframe for the IOUs to achieve both measurable energy savings and strategic plan objectives. Ensuring that utilities' energy efficiency program portfolios for the 2010–2012 cycle are consistent with the goals and strategies in the Strategic Plan represents a large portion of the CPUC's role in ensuring the realization of the Strategic Plan objectives. Central to this large effort was PG&E's specific proposal to foster ZNE.

TABLE 2 California Long Term Energy Efficiency Strategic Plan Programmatic Goals, or the "Big, Bold Energy Efficiency Strategies" (CEESP, 2008)

Sector/ Market	Goal
New Residential Construction	All new residential construction in California will be zero net energy by 2020.
New Commercial Construction	All new commercial construction in California will be zero net energy by 2030.
Heating, Ventilation, and Air Conditioning (HVAC) Industry	HVAC industry and market will be transformed to ensure that its energy performance is optimal for California's climate.
Low Income Customers	All eligible low-income customers will be given the opportunity to participate in low-income energy efficiency programs by 2020.

2.2. Cost to Achieve ZNE

During the development of the Strategic Plan, ZNE was defined broadly as "the implementation of a combination of building energy efficiency design features and on-site clean distributed generation that result in no *net* purchases from the electricity or gas grid, at the level of a single 'project' seeking development entitlements and building code permits." (CPUC Decision 07-10-032, 2007, emphasis added). This broad definition was adopted by the CPUC so that the ZNE programs could investigate a wide variety of options to achieve ZNE in residential and commercial buildings. These options might include community scale renewable energy systems in addition to the more traditional on-building photovoltaics (PVs). Conceptually ZNE goals can be achieved at the scale of a building, collection of buildings, campus, complex, a development, or even a new city. Larger scale can provide significant new opportunities. For example, on a specific building basis, it might prove difficult to design a new multifamily structure to be ZNE, while if the entire site was considered, additional approaches can be employed to drive further to ZNE at lower cost. Some new technologies might be through superior design (optimizing solar gain/shading, high-performance window and shell construction including "cool" roofs) and the renewables integrated into weather protection on top of parking structures. Moving to a larger scale can open progressively attractive opportunities.

Moving beyond on-building energy production is important because a recent report by the National Renewable Energy Laboratory (NREL) concluded that using today's technologies and practices, only 64% of the commercial building stock in the United States could become ZNE (Griffith et al., 2006). This low technical potential is due to the high energy use intensity (EUI) of many building types and a low amount of roof area that could be devoted to renewable energy production. Therefore, having the option of on-site or near-site PV energy production is necessary to be able to achieve penetration through the entire commercial building market.

Having options beyond the installation of PVs on each building is also important to keep the cost of ZNE as low as possible. For example, as part of program planning, PG&E estimated the cost to achieve ZNE in a new 2100 square foot home in four cities that PG&E serves. Single-family homes are widely considered to be the easiest building type in which to achieve ZNE, and several demonstration projects such as the Solar Decathlon[16] have shown that ZNE can be readily achieved in the residential sector. The results of the estimation are presented in Figure 1. In all four cities, the largest portion of the

16. The U.S. Department of Energy Solar Decathlon challenges 20 collegiate teams to design, build, and operate solar-powered houses that are cost-effective, energy-efficient, and attractive. The winner of the competition is the team that best blends affordability, consumer appeal, and design excellence with optimal energy production and maximum efficiency. For more information see *www.solardecathlon.gov*.

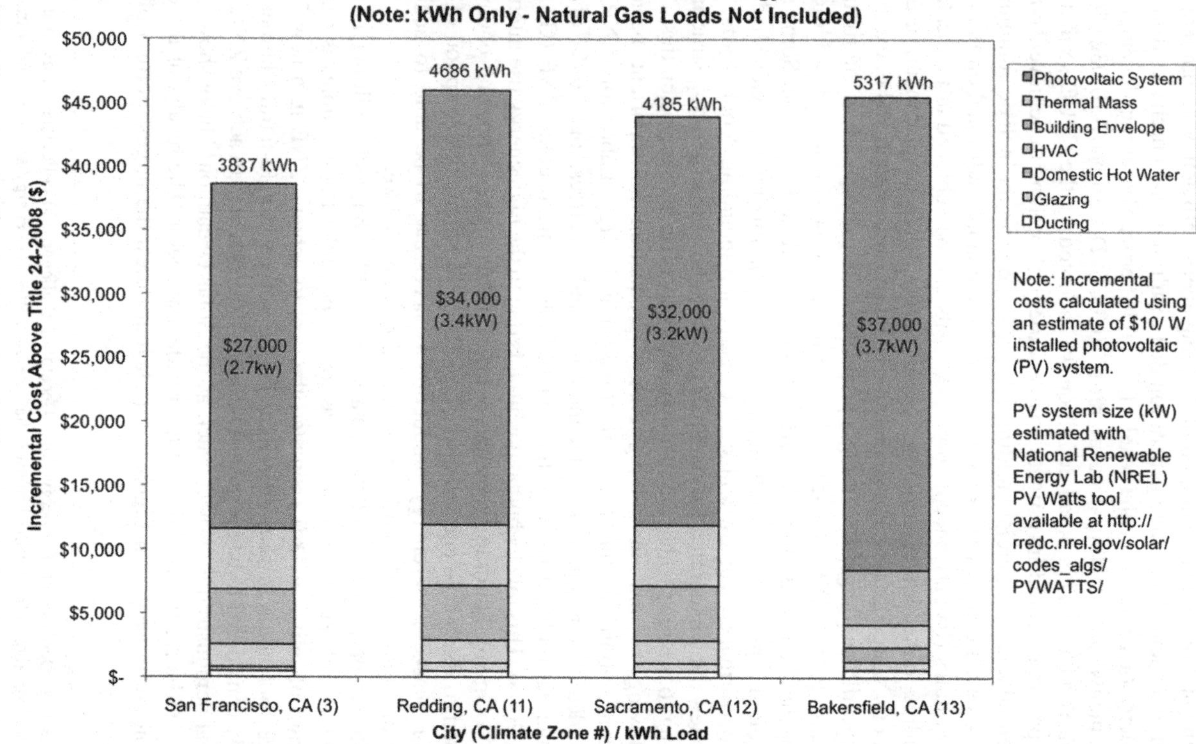

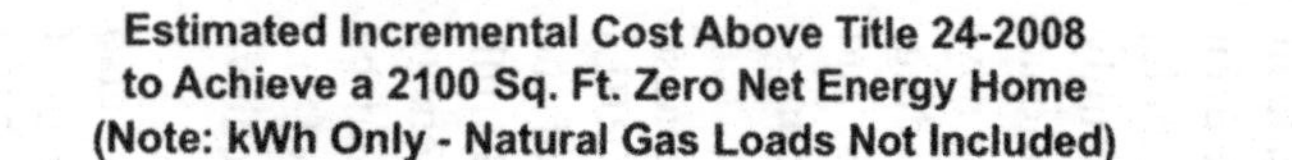

FIGURE 1 Estimate of the Incremental Cost to Achieve ZNE in a 2100 Square Foot Home in four California Communities. Note that the highest proportion of the cost is to provide for a photovoltaic system.

incremental cost was the installation of a PV system on the roof of a house, as high as $37,000 for a home in Bakersfield in California's central valley.[17]

While the total incremental cost of $40,000 to achieve ZNE in a home in San Francisco may only represent a 5% increase in the total cost of the home, in areas such as Sacramento, with lower median home prices, the incremental cost may be 25% of the value of the home.[18] While some of the cost can be recovered through a higher sale price (2%–5%), it is unlikely that the market will support such significant incremental costs in all markets without incentives. These figures will vary regionally with construction costs, the performance of local renewables, and so on.

While ZNE buildings hold the promise of no energy bills, with high incremental costs, the total benefit/cost ratio is less than one, and will remain less than one until either the price of carbon is approximately $65/ton or the incremental cost of the ZNE measures is reduced to approximately $10,000 per home (Goetzler, 2008). Similar results are expected for commercial buildings, though the custom nature of the commercial market makes an "apples to apples" comparison impossible. This lack of cost-effectiveness has framed the approach the utilities took to start to achieve ZNE in California—reducing the first cost of technologies, reducing the cost of design and construction, and increasing the knowledge of ZNE in the general public to increase demand for the products and services.

3. PG&E'S ZERO NET ENERGY PILOT PROGRAM

The PG&E ZNE Pilot Program is one example of how the IOUs are supporting the implementation of the Strategic Plan. Beginning in late 2010, the program will initiate research, develop design guidelines, and identify and initiate demonstration projects around ZNE buildings and developments.

As PG&E designed the ZNE Pilot Program, it identified four key gaps in the current body of knowledge that are significant barriers to ZNE development:

- **First Cost:** The issue of first cost can be overcome by reducing the capital cost of a technology, or, as part of a whole building solution, by reducing other costs of construction to "buy" room in a project budget for deeper levels of energy efficiency. Cost reduction may also be accomplished through better design integration, real-time construction cost estimations during the design process, or utilizing new software tools such as *Building Information Modeling* or BIM. In addition, cost-effectiveness can be

17. In this book's Chapter 9, Gray and Zarnikau also discuss the cost implications of ZNE.

18. Incremental cost to achieve ZNE in the Sacramento region is estimated at $43,925. According to the National Association of Realtors (*www.realtor.org/research/research/metroprice*), the 2009 median house price in the Sacramento region is $180,500.

attained by reducing the amount of time spent identifying, learning about, and locating energy efficiency measures.

- **Integration:** A major issue is the need to have an integrated approach to demand-side management very early in the design process. The ZNE Pilot Program would work with building owners and design teams to affect the form of a building, orientation, and decisions regarding energy consuming equipment in predesign or the schematic design phases of a project.
- **Knowledge:** There is a widespread lack of knowledge regarding ZNE in the-general public. Through the demonstration projects and case studies, future building owners, design teams, and the general public will be able to view ZNE buildings and "kick the tires" on new technologies. This will begin to create demand in the market for buildings and developments with better demand-side management capability and specifically for buildings on the path to ZNE. Increasing public awareness may also help to overcome institutionalized behavior by companies, departments, professional groups, and government entities that may otherwise discourage forward thinking and proactive implementation of energy efficiency measures.
- **Scaling Up:** Until there is a critical mass of ZNE homes and commercial buildings, the perceived difficulty of acquiring the information needed to evaluate measures during design, construction, and operation of a ZNE building may cause design teams and building owners to not pursue ZNE design.

The ZNE Pilot Program was designed to address these barriers through integration, cooperation, and collaboration with a wide range of market actors. The Program plans to engage the publicly owned and IOUs; developers, architects, builders, municipalities, and redevelopment agencies; the California Energy Commission; the U.S. Department of Energy National Laboratories including National Renewable Energy Laboratory (NREL) and Lawrence Berkeley National Laboratory (LBNL); professional building and trade associations; research institutions; and state, federal, regional, and local agencies. With the collaboration of the above organizations, the program will continue to engage long-term strategic planning and also identify near-term steps toward the Strategic Plan goals (Figure 2).

The first part of the program, related to community design, would offer design assistance and technical support furthering advanced community design. This Communities Subprogram will target mixed-use complexes, multifamily complexes, advanced residential new construction, advanced commercial new construction, compact development, and transit-oriented development at the early stages of the entitlement and design process, helping to capture energy and resource savings that would normally fall outside of the scope of a typical project.

To be eligible to participate in the Communities Subprogram, it is expected that a project would need to be in the early stages of entitlement, planning, or

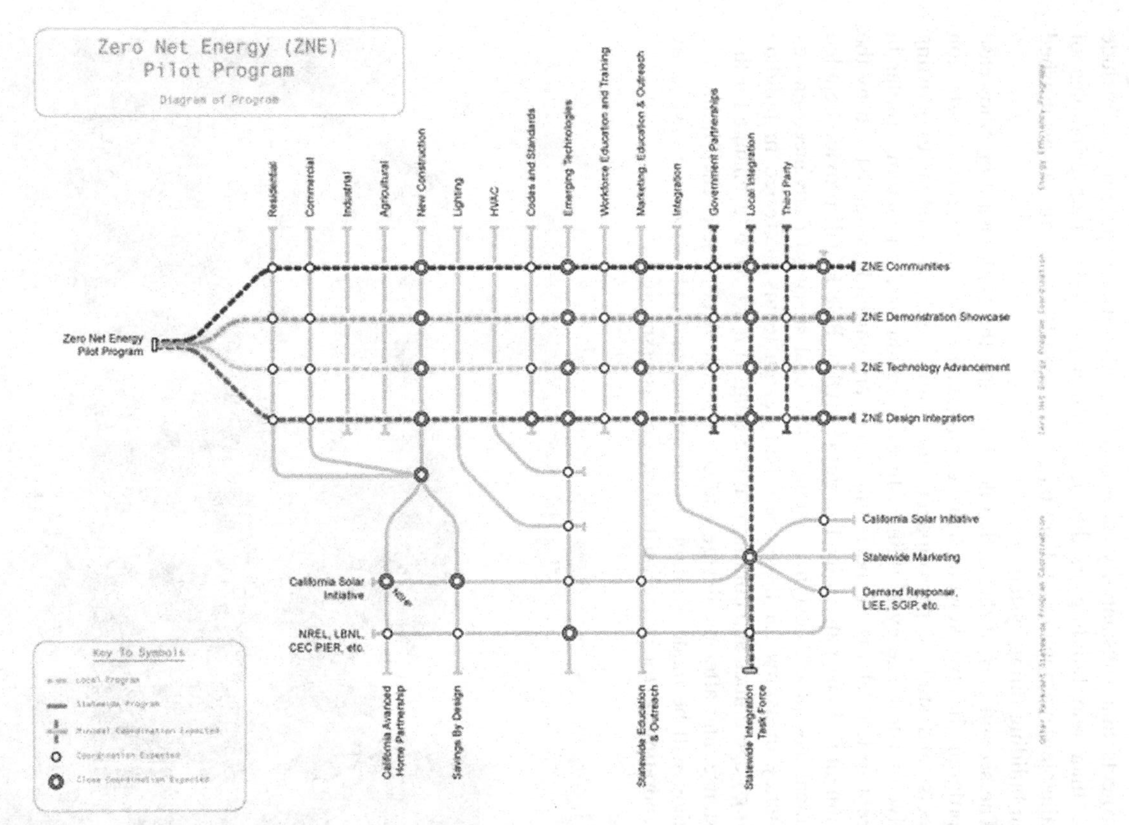

FIGURE 2 PG&E Zero Net Energy Pilot Program Diagram Showing Integration Efforts with Other Energy Efficiency Efforts, from PG&E Program Implementation Plan, 2009.

design; be primarily a residential or commercial development; plan to exceed the California building energy code by at least 30%, preferably much more; and plan to include on-site clean distributed generation. Preference would be given to projects that substantially exceed these minimum requirements, include more than one building in the development, include other principles of sustainable development, and are targeting a certification from an established green building rating system.

The second part of the ZNE Pilot Program, the Demonstration Showcase Subprogram, has two key elements: commercial and residential demonstration projects; and studies, performance monitoring, and assessment of existing passive and low-energy buildings. In this portion of the program, similar in concept to the "Home of the Future" program currently administered by the Sacramento Municipal Utility District (SMUD), PG&E will provide detailed technical assistance, design assistance, and cost sharing of advanced energy efficiency measures for developers and design teams interested in building cutting-edge homes and commercial buildings (Figure 3). In exchange for this assistance, and after the design and construction is complete, each home and building will be made available to the public, and published as a case study incorporating performance verification and assessment.

FIGURE 3 The Sacramento Municipal Utility District (SMUD) "Home of the Future" in Folsom, California. Completed in 2008, the home incorporates a number of energy efficiency and renewable energy features and was open to the public for tours. For more information on the SMUD Home of the Future program, see www.smud.org. Photo Credit: Nicholas Rajkovich.

These demonstration projects will give teams the opportunity to design, build, and construct a near ZNE building at a lower direct cost and in a lower risk environment than if they were to undertake the projects on their own. This will be accomplished by partnering with research organizations such as CEC PIER, the NREL, and the LBNL[19] to provide lessons learned from other demonstration project programs and to offer design and technical assistance for the construction of several homes and commercial projects.

These demonstration projects will also give teams the opportunity to *practice* ZNE design and construction techniques and to have access to research expertise and "lessons learned" from past projects that are not normally available to a design team. In addition, the showcases will engage the public in on-site activities, while ongoing performance assessment and verification will allow teams to refine their techniques for future design projects.

In the third part of the ZNE Pilot Program, the Technology Advancement Subprogram, PG&E plans to deliver information, insights, analytical tools, and resources to accelerate and expand the commercialization of innovative technologies. To accomplish this, the ZNE Pilot Program plans to work with the existing Emerging Technologies Program (ETP) in California to deliver information and insights on customer and community planning needs, as well as technology integration opportunities to help the ETP screen and assess potential technologies. In turn, the ETP will provide insights on technology evolution and trends, market potential, adoption rates, participation in vendor technology evaluations, implementation and management of pilot programs, and design specifications for needed technology to support the ZNE Pilot Program.

In the fourth and final part of the ZNE Pilot Program, the Design Integration Subprogram, PG&E will develop and disseminate information on the best practices for the design of ZNE communities, buildings, and homes by engaging organizations such as the American Institute of Architects California Council (AIACC), the U.S. Green Building Council (USGBC), and the American Society of Heating, Refrigerating, and Air-Conditioning Engineers (ASHRAE).

In addition, to close the loop and allow for the evaluation of proposed ZNE communities, buildings, and homes, assistance will be offered to planning and code officials who are in the process of reviewing proposed ZNE buildings and developments. The results of the Design Integration Subprogram will include best practice guidelines and software tools to design and evaluate "beyond-code" projects. The goal will be to close the loop from design through occupancy, including project phases such as code review and on-site code related inspections. This will require convening and coordinating the ongoing efforts of

19. The U.S. Department of Energy has a research objective of enabling ZNE buildings. For commercial buildings, see *www1.eere.energy.gov/buildings/commercial_initiative/goals.html* and for residential activities see *www1.eere.energy.gov/buildings/goals.html*.

national energy experts, software developers, regulatory bodies, and code officials to adopt a common language for the design, construction, and evaluation of ZNE buildings.

4. REFRAMING ASSUMPTIONS FOR ZERO NET ENERGY

As California and other states contemplate a move to ZNE, it is timely to examine how the legacy of energy efficiency program planning and evaluation may affect the success of this new type of program. Historically, programs promoting energy efficiency have taken a "rational actor" approach to transform patterns of energy use. The goal of most energy efficiency programs has been to *effect* a change in energy consumption by *causing* an end user to purchase a piece of equipment with a higher level of efficiency. This has typically been done by providing the end user with information on the cost of the energy efficiency equipment and/or providing a financial incentive to make the investment in the equipment economically attractive. This policy paradigm is widely called the Physical Technical Economic Model (PTEM).

The PTEM approaches most energy efficiency problems through a methodology called *DMAIC*, defined through the following five steps (Huesing, 2008):

- *Defining* energy savings goals and current processes in use.
- *Measuring* key aspects of current processes and collecting energy and economic data related to the process.
- *Analyzing* the data to determine cause-and-effect relationships and hypothesizing what economic incentive levels are necessary to effect change in the market.
- *Improving* or optimizing an end user's process based upon the data provided.
- *Controlling* to ensure that any deviations from the original program goals are corrected as the program scales up to the entire market.

The PTEM has been effective in causing an increase in investment in specific energy efficiency technologies such as high-efficiency motors. The PTEM has been favored over other approaches primarily because energy savings achieved through these programs can be *clearly defined* and are subject to widely accepted *measurement* or *verification* approaches. However, PTEM-type programs have not been totally effective. Despite substantial efforts and financial incentives to cause investment in energy-efficiency technologies, significant gaps remain between the *potential* and the level of *investment* actually occurring in the market (e.g., Granade et al., 2009). This gap is receiving increased focus because of the substantial goals proposed by the Strategic Plan and because of the vital contribution energy-efficiency programs can make to reducing worldwide GHG emissions.

Because the PTEM relies on the assumption of a rational actor, PTEM-based planners most often rely on two arguments to explain away the gaps:

- The first argument is that there are significant institutional and other barriers in the market that prevent consumers from behaving rationally and therefore prevent the market from performing to its full potential. These barriers include limited availability of capital, predatory pricing, regulatory distortions, transaction costs, and inseparability of energy efficiency from other product attributes (Sullivan, 2009, p. iv).
- The second argument is that the gap is more apparent than is actually real. Under this explanation PTEM-based planners argue that the gap is actually a consequence of the normal operation of an efficient market, and that adoption rates for any new technology should lag behind the level of economic potential at any point in time (Sullivan, 2009, p. iii).

Neither of these explanations is totally satisfying, and recent research has shown that PTEM-based programs tend to oversimplify consumer and business decision-making, leading to an overemphasis on cost-benefit tests and ineffective marketing (Sullivan, 2009, p. iv).

While the PTEM has been useful in building a logical foundation for energy efficiency in the past, its use has also limited its scope. Because energy efficiency as defined under the PTEM uses strict neoclassical engineering and economic models and is generally defined as the total reduction in energy usage from a baseline condition to a "better" condition after a measure is installed, PTEM-based programs inadvertently reward buildings that have a higher initial EUI—for example, single-family housing and multistory office buildings—with greater levels of financial incentives. This occurs because the baseline EUI for these building types is higher, and therefore the potential for total energy savings, measured in kilowatts (kW), kilowatt-hours (kWh), or therms (thm) of natural gas, is also higher.

While there is no causal link between energy efficiency programs and settlement patterns like urban sprawl, it is worth noting that since the 1970s, these programs based on the PTEM have provided millions of dollars to developers of relatively dispersed single family developments and office parks, a pattern of settlement with a high level of per capita GHG emissions, compared to denser, transit-oriented, and/or multifamily housing-related development. While energy efficiency programs have mitigated the energy impact of these developments, they have been unable to effect a change in the pattern and type of development. A PTEM-based program planner might counter the above statement by replying that they must remain agnostic to development types as engineers and economists; their role is only to drive "whatever the market wants" to a higher level of efficiency. But the persistent gap between potential and actual market conditions, as well as a feeling that engineering and economics left much unexplained about observed energy usage, drove a number of researchers in the early 1990s to push for a theory for energy efficiency grounded more broadly in the social sciences.

Beginning in the early 1990s, researchers such as Loren Lutzenhiser showed that the role of human social behavior has been largely overlooked in energy policy, despite the fact that social impacts of energy use have a significant effect on the results of any energy program. He argued that although a social theory of energy was scattered across a number of social science disciplines, a body of research concerned with human factors in energy use did exist and that we should apply these techniques to achieve deeper results (Lutzenhiser, 1993).[20]

One method advocated by Lutzenhiser was segmentation, or "identifying homogenous sub-populations within larger heterogeneous populations" (Moss, 2008, p. 3). This technique, used widely in the field of marketing, is an effort to effectively communicate with and motivate to action an increasingly diverse population of individuals, families, and businesses by understanding their needs and patterns of energy use (ibid., p. 3). Methodologies advocated by this newly minted "social approach" included ethnographic studies to characterize patterns of energy use in specific demographic groups, focus groups, and longitudinal studies of energy usage. The primary focus of the social approach was reducing energy usage in the residential sector, where the PTEM had previously failed to achieve penetration and results. Under the new social approach, the PTEM failure to penetrate into a market was defined as a failure to understand the needs (especially the noneconomic needs) of a market.

However, the social model remains as agnostic on settlement patterns as the PTEM. While perhaps better equipped to understand the needs of the greater population by determining attitudes on sprawl or types of development, it also lacks an ability to inform a value judgment favoring one form of development over another. For instance, if equal percentages of a surveyed subpopulation were to prefer single-family residential and multifamily housing, the social model would fail to reject either and could not provide a clear rationale for program activities focused on one type of housing over the other. Both types of development would remain. In the end, we are seeing socially-based energy planners following the same path as their PTEM brethren, driving "whatever the market was surveyed to want" to a higher level of efficiency. This approach, therefore, is as flawed as the PTEM.

Over time, the original PTEM has been revised by the social approach, resulting in the current "sociotechnical" model used in energy efficiency practice (e.g., Bernstein, 1966; Janda, 2002). The current energy efficiency planning process used in California and other U.S. jurisdictions is an amalgamation of the PTEM and segmentation methodologies. The authority to

20. In this book's Chapter 11 Prindle and Finlinson make similar arguments, as do Ehrhardt--Martinez et al. in Chapter 10.

operate certain programs that are "socially good" but produce little energy or economic savings is granted by lumping all of the energy efficiency programs together into one large program, called a portfolio. Within this portfolio, planners are basically able to average the cost-effectiveness from high-producing programs that have a benefit—cost ratio much greater than one with other programs that produce social good but would otherwise not pass a cost-effectiveness test (Sullivan, 2009, p. vi).

Unfortunately, the current approach is unlikely to allow ZNE to stand on its own within energy efficiency programs because the total portfolio cost-effectiveness remains the overall concern. Cost-effectiveness is highly valued in the political process as a means to prove the efficacy of a program. When the economy falters, or other programs are not meeting energy savings targets, the programs that do not provide measured energy resource savings will be at a disadvantage relative to programs with measurable energy savings potential.

CONCLUSIONS

Any strategic approach to achieve long-term, energy-efficiency savings will require wide stakeholder support, an ability to quantify externalities, an ability to count nonenergy benefits from these programs, and a process for mediating, assessing, and prioritizing new information in the context of shifting current social norms and priorities. One of the current handicaps in California is a lack of reflection and open dialogue among principal stakeholders that would allow them to discuss the shortfalls in current approaches to cost-effectiveness and openly collaborate to resolve the issues. Because a national effort is underway to achieve ZNE in the built environment by the year 2030, and programs such as PG&E's are underway or being considered by other jurisdictions, this effort is timely and necessary.

The effort to expand the foundation on which strategic energy efficiency programs is built will require significant time and resources. The payoff could be well worth it: a sound and agreed upon basis to assess and justify such programs. This might enable expanded energy efficiency program options, like ZNE, to support sustainable development in a way that has not been feasible in the past. Research from the social sciences, documenting that we need to create developments that are socially inclusive and sustainable with regard to land use, energy use, and transit, could finally be brought to bear on energy-efficiency problems and be fundamentally connected with overall programmatic efforts to reduce overall GHG emissions (Blanco et al., 2009). The "payoff" would be incalculable: Energy use in new homes and businesses would trend to zero, flattening energy growth in California. Increased energy efficiency in existing buildings would no longer be offset by additional energy use in new buildings, and overall real energy usage would decrease, hastening progress to sustainability.

ACKNOWLEDGMENTS

We would like to thank Marlene Vogelsang of the PG&E Pacific Energy Center for assisting with our literature search. We would also like to thank Professor Richard K. Norton and Stacey Kartub for reviewing earlier versions of this document and providing valuable feedback.

BIBLIOGRAPHY

Bernstein, R. J. (1966). *John Dewey*. New York: Washington Square Press.

Bernstein, R. J. (1976). *The restructuring of social and political theory*. Harcourt Brace Jovanovich, Inc.

Blanco, H., Alberti, M., et al. (2009). Hot, congested, crowded and diverse: Emerging research agendas in planning. *Progress in Planning, 71*, 153–205.

California Public Utilities Commission (CPUC). (1981). CPUC. San Francisco, CA, CPUC. 93887.

——. (1990). CPUC. San Francisco, CA, CPUC.A.90-04-037.

——. (1995). CPUC. San Francisco, CA, CPUC.D.95-12-063.

——. (1996). CPUC. San Francisco, CA, CPUC.D.96-01-009.

——. (2001). California standard practice manual: Economic analysis of demand-side programs and projects. San Francisco, California, CPUC.

——. (2002). CPUC. San Francisco, CA, CPUC.D.02-03-056.

——. (2003). CPUC. San Francisco, CA, CPUC.D.03-04-055.

——. (2004). CPUC. San Francisco, CA, CPUC.D.04-02-059.

——. (2007). CPUC. San Francisco, CA, CPUC.D.07-10-032.

——. (2008). *Decision approving 2010 to 2012 energy efficiency portfolio and budgets*. San Francisco, CA, CPUC. 08-07-031.

——. (2008). *California energy efficiency. www.californiaenergyefficiency.com*

U.S. Department of Energy (DOE). (2008). *Building technologies program (BTP) multi-year program plan. www1.eere.energy.gov/buildings/mypp.html.*

——. (2009). *Net-zero energy commercial buildings initiative. www1.eere.energy.gov/buildings/commercial_initiative/.*

Dykman, J. (1983). Reflections on planning practice in an age of reaction. *Journal of Planning Education and Research, 3*(1), 5–12.

English, M. R., Schweitzer, M., et al. (1994). Interactive efforts between utilities and non-utility parties: constraints and possibilities. *Energy, 19*(10), 1051–1060.

Feyerabend, P. (1975). How to defend society against science. *Radical Philosophy, 11*, 3–8.

Forester, J. (1993). *Critical theory, public policy, and planning practice*. Albany, NY: State University of New York Press.

Goetzler, W. (2008). *Zero energy homes: Implications for California*. San Diego, CA: 2008 Emerging Technologies Summit.

Granade, H. C., Creyts, J., et al. (2009). *Unlocking energy efficiency in the U.S. economy*. McKinsey & Company.

Griffith, B., Torcellini, P., et al. (2006). *Assessment of the technical potential for achieving zero-energy commercial buildings. www.nrel.gov/docs/fy06osti/39830.pdf.*

Hawthorne, C. (2003). Turning down the global thermostat. *Metropolismag.com. www.metropolismag.com/story/20031001/turning-down-the-global-thermostat.*

Hoch, C. (1984). Doing good and being right: The pragmatic connection in planning theory. *Journal of the American Planning Association, 50*(3), 335–343.

Huesing, T. (2008). *Six sigma through the years. www.motorola.com/content.jsp?global ObjectId=3070-5788.*

IPCC. (2007). Climate change 2007: The physical science basis. *Contribution of Working Group I to the Fourth Assessment Report of the Intergovernmental Panel on Climate Change.* Cambridge, U.K. and New York: Cambridge University Press.

Janda, K. B. (2002). Improving efficiency: A socio-technical approach. In A. Jamison & H. Rohracher (Eds.), *Technology Studies & Sustainable Development* (pp. 343–364). Munich and Vienna: Profil Verlag GmbH.

Klemke, E. D., & Hollinger, R.et al. (Eds.). (1998). *Introductory readings in the philosophy of science.* Amherst, New York: Prometheus Books.

Lutzenhiser, L. (1993). Social and behavioral aspects of energy use. *Annual Review of Energy and Environment, 18*, 247–289.

Moss, S. J. (2008). *Market segmentation and energy efficiency program design.* Berkeley: California Institute for Energy and Environment.

Popper, K. (1959). *The logic of scientific discovery.* New York: Routledge.

Sullivan, M. J. (2009). *Behavioral assumptions underlying energy efficiency programs for businesses.* Berkeley: California Institute for Energy and Environment.

Vine, E., Rhee, C. H., et al. (2006). Measurement and evaluation of energy efficiency programs: California and South Korea. *Energy, 31*, 1000–1113.

Chapter 18

Toward Carbon Neutrality: The Case of the City of Austin, TX

Jennifer Clymer[1]
ICF International[2]

Chapter Outline

1. INTRODUCTION

In February 2007, the Austin City Council enacted the Austin Climate Protection Plan (ACPP) to reduce greenhouse gas (GHG) emissions from the City of Austin's, 2009 municipal operations and the larger community. Austin is the capital of Texas, the fourth largest city in the state, and the sixteenth most populous city in the U.S. The city has a population of 783,000 and comprises 300 square miles. In addition to conventional municipal services, the municipal government (the "City") provides electric, water, wastewater, and solid waste management services. Each of these functions contributes a sizeable portion of Austin's carbon footprint and offers significant opportunities for GHG reductions.

1. The author wishes to thank Ester Matthews, Karl Rábago, and Ed Clark for their review and contributions to this chapter. Acknowledgment is also given to Austin Energy and the City of Austin for their permission to share information about Austin's climate protection efforts.

2. The work described in this chapter was performed while the author was with the City of Austin Climate Protection Program, with acknowledgment for their support.

This chapter highlights the City of Austin's successes and challenges in implementing the ACPP and provides comparisons to other cities that have taken similar action to address climate change. Section 2 sets the stage for Austin's and other cities' climate protection efforts. Section 3 outlines the ACPP. Section 4 describes specific measures Austin is taking to reduce the carbon footprint of its municipal operations and of the community. A comparison of Austin's approach to climate protection to other peer cities is provided in Section 5. Section 6 outlines lessons learned that could be applied to other climate protection programs, and the chapter concludes with a forward-looking view of a world in which all cities achieve, or exceed, Austin's goal of carbon neutrality.

2. WHY CLIMATE ACTION?

Urban areas represent less then 1% of the global land area but house roughly half of the global population. These 3.4 billion urban dwellers currently consume roughly 75% of worldwide energy and emit 75% of global GHG emissions. This urbanization trend is projected to continue, with roughly 60% of the world's population living in urban areas by 2030 (C40 Cities, 2009). A massive amount of energy is required to support this growth and the resultant GHG output if the world continues its projected pace of energy consumption and reliance on fossil fuels. In the absence of climate action, the Intergovernmental Panel on Climate Change (IPCC) projects global GHG emissions will reach somewhere between 25 to 90 billion metric tons of carbon dioxide-equivalents (CO_2-eq.) per year by 2030 (IPCC, 2007).

In addition to their large contribution to global GHG emissions, urban areas are highly susceptible to the impacts of climate change. The effects of elevated temperatures are exacerbated in urban environments due to lack of vegetative cover; concentration of asphalt and concrete structures that absorb—rather than reflect—heat; and excess heat contributed by a denser concentration of vehicles, power plants, air-conditioners, and so on. The combination of these factors can cause urban temperatures to be 2° to 22°F (1° to 12°C) warmer than surrounding rural and suburban environments (U.S. Environmental Protection Agency, 2009), a phenomenon known as the *urban heat island effect*. Warmer urban temperatures contribute to a risk of more extreme heat days that contribute to heat- and air-pollution–related illnesses and deaths.[3]

3. The urban heat island effect contributes to elevated air pollution levels in two ways. First, elevated ambient temperatures increase the need for air-conditioning (where available), which increases the demand for electricity, which in turn increases emissions from local power plants. Second, ozone, a common urban air pollutant, is formed when nitrogen oxides and volatile organic compounds react in the presence of sunlight. Hot, stagnant air provides ideal conditions for ozone formation.

Beyond rising temperatures, other types of extreme weather events are expected to increase due to climate change (IPCC, 2007). The U.S. alone has experienced 96 extreme weather events since 1980 that have individually carried a price tag of $1 billion or more. The number of $1 billion plus events has trended upwards, with a notable increase in the past decade (National Climatic Data Center, 2009). Flooding, hurricanes, and tsunamis are of particular concern due to the coastal location of many urban areas. In the U.S., the Gulf Coast and southeastern Atlantic Coast states have sustained the most damage due to the increased frequency and intensity of hurricanes (National Climatic Data Center, 2009). Coupled with rising sea levels, coastal residents and ecosystems are particularly vulnerable.

Inland areas are also impacted due to drought, flooding, tornadoes, and other weather events that may be exacerbated by climate change. Even in the absence of direct climate or weather events, inland cities that house coastal refugees during events can be impacted. Typically during evacuation events, those who have transportation, adequate financial resources, and alternative housing options leave. Those without such means are provided transportation to sister cities. These visitors often become permanent residents who strain the receiving community's social services as they seek assistance finding housing, jobs, food, and medical care. Local, state, and federal government assistance programs are left to fill the gap.

Given urban areas' disproportionate contribution to climate change and climate change's likely disproportionate impact on urban areas, local governments have become leaders in climate protection. The Kyoto Protocol is driving climate action at all levels of government among its 187 participating member countries. In the U.S., which has not yet ratified the Kyoto Protocol nor taken federal action to limit GHG emissions, several local government climate action campaigns have taken root over the past few years. Two of the most successful campaigns in the U.S. have been the U.S. Conference of Mayors' Climate Protection Campaign (*http://usmayors.org/climateprotection/agreement.htm*) and ICLEI-Local Governments for Sustainability's Cities for Climate Protection Campaign (*www.icleiusa.org/*). As of December 2009, 1016 local U.S. governments had signed onto the U.S. Conference of Mayors' Climate Protection Agreement, and greater than 600 U.S. cities and nearly 2900 cities globally had joined ICLEI's Climate Protection Campaign.

3. TAKING ACTION: AUSTIN'S CLIMATE PROTECTION PLAN

As a result of climate change, Austin may experience more intense precipitation events followed by more intense, and perhaps more prolonged, drought periods. Austin may also experience warmer weather year-round, with fewer freezes during the winter and a prolonged "warm" season (Union of Concerned Scientists, 2009). As a result, Austin stands to suffer from heat and disease-related health concerns, increased water and energy demands, and worsened air quality, among other climate change-induced threats.

In the face of these challenges, Austin is taking action to mitigate climate change and its effects in an effort to preserve Austinites' high quality of living. Austin has adopted local adaptations of the U.S. Conference of Mayors, Climate Protection Agreement and is a member of the ICLEI Cities for Climate Protection Campaign. In 2007, the Austin City Council adopted the ACPP, arguably one of the nation's most aggressive climate action plans.[4,5]

The ACPP strives to make municipal government operations carbon-neutral by 2020, whereas the larger community is working toward carbon neutrality by 2050. The ACPP is divided into five subplans to help the community progress toward these goals:

- ***Municipal Plan*** – Make all City of Austin facilities, vehicles, and operations carbon-neutral by 2020.
- ***Utility Plan*** – Cap CO_2 emissions from existing power plants; expand conservation and renewable energy programs; and make any new electricity generation carbon-neutral.
- ***Homes and Buildings Plan*** – Update building codes for new buildings to be the most energy-efficient in the nation; pursue energy efficiency upgrades for existing buildings; and enhance local Green Building program.
- ***Community Plan*** – Engage Austin citizens, community groups, and businesses to reduce the community's carbon footprint.
- ***"Go Neutral" Plan*** – Provide tools and resources for citizens, businesses, organizations, and visitors to measure and reduce their carbon footprint.

What Are Greenhouse Gases?

GHGs are gases that absorb outgoing radiation and warm the Earth's atmosphere. The accumulation of these GHGs in the atmosphere contributes to global climate change. Austin measures human-induced emissions of the following long-lived GHGs:

- **Carbon dioxide (CO_2)** – Primarily emitted from the combustion of fossil fuels.
- **Methane** – Primarily emitted from the combustion of fossil fuels; the decay of organic materials, for example, in landfills and feedlots; digestion of ruminant animals (e.g., cows); and wastewater treatment.
- **Nitrous oxide** – Primarily emitted from fossil fuel combustion, wastewater treatment, the application of fertilizers, and feedlots.

(Continued)

4. Most governments and other entities refer to "climate action plans." Austin prefers the term "climate protection plan." The two phrases are used interchangeably throughout this chapter.

5. The text of the Austin Climate Protection Plan is available at *http://www.ci.austin.tx.us/acpp/downloads/acpp_res021507.pdf*

What Are Greenhouse Gases?—cont'd

- **Perfluorocarbons** – Primarily emitted from aluminum smelting, semiconductor manufacturing, and systems that use refrigerants and fire suppressants.
- **Hexafluorocarbons** – Primarily emitted from semiconductor manufacturing and systems that use refrigerants and fire suppressants.
- **Sulfur hexafluoride** – Primarily emitted from electric transmission and distribution equipment, semiconductor manufacturing, and magnesium production.

CO_2 is emitted in the largest quantities, but the other GHGs have a larger per-unit impact in causing global warming. To account for these differences in "global warming potential" among the gases, cumulative GHGs are reported in CO_2-equivalents (CO_2-eq.).

3.1. Measuring Austin's Carbon Footprint

Measuring the sum of GHGs an entity contributes to the atmosphere –referred to as a "carbon footprint" or a "GHG emissions inventory" –is a critical first step for taking action to reduce an entity's climate impact.". To better understand the community's contribution to local GHGs, ACPP staff prepared an inventory of the major GHG emission sources in Travis County, the county in which Austin resides. The county was selected as the boundary for the community's carbon footprint because the City provides services outside the city limits and because Travis County captures the majority of the city limits and extended service area. The City of Austin (highlighted in Figure 1) is shown in relation to the municipal utility's electric service delivery area (bordered by the solid line) and Travis County (bordered by the broken line) in Figure 1.

2007, the year in which the ACPP was adopted, was selected as the base year against which future changes in the community's carbon footprint will be measured. Total 2007 GHG emissions for Travis County were approximately 15.7 million metric tons of CO_2-eq. Ninety percent of the emissions came from energy use, with nearly 50% from building energy use and another 40% from transportation. Waste contributed the remaining 11%. Building energy use includes emissions from electricity[6] and natural gas consumption by residential, commercial, and industrial sectors. Transportation includes emissions from on- and off-road vehicles, passenger and freight trains, and air travel in the region. Waste includes emissions from landfills and wastewater treatment plants. Figure 2 shows the breakdown of these emissions in the community.

The average Travis County resident emitted 15 metric tons of CO_2 emissions from energy use (electricity, natural gas, and transportation fuel) in 2007.

---

6. Electricity emissions also capture the embodied energy use for treating, transporting, and heating water.

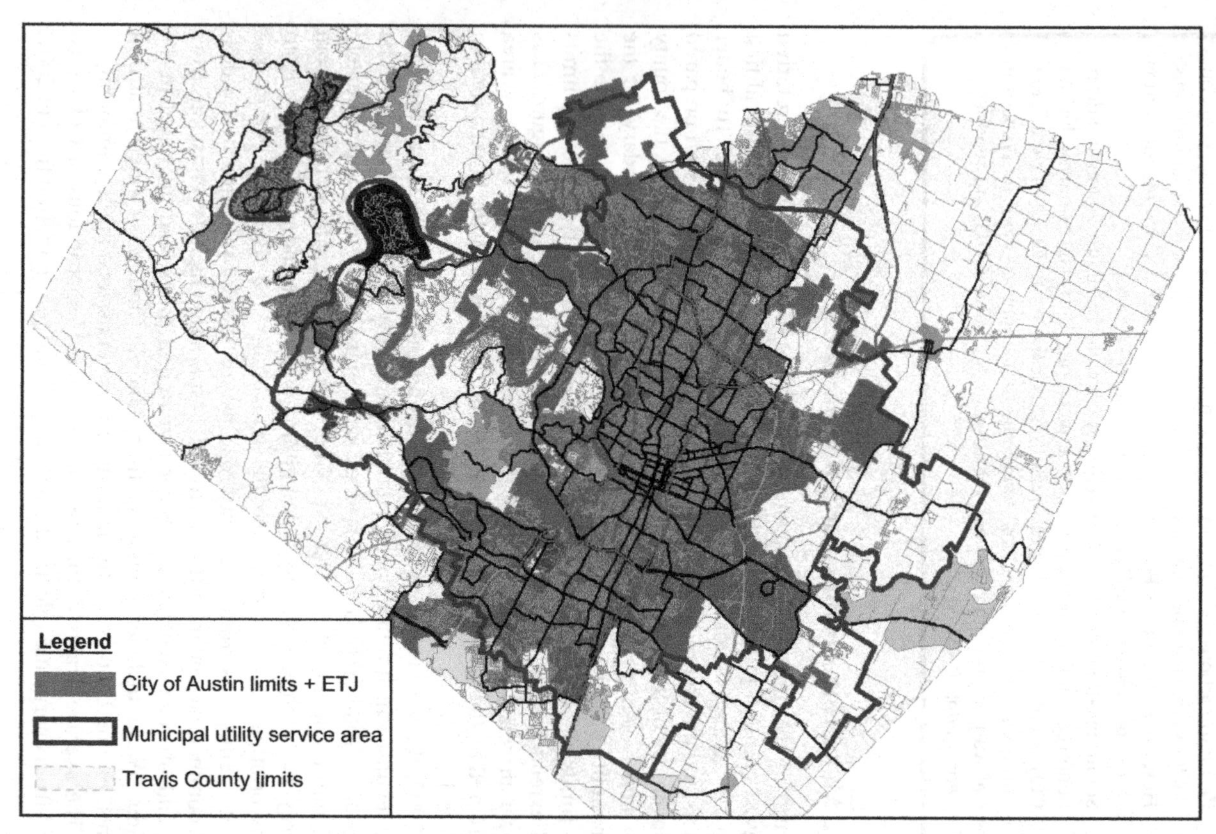

FIGURE 1 Map of City of Austin, Municipal Utility Service Area, and Travis County. *Source: City of Austin*

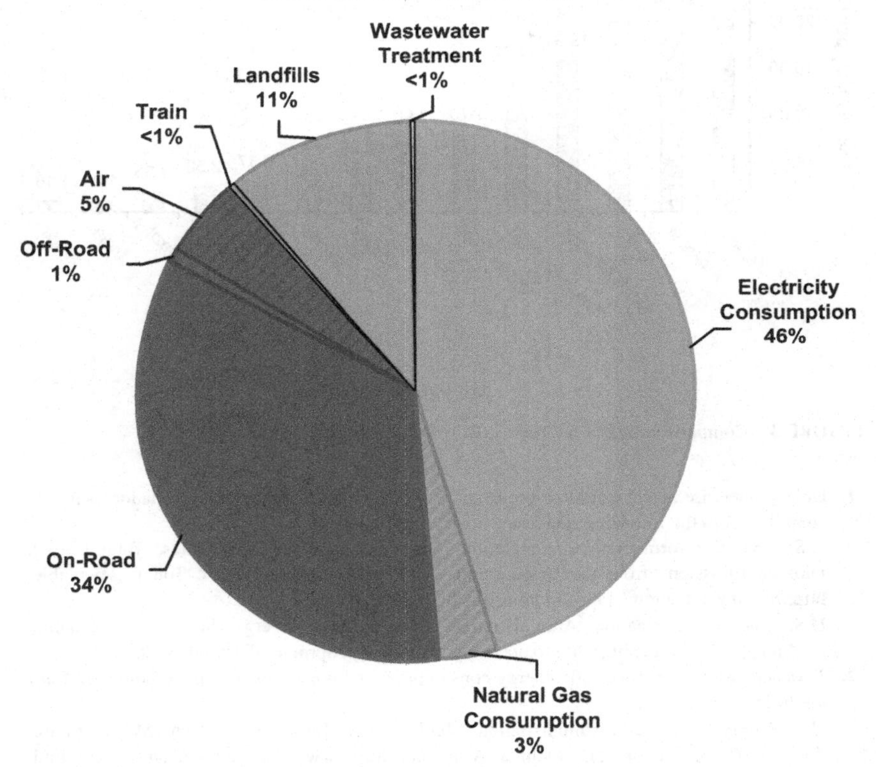

FIGURE 2 Austin Area (Travis County) Carbon Footprint, 2007. *Source: City of Austin*

Figure 3 compares the average energy-related per capita carbon footprint for Travis County to that of other geographic jurisdictions. Because of the region's lack of heavy industry, a fairly energy conscious community, and a relatively clean electricity fuel mix, Travis County's average energy-related per capita carbon footprint is half of the Texas average (27 metric tons of CO_2) and only three-quarters of the U.S. average (20 metric tons of CO_2), but is over three times the world average (5 metric tons of CO_2).

3.1.1. The City of Austin Leading the Charge

Because the City of Austin provides electric, water and wastewater, and solid waste collection services to the community, the government has a sizeable role to play in helping the community reduce its carbon footprint. For this reason, the City began taking action to address its internal operations before asking the

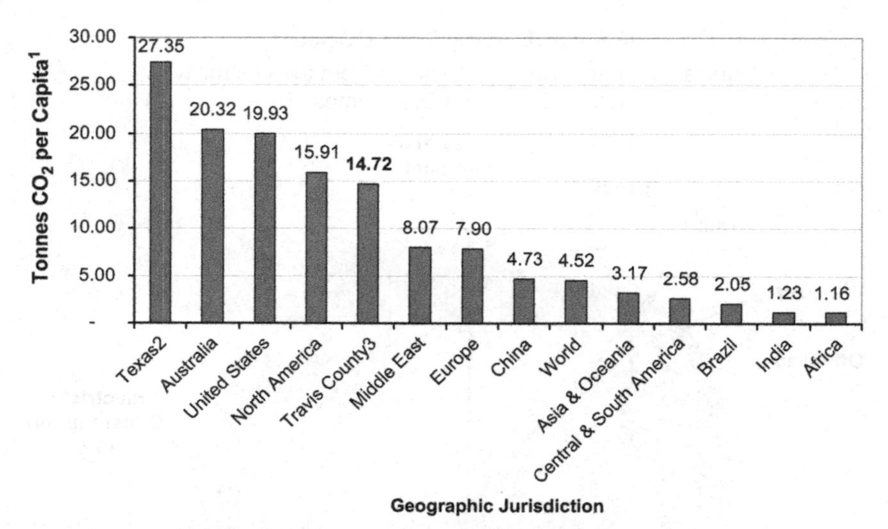

FIGURE 3 Comparison of Energy-related CO_2 Emissions per Capita.
Notes:

1. Unless otherwise noted, emissions are from 2007 energy consumption and population estimates compiled from the following sources:
U.S. Energy Information Administration, "International Energy Statistics, Total Carbon Dioxide Emissions from the Consumption of Energy (Million Metric Tons)." Available: http://tonto.eia.doe.gov/cfapps/ipdbproject/IEDIndex3.cfm.
U.S. Energy Information Administration, "International Energy Statistics, Population (Millions)." Available: http://tonto.eia.doe.gov/cfapps/ipdbproject/IEDIndex3.cfm.
2. Texas emissions are from 2005 energy consumption and population estimates compiled from the following sources:
U.S. Energy Information Administration, "Table 3. State Emissions by Year (Million Metric Tons of Carbon Dioxide)." Online. Available: http://www.eia.doe.gov/environment.html. Accessed: April 21, 2009.
U.S. Census Bureau, "Table 1: Annual Estimates of the Population for the United States and States, and for Puerto Rico: April 1, 2000 to July 1, 2005," (NST-EST2005-01). Online. Available: http://www.census.gov/popest/states/NST-ann-est2005.html. Accessed: April 21, 2009.
3. Travis County emissions are from 2007 energy consumption and population estimates prepared by the Austin Climate Protection Program.

Source: City of Austin

community to do the same. The City is taking a top-down and bottom-up approach to embedding climate change as a key consideration in the day-to-day activities of the City's 13,000 employees.

To spur individual employee action, the City offers an interactive employee training seminar on ways individuals can reduce their climate change impact at work and at home. In addition, departments are taking charge through the creation of department-level climate protection plans. Most governments have climate action plans that span departments. However, Austin is taking a unique approach in which it is empowering each department to develop customized plans that take into account its unique scope of services, emission sources, and logistical/budgetary concerns. In some cases, building climate protection plans have been developed to guide building occupants' operations where no single department has significant operational control over its energy and water use or waste disposal options.

Reducing Austin's Carbon Footprint One Dollar at a Time

Collectively, the quantifiable goals set by the 23 departments and five buildings have the potential in the first year (2010) to reduce:

- 5.8 million kilowatt-hours (kWh) of electricity, avoiding $462,000 in electricity costs
- 48,000 hundred cubic feet (ccf) of natural gas, avoiding $32,000 in heating costs
- 53.6 million gallons of water, avoiding $225,000 in water costs
- 194,300 gallons of transportation fuel, avoiding $619,800 in fuel costs
- 337,300 metric tons of GHG emissions (relative to CY2007 baseline)

The 337,300-metric ton reduction in GHGs in 2010 reduces the City's total municipal operations' carbon footprint by 5% compared to 2007 levels and potentially avoids $1.3 million in operating costs. If this level of reduction is achieved each year between 2010 and 2020, the City can expect to cut its carbon footprint by nearly 50%.

The departmental and building climate protection plans seek to reduce GHG and other air pollutant emissions from the following sources: energy, water, and transportation fuel use; procurement and materials management; and waste. Employee education is also a critical component to ensure the long-term success of the plans in engraining climate-conscious decision-making and behavior change within the departments. The plans are available at: *www.coolaustin.org*.

While the departmental and building climate protection plans provide a bottom-up approach to climate protection, a City-wide perspective is being maintained to identify and address universal barriers hindering the City's ability to become carbon-neutral by 2020. The remainder of this section highlights emission reduction projects for each emissions source category addressed in the departmental and building climate protection plans.

3.2. Energy

To minimize energy use in new buildings, the City requires all new municipal buildings to achieve a Leadership in Energy and Environmental Design (LEED) silver rating. Additional sustainability requirements apply for buildings that are not eligible for LEED rating, as well as for renovation projects and ongoing building operations and maintenance.[7]

To further reduce energy use from existing buildings, the City's municipally owned electric utility, Austin Energy, 2008 and 2009, operates a Municipal Energy Conservation Program that has avoided greater than 46 million kWh and 25,600 metric tons CO_2-eq. over the past two decades. Austin Energy is investing $7.5 million in federal Energy Efficiency and Conservation Block Grant funding in additional energy efficiency and energy management improvements at multiple City facilities over the next two years. Projects include lighting and lighting control retrofits and building commissioning to optimize energy use in buildings. These projects are expected to avoid 5 million kWh; 2800 metric tons CO_2-eq.; and $400,000 in electricity costs per year.

Some departments have uncovered additional energy savings through in-house projects. For example, the City's Information Technology and Telecommunications (ITT) department is installing power management software in all noncritical computers and monitors. Power management software allows ITT personnel to automatically set computers and monitors to a low-power state during periods of inactivity without interfering with their ability to install system updates and patches. The software is expected to annually avoid 3.5 million kWh; 1950 metric tons of CO_2-eq.; and $280,000 in electricity costs. In addition, the City is undertaking a number of energy efficiency upgrades at its main data center. The upgrades are expected to annually avoid 393,000 kWh; 220 metric tons of CO_2-eq.; and $32,000 in electricity costs.

For energy use that cannot be avoided, the ACPP requires the use of emissions-free renewable energy in all municipal facilities by 2012. As of December 2009, nearly 20% of the City's total electricity use came from renewable resources, avoiding 40,100 metric tons CO_2-eq. per year. Obtaining 100% of the City's electricity from renewable energy will reduce the community's collective carbon footprint by 238,000 metric tons CO_2-eq. per year (assuming 2009 energy consumption levels).

3.3. Water

The City has installed water-saving technologies, including low-flow faucets, toilets, urinals, and showerheads, in many of its municipal buildings. Water conservation provides the dual benefit of preserving a limited resource in a frequently drought-affected region and saving energy due to the interrelated nature of energy and water use.[8]

---

7. Austin's sustainable building requirements are available at *www.ci.austin.tx.us/publicworks/sustainability/default.htm*.

8. Treating and transporting water and wastewater requires significant amounts of energy. Roughly 50% of the City of Austin's electricity use is used to treat and pump water.

One way the City is cutting back on water demand is through the use of green infrastructure. Green infrastructure refers to the "use of natural systems and engineered systems [that] mimic natural processes to enhance environmental quality and provide utility services" (City of Austin, *Green infrastructure*, 2009). Drought tolerant plants are employed for landscaping of public property, rain gardens are being incorporated into City parks to naturally treat runoff and protect water quality, and untreated river water or reclaimed water is being used for irrigation. These activities avoid GHG emissions by reducing demand for treated water.[9]

3.4. Transportation

Achieving carbon neutrality will require all City vehicles to be powered by biofuels or electricity from emissions-free renewable energy. Until such options are available for all of the City's transportation needs, the City is striving to downsize the existing fleet, select new vehicles with the lowest carbon footprint available through fuel-efficient design and use of alternative fuels, and promote environmentally responsible driver behavior.

One innovative program to help downsize the fleet is a car-sharing program called Car2Go. The City has partnered with Daimler to offer 200 fuel-efficient Smart Cars to City employees during a six-month trial period. Halfway through the trial, use of the Smart Cars had avoided an estimated 680 gallons of fuel use and 6 metric tons of CO_2-eq. Similarly, City Cycle, an employee bike-sharing program, helped the City of Austin achieve designation as a "Bicycle Friendly Business."

The City is also increasing its portfolio of alternative fuels. In 2009, 55% of City vehicles were alternative fuel-capable. The City provides E85, which is a blend of 85% ethanol and 15% gasoline; B20, which is a blend of 20% plant-based diesel and 80% petro-diesel; compressed natural gas (CNG); propane; and electricity[10] as transportation fuels. Figure 4 shows the distribution of vehicles by fuel type.

Fuel-efficient driving tips are included in the City's mandatory driver training classes, and additional online training opportunities are in development. Similarly, some departments have initiated fuel conservation competitions among work groups, in some cases leading to fuel savings of up to 10%.

9. In addition to the energy intensity of water purification and transport, the water treatment process emits fugitive emissions of methane and nitrous oxide, which respectively have 21 and 310 times the heat-trapping ability of CO_2.

10. Austin Energy, the City's municipally owned electric utility, initiated a campaign, Plug-In Partners (*www.pluginpartners.org/*), in 2005 to encourage U.S. auto makers to develop plug-in hybrid-electric vehicles (PHEVs) as a transition to all electric vehicles. The City has committed to purchase PHEVs as soon as they are economically available and currently operates greater than 50 all electric vehicles, ranging from flatbed trucks to forklifts.

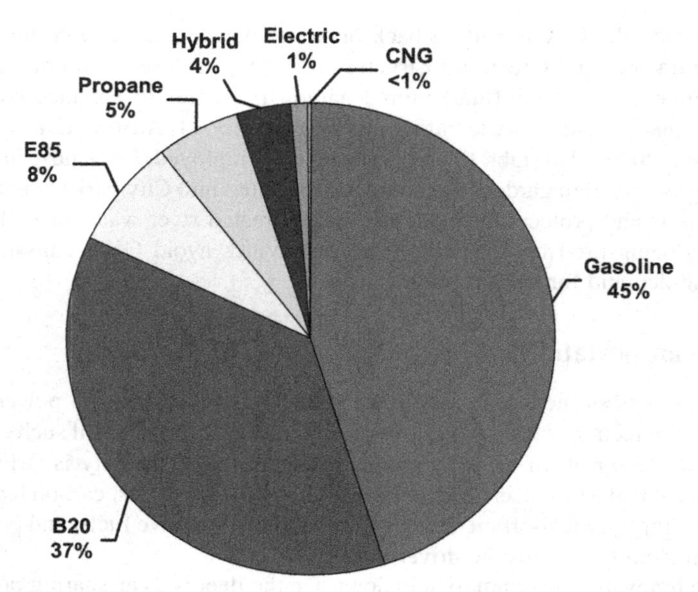

FIGURE 4 City of Austin Vehicles by Fuel Type. *Source: City of Austin*

3.5. Procurement and Materials Management

The City began developing a Responsible Purchasing Program in late 2008. The Purchasing Office is updating contract language to incorporate "Best Value" evaluation criteria that allow for the consideration of environmental impact and other nonfiscal measures in addition to cost. Two major successes of the Responsible Purchasing Program are use of alternative fuel landscaping services contracts and purchasing protocols for custodial supplies and services that outline cleaning methods and products that are protective of human and environmental health.

3.6. Waste

Austin's Zero Waste Initiative seeks to reduce or divert the amount of waste that is sent to area landfills by 90% by 2040. The City is contributing to this effort internally by colocating all-in-one recycling bins next to trash cans (labeled "Landfill Trash") throughout City facilities and at City-sponsored events. Results from a waste audit before and after implementation of this colocation effort at Austin's City Hall revealed a 23% reduction in total daily waste volume and a 42% increase in the daily volume of waste recycled.

In addition, the City is trying to cut down on paper use by replacing plan and permit submittals with an electronic system. In 2007, less than 400 U.S. state and local governments accepted electronic plans and fewer than 50 reviewed plans electronically (Alliance for Building Regulatory Reform in the Digital Age, 2007). The City's Public Works Department is trying to increase this number by piloting a program to convert its historically paper-based project bid system to an electronic process. Moving to an electronic bid system is conservatively estimated to reduce GHG emissions by 9 metric tons of CO_2-eq. per public works project.

The City is also revising its contracts to show preference for vendors who promote product stewardship through the use of recycled content, minimized packaging, and product take-back programs.[11]

4. ENGAGING THE COMMUNITY TO ACHIEVE CARBON NEUTRALITY

The City is promoting a parallel effort within the community to promote regional GHG reductions, building on the combination top-down, bottom-up approach modeled by the municipal government. Support is being built from the bottom-up through a community-driven Climate Action Plan that will identify actions the community can take to reduce its climate impact in the areas of energy, water, transportation, materials management, land use, and climate preparedness. The Community Climate Action Plan, which is expected to be finalized in early 2011, will focus on short-term and medium-term mitigation and adaption strategies, while longer-term strategies will be incorporated in the City's Comprehensive Plan, currently under development.[12]

From a top-down perspective, the municipal government is spearheading a number of climate protection initiatives. This section highlights the City's initial efforts to help the community reduce its impact in the areas of energy, transportation, and waste.

4.1. Energy

Building energy use comprises half of the Austin area's carbon footprint (Figure 2). Electricity use, rather than natural gas or other heating fuel, is the primary driver of energy-related GHG emissions in Austin. Fortunately, Austin owns its electric utility, giving the citizens direct control over their power source.[13]

---

11. Information about Austin's Zero Waste Initiative is available at *http://www.ci.austin.tx.us/sws/zerowaste.htm*.

12. Information about the development of the City's comprehensive plan is available at *www.ci.austin.tx.us/compplan/default.htm*.

13. Two other electric utilities provide electricity to a small portion of customers in Travis County, which is the boundary for measuring Austin's carbon footprint. The City of Austin does not have influence over these utilities; therefore, this chapter focuses on efforts Austin Energy, as a department of the City of Austin, and members of the community are undertaking to reduce their energy-related carbon footprint.

Austin Energy provides power through a diverse generation mix, consisting of coal, natural gas, nuclear, and renewable energy sources. Austin Energy has the top-performing renewable energy program in the nation, the nation's first and largest green building program, and is home to one of the nation's most comprehensive residential and commercial energy-efficiency programs. These programs each have active roles in meeting Austin Energy's climate protection goals.

The contribution of electricity use to the community's carbon footprint can be modified through changes in customer demand for energy, Austin Energy's generation mix, and how efficiently Austin Energy produces and delivers power to its customers. Austin Energy hosted a year-long public participation process in which Austin Energy customers were invited to weigh in on how the utility plans to reduce its GHG emissions in the first two areas (energy conservation and renewable energy) between 2010 and 2020, which is Austin Energy's current planning horizon.

The year-long process culminated with the adoption of Austin Energy's Resource, Generation, and Climate Protection Plan.[14] The plan outlines a generation mix that will enable the utility to reduce its GHG emissions to 20% below 2005 levels in 2020, while meeting the community's growing energy needs.

The proposed plan would add 1012 net MW of new capacity to Austin Energy's existing 2925-MW portfolio. All but 300 MW of load-balancing natural gas generation are proposed to be carbon-neutral. Table 1 shows Austin Energy's current installed capacity and outlines proposed capacity additions by energy source through 2020. This plan relies on 800 MW of peak demand reduction through new and expanded demand side management programs and energy code changes. This 800 MW goal exceeds the original ACPP goal by 100 MW of additional demand savings. Austin Energy responded to the public's request for a larger investment in energy conservation and solar and wind energy and a reduced reliance on coal. If this plan is implemented, Austin Energy's renewable energy portfolio would rise from 11% in 2009 to 35% in 2020, surpassing the ACPP's 30% renewable energy requirement. The Resource, Generation, and Climate Protection Plan additionally raises Austin Energy's solar capacity goal from 100 to 200 MW. Coal's contribution to the utility's resource mix would fall from 32% in 2009 to 23% in 2020. This percent share reduction would be achieved by reducing the capacity factor[15] of Austin Energy's jointly owned power plant to 60% from 90%–95% currently. Reducing output at its coal plant is critical as it represents over 70% of the

14. Austin Energy's Resource, Generation, and Climate Protection Plan is available at *www.austinenergy.com/*. Details about the year-long public participation process to develop the plan is available at *www.austinsmartenergy.com/*.

15. Capacity factor is the energy a generating unit produces in a year divided by the total amount of energy it could produce if it ran at maximum output.

TABLE 1 Austin Energy's Proposed Generation Resources, 2009 Through 2020

Year	Coal (MW)	Gas (MW)	Nuclear (MW)	Biomass (MW)	Wind[1] (MW)	Solar (MW)	Renewable portfolio
2009	607	1444	422	12	439	5	10.5%
2010		100				30	12.5%
2011					(77) / 200		17.7%
2012				100			22.2%
2013					150		26.2%
2014						30	26.4%
2015		200			100		28.7%
2016				50		20	31.6%
2017					(126) / 200	30	35.0%
2018						20	33.6%
2019						30	33.7%
2020					115	40	36.7%
TOTAL	**607**	**1744**	**422**	**162**	**1001**	**205**	**N/A**

[1]*Notes: Values in parentheses represent wind contracts that expire.*
Source: Austin Energy, *Resource & Climate Protection Plan to 2020, Recommendations & Plan.* Presentation by Roger Duncan to Austin City Council, August 18, 2009, slide 15.

utility's GHG emissions. Figure 5 shows Austin Energy's resulting energy portfolio in 2020, if the plan is fully implemented, compared to its 2009 portfolio.

If fully implemented, the Resource, Generation, and Climate Protection Plan would reduce Austin Energy's carbon footprint from 5.6 million metric tons of CO_2-eq. in 2005 to 4.5 million metric tons of CO_2-eq. in 2020 (Figure 6). This 20% decrease in emissions would be achieved while meeting the energy needs of a 30% larger population, reflecting a nearly 40% decline in per capita GHG emissions.

4.1.1. Reducing Customer Demand for Energy

Austin Energy has avoided the need to build a 600 MW coal power plant through energy efficiency and conservation over the past two decades. The utility is seeking to reduce its customers' peak demand for energy by an

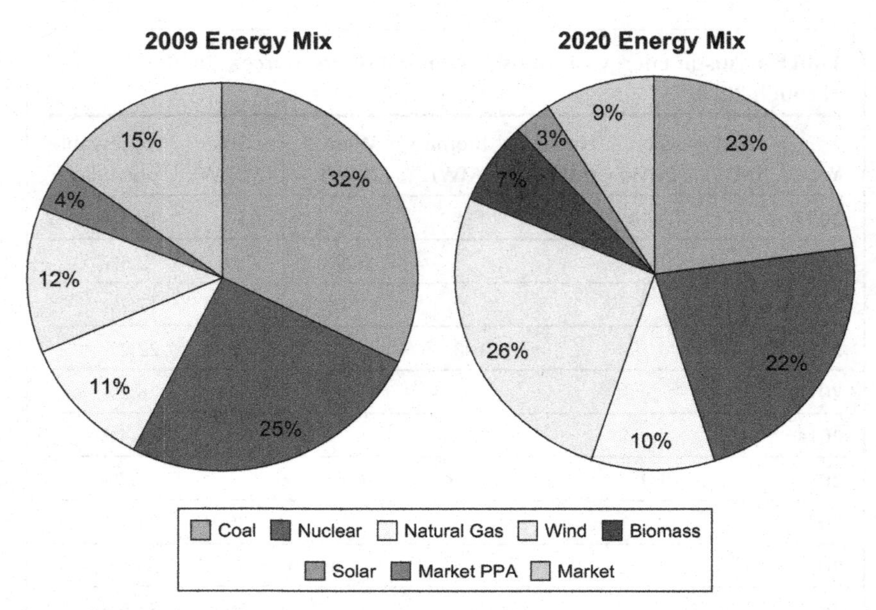

FIGURE 5 Austin Energy's Energy Portfolio, 2009 and 2020. *Source: Austin Energy, Resource & Climate Protection Plan to 2020, Recommendations & Plan. Presentation by Roger Duncan to Austin City Council, August 18, 2009, Slide 17.*

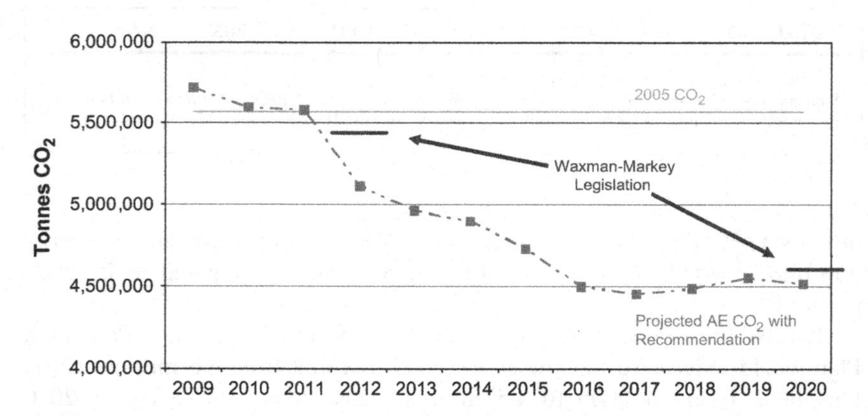

FIGURE 6 Projected Annual Decrease in Austin Energy's GHG Emissions, 2009 through 2020. *Source: Resource & Climate Protection Plan to 2020, Recommendations & Plan. Presentation by Roger Duncan to Austin City Council, August 18, 2009, Slide 18.*

additional 800 MW by 2020. From 2007 through 2009, Austin Energy has reduced peak demand by 184 MW, which accounts for 23% of the 800 MW goal. This reduced demand translates to 352,000 MWh of avoided energy use and 193,200 metric tons of avoided CO_2-eq.

If the 800 MW demand savings goal is reached in 2020, 1.6 million MWh and 499,000 metric tons of CO2-eq. could be avoided. Austin Energy's Energy Efficiency and Green Building programs are implementing a number of innovative programs for new and existing buildings to achieve these savings.

The ACPP established a goal of making all new single-family homes "zero net energy capable" by 2015 and provided energy-efficiency targets for other buildings within Austin.[16] These goals are being achieved through local building energy codes. With the addition of on-site renewable energy generation, a zero net energy home will use only as much energy as it generates over the course of a year. To be energy self-sufficient, Austin has determined that a single-family home must be 65% more efficient than a home built in Austin prior to the first round of code changes in 2007. New homes built after April 2010, when the second round of code changes were adopted, are estimated to be 34% more energy-efficient compared to homes built to code in 2007. Additional code changes will be introduced in 2012 and 2015.

By 2015, other new buildings in Austin will be 75% more energy-efficient by code than they would have been in 2007. The combined first and second round code changes are expected to increase the overall efficiency of new multifamily and commercial buildings by 32% compared to similar buildings built to code in 2007.

Austin Energy seeks to reduce energy use in Austin's existing buildings through its demand-side management programs, which drive down citizens' energy use during peak demand periods in the late afternoon and evening, and through a new ordinance requiring existing buildings to receive an energy audit within a certain timeframe or upon sale of the building. The Energy Conservation Audit and Disclosure (ECAD) Ordinance took effect June 1, 2009 and requires mandatory energy audits at the time of sale for single-family homes and within two years of the ordinance's effective date for commercial and multifamily residential buildings. The intent of the ordinance is to provide information on building energy use to prospective tenants and buyers and to encourage energy efficiency improvements.

Single-family and multifamily homes must undergo a standardized energy audit to assess the efficiency of the building envelope, including insulation, ducts, air infiltration, and windows and doors. The audit also makes recommendations for areas where the property's energy efficiency could be improved and provides information on local incentives to facilitate the upgrades. As of December 2009, 2600 audits of single-family homes have been performed, with over 100 properties participating in Austin Energy's rebate programs after completing the single-family audit. Multifamily audits are required by June 2011. Multifamily properties that exceed 150% of the average energy use per

16. This book's Chapter 9 by Gray and Zarnikau and Chapter 17 by Rajkovich et al. cover the topic of zero net energy capable buildings.

square foot of multifamily properties within Austin Energy's service territory must reduce energy use to bring the property's energy intensity rating to within 110% of the area average.

Commercial buildings are required to obtain an ENERGY STAR® rating using the U.S. Environmental Protection Agency's Portfolio Manager or comparable tool. Portfolio Manager compares a building's energy use to the energy use of similar building types. A rating of 50 (out of 100) indicates that the building uses energy more efficiently than 50% of buildings in its peer group. Commercial building ratings have ranged from a low of 20 to a high of 95. Austin Energy is encouraging property owners to improve the operating efficiency of their buildings to strive toward a 75 rating, which would make the property 75% more energy-efficient than similar property types.

In addition to its energy use and emission reduction benefits, the ordinance has been successful in creating and sustaining local green jobs. Austin Energy has registered 184 individual auditors, representing 142 participating local audit firms. At least a third of the auditors started in the energy audit business as a result of the ECAD Ordinance.[17]

4.1.2. Ramping Up Renewable Energy Use

Austin Energy plans to obtain 35% of its energy needs from renewable resources by 2020. A related goal seeks to expand installed solar capacity to 200 MW by 2020. Renewable energy emits no climate-forcing GHG emissions, thereby reducing Austin Energy's carbon footprint. If the 35% renewable energy goal is attained through purchases or direct ownership, renewable energy consumed by Austin Energy customers will reduce the community's emissions by 1.7 million metric tons of CO_2-eq., or 20% below 2005 emissions, in 2020.

Wind is the dominant renewable energy resource in Texas and comprises 95% of the total renewable energy sources brought into Austin Energy's system. However, wind is a variable resource that provides the greatest amount of energy when it is least needed (overnight),[18] so it does little to meet peak demand. Additionally, wind energy supply has outpaced the state's transmission system's ability to deliver the wind energy from sparsely populated West Texas where the bulk of the wind farms are located to the populated eastern portion of the state, including Austin. This relationship has caused a bottleneck in transporting wind energy, leading Austin Energy to pay costly congestion fees to get the wind energy to its customers.

---

17. The ordinance is further described at *www.austinenergy.com/About%20Us/Environmental%20Initiatives/ordinance/index.htm.*

18. Austin Energy currently purchases wind power from West Texas wind farms, where the wind blows strongest during the nighttime. However, Austin Energy is researching the feasibility of also obtaining some of its wind power from offshore wind farms along the Gulf Coast, which could provide a better match of wind power supply and customer demand for power during the day.

Because of the variability and congestion issues associated with wind, as well as concerns about being too dependent on any one resource, Austin Energy is seeking to diversify the type and geographic location of its renewable energy resources. The Austin City Council approved power purchase agreements from two large, privately owned renewable energy projects in 2008. The first is a 100 MW biomass power plant, the largest wood waste-burning power plant in the nation. The biomass plant is expected to be online in 2012, providing base load power to Austin Energy's customers 24/7. A 30 MW solar photovoltaic (PV) facility is expected to be online by the end of 2011 and will provide energy sufficient to power about 5000 homes each year. Additional new generation projects outlined in Austin Energy's proposed Resource, Generation, and Climate Protection Plan will have to be approved by the Austin City Council.

In addition to utility-scale solar applications, Austin Energy's solar PV rebate program is a key mechanism for achieving the utility's 100 MW to 200 MW solar goal. From 2004 through 2009, Austin Energy solar rebate program participants installed 4.6 MW of solar PV capacity, avoiding 11,500 MWh of "brown" energy and 5700 metric tons of CO_2-eq. emissions.

The solar rebate program's $4 million annual budget has not been able to keep up with Austin's skyrocketing demand for solar. Nearly 2 MW, or 40% of the total installed capacity, was added in 2009 alone, causing the program to lower the residential customer rebate amount from a high of $4.50 per watt to $2.50 per watt. Commercial customers are now offered a performance-based incentive rather than an upfront rebate. Performance-based incentives pay customers for energy produced over a fixed time period rather than paying for the upfront installation costs. This design ensures Austin Energy is paying for actual renewable energy produced (and therefore avoiding the need for fossil fuel generation), and because the incentive rate decreases over time, production-based models can provide more sustainable long-term funding.

The Austin City Council also recently adopted an ordinance that allows Austin Energy to develop a property assessed solar financing mechanism. Under this model, the City issues bonds to support solar installations, securing a lower interest rate and helping to defray the upfront capital costs of the system. Participating citizens then repay the amount the City borrowed on their behalf through property tax assessments.

4.2. Transportation

Transportation energy use comprises nearly 40% of the Austin area's carbon footprint (Figure 2), with on-road vehicles being the biggest transportation contributor. Figure 7 shows the breakdown of each transportation mode's contribution to area GHG emissions.

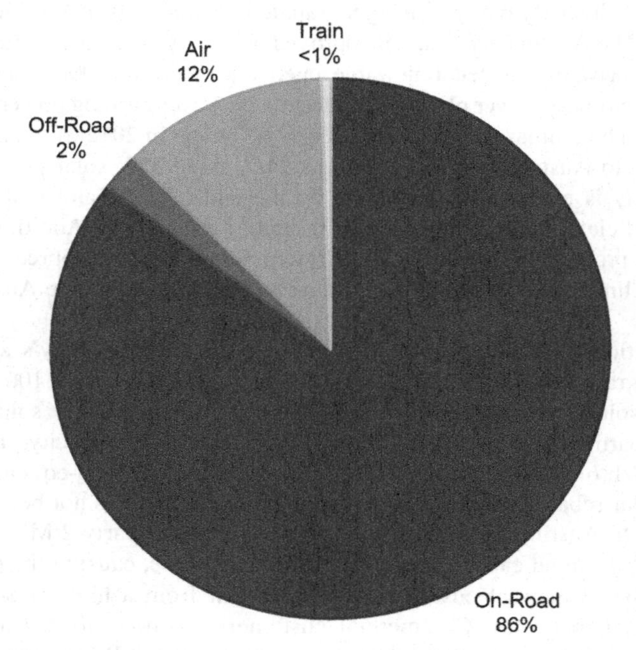

FIGURE 7 Austin Area (Travis County) Transportation Carbon Footprint, 2007. *Source: City of Austin*

The City of Austin and its regional partners are focused on reducing GHG emissions from the transportation sector through the following methods:

- Reducing vehicle miles traveled
- Increasing the use of alternative fuels
- Discouraging idling

The remainder of this section discusses Austin's efforts in each of these areas.

4.2.1. Reducing Vehicle Miles Traveled

The City of Austin promotes high density, mixed use, and transit-oriented development through zoning and incentives.[19] The City also partners with the local metropolitan planning organization and air quality entities to develop and

19. Learn more about Austin's compact, mixed-use development efforts at *www.ci.austin.tx.us/planning/tod/*.

promote alternative transportation options, such as a ride-matching service[20] and transit training classes, throughout the Austin region. The local transit authority opened the region's first commuter rail line in March 2010, and the City is exploring additional in-town rail options. The City also promotes a car-share program through the allocation of a growing number of on-street parking spaces to car-share vehicles.[21]

To promote nonmotorized modes of travel, the City administers a Bicycle and Pedestrian Program that promotes walking and biking through a safe, interconnected network of sidewalks, trails, and bicycle lanes.[22] As a League of American Bicyclists' Bicycle Friendly Community, Austin is striving to make Austin a "world-class bicycling city" (Austin 2009 Bicycle Plan Update, 2009). Austin has set a goal to increase the percent of Austin commuters who bicycle to work to 2% by 2015 and 5% by 2020, with higher targets for central city commuters.

If the City is successful in increasing bicycle commute share to 5% in 2020, 35,000 metric tons of CO_2-eq. could be avoided in each subsequent year. In comparison, Copenhagen residents make 37% of commute trips, and 55% of total trips, by bicycle. Table 2 lists the top 30 bicycle-friendly cities in the developed world, as measured by the percent of trips taken by bicycle. All but three of the top 30 are in Europe, with only one U.S. city (Davis, California) reporting a double-digit bicycle ridership percentage.

4.2.2. Increasing the Use of Alternative Fuels and Discouraging Idling

The City of Austin sponsors a local chapter of the U.S. Department of Energy's Clean Cities Program. The Clean Cities Program develops public-private partnerships to promote fuel conservation, alternative fuels, and alternative fuel vehicles. Locally, the program has helped spur the operation of greater than 20 alternative fuel stations and thousands of alternative fuel and hybrid-electric fleet vehicles. The program also promotes fuel conservation through anti-idling awareness and adoption of idle reduction technology.

4.3. Waste

City of Austin operations represent a fraction of the solid waste generated by the community. Therefore, the City has focused its education and outreach efforts on promoting zero waste among residents, local businesses, and other

---

20. The region operates a ride-matching service at *www.rivercitiesrideshare.com/en-US/*.

21. The car-share program Car2Go (*www.car2go.com/austin/en/*) currently operates in the Austin area.

22. Learn more about Austin's bicycle and pedestrian program at *www.ci.austin.tx.us/publicworks/ncd.htm*.

TABLE 2 Top 30 Most Bicycle-Friendly Cities

City	% of Trips Taken by Bicycle
Copenhagen, Denmark	55%
Gronningen, Netherlands	55%
Greifswalk, Germany	44%
Assen, Netherlands	40%
Amsterdam, Netherlands	40%
Münster, Germany	40%
Utrecht, Netherlands	33%
Ferrara, Italy	30%
Malmö, Sweden	30%
Linköping, Sweden	30%
Västerås, Sweden	30%
Odense, Denmark	25%
Basel, Switzerland	25%
Osaka, Japan	25% (est)
Parma, Italy	25%
Bologna, Italy	20%
Oulu, Finland	20%
Rotterdam, Netherlands	20%
Berne, Switzerland	20%
Tübingen, Germany	20%
Aarhus, Denmark	20%
Tokyo, Japan	20% (est)
Pardubice, Czech Republic	18%
York, UK	18%
Dresden, Germany	17%
Munich, Germany	15%
Davis, USA	15%

TABLE 2 Top 30 Most Bicycle-Friendly Cities—cont'd

City	% of Trips Taken by Bicycle
Cambridge, UK	15%
Berlin, Germany	12%
Turku, Finland	11%

Source: Copenhagenize.com, "The World's Most Bicycle Friendly Cities." July 1, 2009.

members of the community. The City of Austin incentivizes waste minimization through a tiered rate structure for waste collection services, a free waste reduction consulting program for local businesses, and promotion of product stewardship. The City also offers a single-stream recycling program, which accepts comingled recyclables with the hope that simplifying recycling will increase recycling rates. The City further requires businesses with greater than 100 employees and multifamily complexes with greater than 100 units to provide recycling for building occupants. The City also collects yard trimmings and brush and composts these with treated sewage sludge to create Dillo Dirt™, a high quality organic fertilizer. Through these programs, the City has diverted 36% of the waste it collects from area landfills.

4.4. Community Outreach and Education

A critical component to helping the community reach carbon neutrality by 2050 is promoting behavior change through education and outreach. The primary tool available to Austin residents to help them understand the impact their daily activities have in causing climate change is the Austin Carbon Footprint Calculator.[23] The calculator offers a number of unique features not currently available in the majority of carbon footprint calculators. First, the calculator gives City of Austin utility customers the ability to automatically upload electric and water utility account information and garbage cart size. The calculator will also allow users to calculate emissions from water and wastewater usage along with emissions from energy use, solid waste generation, travel, and food consumption. The calculator provides tips and links to local programs and incentives that can assist residents in reducing their carbon footprint. A social networking component allows residents to build online communities with common goals for reducing GHGs.

23. The Austin Carbon Footprint Calculator is available at *www.ci.austin.tx.us/acpp/co2_footprint.htm*.

5. COMPARISON TO OTHER LOCAL GOVERNMENT PROGRAMS

It is difficult to compare one local government to another in the area of climate protection as they are all at different stages of development and have unique structures and program management approaches. That said, in the United States, ICLEI provides the most useful framework for comparison. ICLEI advocates a five-step approach to local climate action.

1. Develop a baseline GHG emissions inventory and forecast.
2. Establish an emissions reduction target based on historical and forecasted emissions.
3. Adopt a climate action plan to achieve the emissions reduction target.
4. Implement the climate action plan.
5. Monitor and report on climate action plan progress and reevaluate plan as needed.

ICLEI has greater than 600 member cities in the U.S. and thousands worldwide. A third of ICLEI's U.S. members have completed at least the first ICLEI climate action milestone of completing a GHG emissions inventory.[24] Three-quarters of the 200 member cities who have completed an inventory have also established an emissions reduction target, and nearly as many have developed climate action plans. Austin is one of 32 U.S. cities, as of November 2009, to have begun implementing its climate action plan (ICLEI, 2009).

The remainder of this section compares Austin's climate action approach to other ICLEI cities for each of ICLEI's five milestones.

5.1. Step 1—Develop GHG Emissions Inventory

ICLEI calls for measurement of the local government's emissions prior to setting GHG reduction goals or outlining a plan for achieving those goals. However, the ACPP adopted by the Austin City Council included a mix of GHG reduction targets and implementation strategies, which were not explicitly based on historical or projected emissions. Only after the plan was adopted was staff allocated to conduct an emissions analysis and otherwise implement the plan.

The City of Austin and other local governments worked with ICLEI, The Climate Registry, the California Climate Action Registry, and the California Air Resources Board to develop a standardized reporting protocol and verification procedure that will enable local governments of similar size and scope to compare their emissions using common metrics (e.g., metric tons

24. ICLEI recommends that member cities complete climate action planning for their municipal operations and community-wide greenhouse gas emissions. However, it includes members who have not completed any of its steps for either municipal operations or community emissions.

CO_2-eq./employee, metric tons CO_2-eq./$ spent).[25] Since the Local Government Operations Protocol was adopted in 2008, many local governments, including Austin, have used it in creating their municipal operations inventories.[26,27]

5.2. Step 2—Establish GHG Emissions Reduction Target

Most local governments (and other entities) commit to reduce their GHG emissions to some percent below a base year's emissions level. This model is based on the approach taken by the Kyoto Protocol, which committed signatory developed countries to reduce their emissions to *X* percent below 1990 levels. Austin's Climate Protection Plan is unique because its emission reduction target is not tied to a base year (although 2007 is being used as the base year for inventory quantification purposes). The ACPP established an emissions reduction goal of zero CO_2-eq. (a.k.a., carbon-neutral) by 2020 for the City's operations, and the Austin community adopted a goal of becoming carbon-neutral by 2050. Austin's commitment to carbon neutrality places it in a small group of committed cities, ranging in size from Lincoln City, Oregon, a tourist-centric coastal city with nearly 8000 residents (Haight, 2008), to Vancouver, British Columbia, home to approximately 600,000 residents (Climate Neutral Network, 2010).[28]

The Carbon Disclosure Project's survey of U.S. cities undertaking climate action plans revealed that many cities' reduction targets may be more symbolic than based on actual science or understanding of an achievable target (Carbonsense, 2008). Austin's carbon neutrality goal aligns with these findings. Governments at all levels find it challenging to strike a balance between pushing the envelope to progress toward the magnitude of GHG reductions called for by the IPCC and other scientists to protect the planet from the worst consequences of climate change and setting targets that are realistically achievable within the timeframe specified. The Carbon Disclosure Project survey found that even the most aggressive emissions reduction targets called for less than a 3% reduction per year (assuming an equivalent percentage is achieved each year). If Austin is to make its government operations carbon-neutral by 2020, it must halve its emissions each year between 2010 and 2020. Even with the City obtaining 100% of its electricity (which provides two-thirds

25. However, one should not compare local governments' carbon footprints as governments have unique structures, scope of services, and reporting preferences.

26. California cities are required to use the Local Government Operations Protocol as it is the method adopted by California for implementing the greenhouse gas reporting requirements of the state's greenhouse gas regulations.

27. The Local Government Operations Protocol can be viewed at *www.icleiusa.org/programs/climate/ghg-protocol*.

28. Lincoln City has not specified the year by which it will achieve carbon neutrality. Vancouver plans to be carbon-neutral by 2012.

of the City's GHG emissions) from renewable energy, a 50% reduction year after year does not seem feasible without the use of carbon offsets.

The Austin community is dependent on the municipal government (namely Austin Energy) to help neutralize its GHG emissions, so the community may be challenged to meet its carbon neutrality goal. However, 2050 is a more feasible timeline for achieving such an aggressive, but necessary, goal.

5.3. Step 3—Adopt Climate Action Plan

Austin's climate action approach is unique in that its Climate Protection Plan was adopted as the first step rather than the third in ICLEI's five-step model. ACPP staff is working with others throughout the City to develop more specific GHG reduction targets and detailed climate protection plans for each of the City departments and the community. These additional emissions reduction targets and action plans are based on the departments' and community's carbon footprints and a more in-depth understanding of barriers and opportunities available to each sector of the City and the community.

Growing awareness of climate change science and the urgency of taking action, as well as increased availability of resources such as local climate action toolkits, best practices guides, and a network of sustainability and energy management staff, have contributed to an increase in the number of local governments adopting climate action plans. Figure 8 illustrates the growth in the number of ICLEI member cities in the U.S. that have adopted climate action plans (shown as Milestone 3 in the light blue bars) through 2009. Prior to 2000, only a handful of cities had completed action plans. By 2005, greater than 50 cities had plans in place, and that number had more than doubled by 2009 (ICLEI, 2009).

5.4. Step 4—Implement Climate Action Plan

Austin is fortunate to have staff devoted specifically to climate protection, municipally owned electric and water utilities, and a network of related energy and environmental programs dispersed throughout the City. The ACPP provides a common thread among the City's sustainability programs. City programs with a climate protection component include those designed to reduce energy consumption in City facilities (Austin Energy's Municipal Energy Management Program) and throughout the community (Austin Energy's energy efficiency and Green Building programs);[29] promote renewable energy development and use (Austin Energy's GreenChoice program);[30]

29. Learn more about the City's energy efficiency and green building programs at *www.austinenergy.com/energy%20Efficiency/index.htm*.

30. Learn more about the City's retail renewable energy program at *www.austinenergy.com/Energy%20Efficiency/Programs/Green%20Choice/index.htm*.

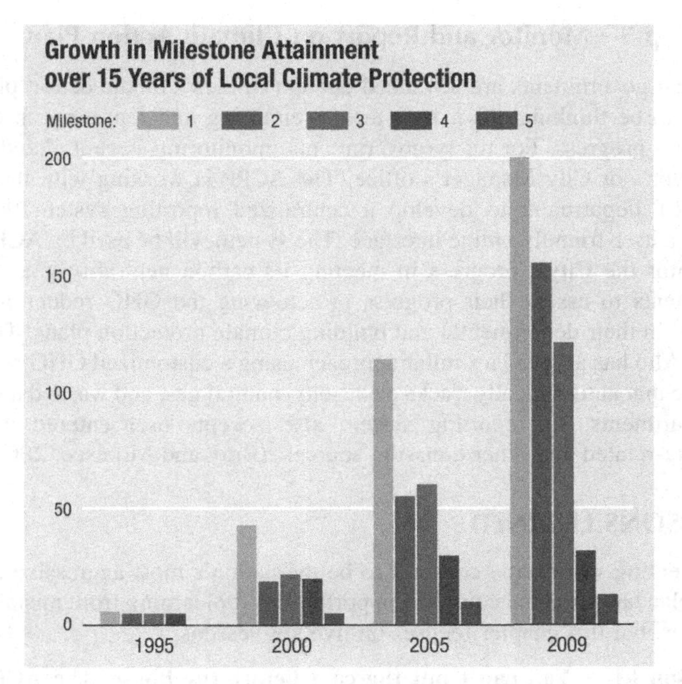

FIGURE 8 ICLEI Member Cities with Climate Action Plans, 1995 through 2009. *Source: ICLEI, Measuring Up, 2009, p. 54.*

minimize water consumption (Austin Water Utility's Water Conservation program);[31] conserve fuel and promote the use of alternative fuels (Austin Energy's Air Quality[32] and Central Texas Clean Cities[33] programs); encourage waste reduction, reuse, and recycling (Solid Waste Services' Zero Waste Initiative);[34] and promote environmentally responsible purchasing practices (Purchasing Office's Responsible Purchasing Program). These resources have allowed the City to take on the broad array of projects and goals outlined in the ACPP.

However, limited staff, funding, and institutional support are often cited as primary roadblocks to the success of local government climate protection and sustainability programs (Urban Sustainability Directors Network, 2009). Many cities are limited to one or two staff housed out of the Mayor's or City Manager's office and, as a result, are constrained in their ability to achieve all facets of their climate action plans.

---

31. Learn more about the City water conservation program at *www.ci.austin.tx.us/watercon/*.

32. Learn more about the City's air quality program at *www.ci.austin.tx.us/airquality/*.

33. Learn more about the City's alternative fuels program at *www.ci.austin.tx.us/cleancities/*.

34. Learn more about the City's zero waste program at *http://www.ci.austin.tx.us/sws/zerowaste.htm*.

5.5. Step 5—Monitor and Report on Climate Action Plan

Few local governments are advanced enough in their climate action planning process to be thinking about long-term monitoring and reporting on climate protection progress. For many governments, monitoring is centralized out of the Mayor's or City Manager's office. The ACPP is working with the City's main ITT department to develop a centralized reporting system that will provide a user-friendly online interface. The system will be used by ACPP staff to monitor the City's progress in meeting its carbon neutrality goal and by departments to assess their progress in achieving the GHG reduction goals included in their departmental and building climate protection plans. The City of Palo Alto has adopted a similar approach using a customized GHG reporting software that automatically tracks electricity, natural gas, and waste data across 13 departments. Its reporting system also accepts user entered data for commute-related and other emission sources. (Bray and Moresco, 2009)

6. LESSONS LEARNED

Implementing what some consider to be the nation's most aggressive climate action plan has provided countless opportunities for learning from mistakes and successes, but this chapter focuses on five key lessons.

- **Lesson #1 – You can't put the cart before the horse**. The ACPP was adopted prior to conducting an emissions inventory of all municipal and community GHG sources. A post-plan assessment of the City's baseline municipal operations GHG emissions inventory revealed the challenges associated with achieving carbon neutrality within a 12-year (now 10-year) window.
  - Electricity use represents two-thirds of the City's municipal operations' carbon footprint. Therefore, powering 100% of the City's electricity with renewable energy would eliminate the majority of the City's carbon footprint. However, Austin Energy has uncovered barriers to converting the City to renewable energy, including uncertainty about the quantity of energy actually consumed by all municipal operations (for example, electricity used by street lights, traffic signals, and security lighting is estimated, not metered) and paying for more expensive renewable energy during a time of budget cuts.
  - Transportation accounts for another quarter of the City's carbon footprint. To make the City's vehicle fleet carbon-neutral, 100% of vehicles must be powered by electricity from renewable resources or from biofuels. The City is actively phasing in electric and biofuel-capable vehicles and the fueling infrastructure to support them, but it remains to be seen whether electric vehicle technology (especially for medium- and heavy-duty vehicles) and biofuels will mature fast enough to meet the 2020 deadline.

- The remaining contribution from building heating, waste, water, and refrigerants may never be made carbon-free, but offsets may be a viable option for the relatively small remaining carbon footprint. The lesson for Austin and others seeking carbon neutrality is that 2020 may not be the most achievable deadline for such a technology-dependent goal.

- **Lesson #2 – Target practice is necessary**. GHG emission reduction targets need to be aggressive as called for by climate science, but a phased approach may be useful for testing the waters of what magnitude of reduction is realistic given technological, fiscal, political, and other constraints over a given time period. Using a stepped approach allows governments to claim a "win" to bolster support for further climate action and can be used to reevaluate future targets in light of how well the earlier target(s) was (were) achieved. Annapolis, Maryland provides a good example for taking steps toward carbon neutrality, with goals to reduce its emissions to 50% below 2006 levels in 2012, 75% below 2006 levels by 2025, and 100% below 2006 levels (carbon-neutral) by 2050.
- **Lesson #3 – Peer pressure is not enough**. Central management and coordination are paramount to institutionalizing climate protection as a core mission of a government's operations. So far, the ACPP has found it to be most successful to take a top-down and bottom-up approach to carbon management. The departmental and building climate protection plans provide a bottom-up approach that allows departments and building management to outline GHG mitigation strategies that work best for their scope of services, resources, and constraints. From the top-down perspective, ACPP staff is working with the City Manager's Office and key departments to develop or enhance City-wide policies that facilitate implementation of the department and building climate protection plans.
- **Lesson #4 – Government is only one piece of the puzzle**. The City of Austin represents a relatively large piece of the community's carbon footprint given the lack of heavy industry in Austin and City ownership of power plants, water and wastewater treatment plants, and a landfill. However, most local governments comprise merely a fraction of the community's total carbon footprint. While government has significant control over setting policies that enable or support emission reductions, many of the most far-reaching climate protection strategies need to be implemented by individuals, businesses, and organizations in the community. Therefore, it is critical to involve the community from the beginning to build trust and set a collaborative tone.
- **Lesson #5 – Mitigation and adaptation go hand-in-hand**. Laying a foundation for climate preparedness does not have to mean that governments and the communities they serve have given up on lessening the impact of climate change. Local governments can continue to reduce GHG emissions, while planning for how they will adapt to a locally changing climate. For example, targeting mitigation strategies, such as residential weatherization

and other energy efficiency upgrades, to low-income populations who devote a larger proportion of their monthly income to utilities not only reduces community GHG emissions, but also helps ease the burden climate change may impose on them through increased utility bills.

CONCLUSIONS

The ACPP provides a viable model for local climate action planning. The plan has found success by leveraging the environmentally responsible mindset of the local community and support of other environmental sustainability programs throughout the City. Austin is on track to meet its bold energy conservation and renewable energy goals, and a solid foundation has been laid for institutionalizing climate protection in the City's operations. Austin's challenge will now be engaging the local and global community in a meaningful dialogue to further reduce GHG emissions.

The beginning of this chapter outlined the stark consequences for urban areas, which house roughly half the world's population, if GHG emissions continue unabated. Cities like Austin are refusing to accept this fate and instead are tackling the problem of climate change head-on through climate protection and sustainability planning initiatives. If all 600 plus members of the U.S. ICLEI network achieve their climate protection goals, these local governments will have avoided 1.4 billion metric tons of CO_2-eq.—the equivalent of shutting down 350 coal-fired power plants—by 2020 (ICLEI, 2009). If this climate change mitigation effort is adopted worldwide, the results could be staggering.

Imagine a world powered with 100% renewable energy, an all-electric vehicle fleet, and a closed-loop cycle in which the products we consume are continually repurposed into productive reuse. Imagine a world in which all food, all energy sources, and all products consumed by a city are produced within 100 miles. Imagine a world where every city balances its GHG emissions with native vegetation and healthy soils that store as much carbon as its residents emit. If all cities worldwide pursue Austin's climate protection goals, this could be the legacy we pass on to future generations.

BIBLIOGRAPHY

Alliance for Building Regulatory Reform in the Digital Age at Fiatech. (2007). *White paper: On best practices in electronic plan submittal, review, tracking and storage. www.natlpartnerstreamline.org/documents/WhitePaper_ElectPlan_092107.pdf.*

Austin Climate Protection Program. (2009). *Austin climate protection program annual report 2009. http://www.ci.austin.tx.us/acpp/downloads/ACPP_Annual_Report_5.20.09_FINAL.pdf.*

Austin Energy. (2008). *Resource guide, planning for Austin's future energy resources. www.austinsmartenergy.com/downloads/AustinEnergyResourceGuide.pdf.*

Austin Energy. (2009). Resource and climate protection plan to 2020, recommendations and plan. *Presentation by Roger Duncan to the Austin City Council. www.austinsmartenergy.com/downloads/RecommendationCO2Plan.pdf.*

Bray, D., & Moresco, J. (2009). *Palo Alto cuts carbon with Hara*. Solution case study, AltaTerra Research. *www.hara.com/request_altaterra.html*.

C40 Cities. (2009). *Cities and climate change*. *www.c40cities.org/climatechange.jsp*.

Carbonsense. (2008). *Carbon disclosure project, cities pilot project 2008*. Carbon Disclosure Project. *www.cdproject.net/CDPResults/65_329_216_CDP-CitiesReport.pdf*.

City of Austin. (2009). *Austin 2020 bicycle plan update*. Department of Public Works, Neighborhood Connectivity Division, Bicycle Program. *www.ci.austin.tx.us/publicworks/bicycle-plan.htm*.

City of Austin. (2009). *Green infrastructure*. *www.ci.austin.tx.us/publicworks/sustainability/green.htm*.

Climate Neutral Network. (2010). *Vancouver*. *www.unep.org/climateneutral/Default.aspx?tabid=207*.

Copenhagenize.com. (2009). *The world's most bicycle friendly cities*. *www.copenhagenize.com/2009/07/worlds-most-bicycle-friendly-cities.html*.

Haight, A. (2008). Lincoln city crimps its carbon. *The Oregonian*, 12 December 2008. *www.oregonlive.com/environment/index.ssf/2008/12/lincoln_city.html*.

ICLEI. (2009). *Measuring up*. *www.icleiusa.org/blog/announcing-iclei-usas-new-measuring-up-report*.

IPCC/Pachauri, R.K., & Reisinger, A. (Eds.). (2007). *Climate change 2007: Synthesis report. Contribution of working groups I, II, and III to the fourth assessment report of the intergovernmental panel on climate change*. Geneva, Switzerland: IPCC. *www.ipcc.ch/publications_and_data/publications_ipcc_fourth_assessment_report_synthesis_report.htm*.

National Climatic Data Center. (2009). *Billion dollar U.S. weather disasters*. National Environmental Satellite, Data, and Information Service, National Oceanic and Atmospheric Administration, U.S. Department of Commerce. *www.ncdc.noaa.gov/oa/reports/billionz.html#narrative*.

Union of Concerned Scientists. (2009). *Texas, climate projections*. *www.ucsusa.org/gulf/gcstatetex_cli.html*.

Urban Sustainability Directors Network. (2009). Personal communication.

U.S. Census Bureau. (2005). *Table 1: Annual estimates of the population for the United States and States, and for Puerto Rico: April 1, 2000 to July 1, 2005*. (NST-EST2005-01). *www.census.gov/popest/states/NST-ann-est2005.html*.

U.S. Energy Information Administration. (Undated). *International energy statistics*. *http://tonto.eia.doe.gov/cfapps/ipdbproject/IEDIndex3.cfm*.

U.S. Energy Information Administration. (Undated). *Table 3. State emissions by year (million metric tons of carbon dioxide)*. *www.eia.doe.gov/environment.html*.

U.S. Environmental Protection Agency. (2009). *Heat island effect*. *www.epa.gov/hiri/*.

Chapter 19

Rising to the Challenge of Sustainability: Three Cases of Climate and Energy Governance

Benjamin K. Sovacool
National University of Singapore

Chapter Outline

1. INTRODUCTION

In the past few years, governments at a variety of scales have implemented programs and policies aimed at creating more sustainable forms of transport, energy use, forestry, urban planning, and agriculture. Consider the following examples:

- The city of Vauban, Germany, has begun charging residents $40,000 per permanent parking space and prohibiting motorized transport on most of its roads to lessen the unsustainable aspects of motorized transportation (Rosenthal, 2009).
- In Ellensburg, Washington, in the United States, the local government accepted hundreds of thousands of dollars of contributions from the community to fund an array of solar panels that currently sell electricity to the municipal utility (Coughlin and Cory, 2009).
- The Private Forest Project in Costa Rica collects a 5% tax on gasoline and distributes the funds to encourage plantation owners and forest managers to preserve their lands, which act as important carbon sinks (Brown and Sovacool, 2010).
- In São Paulo, Brazil, planners built the Bandeirantes Landfill Gas to Energy Project to capture methane and convert it into electricity used by

400,000 homes, preventing the release of 800,000 tons of carbon dioxide (CO_2) equivalent per year (Sovacool and Brown, 2009).

- In Beijing, China, city planners wielded combined heat and power, energy efficiency, fuel substitution programs, electric bicycles, and ring roads to reduce greenhouse gas (GHG) emissions (Sovacool and Brown, 2009).
- In South Korea, the national government passed an aggressive stimulus package that seeks to reinvest more than 2% of the country's Gross Domestic Product (GDP) into the construction of one million green homes, research and development on low carbon energy supply, and clean transport infrastructure and bicycle expressways (Watts, 2009).
- Israel has begun subsidizing the purchase price of Plug-in Hybrid Electric Vehicles to make them equivalent to conventional ones and committed $200 million to build recharging and maintenance stations throughout the country (Erlanger, 2008).
- In Austin, Texas, city officials have launched an ambitious Climate Protection Plan aimed at achieving carbon neutrality by 2020 (see this book's Chapter 18 by Clymer).

These cases, and the countless others that could have been mentioned (such as Berkeley's "financing initiative for renewable and solar technology" [Fuller et al., 2009], Güssing, Hungary's plan to meet all of its energy needs from renewable resources by 2015 [Girardet and Mendonca, 2009], or vehicle moratorium and congestion road pricing in Singapore [Barter, 2008]) demonstrate that communities and countries can rely on an assortment of tools and options, in diverse sectors, to promote sustainability. Many experts also now believe that these types of community-oriented projects may be the best short-term options for addressing sustainability problems especially in view of the failure of the Copenhagen summit on climate change in December 2009, utter lack of progress in U.S. Congress, and the apparent dearth of international interest in making any significant commitments (Sovacool and Brown, 2010; Ostrom, 2009).

One lingering question, however, remains: what makes a particular project or program aimed at improving sustainability successful? To answer that question, this chapter explores the efforts of the Clinton Climate Initiative (CCI), the corporate social responsibility of Motorola, and the proposed public–private partnership of Masdar City in the United Arab Emirates (UAE). The chapter utilizes a three-part analytical framework for each case study, describing each project, elaborating its benefits, and discussing its challenges.

2. CLINTON CLIMATE INITIATIVE

2.1. Description

The William J. Clinton Foundation is a nonprofit organization created in 1997 by the former U.S. President Bill Clinton, who launched the CCI in 2006. CCI is funded by private individuals and foundations and focuses on three strategic program areas: cities, clean energy, and forestry.[1] In August 2006 it became the action arm of the C40 "Climate Leadership Group," a consortium of large cities committed to reducing GHGs.[2] The organization has program staff in many of these 40 cities (called "partner" cities).

The *Cities* program is the most developed to date. CCI city directors deliver three types of capacity building:

- Technical and analytical assistance, such as information about energy technologies and market dynamics, specifications for pilot projects, and emissions abatement analysis
- Project assistance, such as coordination of different industry, financial, regulatory, and electric utility stakeholders
- Purchasing and financial assistance, such as introductions to vendors, discounts on bulk purchases, life-cycle cost, and payback analyses (Clinton Climate Initiative, 2009a)

CCI's Cities program targets this assistance toward six subprograms: building retrofits, outdoor lighting, waste management, GHG emissions measurement, transportation, and a new program in urban development called "climate positive." Table 1 presents an overview of some of the program's accomplishments to date in its most developed subprograms of buildings, lighting, waste, and transportation.

The building retrofit program coordinates public and private building owners, ESCOs, and financial institutions to undertake energy efficiency building retrofit projects.[3] CCI has signed agreements with a number of ESCOs to increase the delivery of building retrofits through energy performance contracting (EPC). The ESCOs have agreed to execute projects under a clear set of contracting terms and conditions, including streamlined procurement, transparency in pricing. These ESCOs contractually guarantee energy savings

---

1. Two other program areas, water and ports, are currently "under development." See Clinton Climate Initiative. (2009b). *Our challenge, our work.* CCI fact sheet, updated May 2, 2009.

2. The C40 cities are Addis Ababa, Ethiopia; Athens, Greece; Bangkok, Thailand; Beijing, China; Berlin, Germany; Bogotá, Colombia; Buenos Aires, Argentina; Cairo, Egypt; Caracas, Venezuela; Chicago, United States; Delhi, India; Dhaka, Bangladesh; Hanoi, Vietnam; Houston, United States; Hong Kong, China; Istanbul, Turkey; Jakarta, Indonesia; Johannesburg, South Africa; Karachi, Pakistan; Lagos, Nigeria; Lima, Peru; London, United Kingdom; Los Angeles, United States; Madrid, Spain; Melbourne, Australia; Mexico City, Mexico; Moscow, Russia; Mumbai, India; New York, United States; Paris, France; Philadelphia, United States; Rio de Janeiro, Brazil; Rome, Italy; São Paulo, Brazil; Seoul, South Korea; Shanghai, China; Sydney, Australia; Tokyo, Japan; Toronto, Canada; Warsaw, Poland.

3. An initial group of 15 cities participated in the retrofit program and offered municipal buildings for the first round of energy retrofits in 2007: Bangkok, Berlin, Chicago, Houston, Johannesburg, Karachi, London, Melbourne, Mexico City, New York, Rome, São Paulo, Seoul, Tokyo, and Toronto.

TABLE 1 Clinton Climate Initiative Accomplishments as of February, 2010

Sub-Program	Location	Description
Buildings	Various	Retrofitted 500 million square feet of real estate in 30 cities including 300 municipal buildings in Seoul, Johannesburg, Houston, London, and Melbourne and commercial buildings in Bangkok, Mumbai, Chicago, and New York
	New York, United States	Retrofitted the Empire State Building to reduce its energy use by 15 percent
Lighting	Los Angeles, California	Replaced conventional streetlights with LED units to reduce CO_2 emissions by 40,500 tons and save $10 million annually
Waste	Delhi, India	Implemented an integrated solid waste management system covering door-to-door collection, transportation, recycling, and composting to prevent 96,000 tons of CO_2 from being emitted each year
	Lagos, Nigeria	Replaced open air dumps with composting, recycling facilities, and a sanitary landfill
Transportation	Various	Invested in bus rapid transit systems in Johannesburg, Mexico City, and Bogota
	Various	Promoted bicycle lanes in Sao Paulo and Buenos Aires

Source: Author.

of 15%–25% per year and maximum project costs, agreeing to compensate the building owner financially for savings shortfalls or to make additional product retrofits at no cost to ensure that performance targets are reached.

How does the system work? Banks lend capital for the retrofits—sometimes enough to cover 100% of the project—and are then repaid using the revenue generated from the energy savings. Honeywell, Johnson Controls, Siemens, and Trane conduct the initial energy audits, perform building retrofits, and guarantee the energy savings of the retrofit projects. ABN AMRO, Citi, Deutsche Bank, JPMorgan Chase, and UBS have pledged $1 billion each to finance city and private efforts to undertake these retrofits, with paybacks for the loans plus interest coming from the energy savings spread across many years.

Energy savings delivered by building retrofits amplify the return on investment of projects. In short, everyone wins: building owners get lower energy bills, ESCOs get to implement projects, and banks get interest on their loans. So far more than 250 building retrofits have been initiated in 30 cities on a variety of different buildings, including municipal offices, commercial firms, schools, universities, and public housing complexes—investments that might not otherwise occur due to lack of information and transaction costs.

One such CCI project is the retrofit of the Empire State Building, which is designed to reduce energy use by 38% and reduce CO_2 emissions by 105,000 metric tons over 15 years. Another project working with the New York City Housing Authority has installed 10,000 compact fluorescent lights to reduce electricity costs by 17% with yearly savings of $367,000 (and a reduction of GHG emissions by 1400 tons of CO_2-equivalent per year). CCI has also formed a purchasing alliance with major providers of energy efficient equipment, heating and cooling products, indoor and outdoor lighting, and materials, so that technology can be purchased at reduced cost (Clinton Climate Initiative, 2007, 2009d).

The outdoor lighting program replaces street and traffic lights with more efficient luminaire technologies and control systems. Street lighting accounts for about 1.3% of all electricity used globally and 5%–37% of a municipal government's electricity consumption and carbon footprint (Koenig, 2009). Advanced luminaries (such as light emitting diodes [LEDs] and induction technologies) often consume 40% less electricity than normal streetlights and 90% less than ordinary traffic lights with no reduction in performance. In Ann Arbor, Michigan, CCI officials worked with the city to replace 100 high pressure sodium lights with 1400 LED streetlights that resulted in energy savings greater than 50% and payback in 4.4 years. In Oslo, Norway, CCI planners equipped 55,000 high pressure sodium streetlights with an intelligent control system and electronic ballasts that reduced energy consumption by 62% (Koenig, 2009). And in Los Angeles, California, CCI worked with city planners to replace 140,000 streetlights with LED models that will save the city $10 million annually and reduce CO_2 emissions by 40,500 tons through reduced maintenance costs and energy savings.

The waste management program assists cities in managing their waste and sanitation systems and provides recommendations for improvement. The program helps cities design the infrastructure, identify the technologies, and adopt the policies to improve their waste management systems and reduce reliance on landfills. Key areas of focus are diverting waste streams from landfill disposal and converting waste into energy, for example by capturing methane gas from landfills or creating biogas from organic waste. CCI works with city planners to evaluate the city's waste management systems, potential technologies, and draft public tender documents. In addition, the program teaches city planners how to recycle construction and demolition waste, and recover commodities such as metals, glass, paper, plastic, and electronic waste (Clinton Climate Initiative, 2009c).

CCI's emissions measurement software, Project 2°, enables cities to track, monitor, and report their GHG emissions. This multilingual, easy to use, common measurement system synthesizes data on fuel and electricity consumption, vehicle traffic, waste production, and industrial processes, and converts them into tons of CO_2 equivalent. Once data are compiled, cities can establish a baseline of their GHG emissions, then manage inventories, create action plans and customized reports, and share best practices. Feedback is still being collected on a trial release of the software, so Project 2° is essentially too new to evaluate.

The transportation program promotes public transportation and nonmotorized transit such as walking and cycling. The global partnership between CCI and the Institute for Transportation and Development Policy (ITDP) focuses on the implementation and improvement of Bus Rapid Transit (BRT) systems and bicycle networks. Projects in cities including Johannesburg, Bogota, São Paulo, and Mexico City are addressing route optimization, operational planning, and, for BRT, fuel substitution. CCI also helps cities adopt proven and emerging technologies to reduce carbon in their transportation sector. It looks at vehicles and propulsion systems, fuel options, and fuel distribution and dispensing infrastructure. The CCI Hybrid Bus Test Program aims to create a market for hybrid bus technology in Latin America. Its Electric Vehicle Working Group brings together 12 cities in the C40 network to organize a coordinated procurement of electric vehicles by owners of public and private sector fleets.

In May 2009 CCI's Cities program launched the Climate Positive Development Program working in concert with the U.S. Green Building Council. It aims to develop large-scale urban projects that are "climate positive" by having zero or below zero GHG emissions (Clinton Climate Initiative, 2009e). The program intends to integrate the CCI Cities subprograms—building retrofits, outdoor lighting, waste management, carbon measurement, and transport programs—into a synergistic project that focuses on:

- Developing high-performance and energy-efficient buildings
- Supplying 100% renewable electricity
- Creating an interconnected transport system based on walking, bicycling, and mass transit
- Facilitating waste streams so that everything is recycled
- Maximizing the efficiency of grey-water systems
- Optimizing environmental performance through the use of local materials
- Sequestering CO_2 and closing the emissions cycle "on-site"

Sixteen communities in six continents have agreed to participate in Climate Positive, and once completed more than one million people could live and work in communities that have no net GHG emissions.[4] Like Project 2°, however, the Climate Positive program is still being developed.

---

4. These locations are: Melbourne, Australia; Palhoça, Brazil; Toronto, Canada; Victoria, Canada; Ahmedabad, India; Jaipur, India; outside Panama City, Panama; Pretoria, South Africa; Johannesburg, South Africa; Seoul, South Korea; Stockholm, Sweden; London, U.K.; San Francisco, California, U.S.; and Destiny, Florida, U.S.

CCI's Clean Energy Program works from the premise that we must reduce emissions from the use of fossil fuels as urgently as we must develop viable renewable energy solutions. CCI therefore focuses on two low-carbon technologies in the power sector: Carbon Capture and Storage (CCS), which isolates CO_2 emissions from power plants and other industrial facilities, and CSP, which harnesses the sun's heat to run conventional turbines or engines. Working with governments and other key stakeholders, CCI aims to deliver large-scale demonstration projects around the world.

The CCS program adopts a "network" approach that connects multiple capture facilities to a common pipeline and storage system sized to support long-term, commercial-scale use. The CCS team works in strategic locations with a concentration of emission sources and sufficient storage capacity, helping governments anticipate and resolve a complex host of critical issues.

The Clean Energy program is also working to develop the market for CSP, a low-emissions renewable technology with strong potential to become a significant global energy source. Here CCI's approach centers on a model for solar parks in which multiple companies lower capital and construction costs by building optimally-scaled CSP plants on a common parcel of land. CCI works in three principal ways: cost analysis, relationship building, and strategic planning and assessment. Projects under development in India, South Africa, Australia, and the U.S. will demonstrate the technology and stimulate government investment and incentives to drive down the cost of the technology.

Finally, the CCI Forestry Program (also known as the "Carbon and Poverty Reduction Program") focuses on improving land use practices to reduce poverty and curb GHG emissions. It currently works with the national governments of Cambodia, Guyana, Kenya, Indonesia, and Tanzania. The program attempts to build capacity to protect and manage forests in these countries to mitigate climate change, but also to ensure that local communities still have viable sources of income.

Efforts are centered on the development of a National Carbon Accounting System (NCAS) for forests to give developing countries "unprecedented accounting rigor" to measure and monitor forest carbon sequestration and emissions. CCI has convened the Carbon Measurement Collaborative,[5] a network of scientists and forestry experts in carbon modeling, land use change, and satellite imaging to design and implement the NCAS in partner countries. CCI is equally focused on supporting both avoided deforestation and reforestation projects. The CCI Forestry team has contributed to innovating and validating an REDD methodology for aggregating dispersed forest communities as one entity so they are able to receive benefits from the sale of carbon

5. As of mid-2009, partners and participants in the CMC included the Australian government, the Environmental Systems Research Institute, the U.S. National Aeronautics and Space Administration, Intergovernmental Group on Earth Observations, Woods Hole Research Center, H. John Heinz III Center, World Resources Institute, and the Green Belt Movement.

credits. Facilitating project financing, design and implementation, CCI aims to reduce poverty among forest-dependent communities by attracting financing, providing jobs, creating new sources of revenue—and safeguarding critical "environmental services" like protecting watersheds and preventing soil erosion (Clinton Climate Initiative, 2009f).

2.2. Benefits

One benefit is that by overcoming initial cost and information hurdles, CCI programs facilitate investment in energy efficiency projects that tend to save large amounts of energy and pay for themselves quickly. The energy performance contracting model for building retrofits is premised entirely on using energy savings to pay the upfront cost of installing more efficient equipment. With more than 250 projects covering over 2500 buildings initiated to date, these projects work because 75% of a typical building's life-cycle cost comes from energy use and operations (a mere 11% comes from design and construction and 14% from finance). With so much energy being used and wasted, cost savings from retrofits can accrue rapidly. We see the same degree of potential savings with outdoor lighting systems: Investment in LEDs, induction systems, and better controls for streetlights and traffic lights can have a payback of three to seven years, demonstrated most clearly with the success of the retrofit program in Los Angeles (Koenig, 2009).

A second benefit is that by aggregating demand across cities, CCI can negotiate lower prices with suppliers of energy efficient products and technologies in the building, lighting, waste, and transportation sectors (de Raaf, 2008). Pricing agreements with manufacturers yield discounts that help public and private buyers by lowering the investment barriers on emerging and mature products and technologies, which offer significant energy efficiency improvement or fuel switching potential.

2.3. Challenges

Despite the progress made so far, CCI programs do face some persistent challenges. Many cities' programs rely on new technology that most people remain uninformed about. The experiences of early adopters of LED lighting systems or fuel-cell vehicles are not widely known, and some officials and property owners can express an understandable distaste toward technologies they have never heard of. In many cities, incentives are still misaligned against energy efficiency because lenders are unfamiliar with energy service providers and/or their technologies, because of the principal agent problem involving landlords and tenants, or because people remain unsure about how to calculate the expected revenue from potential energy savings. Others have limited budgets for energy efficiency projects, or may need to spend scarce time and resources on other more pressing matters (Koenig, 2009). In essence, while the

CCI shows immense promise, it is too early to tell if it can make a lasting impact toward reducing GHG emissions (its primary goal) and improving sustainability.

3. MOTOROLA AND CLIMATE CHANGE

3.1. Description

Paul and Joseph Galvin founded Galvin Manufacturing Corporation in 1928 and renamed it Motorola Incorporated in 1947 after becoming known for their brand name radio products. Motorola has a colorful history of invention, including pioneering in-car radios and public-safety radios and inventing the world's first FM portable two-way radio and commercial portable cellular telephone. Motorola is now one of the world's largest manufacturers of mobile phones, digital video systems and monitors, and cable modems and Internet-related infrastructure. In 2008, Motorola employed more than 60,000 people and reported sales of $30.1 billion, 49% of which were concentrated in one market, the United States (Motorola, 2009). Motorola has more than 27,500 suppliers, spending $19.2 billion with suppliers from 126 countries in 2008. Driven largely by need to distinguish itself from its competitors, Motorola began utilizing its considerable influence as a corporation in 2002 to reduce its environmental impact through a variety of different mechanisms, including pledged emissions reductions, improved energy management, product standards, LEED certification, and corporate social responsibility.

For example, Motorola voluntary joined the Chicago Climate Exchange (CCX) as a founding member (and the first global member). The CCX is the only legally binding GHG reduction and trading system in North America, with global affiliates and projects worldwide. Corporations that choose to join the exchange commit to reducing their GHG emissions 6% by 2010 over a year 2000 baseline, and in early 2009 CCX had more than 350 corporate members including Ford and DuPont as well as municipalities such as Chicago, Illinois, and Oakland, California. As part of this commitment, Motorola submits annual emissions reports to be audited and verified by the Financial Industry Regulatory Authority; if they do not meet their targets they are required to purchase CCX credits to offset the difference.

As a way to meet these commitments and save money at the same time, Motorola has implemented a robust energy efficiency strategy at all of its facilities. Motorola has tasked energy managers at each of its facilities to improve controls for heating and air-conditioning, and also install motion sensors to automatically turn off lights in unoccupied rooms. Managers have shifted cleaning and maintenance to occur during the day coincident with normal working hours (rather than at night) to eliminate excess energy consumption. Managers have also installed more energy-efficient lighting and office equipment, and have consolidated data centers to save electricity usage.

Motorola also purchases renewable energy to offset some of the fossil-fuel based electricity the corporation uses. In 2009, for example, 15% of the electricity needed to power Motorola's global fleet of factories, facilities, and offices came from renewable resources (achieved predominately through the purchase of certified renewable energy credits), and its goal is to reach 20% by 2010 and 30% by 2020. Part of Motorola's efforts to promote energy efficiency focus on making employees aware of corporate environmental targets and highlighting best practices through newsletters, on-site posters, and curriculum programs on how to reduce energy use at work and at home (Motorola, 2009).

Another way Motorola minimizes its environmental impact is through the use of product standards and product design. All Motorola phones have a recyclability target of 65%, meaning more than half of the components and materials from these phones can be easily reused in other devices. Since 2000, the company has reduced the average standby power of its mobile phone chargers by at least 70%, and Motorola started a program in 2008 that has collected and recycled 2560 tons of modems, routers, and phones (Motorola, 2008). Motorola is also careful with the materials it uses. Two materials of concern used in cell phones and other electronics are polyvinyl chloride (PVC) and brominated flame retardants (BFR). More than 50 different types of phones in the current Motorola catalog currently have BFR-free circuit boards and the company intends to eliminate PVC and BFRs from all new designs of mobile phones introduced after 2010 (Motorola, 2008).

Improvements in product design have made a notable difference. As one example, Motorola now manufactures the MOTO™ W233 Renew mobile phone made from recycled post-consumer plastic water cooler bottles that it also certifies as "carbon neutral" through the purchase of carbon offsets. The Renew represents a synthesis of different efforts to improve design and reduce pollution, starting with how the phone is made and shipped to how it is used and disposed of. Motorola manufactures the phone's casing from ground-up water cooler bottles and other plastics taken from landfills that are then combined with a unique polycarbonate plastic feedstock that took four years to develop. The phone is unpainted to aid in recycling, shipped in an efficient package one-fifth the size of similar phones, and its manual and packaging utilizes 100% post-consumer recycled paper with vegetable-based inks.

In the U.S., the Renew phone has a battery capable of nine hours talk time, enabled by a standby mode of only 0.1 watts to save energy. The phone is shipped with a postage-paid recycling envelope so consumers can return their previous mobile phone to Motorola free-of-charge to be recycled. To further assist in the recycling and refurbishing effort, Renew contains no PVC or nickel and can be disassembled into separate battery, housing, motherboard, and display in less than ten seconds. Lastly, Renew is the first phone on the market to be certified CarbonFree®. That is, Motorola offsets all of the equivalent CO_2 emitted from materials extraction, manufacturing, distribution, and operation of the phone by investing in renewable energy and reforestation (Reuters News

Service, 2009). These offsets, which mostly occur from a landfill capture plant in Massachusetts and reforestation of native hardwoods forests in Louisiana, are independently verified by Carbonfund.org™ and were chosen based on recommendations from the Environmental Defense Fund.

Apart from improving energy use at existing buildings and making manufacturing processes and products more sustainable, Motorola also uses LEED certification to make new buildings and facilities more environmentally friendly. The company's new manufacturing facility in Sriperumbudur, India, near the urban center of Chennai, has achieved a "silver" rating from LEED and has a rainwater harvesting and collection system that stores 10 million liters of water and recharges an underground aquifer. The facility uses an ultra-efficient heating, ventilation, and air-conditioning system and reuses treated sewage from a local plant for toilets and to water outdoor vegetation. Lastly, more than three-quarters of the construction waste produced by the building were reused in the actual building as scaffolding and raw material. This was the first manufacturing facility to be LEED certified in India.

Finally, Motorola funnels some of its profits back into corporate social responsibility programs that center on renewable energy and energy security. In 2007, Motorola worked with MTC Namibia, Namibia's cellular telephone service provider, and the GSMA Development Fund, a development fund set up by the telecommunications industry, to install a small-scale 6 kW wind-powered and 5 kW solar-powered cellular stations in rural Namibia, where more than 90% of communities have no access to electricity. Motorola is installing hydrogen-powered fuel cells to back up the public safety network in Denmark (SINE) to provide continuous, secure communication to mission-critical operations across 450 radio stations. These fuel cells displaced diesel generators, which were more polluting and noisy, took longer to start, and required more maintenance. Operators estimate that they will produce net gains of $1–2 million per year for SINE.

3.2. Benefits

The most significant benefit from Motorola's combined efforts to cut emissions, improve energy efficiency, and develop more sustainable products has been drastic reductions in energy use. In 2008, actual data covering 78% of total manufacturing floor space (and estimated for the remaining 22% of space) indicated that Motorola used 955 million kWh of electricity and natural gas, a 21% decrease from the 1207 million kWh in 2005 (Figure 1). A related advantage from using less energy is a cleaner environment from reduced GHG emissions. The total carbon footprint for GHGs under the Kyoto Protocol for Motorola in 2005 was 672,000 metric tons of carbon equivalent, but it dropped 20% to 532,000 tons in 2008. Put another way, the amount of total GHGs emitted by Motorola in 2008 was 17.6 metric tons of CO_2 equivalent per million dollars of sales.

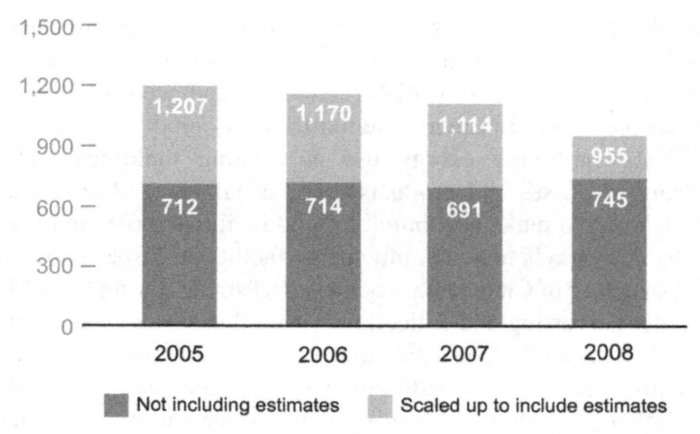

FIGURE 1 Electricity and Natural Gas Use at Motorola (million kWH), 2005 to 2008. *Source: Motorola.*

A secondary benefit from these improvements in environmental performance has been the company's international recognition and improved public image. In 2009 the U.S. Environmental Protection Agency named Motorola a Green Power Leader for purchasing 20% of its U.S.-based electricity (about 78 million kWh) from wind power. The Dow Jones Sustainability Index has named Motorola every year since 2003, and the company has been included in *Corporate Responsibility Officer* magazine's 100 Best Corporate Citizens nine times since 2000, earning fifth place among all of these corporations in the "environment category" for 2009.

3.3. Challenges

While some of Motorola's products are more environmentally responsible than previous versions, and less environmentally damaging than the products of their competitors, many modern telecommunications devices contain potentially hazardous substances, such as PVC and BFRs, that can seep into water supplies or leach into landfills if improperly discarded. These devices also rely on electricity, usually provided from centralized fossil-fueled power stations, in order to operate.

For example, a December 2008 survey of the best "green" phones on the market—published prior to the January 2009 launch of the MOTO™ W233 Renew mobile phone—found that the best designed products, including those from Motorola, still had a serious negative environmental impact. The survey ranked mobile telephones according to four criteria:

- The use of hazardous chemical substances, including PVC and BFRs but also antimony, phthalates, beryllium, and other toxic components
- Power consumption of the products, assessed by comparing them with the most up-to-date ENERGY STAR® efficiency standards

- Product life cycle, including criteria such as recyclability rate, use of recycled plastic in the product, long warranty periods, and good take-back programs
- Availability of data that enable rigorous life-cycle assessments of the energy required to manufacture and use the product

Interestingly, the three best phones—models from Samsung, Motorola, and Nokia, respectively—all scored less than six points, a few tenths past the halfway mark (Rautner and Harrell, 2008). The lesson here appears to be that there is significant opportunity to improve the sustainability of modern electronic devices, including mobile phones.

To extend sales and maximize profits, Motorola is constantly improving its products. Motorola is self-admittedly focused on making people more mobile, placing a premium on mobility and thus the energy consumption facilitating it. Motorola also ceaselessly conducts research to make its products more innovative, something that can be favorable when a Renew phone results but unfavorable if consumers continually purchase new products not when they wear out, but when they no longer are fashionable or contain the latest technology. In this way, Motorola revises its products partially in response to consumer needs, but also partially in an attempt to shape them. It remains driven predominately by the desire to increase profits and returns with sustainability as a laudable but peripheral goal, rather than placing sustainability at the forefront and profits second.

4. MASDAR CITY

4.1. Description

Masdar City, meaning "the source" in Arabic, is a proposed city in Abu Dhabi, UAE, designed to be completely carbon and waste neutral. The city is currently under construction and set to be completed in 2016, and it is being built 17 kilometers from the city of Abu Dhabi and near the Abu Dhabi International Airport (Masdar City, 2009). Current estimates are that Masdar City will cost about $22 billion to complete and will take 8 years to build, the end product being a self-contained city covering 6 square kilometers of reclaimed desert land (depicted in Figure 2) that will eventually be home to 50,000 people and 1500 businesses, along with the infrastructure needed to accommodate a total of 70,000 jobs (including commuters) (*Electricity Journal*, 2009). The project is divided into the five distinct components presented in Table 2, including an industries unit that will invest in local and global renewable energy systems and a university that will train students in sustainability and renewable energy (Reiche, 2010).

Masdar City is also set to be the headquarters of the new International Renewable Energy Agency (IRENA), founded in late 2008, after officials in

FIGURE 2 Artist Depiction of Masdar City[1]
[1]Source: Masdar City, "About Masdar City," http://www.masdaruae.com/en/home/index.aspx, accessed December 2009. *Source: Masdar City.*

TABLE 2 Different Components of the Masdar City Initiative

Masdar City Units	Function
Carbon Management Unit	To develop projects within Masdar City that generate carbon credits that can be sold under the Clean Development Mechanism of the Kyoto Protocol
Industries Unit	To invest both locally and globally in renewable energy systems
Masdar Institute of Science and Technology	To offer advanced graduate degrees focused on the science and engineering of advanced renewable energy systems and sustainable technologies
Property Development	To build the actual carbon-neutral, zero-waste Masdar City
Utilities and Asset Management	To raise money and make investments in technologies that will enable the city to run efficiently and with no net carbon emissions

Source: Author.

Abu Dhabi pledged to give IRENA office space free of charge and an additional $185 million in donations to help get the agency off the ground. Masdar City is being promoted predominately by the government of Abu Dhabi, and not independent or private commercial firms, through the Abu Dhabi Future Energy Company, a seed capital fund owned by the Mubadala Development Company. Construction is currently managed by CH2M Hill.

4.2. Benefits

If Masdar City is built as planned, it promises to have five benefits over other cities:

- First is its predilection for sustainable design. Current plans call for automobiles to be banned within the city, with walking encouraged for short-distance travel and reliance on public mass transit and personal rapid transit systems for longer distances. Architects and engineers also intend to use numerous features of passive design to reduce the city's energy consumption by 75% compared to other cities of comparable size (Fortson, 2009). These include a perimeter city wall that will act as barrier against hot desert winds, shade streets, and facilitate the movement of cooler breezes throughout the city; narrower streets oriented to shield the sun and act as funnels for the wind; shallow pools of water that cool the air through evaporation; and wind towers that push cooler air toward the ground and act as "natural air-conditioners." (*Electricity Journal*, 2009).
- Second are its proposed sources of energy and electricity supply, which are all intended to be low-carbon and renewable. A 60 MW solar power plant currently being built by Conergy is to supply all electricity during construction. One innovative aspect of this facility is that the roof is being built first so that the solar collectors can start producing energy needed to complete the rest of the building before then offering power to construction teams spread across the city (Nader, 2009). This centralized solar plant is to be followed by 130 MW of distributed solar panels, integrated into buildings and hung over alleys where they will double as sun shades to keep temperatures low. Such solar sources of energy are to be supplemented with a 20 MW wind farm installed outside the city wall and a small geothermal power station.
- Third, it is to be designed to maximize the use of resources such as water and minimize the production of waste. A $50 million solar powered desalination plant is to create most of the city's water, 80% of which is to be recycled, with even grey water being reused to irrigate crops and water green spaces and gardens. The City also aims to be zero waste, relying on digestion to convert biological wastes into soil and fertilizer, incineration to convert solid wastes into electricity, and recycling to convert plastics and metals into usable products.

- Fourth, it is to use the sale of carbon credits to partly finance local operations. A very small proportion of the City's operating costs are to be funded by the sale of carbon credits under the Clean Development Mechanism of the Kyoto Protocol. Planners currently expect Masdar City to produce about one million carbon credits per year when operational in 2016, enough to create $30 million in annual revenues at today's prices (Fortson, 2009).
- Fifth and lastly, Masdar City is to host another grand scheme in Abu Dhabi to build one of the world's largest carbon capture and storage systems, able to capture about 6.5 million tons of CO_2 from power plants and industrial facilities nearby by 2013. The captured CO_2 is to be transported and injected into oil reservoirs near Abu Dhabi for enhanced oil recovery (Masdar City, 2009).

4.3. Challenges

Notwithstanding these benefits, Masdar City does face serious challenges. By far the greatest challenge is that very little of the project has actually been built. One reporter that recently visited the construction site noted that all they could see were "a few tractors and a pair of cranes." (Masdar City, 2009) Nothing else yet exists, meaning many of the city's most impressive attributes remain on paper only, and accomplishing them will depend on future variables (Reiche, 2010).

Another challenge is related to the replicability of Masdar City in other areas of the world. One key difference between the political environment in Abu Dhabi and democratic countries is that it is run by appointed emirate rulers. These rulers, along with a Federal National Council, exert complete control over the political system. Political parties and other institutions associated with politics are completely banned (Reiche, 2010). The implication may be that a project such as Masdar City works only where authorities are able to fully exert their control without the interference from other groups—or concerns about commercial feasibility, payback periods, and issues that often plague similar developments elsewhere.

Yet another issue is the very carbon- and energy-intensive environment throughout Abu Dhabi and the UAE that surround Masdar City. Per-capita CO_2 emissions in Abu Dhabi are much higher than the United States—Abu Dhabi emits about 30 metric tons of CO_2 per person compared to 19 metric tons per person in the United States—despite the lack of industry and little need for transportation; and the environmental per capita footprint for Abu Dhabi is the *worst* in the world. That is, residents of the UAE consume more natural resources than anybody else on Earth, even more than those residing in the United States, Europe, or Kuwait (Landais, 2008). As soon as one departs from the yet-to-be-completed Masdar City, noted one editorial, they will likely be "back to the real world of big SUVs driving around aimlessly on cheap oil, high rises, and megamalls chilled with humming air-conditioning in the desert sun" (*Electricity Journal*, 2009). Connecting Masdar City to the rest of the country, moreover, will not be as sustainable as it seems; food will still need to be imported and large car parks are being planned for outside the city for the droves of tourists and residents that want to visit (Fortson, 2009). There is also some concern that the revenues raised from the city, if it ever does turn a profit, will be funnelled back into fossil fuels. Abu Dhabi, for example, receives 70% of its GDP from oil and the emirate owns 95% of the UAE's proven oil reserves (Reiche, 2010).

The global financial downturn is also already curtailing investment in the project. If populated as expected, the city will cost about $440,000 for every resident that lives within its walls—making it exceptionally expensive. The City still needed about $18 billion of future investment as of 2009 yet currency devaluation and financial speculation have made raising that amount close to impossible (Fortson, 2009). There is a tendency for many investors to view Masdar City as more of an "expensive experiment" or "white elephant" instead of a real revenue earning project, contributing to reluctance among financiers to fully embrace it (Reiche, 2010).

Moreover, it is uncertain whether the government can attract the 40,000 people expected to live within the city. Will most of these come from the United States and Europe, and what jobs will be offered other than those at the Masdar Institute for Science and Technology? With apparent difficulties associated with securing working contracts for expatriates, and the lack of environmental awareness among the indigenous population, the City may not be able to convince enough people to actually move there and call it their home (Reiche, 2010).

Taken collectively, these barriers may mean that Masdar City looks fantastic on paper but will be difficult or even impossible to duplicate.

CONCLUSIONS

Four useful lessons may be drawn from these case studies. First, cooperation between governments, civil society, and the private sector can be an elemental component of promoting sustainability, with each of these case studies showing the importance of multiple actors working together at multiple scales and levels. The CCI shows us that targeted action is needed to overcome split incentives in the market, provide information and knowledge, measure emissions credibly, and connect lenders and manufacturers with political leaders in order to address the major sources of GHGs. By operating at a nexus of business, politics, lenders, and experts, CCI is able to bring together the necessary actors to advance projects with supportable business plans and sustainable financing mechanisms that can serve as scalable models for others to follow. Motorola's efforts have enrolled not only corporate leaders, employees, and factories, but also the corporation's vast network of suppliers.

Far from being alone, many other commercial firms are also beginning to promote energy efficiency and sustainability, as documented in this book's Chapter 11 by Prindle on commercial energy use. Although it may be too early to tell for Masdar City, since that project is only in its nascent stages of development, if completed it would demonstrate the utility of relying on a network of venture capital firms and government-owned companies to build a sustainable city from the ground up.

Second, creating more sustainable forms of transport, agriculture, forestry, and energy supply can result in economic gains. The CCI has leveraged $5 billion in energy efficiency funding that will likely provide a return on investment many times over. The simplest lesson from Motorola appears to be that one can manufacture products that are truly better for the environment and profitable at the same time. Masdar City plans to use the proceeds from selling carbon credits on the global market to offset a small amount of the city's operational expenses.

Third, each case study has had to confront at times tenacious challenges. CCI's programs have had to overcome lack of knowledge about energy efficiency practices and technologies, split incentives, the first-cost hurdle, and a misalignment of government incentives. Put another way, energy efficiency efforts have had to swim upstream against other policies and incentives that often encourage economic growth and increased energy consumption. Motorola has taken great strides in making its manufacturing facilities, processes, and products less harmful to the environment and climate, but still relies on toxic substances and GHG-intensive supply chains. As a corporation responsible to its board of directors and shareholders, Motorola is also required to emphasize profits and corporate growth above sustainability. Masdar City faces perhaps the most serious obstacles. Very little of it has been built as of 2010, the UAE remains a carbon-intensive and energy-intensive country, the global financial crisis is turning away lenders and investors, and it remains uncertain whether the City can attract enough professionals and residents.

Fourth, although each case study has its own unique set of problems, they also relied on a multitude of mechanisms and programs simultaneously to overcome impediments and realize benefits. Masdar City, if it does work as planned, would underlie the importance of integrating city planning, passive design, energy supply, transport, water, and recycling efforts so that the entire community is zero-carbon and zero-waste. CCI's programs involve not only building retrofits and lighting systems but also waste management, transportation, concentrating solar power, carbon capture and storage, and forestry.. Motorola's efforts have similarly focused on a variety of areas simultaneously, from implementing corporate-wide energy efficiency practices and pledging GHG reduction commitments to devising innovative products that have fewer hazardous substances and a greater number of recyclable components. In the end, these case studies show that actors can promote sustainability through a variety of channels, with a net gain to people and the planet.

BIBLIOGRAPHY

Barter, P. A. (2008). Singapore's urban transport: Sustainability by design or necessity. In T.-C. Wong, B. Yuen & C. Goldblum (Eds.), *Spatial planning for a sustainable Singapore* (pp. 95–112). New York: Springer.

Brown, M. A., & Sovacool, B. K. (2010). *Climate change and energy security: A global overview of technology and policy options*. Cambridge: MIT Press.

Clinton Climate Initiative. (2007). President Clinton announces landmark program to reduce energy use in buildings worldwide. *CCI press release*. May 16, 2007.

Clinton Climate Initiative. (2009a). *Cities*. CCI fact sheet, updated April 28, 2009.

Clinton Climate Initiative. (2009b). *Our challenge, our work*. CCI fact sheet, updated May 2, 2009.

Clinton Climate Initiative. (2009c). *Waste management program*. CCI fact sheet, updated May 2, 2009.

Clinton Climate Initiative. (2009d). *Energy efficiency: Building retrofit program*. CCI fact sheet, updated May 8, 2009.

Clinton Climate Initiative. (2009e). Clinton Climate Initiative to demonstrate model for sustainable urban growth with projects in 10 countries on six continents. *Business Wire*. May 18, 2009.

Clinton Climate Initiative. (2009f). *Forestry: Carbon and poverty reduction program*. CCI fact sheet, updated July 17, 2009.

Coughlin, J., & Cory, K. (2009). *Solar photovoltaic financing: Residential sector deployment*. Golden, CO: National Renewable Energy Laboratory Technical Report NREL/TP-6A2-44853, pp. 40–41.

de Raaf, W.-J. (2008). *Increasing energy efficiency in buildings and lighting*. Presentation at C40 World Ports Climate Conference, Rotterdam, July 10, 2008.

Electricity Journal. (2009). Abu Dhabi's Masdar project: Dazzling? Or just a mirage? *Electricity Journal, 22*(5), 4–5.

Erlanger, S. (2008). Israel is set to promote the use of electric cars. *New York Times*. 21 January 2008.

Fortson, D. (2009). Green city rises from the desert. *The Sunday Times (London)*, 9, 1 February 2009.

Fuller, M. C., Portis, S. C., & Kammen, D. M. (2009). Toward a low-carbon economy: Municipal financing for energy efficiency and solar power. *Environment, 51*(1), 22–32.

Girardet, H., & Mendonca, M. (2009). *A renewable world: Energy, ecology, equality*. London: Green Books. p. 162.

Koenig, R. (2009). *Introduction to the Clinton Climate Initiative*. Presentation to Pittsburgh City Council, February 9, 2009.

Landais, E. (2008). UAE tops world on per capita carbon footprint. *Gulf News*. 30 October 2008.

Masdar City. (2009). *Welcome to Masdar City. www.masdarcity.ae/en/index.aspx.*

Motorola Corporation. (2008). *Motorola corporate responsibility summary report: How does innovation promote responsibility?* Schaumburg, Illinois: Motorola.

Motorola Corporation. (2009). *Corporate profile*. Schaumburg, Illinois: Motorola.

Nader, S. (2009). Paths to a low-carbon economy – the Masdar example. *Energy Procedia, 1*, 3951–3958.

Ostrom, E. (2009). A polycentric approach for coping with climate change. *Report prepared for the WDR2010 Core Team, Development and Economics Research Group, World Bank*. Bloomington, IN: Indiana University.

Rautner, M., & Harrell, C. (2008). *Green electronics survey December 2008*. Amsterdam: Greenpeace International.

Reiche, D. (2010). Renewable energy policies in the Gulf countries: A case study of the carbon-neutral "Masdar City" in Abu Dhabi. *Energy Policy, 38*, 378–382.

Reuters News Service. (2009). *Motorola phone made from recycled bottles. Publication.* January 6, 2009.

Rosenthal, E. (2009). In German suburb, life goes on without cars. *New York Times*. 11 May 2009.

Sovacool, B. K., & Brown, M. A. (2009). Twelve Metropolitan Carbon Footprints: A Preliminary Global Comparative Assessment. *Energy Policy, 38*, 4856–4869.

Sovacool, B. K., & Brown, M. A. (2010). Addressing climate change: Global vs. local scales of jurisdiction? In F. P. Sioshansi (Ed.), *Generating electricity in a carbon constrained world* (pp. 109–124). New York: Elsevier.

Watts, J. (2009). South Korea lights the way on carbon emissions with its £23bn green deal. *The Guardian*. 21 April 2009.

Epilogue

Can We Get There from Here?

Fereidoon P. Sioshansi
Menlo Energy Economics

The contributors to this volume have done an admirable job in addressing the main theme of the book, namely can we have our cake and eat it too? Can we meet the basic energy requirements of a growing global population with rising aspirations for higher *and* more equitable living standards in a sustainable way? This, as further examined in the various chapters, is a tall order. We are not talking about a continuation of the status quo, where a fraction of the world's inhabitants enjoy high living standards—and use a disproportionately large percentage of global natural resources including energy—while the rest await their turn.

Contributing authors have different views and take different perspectives. Not surprisingly, they come up with different solutions on how we should proceed. As editor, I have encouraged this diversity of opinion. Consequently, it is not easy to summarize the overall message of the book but at the expense of over-simplification I offer the following as my *personal* take on their collective wisdom and insights:

- First is the realization that the status quo does not appear to be sustainable—a view shared by many.
- Second is the recognition of the sheer enormity and scale of the problem—namely the many adjustments, large and small, that have to be made to move toward a more sustainable future path.
- Third is that despite the daunting challenges, there are many degrees of freedom, many opportunities, and many ways in which we can influence the outcome—if we choose wisely, if we act, and if we persevere.

Making appropriate adjustments in the supply side of the energy equation is enormously important and must be pursued with vigor. But I am *personally* convinced that the demand-side offers more rewarding opportunities, a theme frequently repeated in the book. In this context, I am convinced that *incremental* adjustments will not suffice; we need to make bold, radical changes—one step at a time. This requires asking *what we use energy for*, *why we use so much energy*, and *how we can use energy more sparingly.* There are enormous opportunities to achieve far *more* with far *less*, and that—at least for me—is the main message of the book.

A Chinese proverb says that a journey of 5000 miles begins with a single step. It is time to begin the long journey toward a more just, equitable, and sustainable future.

Index

A

B

C

D

E

F

G

J

K

L

M

N

O

S

T

Printed in the United States
By Bookmasters